STUDENT SOLUTIONS MANUAL

FINITE MATHEMATICS

An Applied Approach

Tenth Edition

THE WILEY BICENTENNIAL–KNOWLEDGE FOR GENERATIONS

Each generation has its unique needs and aspirations. When Charles Wiley first opened his small printing shop in lower Manhattan in 1807, it was a generation of boundless potential searching for an identity. And we were there, helping to define a new American literary tradition. Over half a century later, in the midst of the Second Industrial Revolution, it was a generation focused on building the future. Once again, we were there, supplying the critical scientific, technical, and engineering knowledge that helped frame the world. Throughout the 20th Century, and into the new millennium, nations began to reach out beyond their own borders and a new international community was born. Wiley was there, expanding its operations around the world to enable a global exchange of ideas, opinions, and know-how.

For 200 years, Wiley has been an integral part of each generation's journey, enabling the flow of information and understanding necessary to meet their needs and fulfill their aspirations. Today, bold new technologies are changing the way we live and learn. Wiley will be there, providing you the must-have knowledge you need to imagine new worlds, new possibilities, and new opportunities.

Generations come and go, but you can always count on Wiley to provide you the knowledge you need, when and where you need it!

WILLIAM J. PESCE
PRESIDENT AND CHIEF EXECUTIVE OFFICER

PETER BOOTH WILEY
CHAIRMAN OF THE BOARD

STUDENT SOLUTIONS MANUAL

FINITE MATHEMATICS

An Applied Approach

Tenth Edition

Michael Sullivan
Chicago State University

Kathleen Miranda

John Wiley & Sons, Inc.

Bicentennial Logo Design: Richard J. Pacifico

To order books or for customer service, please call 1-800-CALL-WILEY (225-5945).

ISBN-13 978-0-470-24964-2

Printed in the United States of America

10 9 8 7 6 5 4 3 2 1

Printed and bound by Bind-Rite Graphics, Inc.

Table of Contents

Chapter 1

Linear Equations

1.1 Rectangular Coordinates; Lines

1. True

3. True

5. vertical

7. $A = (4, 2)$ $\quad B = (6, 2)$ $\quad C = (5, 3)$ $\quad D = (-2, 1)$

$E = (-2, -3)$ $\quad F = (3, -2)$ $\quad G = (6, -2)$ $\quad H = (5, 0)$

9. The set of points of the form, $(2, y)$, where y is a real number, is a vertical line passing through 2 on the x-axis.

The equation of the line is $x = 2$.

11. $y = 2x + 4$

x	0	−2	2	−2	4	−4
y	**4**	0	**8**	**0**	**12**	**−4**

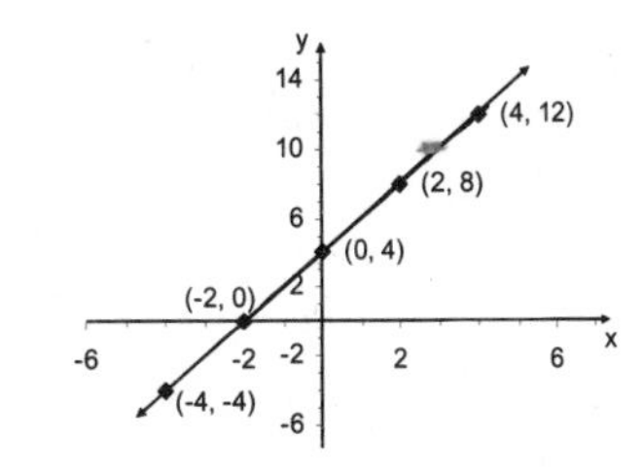

13. $2x - y = 6$

x	0	**3**	2	-2	4	-4
y	**-6**	0	**-2**	**-10**	**2**	**-14**

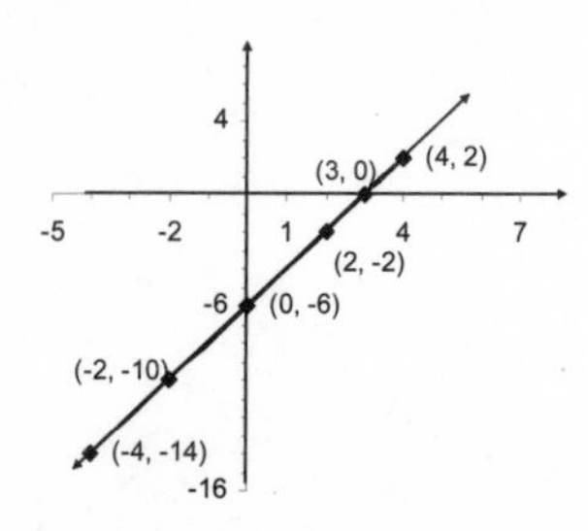

15. (a) The vertical line containing the point (2, –3) is $x = 2$.
(b) The horizontal line containing the point (2, –3) is $y = -3$.
(c)
$$y+3=5(x-2)$$
$$y+3=5x-10$$
$$13=5x-y$$
The line with a slope of 5 containing the point (2, –3) is $5x - y = 13$.

17. (a) The vertical line containing the point (–4, 1) is $x = -4$.
(b) The horizontal line containing the point (–4, 1) is $y = 1$.
(c)
$$y-1=5(x+4)$$
$$y-1=5x+20$$
$$-21=5x-y$$
The line with a slope of 5 containing the point (–4, 1) is $5x - y = -21$.

19. (a) The vertical line containing the point (0, 3) is $x = 0$.
(b) The horizontal line containing the point (0, 3) is $y = 3$.
(c)
$$y-3=5x$$
$$-3=5x-y$$
The line with a slope of 5 containing the point (0, 3) is $5x - y = -3$.

21.
$$m=\frac{y_2-y_1}{x_2-x_1}=\frac{1-0}{2-0}=\frac{1}{2}$$

We interpret the slope to mean that for every 2 unit change in x, y changes 1 unit. That is, for every 2 units x increases, y increases by 1 unit.

23. $$m=\frac{y_2-y_1}{x_2-x_1}=\frac{3-1}{-1-1}=-1$$
We interpret the slope to mean that for every 1 unit change in x, y changes by (–1) unit. That is, for every 1 unit increase in x, y decreases by 1 unit.

25. $$m=\frac{y_2-y_1}{x_2-x_1}=\frac{3-0}{2-1}=\frac{3}{1}=3$$

A slope of 3 means that for every 1 unit change in x, y will change 3 units.

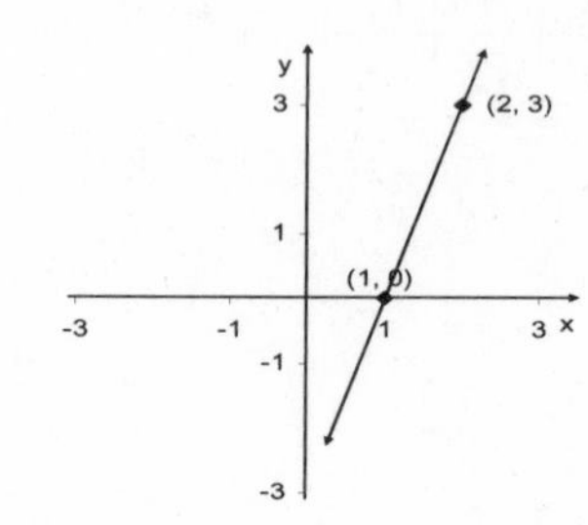

27. $m = \dfrac{y_2 - y_1}{x_2 - x_1} = \dfrac{1-3}{2-(-2)} = \dfrac{-2}{4} = -\dfrac{1}{2}$

A slope of $-\dfrac{1}{2}$ means that for every 2 unit increase in x, y will decrease (-1) unit.

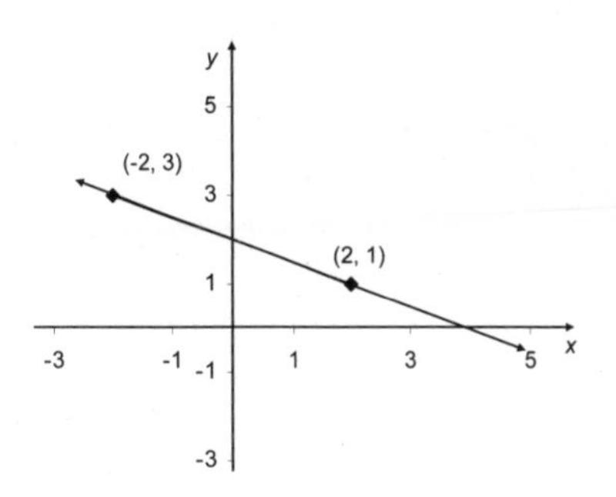

29. $m = \dfrac{y_2 - y_1}{x_2 - x_1} = \dfrac{(-1)-(-1)}{2-(-3)} = \dfrac{0}{5} = 0$

A slope of zero indicates that regardless of how x changes, y remains constant.

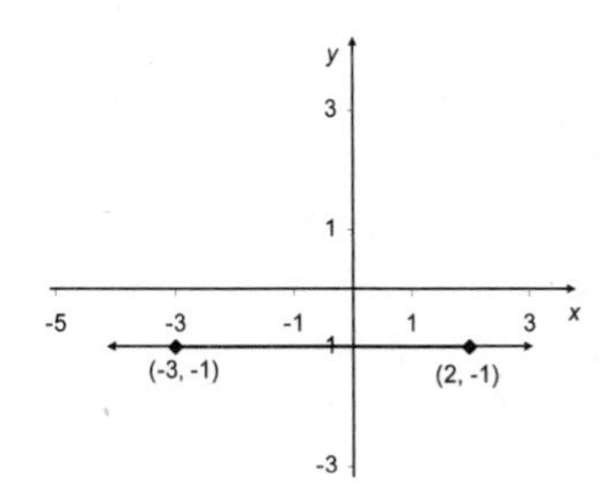

31. $m = \dfrac{y_2 - y_1}{x_2 - x_1} = \dfrac{(-2)-2}{(-1)-(-1)} = \dfrac{-4}{0}$

The slope is not defined.

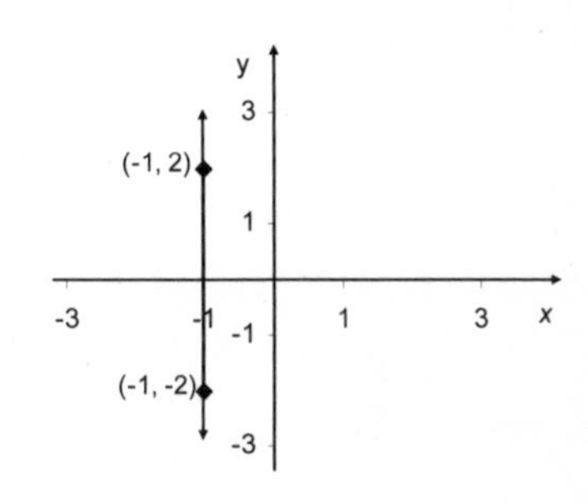

33.

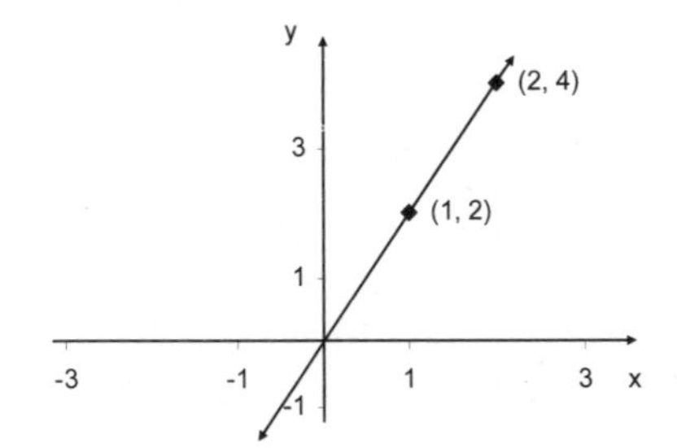

35.

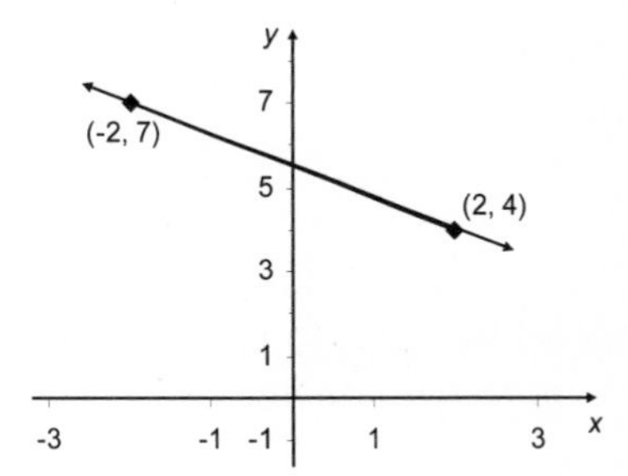

37.

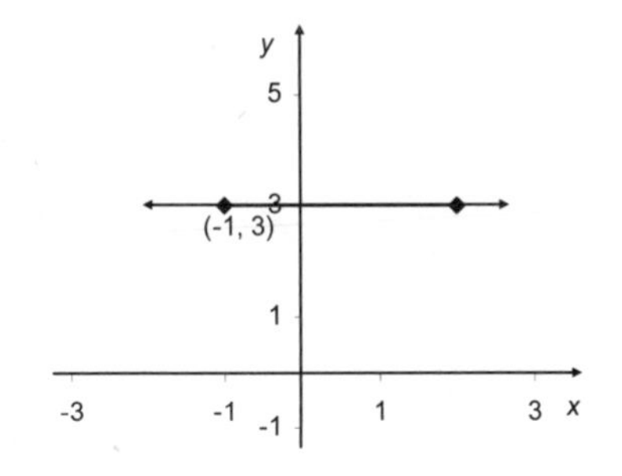

39.

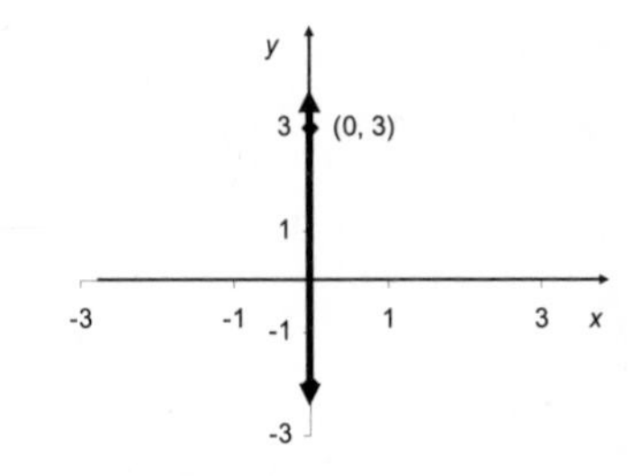

41. Use the points (0, 0) and (2, 1) to compute the slope of the line:

$$m = \frac{y_2 - y_1}{x_2 - x_1} = \frac{1-0}{2-0} = \frac{1}{2}$$

Since the y-intercept, (0, 0), is given, use the slope-intercept form of the equation of the line:

$$y = \frac{1}{2}x + 0$$
$$y = \frac{1}{2}x$$

Then write the general form of the equation: $x - 2y = 0$

43. Use the points (1, 1) and (–1, 3) to compute the slope of the line:

$$m = \frac{y_2 - y_1}{x_2 - x_1} = \frac{3-1}{(-1)-1} = \frac{2}{-2} = -1$$

Now use the point (1, 1) and the slope $m = -1$ to write the point-slope form of the equation of the line:

$$\begin{aligned} y - y_1 &= m(x - x_1) \\ y - 1 &= (-1)(x - 1) \\ y - 1 &= -x + 1 \\ x + y &= 2 \end{aligned}$$

45. Since the slope and a point are given, use the point-slope form of the line:

$$\begin{aligned} y - y_1 &= m(x - x_1) \\ y - 1 &= 2(x - (-4)) \\ y - 1 &= 2x + 8 \\ 2x - y &= -9 \end{aligned}$$

47. Since the slope and a point are given, use the point-slope form of the line:

$$\begin{aligned} y - y_1 &= m(x - x_1) \\ y - (-1) &= -\frac{2}{3}(x - 1) \\ 3y + 3 &= -2(x - 1) \\ 3y + 3 &= -2x + 2 \\ 2x + 3y &= -1 \end{aligned}$$

49. Since we are given two points, (1, 3) and (–1, 2), first find the slope.

$$m = \frac{3-2}{1-(-1)} = \frac{1}{2}$$

Then use the slope, one of the points, (1, 3), and the point-slope form of the line:

$$\begin{aligned} y - y_1 &= m(x - x_1) \\ y - 3 &= \frac{1}{2}(x - 1) \\ 2y - 6 &= x - 1 \\ x - 2y &= -5 \end{aligned}$$

51. Since we are given the slope $m = -2$ and the y-intercept (0, 3), we use the slope-intercept form of the line:

$$\begin{aligned} y &= mx + b \\ y &= -2x + 3 \\ 2x + y &= 3 \end{aligned}$$

53. We are given the slope $m = 3$ and the x-intercept $(-4, 0)$, so we use the point-slope form of the line:

$$\begin{aligned} y - y_1 &= m(x - x_1) \\ y - 0 &= 3\ (x - (-4)) \\ y &= 3x + 12 \\ 3x - y &= -12 \end{aligned}$$

55. We are given the slope $m = \frac{4}{5}$ and the point $(0, 0)$, which is the y-intercept. So, we use the slope-intercept form of the line:

$$\begin{aligned} y &= mx + b \\ y &= \frac{4}{5}x + 0 \\ 5y &= 4x \\ 4x - 5y &= 0 \end{aligned}$$

57. We are given two points, the x-intercept $(2, 0)$ and the y-intercept $(0, -1)$, so we need to find the slope and then to use the slope-intercept form of the line to get the equation.

$$\text{slope} = \frac{0 - (-1)}{2 - 0} = \frac{1}{2}$$

$$\begin{aligned} y &= mx + b \\ y &= \frac{1}{2}x - 1 \\ 2y &= x - 2 \\ x - 2y &= 2 \end{aligned}$$

59. Since the slope is undefined, the line is vertical. The equation of the vertical line containing the point $(1, 4)$ is: $x = 1$

61. Since the slope = 0, the line is horizontal.

The equation of the horizontal line containing the point $(1, 4)$ is: $y = 4$

63. $y = 2x + 3$,

slope: $m = 2$; y-intercept: $(0, 3)$

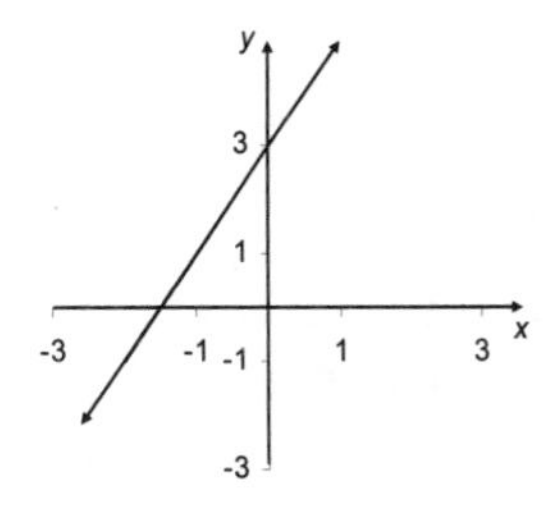

65. To obtain the slope and y-intercept, we transform the equation into its slope-intercept form by solving for y.

$$\frac{1}{2}y = x - 1$$
$$y = 2x - 2$$

slope: $m = 2$; y-intercept: $(0, -2)$

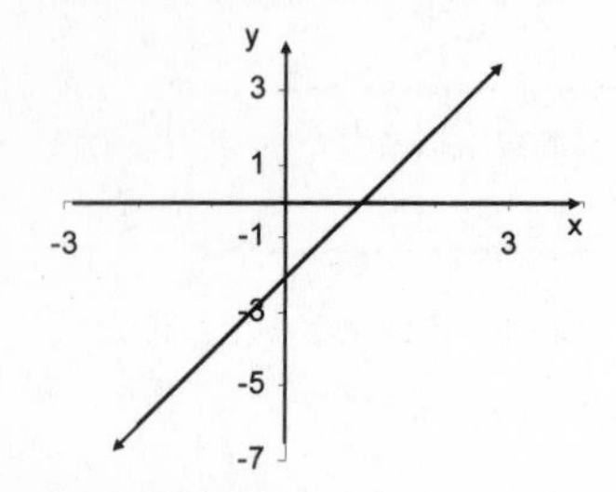

67. To obtain the slope and y-intercept, we transform the equation into its slope-intercept form by solving for y.

$$2x - 3y = 6$$
$$y = \frac{2}{3}x - 2$$

slope: $m = \frac{2}{3}$; y-intercept: $(0, -2)$

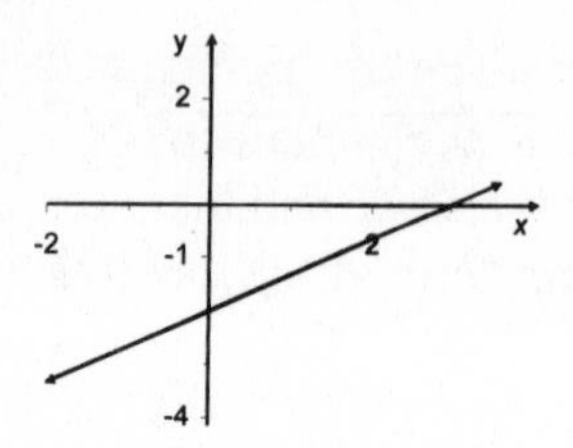

69. To obtain the slope and y-intercept, we transform the equation into its slope-intercept form by solving for y.

$$x + y = 1$$
$$y = -x + 1$$

slope: $m = -1$; y-intercept: $(0, 1)$

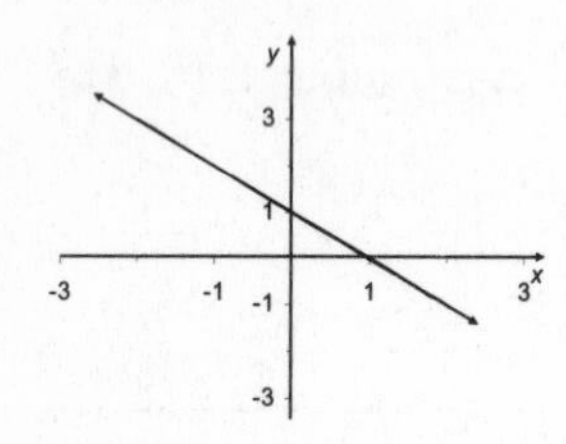

71. The slope is not defined; there is no y-intercept. So the graph is a vertical line.

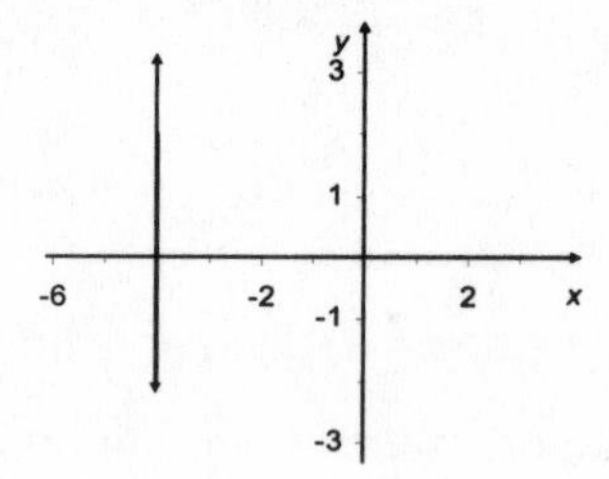

73. slope: $m = 0$; y-intercept: $(0, 5)$

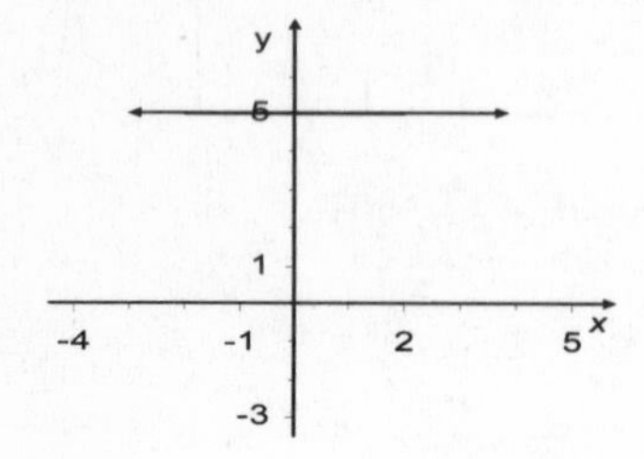

75. To obtain the slope and y-intercept, transform the equation into its slope-intercept form by solving for y.

$$y - x = 0$$
$$y = x$$

slope: $m = 1$; y-intercept $= (0, 0)$

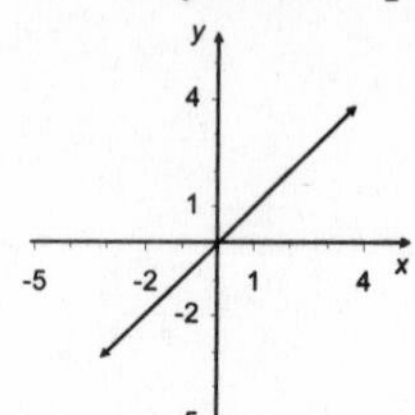

77. To obtain the slope and y-intercept, transform the equation into its slope-intercept form by solving for y.

$$2y - 3x = 0$$
$$2y = 3x$$
$$y = \frac{3}{2}x$$

slope: $m = \frac{3}{2}$; y-intercept = (0, 0)

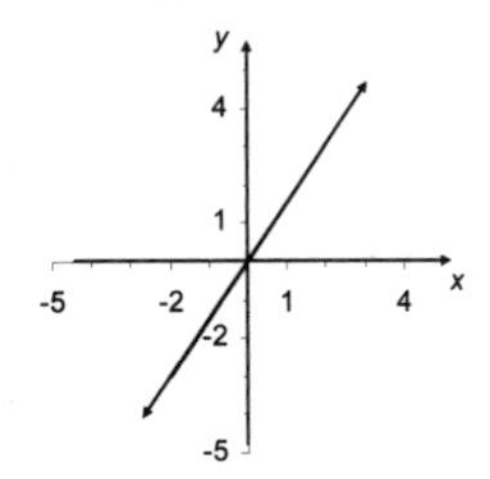

79. To graph an equation on a graphing utility, first solve the equation for y.

$$1.2x + 0.8y = 2$$
$$0.8y = -1.2x + 2$$
$$y = -1.5x + 2.5$$

Window: Xmin = −10; Xmax = 10
Ymin = −10; Ymax = 10

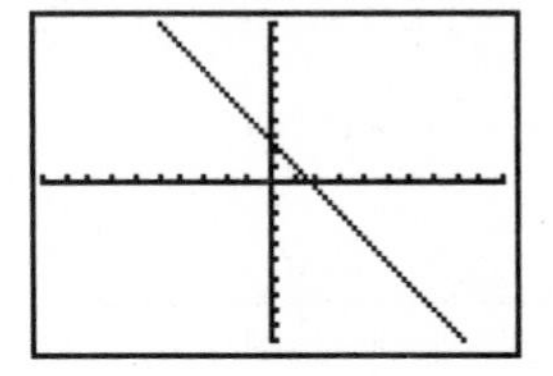

The x-intercept is (1.67, 0).
The y-intercept is (0, 2.50).

81. To graph an equation on a graphing utility, first solve the equation for y.

$$21x - 15y = 53$$
$$15y = 21x - 53$$
$$y = \frac{21}{15}x - \frac{53}{15}$$
$$y = \frac{7}{5}x - \frac{53}{15}$$

Window: Xmin = −10; Xmax = 10
Ymin = −10; Ymax = 10

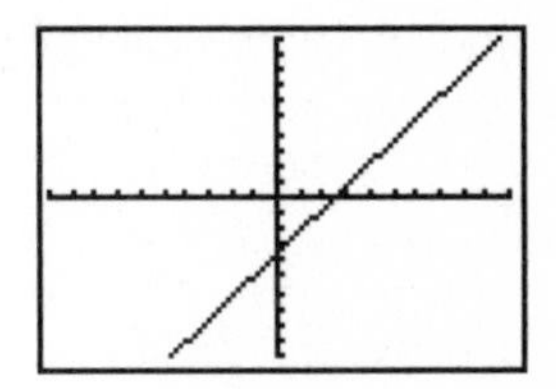

The x-intercept is (2.52, 0).
The y-intercept is (0, −3.53).

83. To graph an equation on a graphing utility, first solve the equation for y.

$$\frac{4}{17}x + \frac{6}{23}y = \frac{2}{3}$$
$$\frac{6}{23}y = -\frac{4}{17}x + \frac{2}{3}$$
$$y = \frac{23}{6}\left(-\frac{4}{17}x + \frac{2}{3}\right)$$
$$y = -\frac{46}{51}x + \frac{23}{9}$$

Window: Xmin = −10; Xmax = 10
Ymin = −10; Ymax = 10

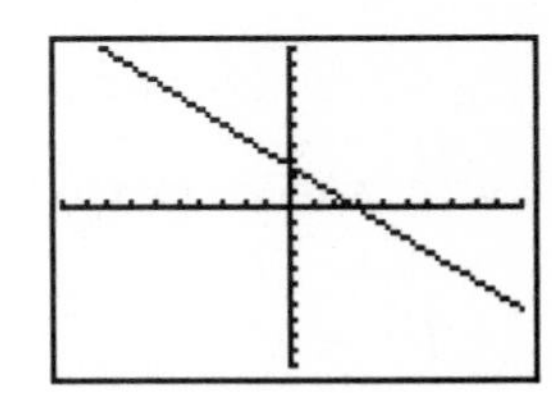

The x-intercept is (2.83, 0).
The y-intercept is (0, 2.56).

85. To graph an equation on a graphing utility, first solve the equation for y.

$$\pi x - \sqrt{3}y = \sqrt{6}$$
$$\sqrt{3}y = \pi x - \sqrt{6}$$
$$y = \frac{\pi}{\sqrt{3}}x - \sqrt{2}$$

Window: Xmin = –10; Xmax = 10
Ymin = –10; Ymax = 10

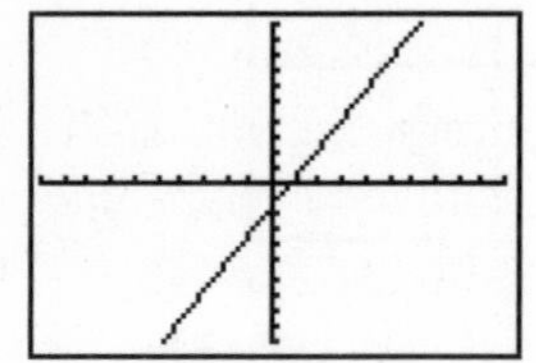

The x-intercept is (0.78, 0).
The y-intercept is (0, –1.41).

87. The graph passes through the points (0, 0) and (4, 8). We use the points to find the slope of the line:

$$m = \frac{y_2 - y_1}{x_2 - x_1} = \frac{8-0}{4-0} = \frac{8}{4} = 2$$

The y-intercept (0, 0) is given, so we use the y-intercept and the slope $m = 2$, to get the slope-intercept form of the line:

$$y = mx + b$$
$$y = 2x + 0$$
$y = 2x$ which is answer (b).

89. The graph passes through the points (0, 0) and (2, 8). We use the points to find the slope of the line:

$$m = \frac{y_2 - y_1}{x_2 - x_1} = \frac{8-0}{2-0} = \frac{8}{2} = 4$$

The y-intercept (0, 0) is given, so we use the y-intercept and the slope $m = 4$, to get the slope-intercept form of the line:

$$y = mx + b$$
$$y = 4x + 0$$
$y = 4x$ which is answer (d).

91. Using the intercepts (–2, 0) and (0, 2),
$$m = \frac{y_2 - y_1}{x_2 - x_1} = \frac{2-0}{0-(-2)} = \frac{2}{2} = 1$$
Slope-intercept form: $y = x + 2$
General form: $x - y = -2$

93. Using the intercepts (3, 0) and (0, 1), $m = \frac{y_2 - y_1}{x_2 - x_1} = \frac{1-0}{0-3} = \frac{1}{-3} = -\frac{1}{3}$

Slope-intercept form: $y = -\frac{1}{3}x + 1$ General form:

$$3y = 3\left(-\frac{1}{3}x + 1\right)$$
$$3y = -x + 3$$
$$x + 3y = 3$$

95. (a) The equation is $C = 0.155x$, where x is the number of miles the car is driven.

(b) $x = 15{,}000$
$C = (0.155) \cdot (15{,}000) = 2325$
It costs \$2325.00 to drive the car 15,000 miles.

(c)

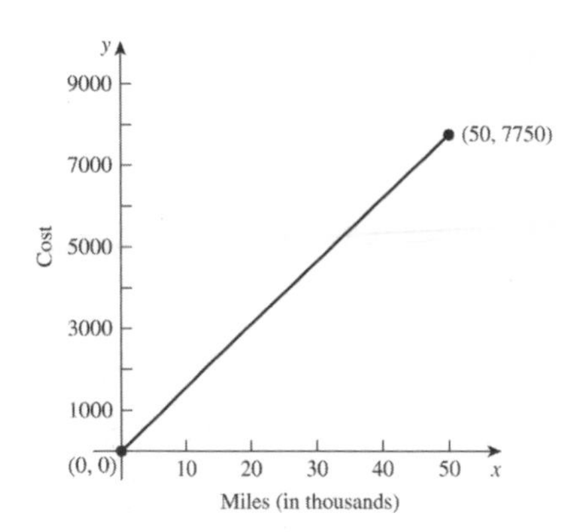

(d) The slope describes the cost of driving a standard-sized car an additional mile.

97. (a) The fixed cost of electricity for the month is \$7.58. In addition, the electricity costs \$0.08275 (8.275 cents) for every kilowatt-hour (KWH) used. If x represents the number of KWH of electricity used in a month, the total monthly is represented by the equation:

$$C = 0.08275x + 7.58, \quad 0 \le x \le 400$$

(b)

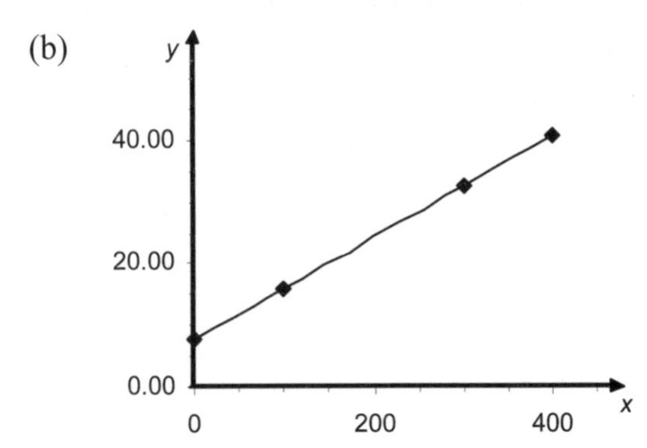

(c) The charge for using 100 KWH of electricity is found by substituting 100 for x in part (a)

$$\begin{aligned} C &= 0.08275(100) + 7.58 \\ &= 8.275 + 7.58 \\ &= 15.855 \\ &= \$15.86 \end{aligned}$$

(d) The charge for using 300 KWH of electricity is found by substituting 300 for x in part (a)

$$\begin{aligned} C &= 0.08275(300) + 7.58 \\ &= 24.825 + 7.58 \\ &= 32.405 \\ &= \$32.41 \end{aligned}$$

(e) The slope of the line, $m = 0.08275$, indicates that for every extra KWH used (up to 400 KWH), the electric bill increases by 8.275 cents.

99. (a) If \$1,252,000 are the fixed cost and \$1.13 the variable cost of delivering the paper, we write $C = 1.13x + 1{,}252{,}000$ where x denotes the number of papers delivered.

(b) If $x = 100{,}000$, then
$C = 1.13 \cdot 100{,}000 + 1{,}252.000$
$C = \$1{,}365{,}000.00$

(c)

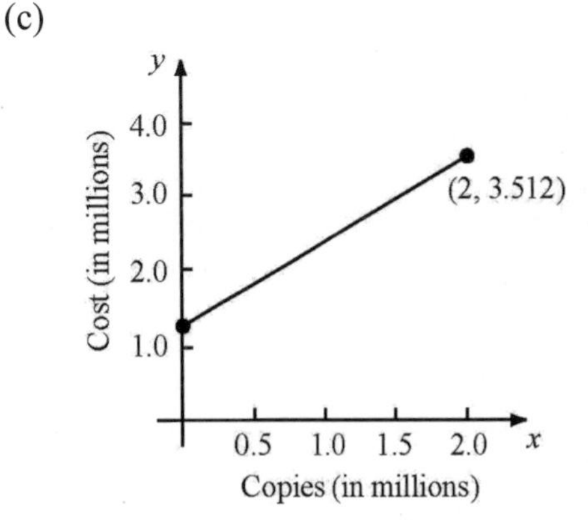

(d) The slope is the cost to the *Chicago Tribune* of delivering one more newspaper.

(e) The y-intercept is the cost to the Chicago Tribune if no papers are delivered.

101. Since we are told the relationship is linear, we will use the two points to get the slope of the line:

$$m = \frac{C_2° - C_1°}{F_2° - F_1°} = \frac{100-0}{212-32} = \frac{100}{180} = \frac{5}{9}$$

We use the point (0, 32) and the fact that the slope is $\frac{5}{9}$ to get the point-slope form of the equation.

$$C° - C_1° = m(F° - F_1°)$$
$$C° - 0 = \frac{5}{9}(F° - 32)$$
$$C° = \frac{5}{9}(F° - 32)$$

To find the Celsius measure of 68 °F we substitute 68 for F in the equation and simplify:

$$C° = \frac{5}{9}(68-32)$$
$$= 20°$$

103. (a) Use the points (6, 38.4) and (36, 37.7) to find the slope of the equation.

$$m = \frac{38.4-37.7}{6-36} = \frac{0.7}{-36} = -0.0233$$

We will use the slope and the point (6, 38.4) to write the point-slope form of the equation.

$$A - 38.4 = -0.0233(t-6)$$
$$A = -0.0233t + 0.14 + 38.4$$
$$A = -0.0233t + 38.54$$

(b) When $t = 20$, $A = -0.023t + 38.54 = 38.07$
On September 20, 2006 there were 38.07 billion gallons of water in the reservoir.

(c) The slope indicates that the reservoir loses 23.3 million gallons of water a day.
(d) When $t = 61$, $A = -0.0233(61) + 38.54 = 37.12$
On October 31, 2006 there were 37.12 billion gallons of water in the reservoir.

(e) The reservoir will be empty when $A = 0$.

$$-0.0233t + 38.54 = 0$$
$$0.0233t = 38.54$$
$$t = 1654.08$$

The reservoir will be empty on day 1654, or after about 4.53 years.

105. (a) When $x = 0$, that is in 2006,

$$S = 5000(0) + 80{,}000$$
$$S = \$80{,}000$$

(b) When $x = 3$, that is in 2009,

$$S = 5000(3) + 80{,}000$$
$$S = \$95{,}000$$

(c) If the trend continues, sales in 2012 should be equal to S when $x = 6$,

$$S = 5000(6) + 80{,}000$$
$$S = \$110{,}000$$

(d) If the trend continues, sales in 2015 should be equal to S when $x = 9$.

$$S = 5000(9) + 80{,}000$$
$$S = \$125{,}000$$

107. (a) Writing 5% as a decimal, 5% = 0.05, we form the equation,

$$S = 0.05x + 400$$

(b) If $x = \$4000.00$, then Dan's earnings will be,

$$\begin{aligned} S &= 0.05(4000) + 400 \\ &= 200 + 400 \\ &= \$600.00 \end{aligned}$$

(c) Let S equal the median earnings, and solve for x.

$$\begin{aligned} 744 &= 0.05x + 400 \\ 344 &= 0.05x \\ 6888 &= x \end{aligned}$$

Dan would need to have sales that generate \$6888.00 in profit to earn the median amount.

109. (a) Since the rate of increase is constant, we use the points to find the slope of the line.

$$m = \frac{S_2 - S_1}{t_2 - t_1} = \frac{499 - 468}{2005 - 1991} = \frac{31}{14} = 2.214$$

Then we choose a point, say (1991, 468), and write the point-slope form of the line.

$$\begin{aligned} S - S_1 &= m(t - t_1) \\ S - 468 &= 2.214(t - 1991) \\ S &= 2.214t - 4408.07 + 468 \\ S &= 2.214t - 3940.07 \end{aligned}$$

(b) If the trend continues, in 2008,

$$S = 2.214(2008) - 3940.07 = 505.64$$

The projected SAT mathematics score is 506.

111. (a) Since we assume the rate of increase is constant, we use the points to find the slope of a line.

$$m = \frac{P_2 - P_1}{t_2 - t_1} = \frac{27.6 - 23.0}{2005 - 1995} = \frac{4.6}{10} = 0.46$$

Now we choose a point, say (23, 1995), and write the point-slope form of the line.

$$\begin{aligned} P - P_1 &= m(t - t_1) \\ P - 23 &= 0.46(t - 1995) \\ P &= 0.46t - 917.7 + 23 \\ P &= 0.46t - 894.7 \end{aligned}$$

(b) To find the predicted percentage of people who will have a bachelor's degree, let $t = 2010$.

$$P = 0.46(2010) - 894.7 = 29.9$$

By 2010 it is predicted that 29.9% of people over 25 years of age will have a bachelor's degree or higher.

(c) The slope is the annual average increase in the percentage of people over 25 years of age who have a bachelor's degree or higher.

113. (a) Since the cost of the houses is linear, we first use the points to find the slope of the line.

$$m = \frac{C_2 - C_1}{t_2 - t_1} = \frac{141{,}200 - 136{,}200}{2005 - 2004} = 5000$$

Next, choose a point, say (2004, 136200), and write the point slope form of the line.

$$C - C_1 = m(t - t_1)$$
$$C - 136{,}000 = 5000(t - 2004)$$
$$C = 5000t - 10{,}020{,}000 + 136{,}000$$
$$C = 5000t - 9{,}884{,}000 \text{ dollars}$$

(b) The projected average cost of a house in 2008 is found by letting $t = 2008$ in the equation.

$$C = 5000(2008) - 9{,}884{,}000$$
$$C = \$156{,}000$$

115. (a) First we find the slope of the line, using $t = 0$ to represent 2004.

$$m = \frac{S_2 - S_1}{t_2 - t_1} = \frac{193.641 - 150.865}{1 - 0} = 42.776$$

Then we use the point (0, 150.865) to write the point-slope form of the equation of the line.

$$S - S_1 = m(t - t_1)$$
$$S - 150.865 = 42.776(t - 0)$$
$$S = 42.776t + 150.865$$

(b) In 2009 $t = 5$, so we get $S = 42.776(5) + 150.865 = 364.745$

In 2009 the total sales and other operating income for Chevron Corporation is predicted to be \$364.745 billion.

117. (a) If the Smith's drive x miles in a year, and their car averages 17 miles per gallon of gasoline, then they will use about $\frac{x}{17}$ gallons of gasoline per year.

In 2006, the Smith's annual fuel cost is given by

$$C = 2.934\left(\frac{x}{17}\right) = 0.1726x$$

(b) In 2005, the Smith's annual fuel cost is given by

$$C = 2.226\left(\frac{x}{17}\right) = 0.1309x$$

(c) Assuming that the Smith's drove 15,000 miles, their fuel cost in 2006 was

$$C = 0.1726(15{,}000) = \$2589.00$$

(d) Assuming that the Smith's drove 15,000 miles, their fuel cost in 2005 was

$$C = 0.1309(15{,}000) = \$1964$$

(e) The difference in the annual costs is \$2589 – \$1964 = \$625. The Smith's spend \$625 more at the 2006 price than at the 2005 price.

119. (a) First we compute the slope of the line.

$$m = \frac{N_2 - N_1}{t_2 - t_1} = \frac{954.7 - 747.3}{4 - 0} = \frac{207.4}{4} = 51.85$$

Then we use the slope and the point (0, 747.3) to write the point-slope form of the line.

$$N - N_1 = m(t - t_1)$$
$$N - 747.3 = 51.85(t - 0)$$
$$N = 51.85t + 747.3$$

(b) The slope indicates that credit and debit cards in force are increasing at an average rate of 51.85 million cards per year.

(c) In 2009, $t = 7$, and $N = 51.85(7) + 747.3 = 1110.25$
There will be an estimated 1.110 billion credit cards in force at the end of the first quarter of 2009.

(d) To estimate the year that the number of credit cards will first exceed 1.5 billion, let $N = 1500$ (since the equation is written in millions), and solve for t.

$$1500 = 51.85t + 747.3$$
$$752.7 = 51.85t$$
$$t = \frac{752.7}{51.85} = 14.52$$

Since $t = 0$ represented the year 2002, $t = 14.52$ is in the year 2016.
The number of credit and debit cards will surpass 1.5 billion in 2016.

121. From the graph we can see that the line has a negative slope and a y-intercept of the form $(0, b)$ where b is a positive number. Put each of the equations into slope-intercept form and choose those with negative slope and positive y-intercept.

(a) $y = -\frac{2}{3}x + 2$ (c) $y = -\frac{3}{4}x + 3$ (f) $y = -2x + 1$ (g) $y = -\frac{1}{2}x + 10$

123. A vertical line cannot be written in slope-intercept form since its slope is not defined.

125. Two lines that have equal slopes and equal y-intercepts have equivalent equations and identical graphs.

127. If two lines have the same slope, but different x-intercepts, they cannot have the same y-intercept. If Line 1 has x-intercept $(a, 0)$ and Line 2 has x-intercept $(c, 0)$, but both have the same slope m, write the equation of each line using the point-slope form then change them to slope-intercept form and compare the y-intercepts:

Line 1:	Line 2:
$y - 0 = m(x - a)$	$y - 0 = m(x - c)$
$y = mx - ma$	$y = mx - mc$
y-intercept is $(0, -ma)$	y-intercept is $(0, -mc)$

Since a and c are different and m is not zero, $-ma$ and $-mc$ must be different.

129. **monter** (*t.v.*) to go up; to ascend, to mount; to climb; to embark; to rise, to slope up, to be uphill; to grow up; to shoot; to increase. (source: *Cassell's French Dictionary*).

We use m to represent slope. The French verb monter means to rise, to climb or to slope up.

1.2 Pairs of Lines

1. parallel

3. To determine whether the pair of lines is parallel, coincident, or intersecting, rewrite each equation in slope-intercept form, compare their slopes, and, if necessary, compare their y-intercepts.

$L: \quad x + y = 10$
$\quad y = -x + 10$

$M: \quad 3x + 3y = 6$
$\quad 3y = -3x + 6$
$\quad y = -x + 2$

slope: $m = -1$; y-intercept: $(0, 10)$ slope: $m = -1$; y-intercept: $(0, 2)$
The slopes of the two lines are the same, but the y-intercepts are different, so the lines are parallel.

5. To determine whether the pair of lines is parallel, coincident, or intersecting, rewrite each equation in slope-intercept form, compare their slopes, and, if necessary, compare their y-intercepts.

$L: \quad 2x + y = 4$
$\quad y = -2x + 4$

$M: \quad 2x - y = 8$
$\quad -y = -2x + 8$
$\quad y = 2x - 8$

slope: $m = -2$ slope: $m = 2$
Since the slopes of the two lines are different, the lines intersect.

7. To determine whether the pair of lines is parallel, coincident, or intersecting, rewrite each equation in slope-intercept form, compare their slopes, and, if necessary, compare their y-intercepts.

$L: \quad -x + y = 2$
$\quad y = x + 2$

$M: \quad 2x - 2y = -4$
$\quad -2y = -2x - 4$
$\quad y = x + 2$

slope: $m = 1$; y-intercept: $(0, 2)$ slope: $m = 1$; y-intercept: $(0, 2)$
Since both the slopes and the y-intercepts of the two lines are the same, the lines are coincident.

9. To determine whether the pair of lines is parallel, coincident, or intersecting, rewrite each equation in slope-intercept form, compare their slopes, and, if necessary, compare their y-intercepts.

$$L: \quad 2x - 3y = -8 \qquad -3y = -2x - 8 \qquad y = \frac{2}{3}x + \frac{8}{3}$$

$$M: \quad 6x - 9y = -2 \qquad -9y = -6x - 2 \qquad y = \frac{2}{3}x + \frac{2}{9}$$

slope: $m = \frac{2}{3}$; y-intercept: $\left(0, \frac{8}{3}\right)$ slope: $m = \frac{2}{3}$; y-intercept: $\left(0, \frac{2}{9}\right)$

The slopes of the two lines are the same, but the y-intercepts are different, so the lines are parallel.

11. To determine whether the pair of lines is parallel, coincident, or intersecting, rewrite each equation in slope-intercept form, compare their slopes, and, if necessary, compare their y-intercepts.

$$L: \quad 3x - 4y = 1 \qquad -4y = -3x + 1 \qquad y = \frac{3}{4}x - \frac{1}{4}$$

$$M: \quad x - 2y = -4 \qquad -2y = -x - 4 \qquad y = \frac{1}{2}x + 2$$

slope: $m = \frac{3}{4}$ slope: $m = \frac{1}{2}$

Since the slopes of the two lines are different, the lines intersect.

13. L: $x = 3$ M: $y = -2$

slope: not defined; no y-intercept slope: $m = 0$; y-intercept: $(0, -2)$

Since the slopes of the two lines are different, the lines intersect.

15. To find the point of intersection of two lines, first put the lines in slope-intercept form.

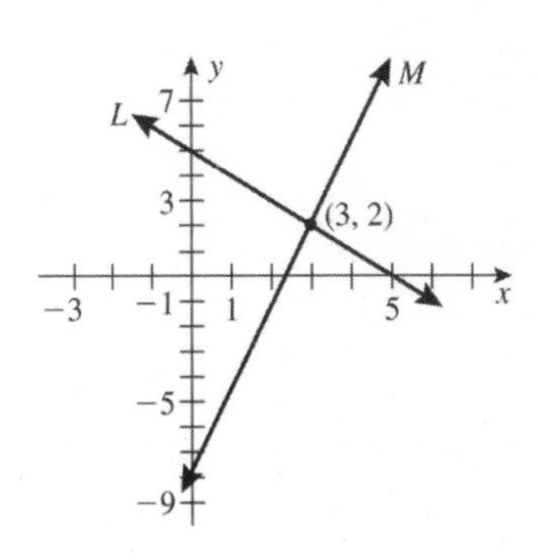

$$L: \quad x + y = 5 \qquad y = -x + 5$$

$$M: \quad 3x - y = 7 \qquad y = 3x - 7$$

Since the point of intersection, (x_0, y_0), must be on both L and M, we set the two equations equal to each other and solve for x_0. Then we substitute the value of x_0 into the equation of one of the lines to find y_0.

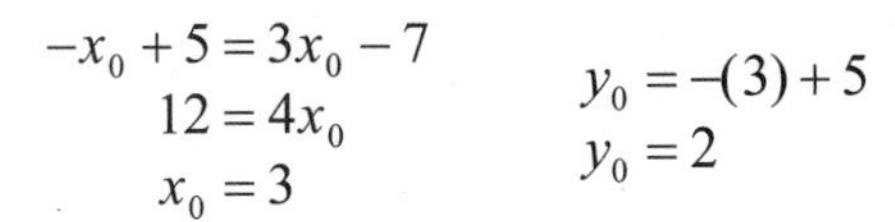

$$-x_0 + 5 = 3x_0 - 7 \qquad 12 = 4x_0 \qquad x_0 = 3$$

$$y_0 = -(3) + 5 \qquad y_0 = 2$$

The point of intersection is $(3, 2)$.

17. To find the point of intersection of two lines, first put the lines in slope-intercept form.

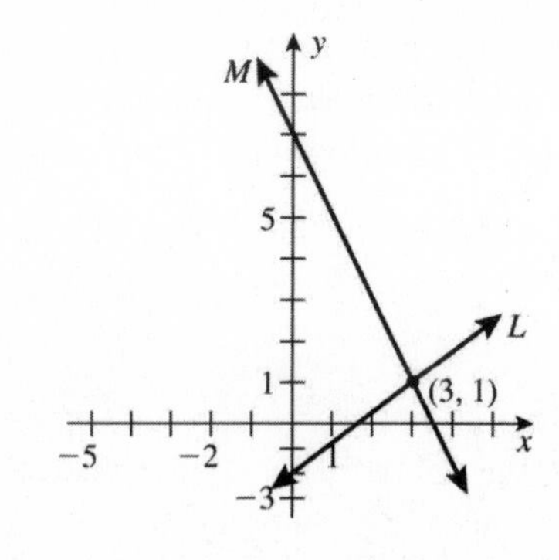

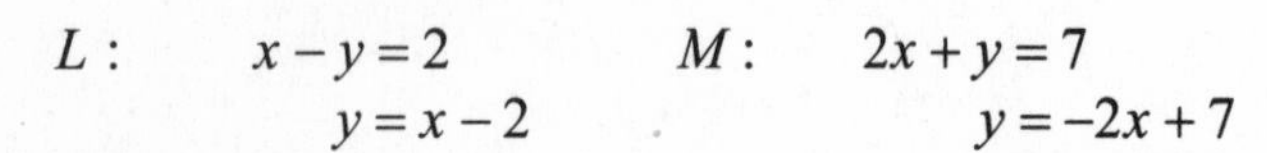

$L: \quad x - y = 2 \qquad\qquad M: \quad 2x + y = 7$

$\qquad y = x - 2 \qquad\qquad\qquad y = -2x + 7$

Since the point of intersection, (x_0, y_0), must be on both L and M, we set the two equations equal to each other and solve for x_0. Then we substitute the value of x_0 into the equation of one of the lines to find y_0.

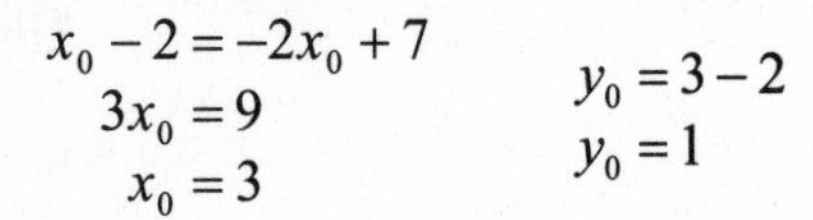

$$\begin{aligned} x_0 - 2 &= -2x_0 + 7 \\ 3x_0 &= 9 \\ x_0 &= 3 \end{aligned} \qquad \begin{aligned} y_0 &= 3 - 2 \\ y_0 &= 1 \end{aligned}$$

The point of intersection is (3, 1).

19. To find the point of intersection of two lines, first put the lines in slope-intercept form.

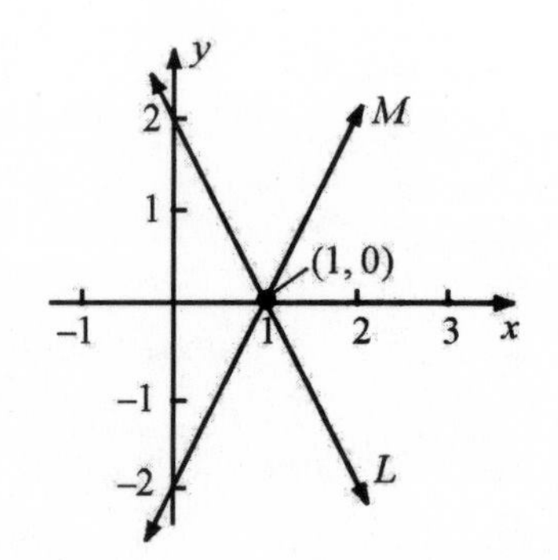

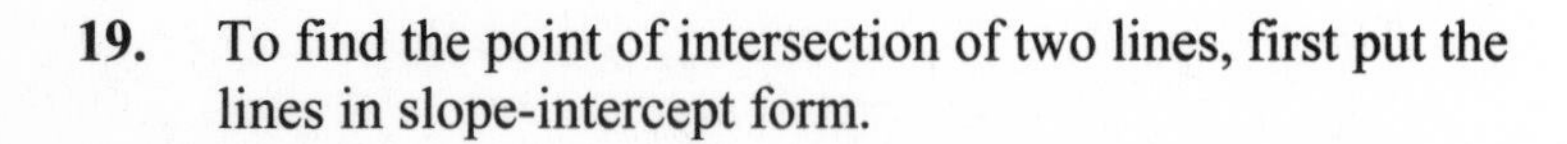

$L: \quad 4x + 2y = 4 \qquad\qquad M: \quad 4x - 2y = 4$

$\qquad y = -2x + 2 \qquad\qquad\qquad y = 2x - 2$

Since the point of intersection, (x_0, y_0), must be on both L and M, set the two equations equal to each other and solve for x_0. Then substitute the value of x_0 into the equation of one of the lines to find y_0.

$$\begin{aligned} -2x_0 + 2 &= 2x_0 - 2 \\ 4 &= 4x_0 \\ x_0 &= 1 \end{aligned} \qquad \begin{aligned} y_0 &= -2(1) + 2 \\ y_0 &= 0 \end{aligned}$$

The point of intersection is (1, 0).

21. To find the point of intersection of two lines, first put the lines in slope-intercept form.

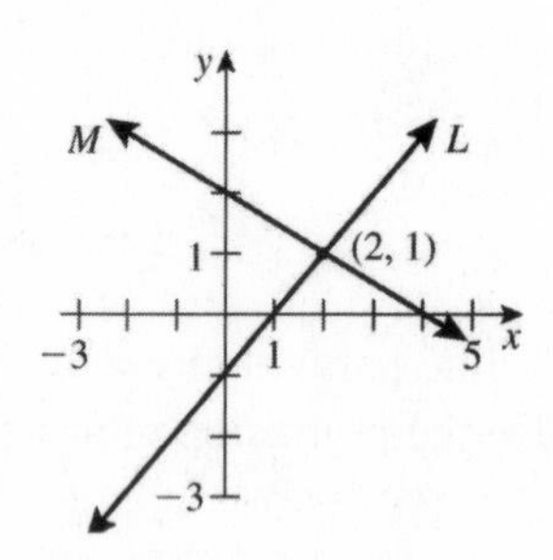

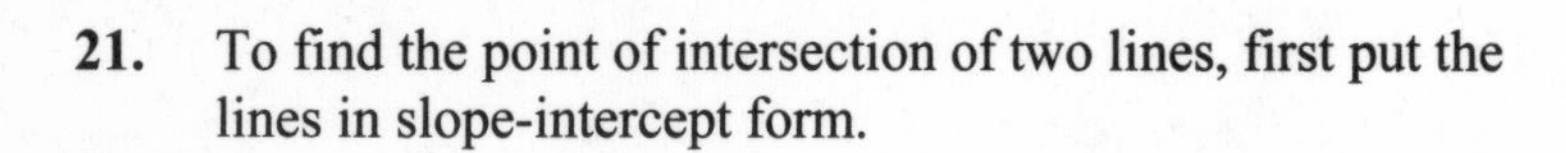

$L: \ 3x - 4y = 2 \qquad\qquad M: \ x + 2y = 4$

$\qquad y = \frac{3}{4}x - \frac{1}{2} \qquad\qquad\qquad y = -\frac{1}{2}x + 2$

Since the point of intersection, (x_0, y_0), must be on both L and M, we set the two equations equal to each other and solve for x_0. Then we substitute the value of x_0 into the equation of one of the lines to find y_0.

$$\begin{aligned} \frac{3}{4}x_0 - \frac{1}{2} &= -\frac{1}{2}x_0 + 2 \\ 3x_0 - 2 &= -2x_0 + 8 \\ 5x_0 &= 10 \\ x_0 &= 2 \end{aligned} \qquad \begin{aligned} y_0 &= -\frac{1}{2}(2) + 2 \\ y_0 &= 1 \end{aligned}$$

The point of intersection is (2, 1).

23. To find the point of intersection of two lines, first put the lines in slope-intercept form.

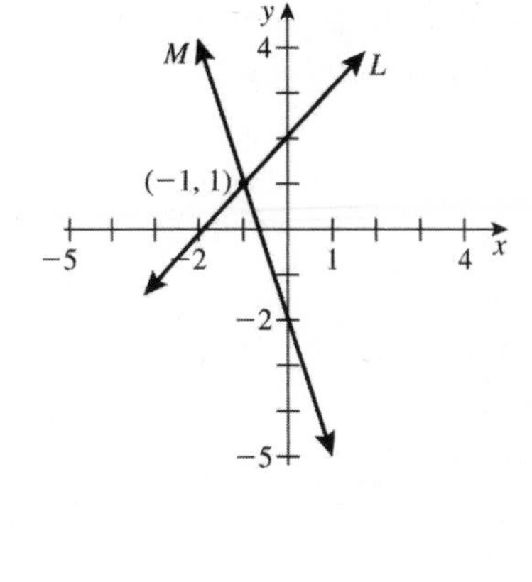

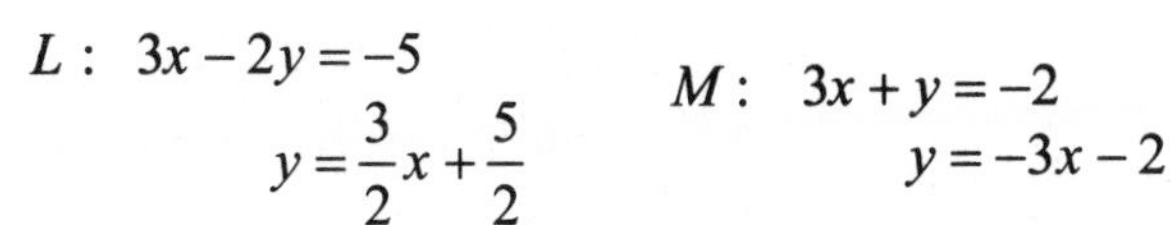

$$L:\ 3x - 2y = -5 \qquad\qquad M:\ 3x + y = -2$$
$$y = \frac{3}{2}x + \frac{5}{2} \qquad\qquad y = -3x - 2$$

Since the point of intersection, (x_0, y_0), must be on both L and M, we set the two equations equal to each other and solve for x_0. Then we substitute the value of x_0 into the equation of one of the lines to find y_0.

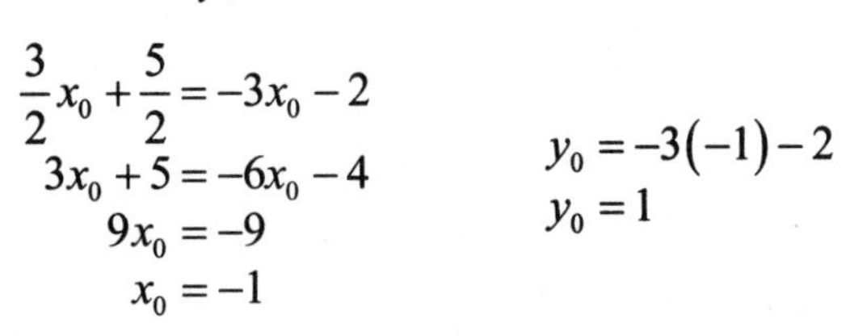

$$\begin{aligned} \frac{3}{2}x_0 + \frac{5}{2} &= -3x_0 - 2 \\ 3x_0 + 5 &= -6x_0 - 4 \\ 9x_0 &= -9 \\ x_0 &= -1 \end{aligned} \qquad \begin{aligned} y_0 &= -3(-1) - 2 \\ y_0 &= 1 \end{aligned}$$

The point of intersection is (−1, 1).

25. L is the vertical line on which the x-value is always 4.

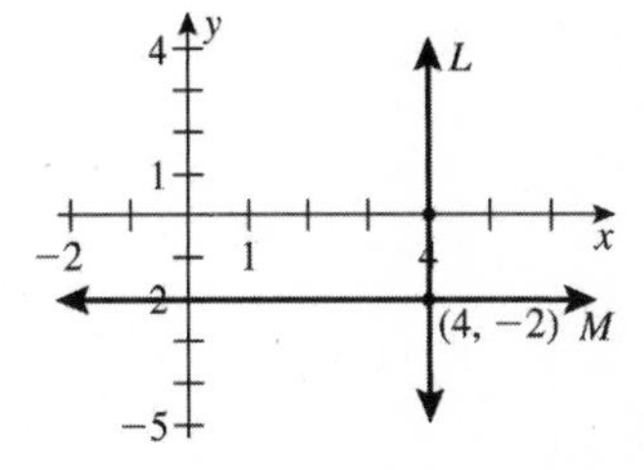

M is the horizontal line on which y-value is always −2.

The point of intersection is (4, −2).

27. L is parallel to $y = 2x$, so the slope of L is $m = 2$. We are given the point (3, 3) on line L. Use the point-slope form of the line.

$$y - y_1 = m(x - x_1)$$
$$y - 3 = 2(x - 3)$$
$$y - 3 = 2x - 6$$

slope-intercept form: $y = 2x - 3$

general form: $2x - y = 3$

29. We want a line parallel to $y = 4x$. So our line will have slope $m = 4$. It must also contain the point (−1, 2). Use the point slope form of the equation of a line:

$$y - y_1 = m(x - x_1)$$
$$y - 2 = 4(x + 1)$$
$$y - 2 = 4x + 4$$

slope-intercept form: $y = 4x + 6$

general form: $4x - y = -6$

31. We want a line parallel to $2x - y = -2$. Find the slope of the line and use the given point, (0, 0) to obtain the equation. Since the y-intercept (0, 0) is given, use the slope-intercept form of the equation of a line.

Original line:

$$\begin{aligned} 2x - y &= -2 \\ y &= 2x + 2 \\ m &= 2 \end{aligned}$$

Parallel line:

$$\begin{aligned} y &= mx + b \\ y &= 2x + 0 \\ y &= 2x \end{aligned}$$

general form: $2x - y = 0$

33. We want a line parallel to the line $x = 3$. This is a vertical line so the slope is not defined. A parallel line will also be vertical, and it must contain the point (4, 2). The parallel line will have the equation $x = 4$

35. To find the equation of the line, we must first find the slope of the line containing the points (–2, 9) and (3, –10).

$$m = \frac{y_2 - y_1}{x_2 - x_1} = \frac{9-(-10)}{(-2)-3} = \frac{19}{-5} = -\frac{19}{5}$$

The slope of a line parallel to the line containing these points is also $-\frac{19}{5}$. Use the slope and the point (–2, –5) to write the point-slope form of the parallel line.

$$\begin{aligned} y - y_1 &= m(x - x_1) \\ y + 5 &= -\frac{19}{5}(x+2) \end{aligned}$$

Solve for y to get the slope-intercept form:

$$\begin{aligned} y &= -\frac{19}{5}x - \frac{38}{5} - 5 \\ y &= -\frac{19}{5}x - \frac{63}{5} \end{aligned}$$

Multiply both sides by 5 and rearrange terms to obtain the general form of the equation:

$$19x + 5y = -63$$

37. (a) Assuming that the rate of growth is constant, we find the slope of the line passing through the points (0, 176) and (365, 624).

$$m = \frac{y_2 - y_1}{x_2 - x_1} = \frac{624-176}{365-0} = \frac{448}{365} = 1.227$$

Next use the slope and the point (0, 176) to write the point-slope form of the line.

$$\begin{aligned} y - y_1 &= m(x - x_1) \\ y - 176 &= 1.227(x - 0) \\ y &= 1.227x + 176 \end{aligned}$$

(b) To find the point of intersection of the line $y = 1.227x + 176$ and the line $y = 700$, set them equal and solve for x.

$$\begin{aligned} 700 &= 1.227x + 176 \\ 524 &= 1.227x \\ x &= \frac{524}{1.227} = 427.058 \end{aligned}$$

There will be 700 HD radio stations 427 days after December 31, 2004 or on March 4, 2006.

39. We will let

$x =$ the number of caramels the box of candy, and
$y =$ the number of creams in the box of candy.

Since there are a total of 50 pieces of candy in a box, we have

$$x + y = 50 \text{ , or}$$
$$y = 50 - x$$

Each caramel costs \$0.10 to make, and each cream costs \$0.20 to make. So, the cost of making a box of candy is given by the equation:

$$\begin{aligned} C &= 0.1x + 0.2y \\ &= 0.1x + 0.2(50 - x) \end{aligned}$$

The box of candy sells for \$8.00. So to break even, we need

$$\begin{aligned} R &= C \\ 8 &= 0.1x + 0.2(50 - x) \\ 8 &= 0.1x + 10 - 0.2x \\ 8 &= 10 - 0.1x \\ 0.1x &= 2 \\ x &= 20 \end{aligned}$$

To break even, put 20 caramels and 30 creams into each box.
If the candy shop owner increases the number of caramels to more than 20 (and decreases the number of creams) the owner will obtain a profit since the caramels cost less to produce than the creams.

41. Investment problems are simply mixture problems involving money. We will use a table to organize the information.

Let x denote the amount Mr. Nicholson invests in AA bonds, and y denote the amount he invests in S & L Certificates.

Investment	Amount Invested	Interest Rate	Interest Earned
AA Bonds	x	$10\% = 0.10$	$0.10x$
S & L Certificates	$y = 150{,}000 - x$	$5\% = 0.05$	$0.05y = 0.05(150{,}000 - x)$
Total	$x + y = 150{,}000$		\$10,000

The last column gives the information we need to set up the equation solve since the sum of the interest earned on the two investments must equal the total interest earned.

$$\begin{aligned} 0.1x + 0.05(150{,}000 - x) &= 10{,}000 \\ 0.1x + 7500 - 0.05x &= 10{,}000 \\ 0.05x + 7500 &= 10{,}000 \\ 0.05x &= 2500 \\ x &= 50{,}000 \end{aligned}$$

Mr. Nicholson should invest \$50,000 in AA Bonds and \$100,000 in Savings and Loan Certificates in order to earn \$10,000 per year.

43. Let x denote the amount of Kona coffee and y denote the amount of Columbian coffee in the mix. We will use the hint and assume that the total weight of the blend is 100 pounds.

Coffee	Amount Mixed	Price per Pound	Total Value
Kona	x	\$22.95	$\$22.95x$
Columbian	$y = 100 - x$	\$ 6.75	$\$6.75y = \$6.75(100 - x)$
Mixture	$x + y = 100$	\$10.80	$\$10.80(100) = \$1080$

The last column gives the information necessary to write the equation, since the sum of the values of each of the two individual coffees must equal the total value of the mixture.

$$\begin{aligned} 22.95x + 6.75(100 - x) &= 10.80(100) \\ 22.95x + 675 - 6.75x &= 1080 \\ 16.2x &= 405 \\ x &= 25 \end{aligned}$$

Mix 25 pounds of Kona coffee with 75 pounds of Columbian coffee to obtain a blend worth \$10.80 per pound.

45. This is a classical mixture problem. Let x represent the amount of Acid A used and let y represent the amount of Acid B used in the solution. We will use a table to organize the information.

Ingredients	Amount Used	Acidity	Solution
Acid A	x	$15\% = 0.15$	$0.15x$
Acid B	$y = 100 - x$	$5\% = 0.05$	$0.05y = 0.05(100 - x)$
Total	$x + y = 100$	$8\% = 0.08$	$0.08(100) = 8$

The last column gives the information necessary to determine how much of each of the two ingredients should be mixed, since the sum of the two individual solutions must equal the total solution.

$$\begin{aligned} 0.15x + 0.05(100 - x) &= 8 \\ 0.15x + 5 - 0.05x &= 8 \\ 0.1x &= 3 \\ x &= 30 \end{aligned}$$

We should mix 30 cubic centimeters of 15% solution with 70 cubic centimeters of 5% solution to obtain a 100 cubic centimeters of 8% solution.

47. (a) The realized gain, y, is the difference between the 2001 value and the 2006 value of the gold.

$$\begin{aligned} y &= 549.86x - 265.79x \\ y &= 284.07x \end{aligned}$$

(b) The point of intersection is the point (x_0, y_o) that satisfies both equation $y = 284.07x$ and equation $y = 10{,}000$.

$$\begin{aligned} 10{,}000 &= 284.07x \\ x &= \frac{10{,}000}{284.07} = 35.20 \end{aligned}$$

The individual would have had to bought and sold 35.2 ounces of gold to realize a gain of \$10,000.00.

49. The two lines in the graph are parallel. Parallel lines have equal slopes, so to determine which set of equations is parallel we must compare their slopes. Do this by writing each equation in slope-intercept form.

Also notice that the slopes of the graphed equations are positive, and one line has a positive y-intercept and the other has a negative y-intercept.

Use all this information to answer the question.

(c) $x-y=-2$ $\quad$ $x-y=1$

$y=x+2$ $\quad$ $y=x-1$

both equations have slope $m=1$. The first has y-intercept (0, 2) and the second has y-intercept (0, –1), so these equations might have the graph illustrated.

1.3 Applications to Business and Economics

1. False

3. The break-even point is the point where the revenue and the cost are equal.

Setting $R=C$, we find

$$30x=10x+600$$
$$20x=600$$
$$x=30$$

That is, 30 units must be sold to break even. The break-even point is $x=30$, $R=30(30)=900$ or (30, 900).

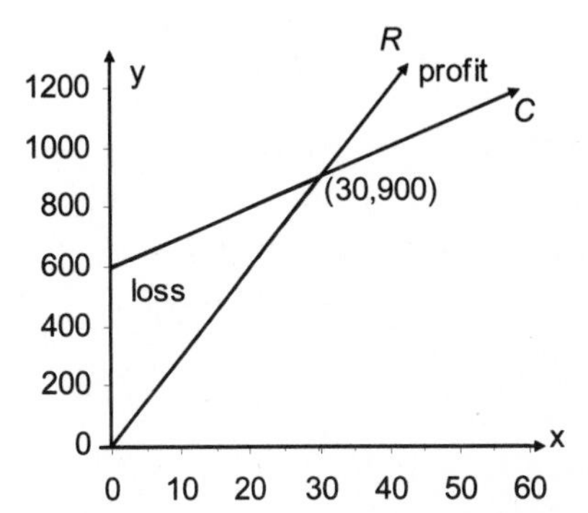

5. The break-even point is the point where the revenue and the cost are equal.

Setting $R=C$, we find

$$0.30x=0.20x+50$$
$$0.10x=50$$
$$x=500$$

That is, 500 units must be sold to break even.
Break-even point: (500, 150)

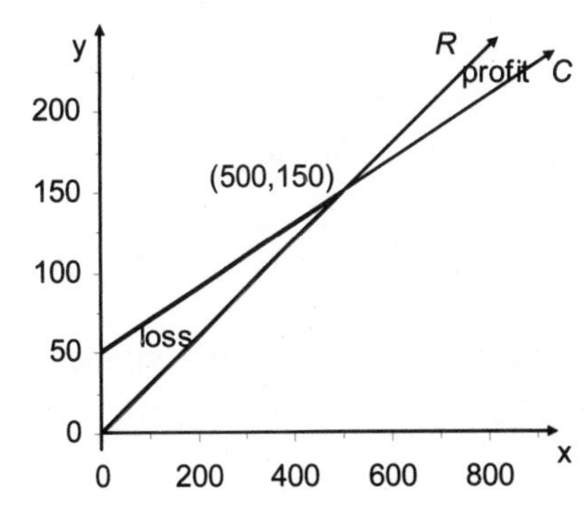

7. The market price is the price at which the supply and the demand are equal.

$$S=D$$
$$p+1=3-p$$
$$2p=2$$
$$p=1$$

At a price of \$1.00 supply and demand are equal, so \$1.00 is the market price.

9. The market price is the price at which the supply and the demand are equal.

$$\begin{aligned} S &= D \\ 20p + 500 &= 1000 - 30p \\ 50p &= 500 \\ p &= 10 \end{aligned}$$

At a price of \$10.00 supply and demand are equal, so \$10.00 is the market price.

11. The break-even point is the point where the revenue and the cost are equal.

Cost is given by the variable cost of producing x pennants at \$0.75 per pennant, plus the fixed operational overhead of \$300 per day.

$$C = \$0.75x + \$300$$

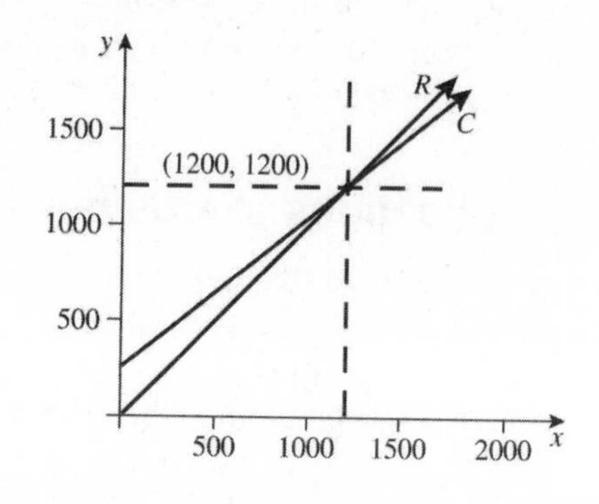

Revenue is the product of price of each pennant (\$1) and the number of pennants sold.

$$R = \$1x$$

Setting $R = C$, we find

$$\begin{aligned} 1x &= 0.75x + 300 \\ 0.25x &= 300 \\ x &= 1200 \end{aligned}$$

1200 pennants must be sold each day to break even.

13. **(a)** The market price is the price at which the supply and the demand are equal.

$$\begin{aligned} S &= D \\ 0.7p + 0.4 &= -0.5p + 1.6 \\ 1.2p &= 1.2 \\ p &= 1 \end{aligned}$$

The market price is \$1.00 per pound.

(b) To find the quantity supplied at market price, let $p = 1$ and solve for S:

$$\begin{aligned} S &= 0.7(1) + 0.4 \\ &= 1.1 \end{aligned}$$

So 1.1 million pounds are demanded at \$1.00.

(c)

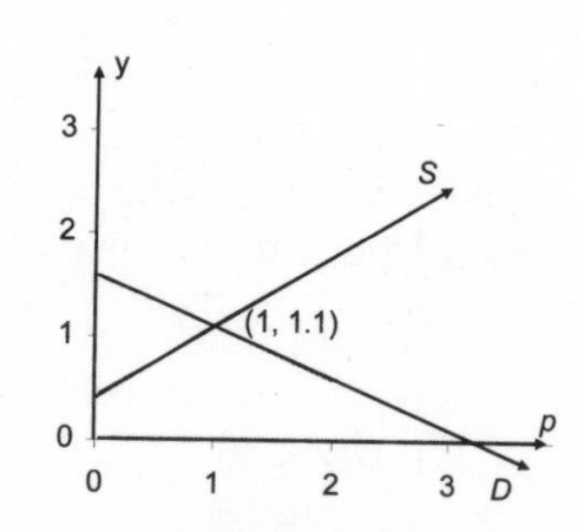

(d) The point of intersection called the market equilibrium. It is the price where the quantity supplied equals the quantity demanded.

15. At the market price of \$3.00, $S = 2(3) + 5 = 11$
so 11 units of the commodity are supplied. Since at the market price the supply and demand are equal, the point (3, 11) satisfies the demand equation. In addition, we are told that at a price of \$1.00, 19 units are demanded. To find the demand equation, use the points (3, 11) and (1, 19) to find the slope.

$$m = \frac{D_2 - D_1}{p_2 - p_1} = \frac{19 - 11}{1 - 3} = \frac{8}{-2} = -4$$

We then use the point (1, 19), the slope $m = -4$, and the point-slope form of the equation.

$$D - D_1 = m(p - p_1)$$
$$D - 19 = -4(p - 1)$$
$$D = -4p + 23$$

17. (a) If R denotes the revenue and p denotes the price per paper, then

$$R = px$$
$$R = 1.79x$$

The revenue from delivering x newspapers is $R = \$1.79x$.

(b) The cost of delivering x newspapers is $C = 1.13x + 1{,}252{,}000$

(c) Profit is the difference between revenue and cost.

$$P = R - C$$
$$P = 1.79x - (1.13x + 1{,}252{,}000)$$
$$P = 0.66x - 1{,}252{,}000$$

The profit from delivering x newspapers is given by $P = \$0.66x - \$1{,}252{,}000$.

(d) The break-even point is the quantity for which the profit is zero.

$$0 = 0.66x - 1{,}252{,}000$$
$$0.66x = 1{,}252{,}000$$
$$x = \frac{1{,}252{,}000}{0.66} = 1{,}896{,}969.697$$

The Tribune must deliver 1,896,970 Sunday papers to break even.

(e)

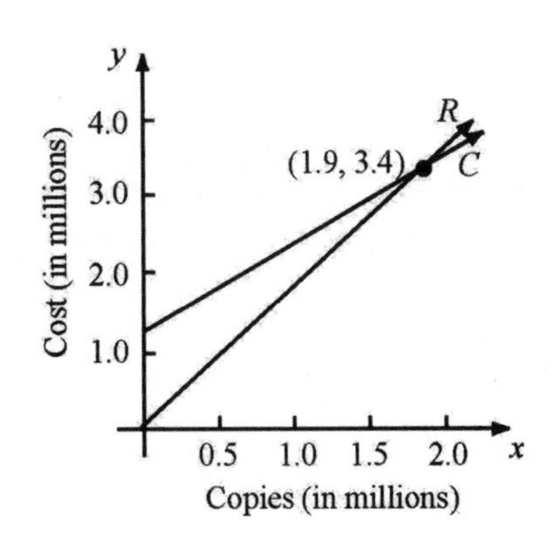

(f)

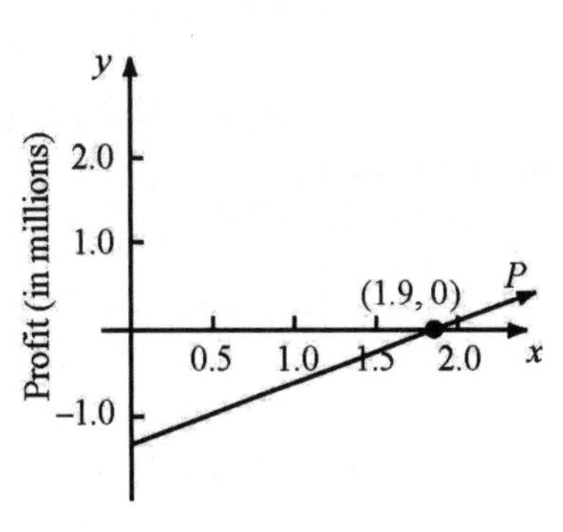

(g) The break-even point is the same as the x-intercept of the profit equation.

19. (a) Since we assume the relationship is linear, we find the equation of a line. We first use the two points to determine the slope.

$$m = \frac{R_2 - R_1}{x_2 - x_1} = \frac{2.407 - 2.153}{2006 - 2005} = 0.254$$

We use the slope and the point (2006, 2.407) to write the point-slope form of the equation.

$$\begin{aligned} R - R_1 &= m(x - x_1) \\ R - 2.407 &= 0.254(x - 2006) \\ R &= 0.254x - 509.524 + 2.407 \\ R &= 0.254x - 507.117 \end{aligned}$$

(b) For the outlay, the slope is

$$m = \frac{O_2 - O_1}{x_2 - x_1} = \frac{2.654 - 2.472}{2006 - 2005} = 0.182$$

We use the slope and the point (2006, 2.654) to write the equation.

$$\begin{aligned} O - O_1 &= m(x - x_1) \\ O - 2.654 &= 0.182(x - 2006) \\ O &= 0.182x - 365.092 + 2.654 \\ O &= 0.182x - 362.438 \end{aligned}$$

(c) The budget will be balanced in the year in which revenue equals outlay.

$$\begin{aligned} R &= O \\ 0.254x - 507.117 &= 0.182x - 362.438 \\ 0.072x &= 144.679 \\ x &= \frac{144.679}{0.072} = 2009.43 \end{aligned}$$

The budget will be balanced during 2009.

21. Let x denote the number of DVDs purchased.
The cost of purchasing x DVDs from the club is

$$\begin{aligned} C &= 0.49(4) + 17.95(x - 4) + 2.31x \\ &= 1.96 + 17.95x - 71.8 + 2.31x \\ &= 20.26x - 69.84 \end{aligned}$$

The cost of purchasing x DVDs from the discount retailer is

$$\begin{aligned} D &= 14.95(1.07)x \\ &= 16.00x \end{aligned}$$

We want to buy as many DVDs as possible from the club, while keeping the cost below the retailers. So we set $C = D$ and solve for x.

$$\begin{aligned} 20.26x - 69.84 &= 16.00x \\ 4.26x &= 69.84 \\ x &= 16.39 \end{aligned}$$

At most sixteen DVDs can be ordered from the club to keep the price lower than that of the discount retailer.

1.4 Scatter Diagrams; Linear Curve Fitting

1. True

3. A relation exists, and it appears to be linear.

5. A relation exists, and it appears to be linear.

7. No relation exists.

9. (a) and (c)

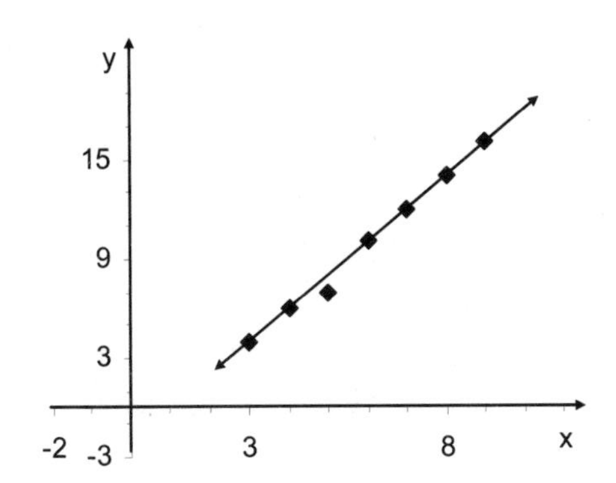

(b) Answers will vary. We select points (3, 4) and (9, 16). The slope of the line containing these points is:

$$m = \frac{y_2 - y_1}{x_2 - x_1} = \frac{16-4}{9-3} = \frac{12}{6} = 2$$

The equation of the line is:

$$y - y_1 = m(x - x_1)$$
$$y - 4 = 2(x - 3)$$
$$y - 4 = 2x - 6$$
$$y = 2x - 2$$

(c) The line on the scatter diagram on the left will vary depending on your choice of points in part (b).

(d) Window: Xmin = −2; Xmax = 10
Ymin = −3; Ymax = 20

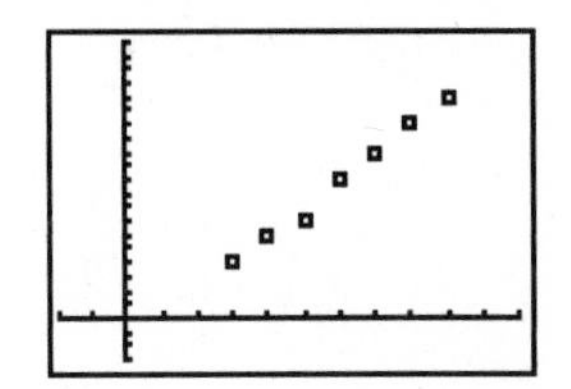

(e) Using the LinReg program, the line of best fit is:

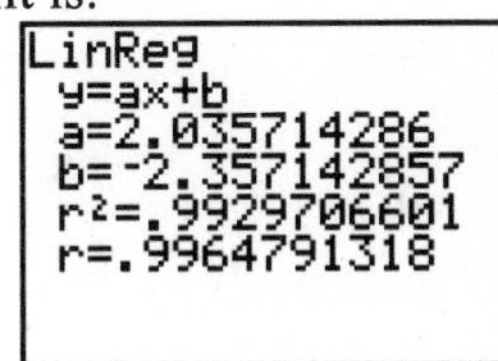

$y = 2.0357x - 2.357$

(f)

11. (a) and (c)

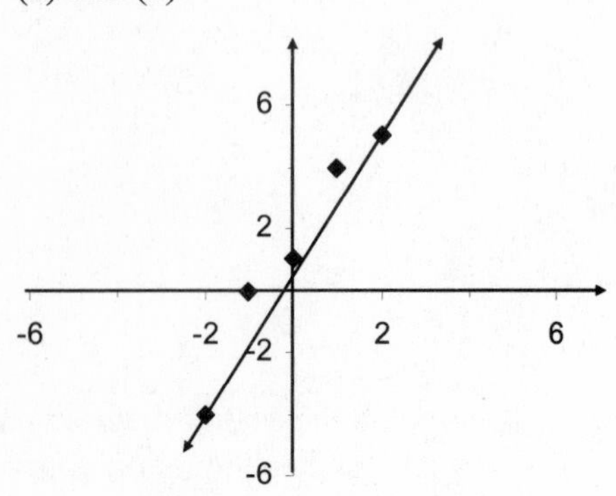

(b) Answers will vary. We select points (−2, −4) and (2, 5). The slope of the line containing these points is:

$$m = \frac{y_2 - y_1}{x_2 - x_1} = \frac{5-(-4)}{2-(-2)} = \frac{9}{4}$$

The equation of the line is:

$$y - y_1 = m(x - x_1)$$
$$y - 5 = \frac{9}{4}(x - 2)$$
$$y - 5 = \frac{9}{4}x - \frac{9}{2}$$
$$y = \frac{9}{4}x + \frac{1}{2}$$

(c) The line on the scatter diagram on the left will vary depending on your choice of points in part (b).

(d) Window: Xmin = −6; Xmax = 6
Ymin = −6; Ymax = 7

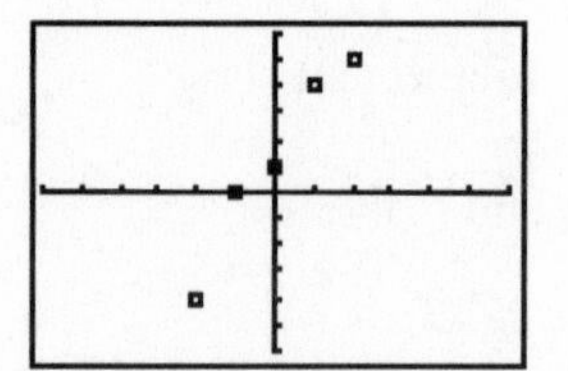

(e) Using the LinReg program the line of best fit is:

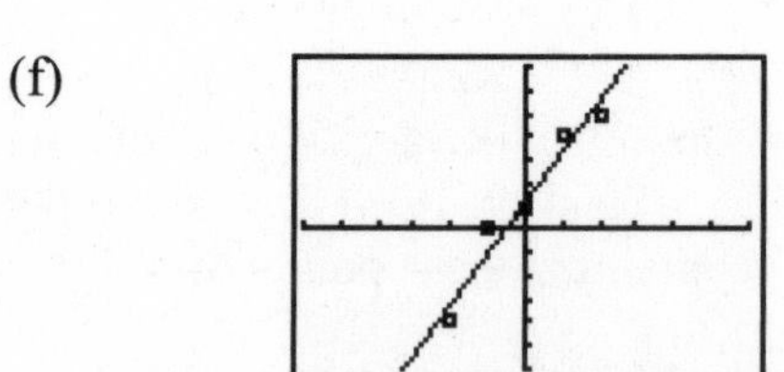

$y = 2.2x + 1.2$

(f)

13. (a) and (c)

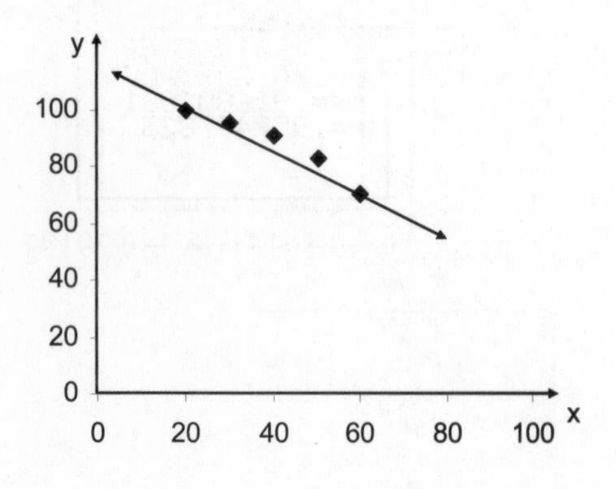

(b) Answers will vary. We select points (20, 100) and (60, 70). The slope of the line containing these points is:

$$m = \frac{y_2 - y_1}{x_2 - x_1} = \frac{70-100}{60-20} = \frac{-30}{40} = -\frac{3}{4}$$

The equation of the line is:

$$y - y_1 = m(x - x_1)$$
$$y - 100 = -\frac{3}{4}(x - 20)$$
$$y - 100 = -\frac{3}{4}x + 15$$
$$y = -\frac{3}{4}x + 115$$

(c) The line on the scatter diagram on the left will vary depending on your choice of points in part (b).

(d) Window: Xmin = 0; Xmax = 100
Ymin = 1; Ymax = 120

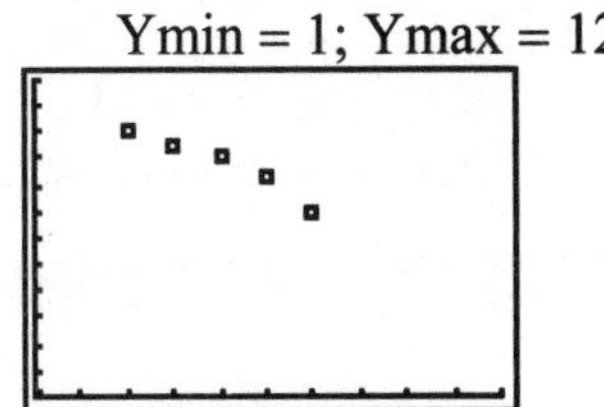

(e) Using the LinReg function the line of best fit is:

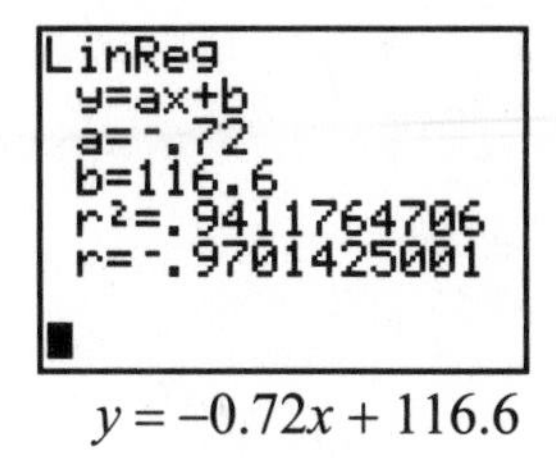

$y = -0.72x + 116.6$

(f)

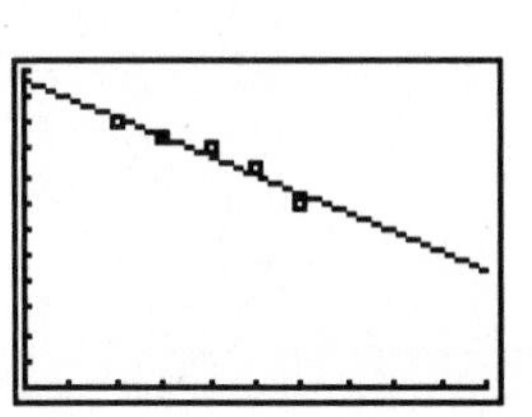

15. (a) and (c)

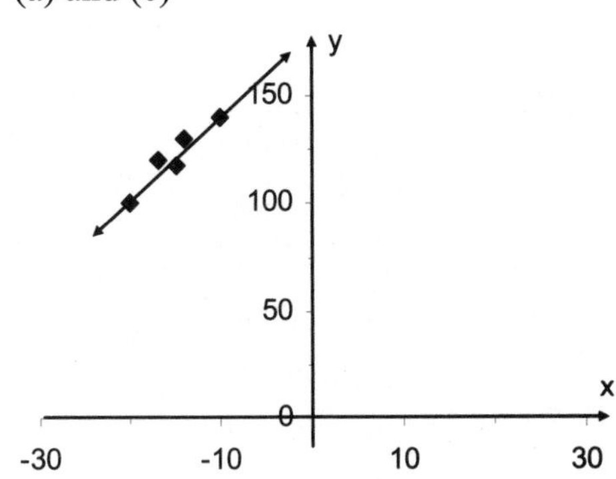

(b) Answers will vary. We select points (–20, 100) and (–10, 140). The slope of the line containing these points is:

$$m = \frac{y_2 - y_1}{x_2 - x_1} = \frac{140 - 100}{(-10) - (-20)} = \frac{40}{10} = 4$$

The equation of the line is:

$$y - y_1 = m(x - x_1)$$
$$y - 100 = 4(x - (-20))$$
$$y - 100 = 4x + 80$$
$$y = 4x + 180$$

(c) The line on the scatter diagram on the left will vary depending on your choice of points in part (b).

(d) Window: Xmin = –30; Xmax = 10
Ymin = 0; Ymax = 160

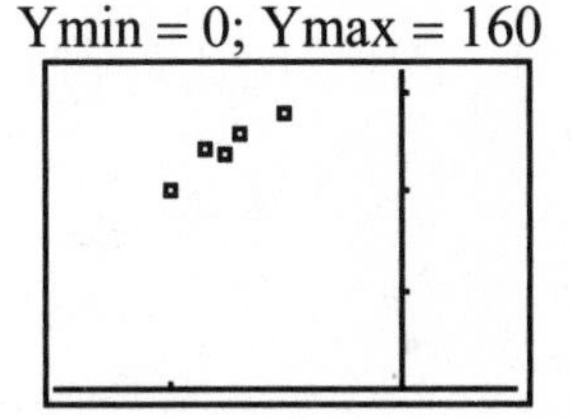

(e) Using the LinReg function, the line of best fit is:

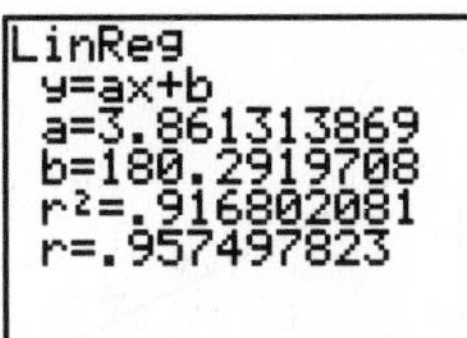

$y = 3.86131x + 180.29197$

(f)

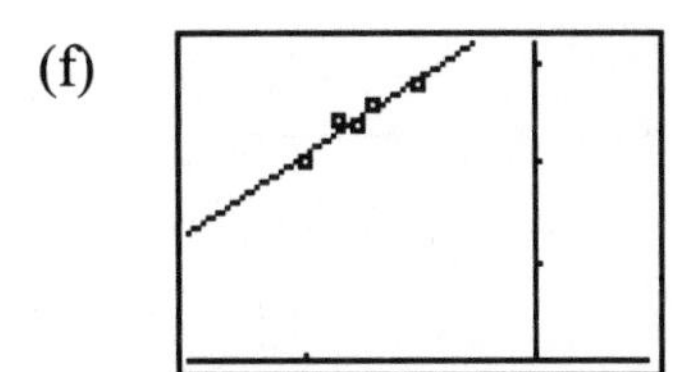

17. (a)

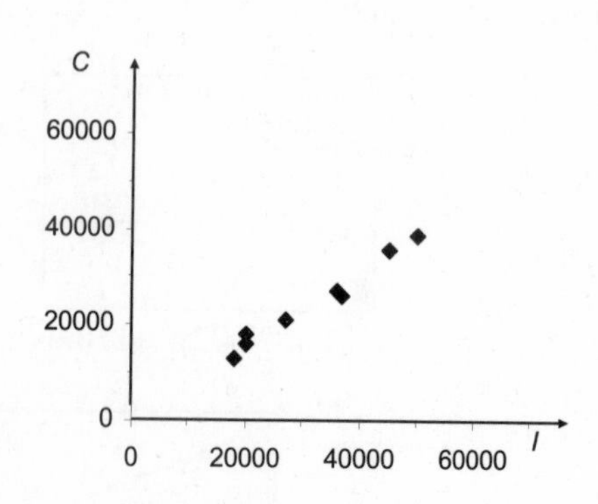

(b) Answers will vary. We select points (20, 16) and (50, 39) numbers in thousands. The slope of the line containing these points is:

$$m = \frac{C_2 - C_1}{I_2 - I_1} = \frac{39-16}{50-20} = \frac{23}{30}$$

The equation of the line is:

$$C - C_1 = m(I - I_1)$$
$$C - 16 = \frac{23}{30}(I - 20)$$
$$C - 16 = \frac{23}{30}I - \frac{46}{3}$$
$$C = \frac{23}{30}I + \frac{2}{3}$$

(c) The slope of this line indicates that a family will spend $23 of every extra $30 of disposable income.

(d) To find the consumption of a family whose disposable income is $42,000, substitute 42 for x in the equation from part (b).

$$C = \frac{23}{30}(42) + \frac{2}{3}$$
$$= \frac{473}{15}$$
$$= 32.86667$$

So the family will spend $32,867.

(e) Using the LinReg function on the graphing utility, the line of best fit is:

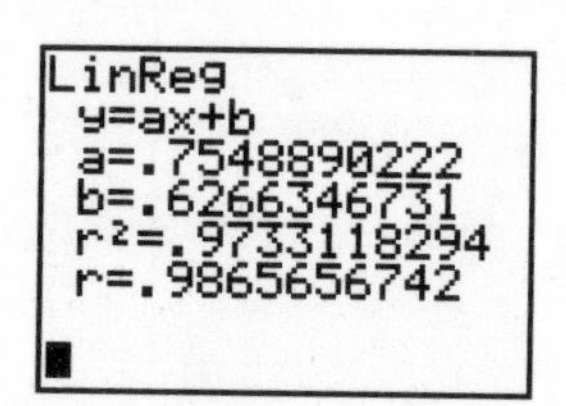

$y = 0.75489x + 0.62663$

19. (a) Window: Xmin = 0; Xmax = 75,500
Ymin = 0; Ymax = 264,678

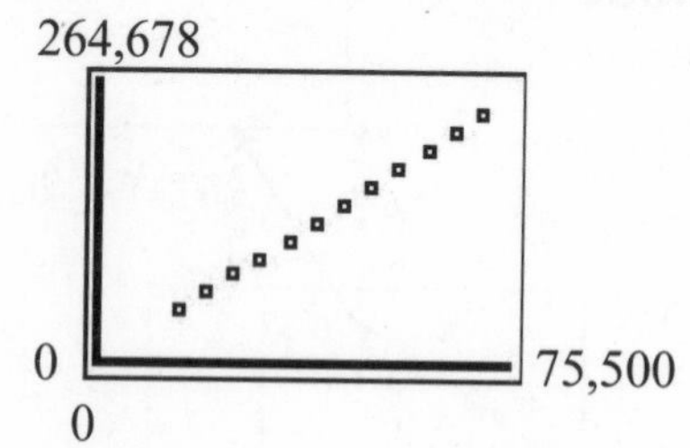

(b) Using the LinReg function and L1 and L2, the line of best fit is:

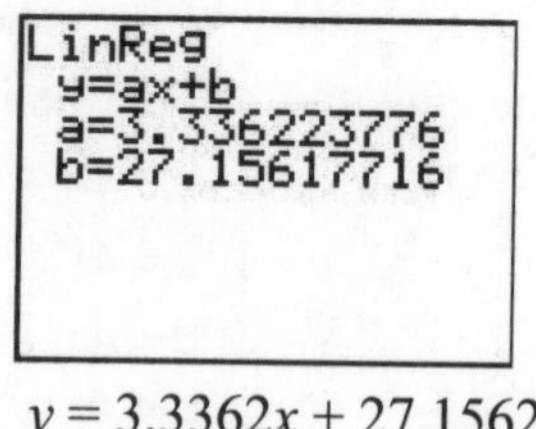

$y = 3.3362x + 27.1562$

(c)

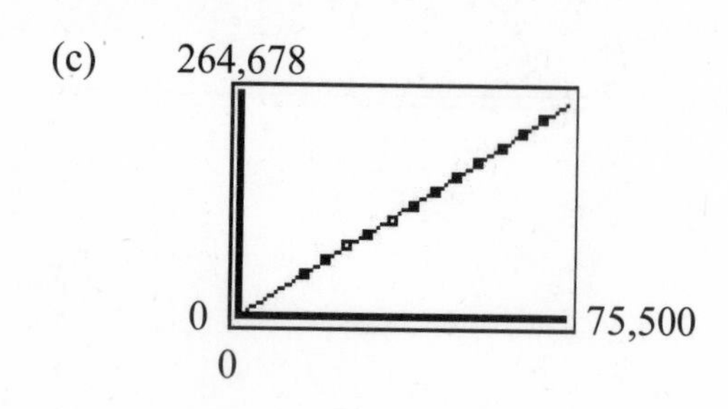

(d) The slope indicates that a person can borrow an additional $3.34 for each additional dollar of income.

(e) Evaluate the line of best fit at $x = 42{,}000$.

$$y = 3.3362(42{,}000) + 27.1563$$
$$= 140{,}148$$

With an income of $42,000 a person can borrow $140,148.

21. (a) Window: Xmin = −10; Xmax = 110
Ymin = 50; Ymax = 70

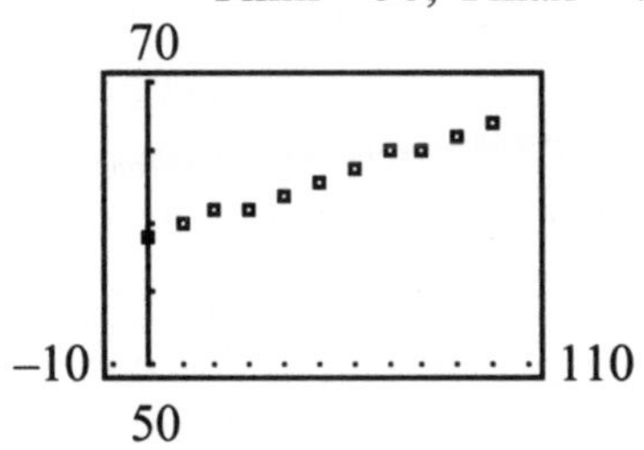

(b) Using the LinReg function, the line of best fit is:

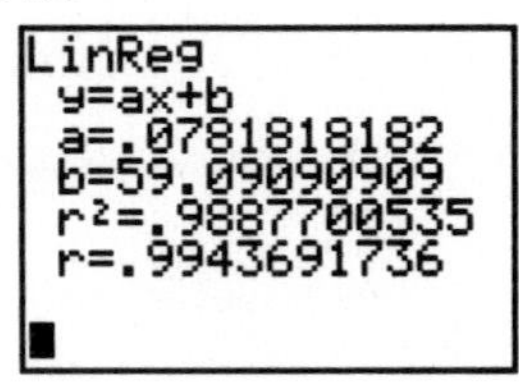

$$y = 0.07818x + 59.0909$$

(c)

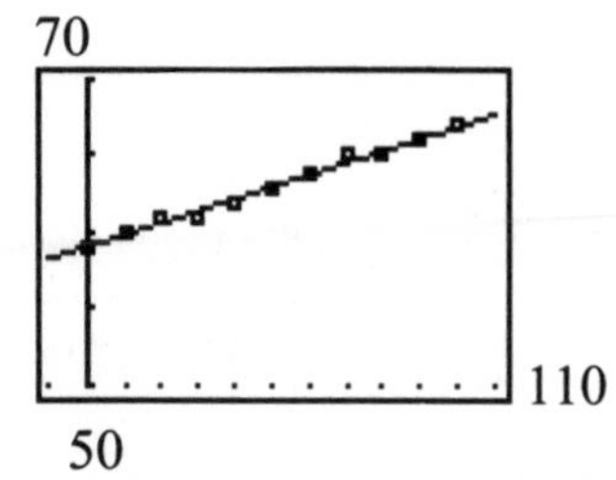

(d) The slope indicates the apparent change in temperature in a 65°F room for every percent increase in relative humidity.

(e) To determine the apparent temperature when the relative humidity is 75%, evaluate the equation of the line of best fit when $x = 75$.

$$y = 0.07818(75) + 59.0909 = 64.95$$

When the relative humidity is 75%, the temperature of the room will appear to be 65°F.

23. (a) Using the LinReg function, the line of best fit is
$$E = 0.4898t + 34.2864$$

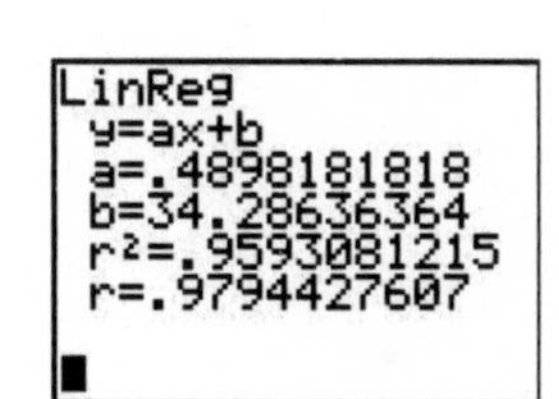

The predicted energy use in 2015, $t = 20$, is

$$E = 0.4898 \cdot 20 + 34.2864 = 44.08$$

quadrillion btu. The predicted use in 2030, $t = 35$, is

$$E = 0.4898 \cdot 35 + 34.2864 = 51.43$$

quadrillion btu.

Chapter 1 Linear Equations; Review Exercises

1.

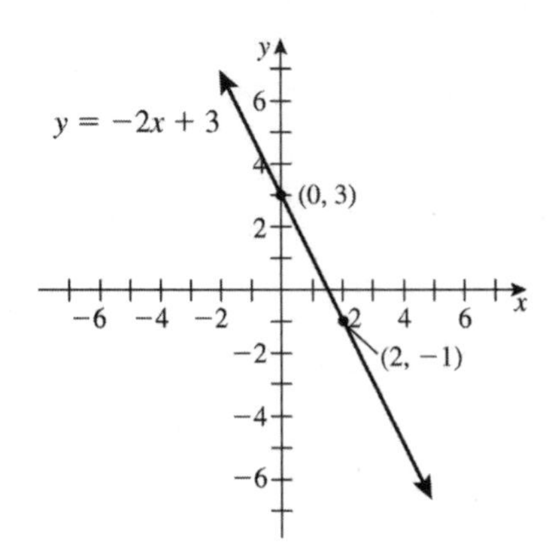

3.

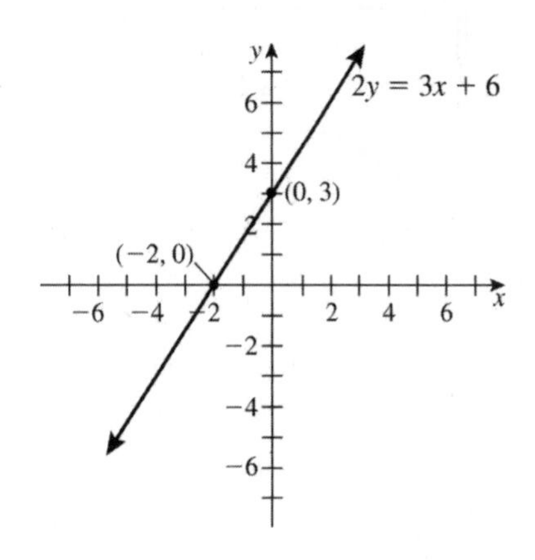

5. (a) $m = \dfrac{y_2 - y_1}{x_2 - x_1} = \dfrac{4-2}{(-3)-1} = \dfrac{2}{-4} = -\dfrac{1}{2}$

A slope of $-\dfrac{1}{2}$ means that for every 2 unit change in x, y will change (-1) unit. That is, for every 2 units x moves to the right, y will move down 1 unit.

(b) Use the point (1, 2) and the slope to get the point-slope form of the equation of the line:

$$y - y_1 = m(x - x_1)$$
$$y - 2 = -\frac{1}{2}(x-1)$$

Simplifying and solving for y gives the slope-intercept form:

$$y - 2 = -\frac{1}{2}x + \frac{1}{2}$$
$$y = -\frac{1}{2}x + \frac{5}{2}$$

Rearranging terms gives the general form of the equation: $x + 2y = 5$

(c)

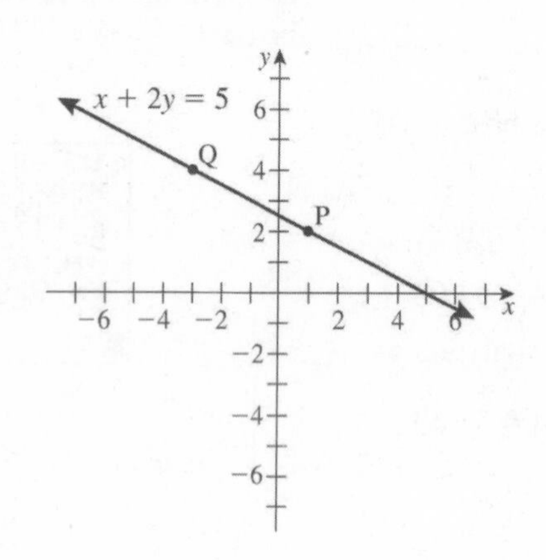

7. (a) $m = \dfrac{y_2 - y_1}{x_2 - x_1} = \dfrac{5-3}{(-1)-(-2)} = \dfrac{2}{1} = 2$

A slope of 2 means that for every 1 unit change in x, y will change by 2 units. That is, for every 1 unit x moves to the right, y will move up 2 units.

(b) Use the point (–1, 5) and the slope to get the point-slope form of the equation:

$$y - y_1 = m(x - x_1)$$
$$y - 5 = 2(x - (-1))$$

Simplifying and solving for y, gives the slope-intercept form:

$$y - 5 = 2x + 2$$
$$y = 2x + 7$$

Rearranging terms gives the general form of the equation: $2x - y = -7$

(c)

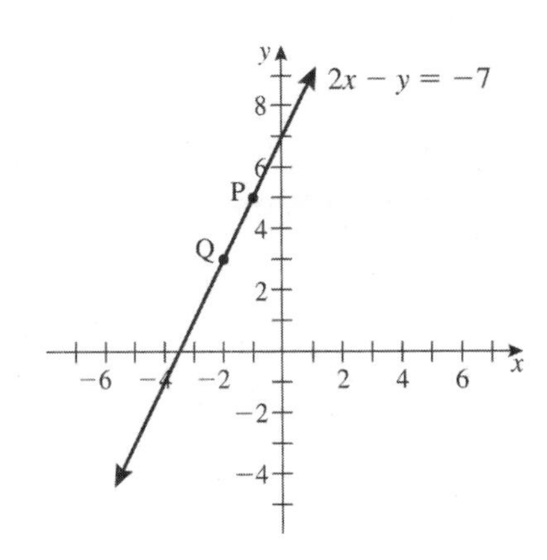

9. Since we are given the slope $m = -3$ and a point, we get the point-slope equation of the line:

$$y - y_1 = m(x - x_1)$$
$$y - (-1) = -3(x - 2)$$

Solving for y puts the equation into the slope-intercept form:

$$y + 1 = -3x + 6$$
$$y = -3x + 5$$

Rearranging terms gives the general form of the equation:

$$3x + y = 5$$

11. We are given the slope $m = 0$ and a point on the line. We either use the point-slope formula or recognize that this is a horizontal line and the equation of a horizontal line is: $y = b$.

$$y = 4$$

This is the general form of the equation.

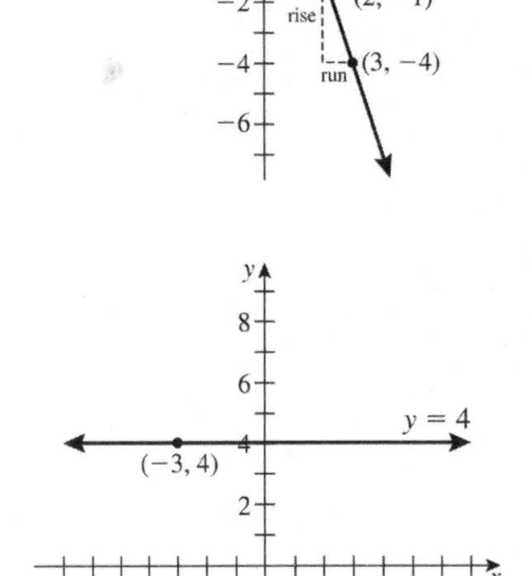

13. We are told that the line is vertical, so the slope is not defined. We also know the line contains the point (8, 5). The general equation of a vertical line is $x = a$

$$x = 8$$

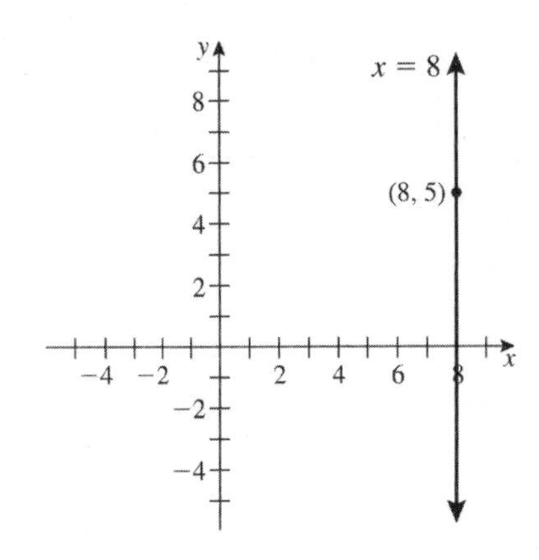

15. We are given the x-intercept and a point. First we find the slope of the line containing the two points.

$$m = \frac{y_2 - y_1}{x_2 - x_1} = \frac{(-5) - 0}{4 - 2} = \frac{-5}{2} = -\frac{5}{2}$$

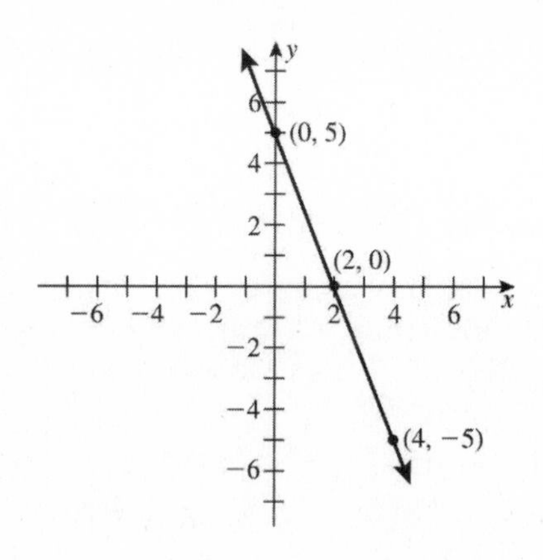

We then use the point (2, 0) and the slope to get the point-slope form of the equation of the line.

$$y - y_1 = m(x - x_1)$$
$$y - 0 = -\frac{5}{2}(x - 2)$$

To get the slope-intercept form, solve for y:

$$y = -\frac{5}{2}x + 5$$

Rearrange the terms for the general form of the equation

$$5x + 2y = 10$$

17. We are given two points, the x-intercept and the y-intercept. Use the two points to find the slope of the line.

$$m = \frac{y_2 - y_1}{x_2 - x_1} = \frac{(-4) - 0}{0 - (-3)} = \frac{-4}{3} = -\frac{4}{3}$$

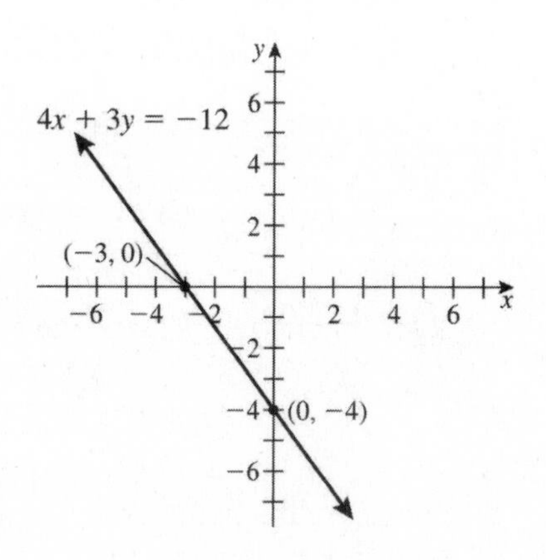

Since one of the points is the y-intercept, use it and the slope to write the slope-intercept form of the equation:

$$y = -\frac{4}{3}x - 4$$

Rearrange the terms to obtain the general form of the equation:

$$4x + 3y = -12$$

19. Since the line we are seeking is parallel to $2x + 3y = -4$, the slope of the two lines are the same. Find the slope of the given line by putting it into slope-intercept form:

$$2x + 3y = -4$$
$$3y = -2x - 4$$
$$y = -\frac{2}{3}x - \frac{4}{3}$$

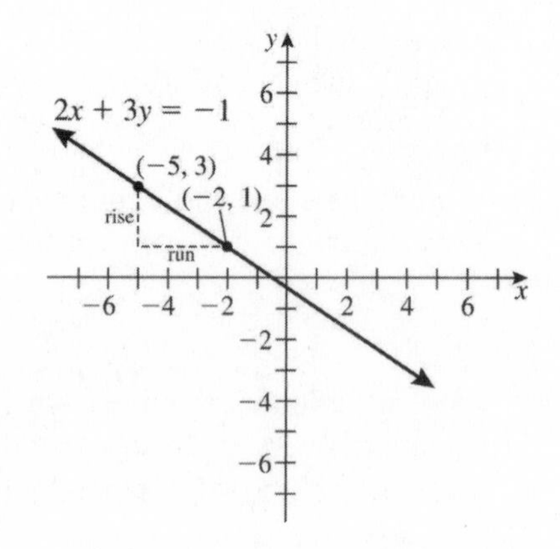

The slopes of the two lines are $m = -\frac{2}{3}$. Use the slope and the point (–5, 3) to write the point-slope form of the equation of the parallel line.

$$y - y_1 = m(x - x_1)$$
$$y - 3 = -\frac{2}{3}(x - (-5))$$
$$y - 3 = -\frac{2}{3}(x + 5)$$

To put the equation into slope-intercept form, solve for y.

$$y-3=-\frac{2}{3}x-\frac{10}{3}$$
$$y=-\frac{2}{3}x-\frac{1}{3}$$

Rearrange the terms to obtain the general form of the equation:

$$2x+3y=-1$$

21. To find the slope and y-intercept of the line, put the equation into the slope-intercept form.

$$9x+2y=18$$
$$2y=-9x+18$$
$$y=-\frac{9}{2}x+9$$

The slope is $-\frac{9}{2}$, and the y-intercept is (0, 9).

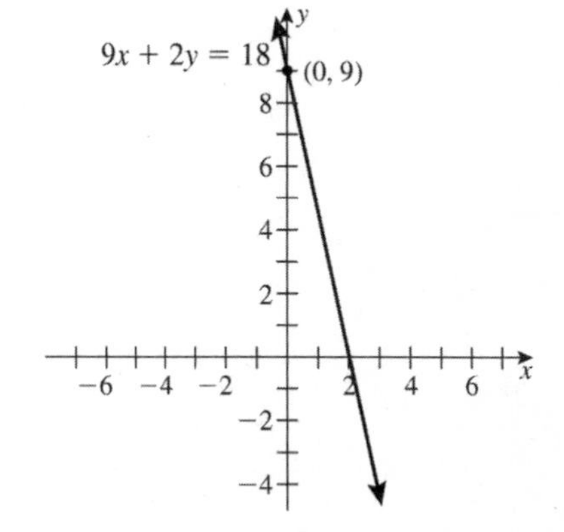

23. To find the slope and y-intercept of the line, put the equation into the slope-intercept form.

$$4x+2y=9$$
$$2y=-4x+9$$
$$y=-2x+\frac{9}{2}$$

The slope is -2, and the y-intercept is $\left(0,\ \frac{9}{2}\right)$.

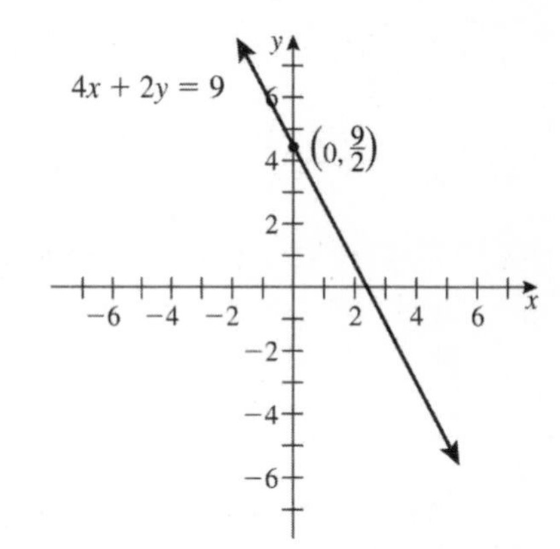

25. To find the slope and y-intercept of the line, put the equation into slope-intercept form.

$$\frac{1}{2}x+\frac{1}{3}y=\frac{1}{6}$$
$$\frac{1}{3}y=-\frac{1}{2}x+\frac{1}{6}$$
$$y=-\frac{3}{2}x+\frac{1}{2}$$

The slope is $-\frac{3}{2}$, and the y-intercept is $\left(0,\frac{1}{2}\right)$.

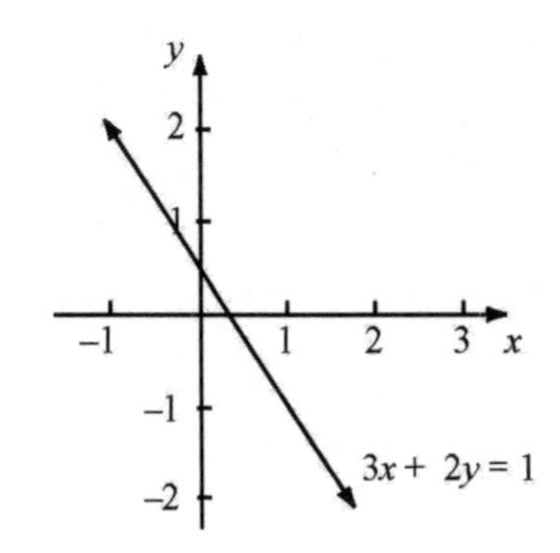

27. We put each equation into slope-intercept form:

$$3x-4y=-12 \qquad 6x-8y=-9$$
$$-4y=-3x-12 \qquad -8y=-6x-9$$
$$y=\frac{3}{4}x+3 \qquad y=\frac{6}{8}x+\frac{9}{8}$$
$$y=\frac{3}{4}x+\frac{9}{8}$$

Since both lines have the same slope but different y-intercepts, the lines are parallel.

29. We put each equation into slope-intercept form:

$$x-y=-2 \qquad 3x-4y=-12$$
$$-y=-x-2 \qquad -4y=-3x-12$$
$$y=x+2 \qquad y=\frac{3}{4}x+3$$

Since the lines have different slopes, they intersect.

31. We put each equation into slope-intercept form:

$$4x+6y=-12 \qquad 2x+3y=-6$$
$$6y=-4x-12 \qquad 3y=-2x-6$$
$$y=-\frac{2}{3}x-2 \qquad y=-\frac{2}{3}x-2$$

Since both lines have the same slope and the same y-intercept, the lines are coincident.

33. To find the point of intersection of two lines, first put the lines in slope-intercept form:

$$L:\ x-y=4 \qquad M:\ x+2y=7$$
$$y=x-4 \qquad y=-\frac{1}{2}x+\frac{7}{2}$$

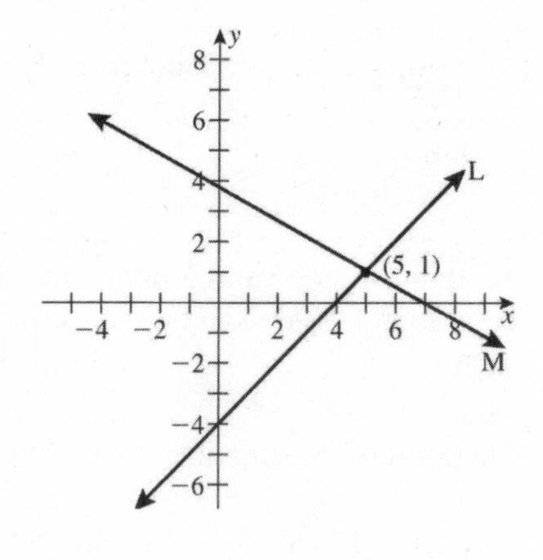

Since the point of intersection, (x_0, y_0), must be on both L and M, we set the two equations equal to each other and solve for x_0. Then we substitute the value of x_0 into the equation of one of the lines to find y_0.

$$x_0-4=-\frac{1}{2}x_0+\frac{7}{2} \qquad y_0=x_0-4$$
$$2x_0-8=-x_0+7 \qquad y_0=5-4$$
$$3x_0=15 \qquad y_0=1$$
$$x_0=5$$

The point of intersection is $(5, 1)$.

35. To find the point of intersection of two lines, first put the lines in slope-intercept form.

$L: \ x - y = -2$ $\qquad$ $M: \ x + 2y = 7$

$y = x + 2$ $\qquad$ $2y = -x + 7$

$y = -\frac{1}{2}x + \frac{7}{2}$

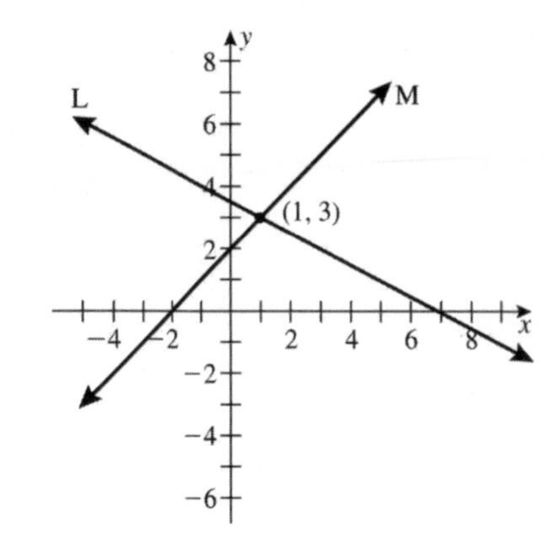

Since the point of intersection, (x_0, y_0), must be on both L and M, we set the two equations equal to each other and solve for x_0. Then we substitute the value of x_0 into the equation of one of the lines to find y_0.

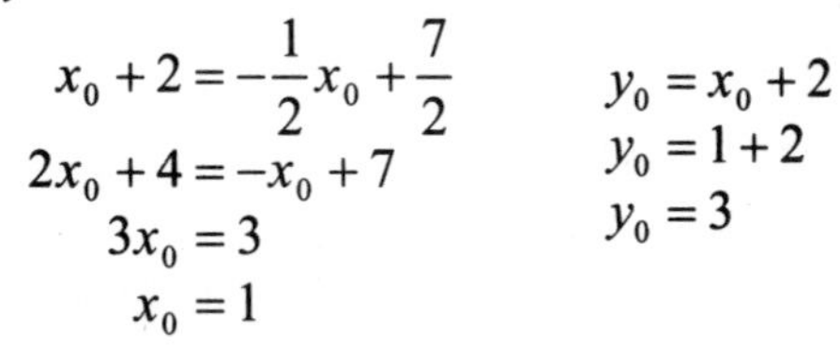

$$x_0 + 2 = -\frac{1}{2}x_0 + \frac{7}{2} \qquad y_0 = x_0 + 2$$
$$2x_0 + 4 = -x_0 + 7 \qquad y_0 = 1 + 2$$
$$3x_0 = 3 \qquad y_0 = 3$$
$$x_0 = 1$$

The point of intersection is (1, 3).

37. To find the point of intersection of two lines, first put the lines in slope-intercept form.

$L: \ 2x - 4y = -8$ $\qquad$ $M: \ 3x + 6y = 0$

$-4y = -2x - 8$ $\qquad$ $6y = -3x$

$y = \frac{1}{2}x + 2$ $\qquad$ $y = -\frac{1}{2}x$

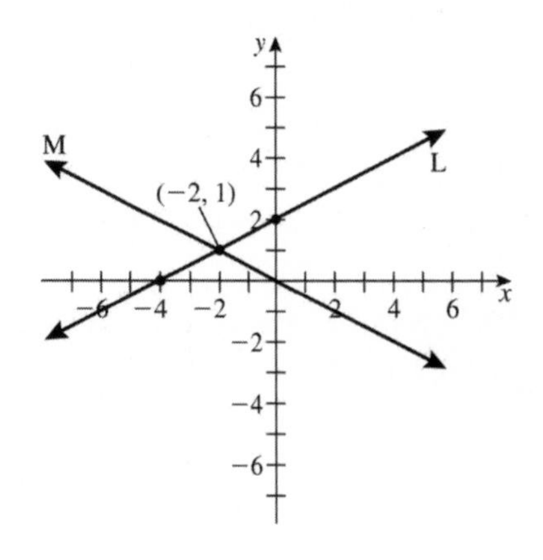

Since the point of intersection, (x_0, y_0), must be on both L and M, we set the two equations equal to each other and solve for x_0. Then we substitute the value of x_0 into the equation of one of the lines to find y_0.

$$\frac{1}{2}x_0 + 2 = -\frac{1}{2}x_0 \qquad y_0 = -\frac{1}{2}x_0$$
$$x_0 + 4 = -x_0 \qquad y_0 = -\frac{1}{2}\cdot(-2)$$
$$2x_0 = -4$$
$$x_0 = -2 \qquad y_0 = 1$$

The point of intersection is (−2, 1).

39. It is often convenient to use a table to organize the information in an investment problem.

	Amount Invested	Interest Rate	Interest Earned
B-Bonds	x	$12\% = 0.12$	$0.12x$
Bank	$y = 90{,}000 - x$	$5\% = 0.05$	$0.05y = 0.05(90{,}000 - x)$
Total	$x + y = 90{,}000$		\$10,000

The last column gives the information needed for the equation since the sum of the interest earned on the individual investments must equal the total interest earned.

$$\begin{aligned}0.12x + 0.05(90{,}000 - x) &= 10{,}000\\0.12x + 4500 - 0.05x &= 10{,}000\\0.07x &= 5{,}500\\x &= 78{,}571.429\end{aligned}$$

Mr. and Mrs. Byrd should invest \$78,571.43 in B-rated bonds and \$11,428.57 in the well-known bank in order to achieve their investment goals.

41. (a) The break-even point is the point where the cost equals the revenue, or when the profit is zero. Before we can find the break-even point we need the equation that describes cost.

We are told the fixed costs, the band and the advertising, and the variable costs. If we let x denote the number of tickets sold, the cost of the dance is described by the equation:

$$\begin{aligned}C &= 500 + 100 + 5x\\&= 5x + 600\end{aligned}$$

The revenue is given by the equation, $R = 10x$, since each ticket costs \$10.
Setting $C = R$, and solving for x, will tell how many tickets must be sold to break even.

$$\begin{aligned}5x + 600 &= 10x\\600 &= 5x\\x &= 120\end{aligned}$$

So 120 tickets must be sold for the group to break even.

(b) Profit is the difference between the revenue and the cost. To determine the number of tickets that need to be sold to clear a profit of \$900, we will solve the equation:

$$\begin{aligned}P &= R - C\\900 &= 10x - (5x + 600)\\900 &= 5x - 600\\1500 &= 5x\\x &= 300\end{aligned}$$

The church group must sell 300 tickets to realize a profit of \$900.

(c) If tickets cost \$12, the break-even point will come from the equation

$$\begin{aligned}R &= C\\12x &= 5x + 600\\7x &= 600\\x &= 85.71\end{aligned}$$

To break even, 86 tickets must be sold.

To find the number of ticket sales needed to have a \$900 profit, solve the equation:

$$\begin{aligned} P &= R - C \\ 900 &= 12x - (5x + 600) \\ 900 &= 7x - 600 \\ 1500 &= 7x \\ x &= 214.29 \end{aligned}$$

So the church group needs to sell 215 tickets at \$12 each to realize a profit of \$900.

43.

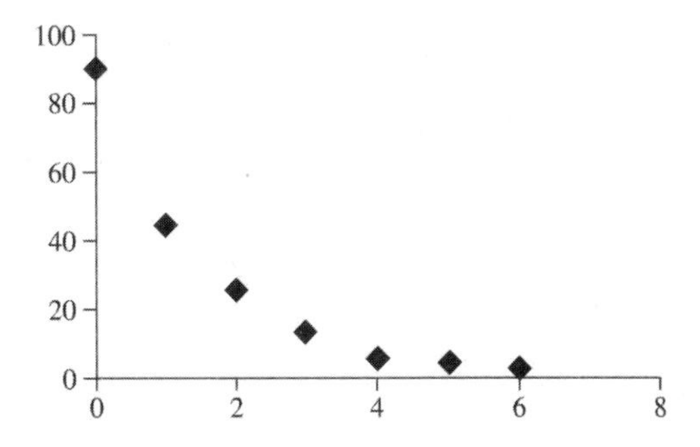

This relation does not appear to be linear.

45. (a) The market price is the price for which supply equals demand. To find market price set $S = D$ and solve for p.

$$\begin{aligned} 0.8p + 0.2 &= -0.4p + 1.8 \\ 1.2p &= 1.6 \\ p &= 1.33 \end{aligned}$$

The market price is \$1.33 per bushel.

(b) When $p = 1.33$, $S = 0.8(1.33) + 0.2 = 1.264$. There will be 1.264 million bushels of corn supplied at the market price of \$1.33.

(c)

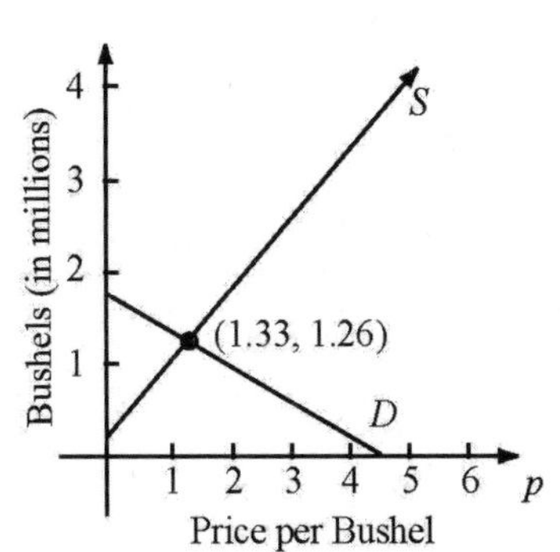

(d) At a price of \$1.33 per bushel the supply and the demand for corn are equal.

47. (a)

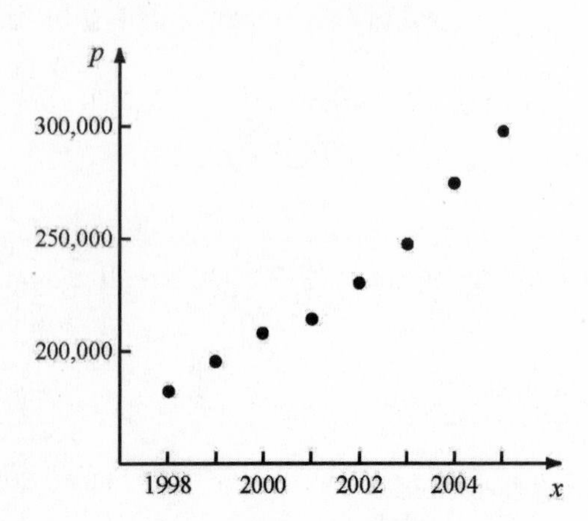

(b) $m = \dfrac{p_2 - p_1}{x_2 - x_1} = \dfrac{228,700 - 181,900}{2002 - 1998} = \dfrac{45,800}{4} = 11,700$

(c) The price of houses sold in the United States increased by an average of \$11,700 per year from 1998 through 2002.

(d) $m = \dfrac{p_2 - p_1}{x_2 - x_1} = \dfrac{297,000 - 228,700}{2005 - 2002} = \dfrac{68,300}{3} = 22,766.67$

(e) The price of houses sold in the United States increased by an average of \$22,766.67 per year from 2002 through 2005.

(f)

```
LinReg
 y=ax+b
 a=15876.19048
 b=-31545670.24
```

The slope of best fit is $a = 15,876.19$

(g) The price of houses sold in the United States increased by an average of \$15,876.19 per year from 1998 through 2005.

(h) The average annual increase in price of houses sold in the United States is not constant. The slope depends on the points used to calculate it.

(i) The average price of houses sold in the United States is increasing.

49. (a)

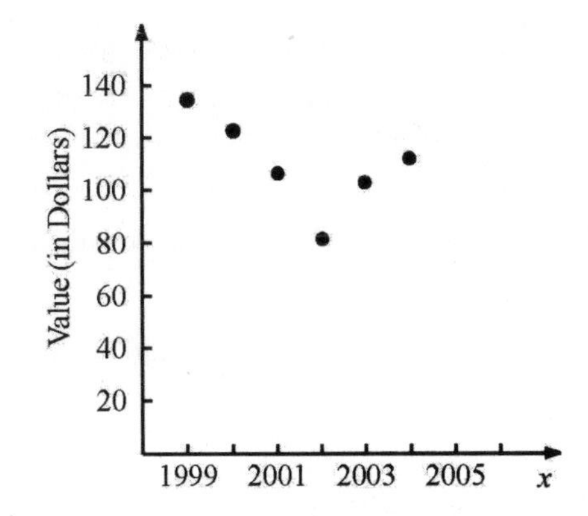

(b) The data do not appear to be linearly related.

(c) Since the data are not linearly related, the prediction in Problem 48 is not valid.

51. (a) This is the equation of a vertical line that includes the point (0, 0). Its graph is the y-axis.

(b) The graph of $y = 0$ is the x-axis. It is a horizontal line and has a slope of zero.

(c) This is an equation of a line with a slope of -1 and a y-intercept of (0, 0). Its graph is a negatively sloped diagonal line through the origin.

Chapter 1 Project

1. Since Avis charges a flat rate that does not depend on the number of miles, x, driven the equation is that of a horizontal line.

$A = 64.99$

3.

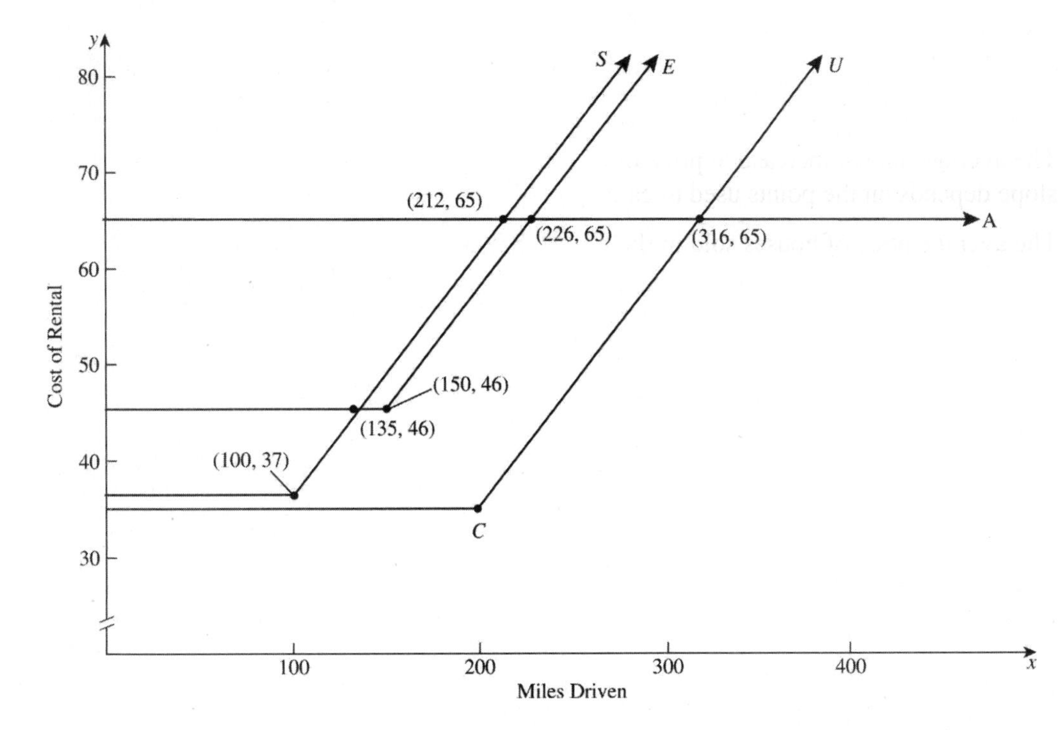

5. By planning your sightseeing carefully and figuring how far you intend to drive, the information in question 4 will help you to decide which car to rent. If you are going to drive more than 226 miles, renting from Avis would be more economical. Otherwise, you would save money by renting from Enterprise.

7. See graph in Problem 3.

9. Usave Car Rental also has a two tiered pricing policy similar to Enterprise and AutoSave. The equations describing the cost of renting a Usave car are:

$$\begin{aligned} U &= 35.99 && \text{when } x \le 200 \\ U &= 35.99 + 0.25(x - 200) && \text{when } x > 200 \\ &= 35.99 + 0.25x - 50 \\ &= 0.25x - 14.01 \end{aligned}$$

Usave Car Rental is always less expensive than Enterprise and AutoSave. Usave's base rate is lower than either of the two, and it offers more free miles than both the other companies. There will be a point, however, where Avis again becomes more economical. To determine at which mileage Avis is the better deal, find the point of intersection of A and U. Consider only $x_0 > 200$.

$$\begin{aligned} U &= A \\ 0.25x_0 - 14.01 &= 64.99 \\ 0.25x_0 &= 79 \\ x_0 &= 316 \end{aligned}$$

Therefore, if you drive more than 316 miles, the Avis car which costs $64.99 is the least expensive to rent.

Mathematical Questions Form Professional Exams

1. The break-even point is the value of x for which the revenue equals cost. If x units are sold at price of $2.00 each, the revenue is $R = 2x$.

Cost is the total of the fixed and variable costs. We are told that the fixed costs are $6000, and that the variable cost per item is 40% of the price. So the cost is given by the equation:

$$\begin{aligned} C &= (0.40)(2)x + 6000 \\ &= 0.80x + 6000 \end{aligned}$$

Setting $R = C$ and solving for x yields,

$$\begin{aligned} R &= C \\ 2x &= 0.8x + 6000 \\ 1.2x &= 6000 \\ x &= 5000 \end{aligned}$$

Answer: (b) 5000 units

3. The break-even point is the number of units that must be sold for revenue to equal cost. Using the notation given, we have:

$$R = SPx \quad \text{and} \quad C = VCx + FC$$

To find the sales level necessary to break even, we set $R = C$ and solve for x.

$$\begin{aligned} R &= C \\ SPx &= VCx + FC \\ SPx - VCx &= FC \\ (SP - VC)x &= FC \\ x &= \frac{FC}{SP - VC} \end{aligned}$$

Answer: (d)

5. Since straight-line depreciation remains the same over the life of the property, its expense over time will be a horizontal line. Sum-of-year's-digits depreciation expense decreases as time increases.
Answer: (c)

7. **Answer:** (b); Y is an estimate of total factory overhead.

Chapter 2

Systems of Linear Equations; Matrices

2.1 Systems of Linear Equations: Substitution; Elimination

1. True – Solve each equation for y and compare their slopes and y-intercepts.

$$3x - 4y = 24 \qquad\qquad 3x - 4y = 12$$
$$-4y = -3x + 24 \qquad\qquad -4y = -3x + 12$$
$$y = \frac{3}{4}x - 6 \qquad\qquad y = \frac{3}{4}x - 3$$

Since the slopes are the same but the y-intercepts are different, the lines are parallel.

3. Since both equations are equal to y, they are equal to each other. First solve for x, they use the result to solve for y.

$$60x - 900 = -15x + 2850 \qquad\qquad y = 60x - 900$$
$$75x = 3750 \qquad\qquad y = 60(50) - 900$$
$$x = 50 \qquad\qquad y = 2100$$

The point of intersection is (50, 2100).

5. False

7. To be a solution to the system of equations, the given values of the variables must solve each equation. So we evaluate each equation at $x = 2$ and $y = -1$.

$$2x - y = 5 \qquad\qquad 5x + 2y = 8$$
$$2(2) - (-1) = 4 + 1 = 5 \qquad\qquad 5(2) + 2(-1) = 10 - 2 = 8$$

Since both equations are satisfied, $x = 2, y = -1$ is a solution to the system.

9. To be a solution to the system of equations, the given values of the variables must solve each equation. So we evaluate each equation at $x = 2$ and $y = \frac{1}{2}$.

$$3x + 4y = 4$$
$$3(2) + 4\left(\frac{1}{2}\right) = 8 \neq 4$$

Since the first equation is not satisfied, $x = 2, y = \frac{1}{2}$ is not a solution to the system.

11. To be a solution to the system of equations, the given values of the variables must solve each equation. So we evaluate each equation at $x = 4$ and $y = 1$.

$$\begin{array}{ll} x - y = 3 & \frac{1}{2}x + y = 3 \\ 4 - 1 = 3 & \frac{1}{2}(4) + 1 = 3 \end{array}$$

Since both equations are satisfied, $x = 4, y = 1$ is a solution to the system.

13. To be a solution to the system of equations, the given values of the variables must solve each equation. So we evaluate each equation at $x = 1, y = -1$, and $z = 2$.

$$\begin{array}{lll} 3x + 3y + 2z = 4 & x - y - z = 0 & 2y - 3z = -8 \\ 3(1) + 3(-1) + 2(2) = 4 & 1 - (-1) - (2) = 0 & 2(-1) - 3(2) = -8 \end{array}$$

Since all three equations are satisfied, $x = 1, y = -1, z = 2$ is a solution to the system.

15. To be a solution to the system of equations, the given values of the variables must solve each equation. So we evaluate each equation at $x = 2, y = -2$, and $z = 2$.

$$\begin{array}{lll} 3x + 3y + 2z = 4 & x - 3y + z = 10 & 5x - 2y - 3z = 8 \\ 3(2) + 3(-2) + 2(2) = 4 & (2) - 3(-2) + 2 = 10 & 5(2) - 2(-2) - 3(2) = 8 \end{array}$$

Since all three equations are satisfied, $x = 2, y = -2, z = 2$ is a solution to the system.

17. Choosing the method of elimination to solve this system,

$$\begin{cases} x + y = 8 & (1) \\ x - y = 4 & (2) \end{cases}$$

$2x = 12$	Add the two equations.
$x = 6$	Solve for x.
$6 + y = 8$	Back-substitute 6 for x in equation (1).
$y = 2$	Solve for y.

The solution to the system is $x = 6, y = 2$, or written as an ordered pair, (6, 2).

19. Choosing the method of elimination to solve this system,

$$\begin{cases} 5x - y = 13 & (1) \\ 2x + 3y = 12 & (2) \end{cases}$$

$15x - 3y = 39$ (1)	Multiply equation (1) by 3.
$17x = 51$	Add (1) to equation (2).
$x = 3$	Solve for x.
$5(3) - y = 13$	Back-substitute 3 for x in equation (1).
$y = 2$	Solve for y.

The solution to the system is $x = 3, y = 2$, or written as an ordered pair, (3, 2).

21. Choosing the method of substitution to solve this system,

$$\begin{cases} 3x = 24 & (1) \\ x + 2y = 0 & (2) \end{cases}$$

$x = 8$ (1)	Solve equation (1) for x.
$8 + 2y = 0$ (2)	Substitute 8 for x in equation (2).
$y = -4$	Solve for y.

The solution to the system is $x = 8, y = -4$, or written as an ordered pair, (8, –4).

23. Choosing the method of elimination to solve this system,

$$\begin{cases} 3x - 6y = 2 & (1) \\ 5x + 4y = 1 & (2) \end{cases}$$

$6x - 12y = 4$ (1)	Multiply equation (1) by 2.
$15x + 12y = 13$ (2)	Multiply equation (2) by 3.
$21x = 7$	Add equations (1) and (2).
$x = \frac{1}{3}$	Solve for x.
$3\left(\frac{1}{3}\right) - 6y = 2$	Back-substitute $\frac{1}{3}$ for x in equation (1).
$y = -\frac{1}{6}$	Solve for y.

The solution to the system is $x = \frac{1}{3}, y = -\frac{1}{6}$, or written as an ordered pair, $\left(\frac{1}{3}, -\frac{1}{6}\right)$.

25. Choosing the method of elimination to solve this system,

$$\begin{cases} 2x + y = 1 & (1) \\ 4x + 2y = 3 & (2) \end{cases}$$

$4x + 2y = 2$ (1) Multiply equation (1) by 2.

$0 = 1$ Subtract (1) from equation (2). This results in a contradiction.

There is no solution to this system of equations. The system is inconsistent.

27. Choosing the method of substitution to solve this system,

$$\begin{cases} 2x - y = 0 & (1) \\ 3x + 2y = 7 & (2) \end{cases}$$

$y = 2x$ Solve equation (1) for y.

$3x + 2(2x) = 7$ Substitute $2x$ for y in equation (2).

$7x = 7$ Simplify.

$x = 1$ Solve for x.

$y = 2$ Back-substitute 1 for x in equation (1) and solve for y.

The solution to the system of equations is $x = 1$ and $y = 2$, or written as an ordered pair, (1, 2).

29. Choosing the method of elimination to solve this system,

$$\begin{cases} x + 2y = 4 & (1) \\ 2x + 4y = 8 & (2) \end{cases}$$

$2x + 4y = 8$ (1) Multiply equation (1) by 2.

$0 = 0$ Subtract equation (1) from equation (2). The equations are dependent.

There are infinitely many solutions to the system of equations. They can be written as $x = -2y + 4$ where y is any real number, or as $y = -\frac{1}{2}x + 2$ where x is any real number, or using ordered pairs, we write $\{(x, y) \mid x = -2y + 4, y \text{ is any real number}\}$ or $\left\{(x, y) \mid y = -\frac{1}{2}x + 2, x \text{ any real number}\right\}$.

31. Choosing the method of substitution to solve this system,

$$\begin{cases} 2x - 3y = -1 & (1) \\ 10x + y = 11 & (2) \end{cases}$$

$y = 11 - 10x$ (2) Solve equation (2) for y.

$2x - 3(11 - 10x) = -1$ Substitute $11 - 10x$ for y in equation (1).

$2x - 33 + 30x = -1$ Simplify.

$32x = 32$

$x = 1$ Solve for x.

$y = 11 - 10(1)$ (2) Back-substitute 1 for x in equation (2).

$y = 1$ Simplify.

The solution of the system is $x = 1, y = 1$, or written as an ordered pair, (1, 1).

33. Choosing the method of elimination to solve this system,

$$\begin{cases} 2x + 3y = 6 & (1) \\ x - y = \dfrac{1}{2} & (2) \end{cases}$$

$2x - 2y = 1 \quad (2)$ — Multiply equation (2) by 2.
$5y = 5$ — Subtract equation (2) from equation (1).
$y = 1$ — Solve for y.
$2x + 3(1) = 6$ — Back-substitute 1 for y in equation (1).
$2x = 3$ — Simplify.
$x = \dfrac{3}{2}$ — Solve for x.

The solution of the system is $x = \dfrac{3}{2}, y = 1$, or written as an ordered pair, $\left(\dfrac{3}{2}, 1\right)$.

35. Choosing the method of elimination to solve this system,

$$\begin{cases} \dfrac{1}{2}x + \dfrac{1}{3}y = 3 & (1) \\ \dfrac{1}{4}x - \dfrac{2}{3}y = -1 & (2) \end{cases}$$

$12x + 8y = 72 \quad (1)$ — Multiply equation (1) by 24.
$3x - 8y = -12 \quad (2)$ — Multiply equation (2) by 12.
$15x = 60$ — Add equations (1) and (2).
$x = 4$ — Solve for x.
$\dfrac{1}{4}(4) - \dfrac{2}{3}y = -1$ — Back-substitute 4 for x in equation (2).
$-\dfrac{2}{3}y = -2$ — Simplify.
$y = 3$ — Solve for y.

The solution of the system is $x = 4, y = 3$, or written as an ordered pair, (4, 3).

37. Choosing the method of elimination to solve this system,

$$\begin{cases} 3x - 5y = 3 & (1) \\ 15x + 5y = 21 & (2) \end{cases}$$

$18x = 24$ — Add equations (1) and (2).
$x = \dfrac{4}{3}$ — Solve for x.
$3\left(\dfrac{4}{3}\right) - 5y = 3$ — Back-substitute $\dfrac{4}{3}$ for x in equation (1).
$4 - 5y = 3$ — Simplify.
$y = \dfrac{1}{5}$ — Solve for y.

The solution of the system is $x = \dfrac{4}{3}, y = \dfrac{1}{5}$, or written as an ordered pair, $\left(\dfrac{4}{3}, \dfrac{1}{5}\right)$.

39.
$$\begin{cases} x - y \quad = 6 & (1) \\ 2x - \quad 3z = 16 & (2) \\ \quad 2y + z = 4 & (3) \end{cases}$$

$2x - 2y = 12 \quad (1)$ — Multiply equation (1) by 2.

$2y + 3z = 4 \quad (2)$ — Subtract equation (1) from (2).

$2z = 0$ — Subtract equation (3) from (2).

$z = 0$ — Solve for z.

$2x - 3(0) = 16$ — Back-substitute 0 for z in equation (2).

$x = 8$ — Solve for x.

$8 - y = 6$ — Back-substitute 8 for x in equation (1).

$y = 2$ — Solve for y.

The solution of the system is $x = 8$, $y = 2$, and $z = 0$, or written as an ordered triple, (8, 2, 0).

41.
$$\begin{cases} x - 2y + 3z = 7 & (1) \\ 2x + y + z = 4 & (2) \\ -3x + 2y - 2z = -10 & (3) \end{cases}$$

$x - 2y + 3z = 7 \quad (1)$

$2x + y + z = 4 \quad (2)$ Multiply by 2

$$\begin{array}{rl} x - 2y + 3z = 7 & (1) \\ 4x + 2y + 2z = 8 & (2) \\ \hline 5x \qquad + 5z = 15 & \text{Add} \end{array}$$

$$\begin{array}{rl} x - 2y + 3z = 7 & (1) \\ -3x + 2y - 2z = -10 & (2) \\ \hline -2x + \qquad z = -3 & \text{Add} \end{array}$$

$$\begin{cases} x - 2y + 3z = 7 & (1) \\ 5x + \quad 5z = 15 & (2) \\ -2x + \quad z = -3 & (3) \end{cases}$$

Working only with equations (2) and (3),

$5x + 5z = 15 \quad (2)$

$-2x + z = -3 \quad (3)$ Multiply by (-5)

$$\begin{array}{rl} 5x + 5z = 15 & (2) \\ 10x - 5z = 15 & (3) \\ \hline 15x \qquad = 30 & \text{or } x = 2 \end{array}$$

Back-substitute 2 for x in equation (3) and solve for z.

$-2x + z = -3 \quad (3)$

$-2(2) + z = -3 \quad (3)$

$z = 1$

Finally back-substitute $x = 2$ and $z = 1$ in equation (1) and solve for y.

$x - 2y + 3z = 7 \quad (1)$

$2 - 2y + 3(1) = 7$

$y = -1$

The solution of the system is $x = 2$, $y = -1$, and $z = 1$, or written as an ordered triple, (2, –1, 1).

43.
$$\begin{cases} x-y-z=1 & (1) \\ 2x+3y+z=2 & (2) \\ 3x+2y=0 & (3) \end{cases}$$

Since (3) has no z term we will eliminate z first.

$$\begin{array}{rl} x-\ y-z=1 & (1) \\ 2x+3y+z=2 & (2) \\ \hline 3x+2y\quad=3 & \text{(Add) (2)} \end{array}$$

We now use the revised (2) and (3).

$$\begin{array}{rl} 3x+2y=3 & (2) \\ 3x+2y=0 & (3) \\ \hline 0=3 & \text{(Subtract) (3)} \end{array}$$

Equation (3) has no solution; the system is inconsistent.

45.
$$\begin{cases} x-\ y-\ z=\ 1 & (1) \\ -x+2y-3z=-4 & (2) \\ 3x-2y-7z=\ 0 & (3) \end{cases}$$

Eliminate x:

$$\begin{array}{rl} x-\ y-\ z=\ 1 & (1) \\ -x+2y-3z=-4 & (2) \\ \hline y-4z=-3 & \text{(Add) (2)} \end{array}$$

$$\begin{array}{rl} x-\ y-\ z=1 & (1) \\ 3x-2y-7z=0 & (3) \end{array} \quad \text{Multiply by } -3 \quad \begin{array}{rl} -3x+3y+3z=-3 & (1) \\ 3x-2y-7z=\ 0 & (3) \\ \hline y-4z=-3 & \text{(Add) (3)} \end{array}$$

We now used revised (2) and (3).

$$\begin{array}{rl} y-4z=-3 & (2) \\ y-4z=-3 & (3) \\ \hline 0=0 & (3) \end{array}$$

The original system is equivalent to a system containing 2 equations, so the equations are dependent and the system has infinitely many solutions. If z represents any real number, then substituting $y=4z-3$ into (1) gives

$$x-\ y-\ z=1 \quad (1)$$

$$x-(4z-3)-z=1 \text{ or } x=5z-2$$

The solution to the system is $\begin{cases} x=5z-2 \\ y=4z-3 \end{cases}$ where z is any real number.

47.
$$\begin{cases} 2x-2y+3z=6 & (1) \\ 4x-3y+2z=0 & (2) \\ -2x+3y-7z=1 & (3) \end{cases}$$

Eliminating x:

$$\begin{array}{ll} 2x-2y+3z=6 & (1) \\ 4x-3y+2z=0 & (2) \end{array} \quad \text{Multiply by } -2 \quad \begin{array}{rl} -4x+4y-6z=-12 & (1) \\ 4x-3y+2z=\ \ 0 & (2) \\ \hline y-4z=-12 & (\text{Add})\ (2) \end{array}$$

$$\begin{array}{rl} 2x-2y+3z=6 & (1) \\ -2x+3y-7z=1 & (3) \\ \hline y-4z=7 & (\text{Add})\ (3) \end{array}$$

We now use the revised (2) and (3).

$$\begin{array}{rl} y-4z=-12 & (2) \\ y-4z=\ \ 7 & (3) \\ \hline 0=\ -5 & (\text{Subtract})\ (3) \end{array}$$

Equation (3) has no solution; the system is inconsistent.

49.
$$\begin{cases} x+\ y-\ z=\ 6 & (1) \\ 3x-2y+\ z=-5 & (2) \\ x+3y-2z=14 & (3) \end{cases}$$

Eliminating x:

$$\begin{array}{ll} x+\ y-z=\ 6 & (1) \\ 3x-2y+z=-5 & (2) \end{array} \quad \text{Multiply by 3} \quad \begin{array}{rl} 3x+3y-3z=18 & (1) \\ 3x-2y+\ z=-5 & (2) \\ \hline 5y-4z=23 & (\text{Subtract})\ (2) \end{array}$$

$$\begin{array}{rl} x+\ y-\ z=\ 6 & (1) \\ x+3y-2z=\ 14 & (3) \\ \hline -2y+\ z=-8 & (\text{Subtract})\ (3) \end{array}$$

We now work with revised (2) and (3) and solve for y.

$$\begin{array}{rl} 5y-4z=23 & (2) \\ -2y+\ z=-8 & (3) \end{array} \quad \text{Multiply by 4} \quad \begin{array}{rl} 5y-4z=\ 23 & (2) \\ -8y+4z=-32 & (3) \\ \hline -3y\quad\ =-9 & (\text{Add})\ (3) \\ y=3 & \end{array}$$

Back-substitute $y = 3$ into (2): $5(3) - 4z = 23$

$15 - 4z = 23$ or $z = -2$.

Then back substitute $y = 3$ and $z = -2$ into (1): $x + 3 - (-2) = 6$ or $x = 1$.

The solution to the system is $x = 1$, $y = 3$, and $z = -2$, or written as an ordered triple, $(1, 3, -2)$.

51.
$$\begin{cases} x+2y-z=-3 & (1)\\ 2x-4y+z=-7 & (2)\\ -2x+2y-3z=4 & (3)\end{cases}$$

Eliminate x:

$$x+2y-z=-3 \quad (1) \qquad \text{Multiply by 2} \qquad 2x+4y-2z=-6 \quad (1)$$
$$2x-4y+z=-7 \quad (2) \qquad\qquad\qquad\qquad 2x-4y+z=-7 \quad (2)$$
$$8y-3z=1 \quad \text{(Subtract)} \quad (2)$$

$$x+2y-z=-3 \quad (1) \qquad \text{Multiply by 2} \qquad 2x+4y-2z=-6 \quad (1)$$
$$-2x+2y-3z=4 \quad (3) \qquad\qquad\qquad\qquad -2x+2y-3z=4 \quad (3)$$
$$6y-5z=-2 \quad \text{(Add)} \quad (3)$$

We now work with revised (2) and (3) and solve for z.

$$8y-3z=1 \quad (2) \qquad \text{Multiply by 3} \qquad 24y-9z=3 \quad (2)$$
$$6y-5z=-2 \quad (3) \qquad \text{Multiply by 4} \qquad 24y-20z=-8 \quad (3)$$
$$11z=11 \quad \text{(Subtract)} \quad (3)$$
$$z=1$$

Back-substitute $z=1$ into (2) to solve for y: $8y-3(1)=1$ or $y=\frac{1}{2}$.

Last back substitute $y=\frac{1}{2}$ and $z=1$ into (1) to solve for x: $x+2\left(\frac{1}{2}\right)-1=-3$ or $x=-3$.

The solution to the system is $x=-3$, $y=\frac{1}{2}$, and $z=1$ or $\left(-3,\frac{1}{2},1\right)$.

53. We let l represent the length of the floor and w represent its width. We are told the perimeter is 90 feet and that the length is twice the width. We need to solve the system of equations

$$\begin{cases} 2l+2w=90 & (1)\\ l=2w & (2)\end{cases}$$

We solve this system by substitution. Substituting $2w$ for l in (1), we get

$$2(2w)+2w=90 \quad (1)$$
$$6w=90$$
$$w=15$$

Substituting 15 for w in (2) gives $l=2(15)=30$. The floor has a length of 30 feet and a width of 15 feet.

55. We let x denote the number of acres of corn planted and y denote the number of acres of soybeans planted. We solve the system of equations

$$\begin{cases} x+y=298 & (1)\\ 333x+227y=85{,}337 & (2)\end{cases}$$

We solve this system by substitution. First, solve equation (1) for y, $y=298-x$, then substitute $298-x$ for y in equation (2).

$$333x+227y=85{,}337 \quad (2)$$
$$333x+227(298-x)=85{,}337$$
$$333x+67{,}646-227x=85{,}337$$
$$106x=17{,}691$$
$$x=166.896 \quad \text{and} \quad y=298-166.896=131.104$$

The farmer should plant 166.9 acres of corn and 131.1 acres of soybeans.

57. Let x represent the number of cashews in the mixture and let y represent the total weight of the mix. To find the amount of cashews needed for the mixture, we solve the system of equations

$$\begin{cases} 30 + x = y & (1) \\ 5.00x + 1.50(30) = 3.00y & (2) \end{cases}$$

We solve this system by substitution, replacing y in (2) with $30 + x$.

$$\begin{aligned} 5x + 1.5(30) &= 3(30 + x) \\ 5x + 45 &= 90 + 3x \\ 2x &= 45 \\ x &= 22.5 \quad \text{and} \quad y = 30 + x = 55.5 \end{aligned}$$

The manager should add 22.5 pounds of cashews to the peanuts to make the mixture.

59. Let x represent the cost of liter of milk and y represent the cost of a 330 gram package of tofu. To find the cost of each, we solve the system of equations

$$\begin{aligned} 3x + 2y &= 1002 \quad (1) \\ 3y &= 8 + 2x \quad (2) \end{aligned}$$

We solve this system by substitution, replacing $\frac{8+2x}{3}$ for y in equation (1).

$$\begin{aligned} 3x + 2\left(\frac{8+2x}{3}\right) &= 1002 \\ 9x + 2(8 + 2x) &= 3006 \\ 9x + 16 + 4x &= 3006 \\ 13x &= 2990 \\ x &= 230 \quad \text{and} \quad y = \frac{8+2(230)}{3} = 156 \end{aligned}$$

A liter of milk costs 230 yen, and a 330 gram package of tofu costs 156 yen.

61. Let x represent the cost of a pound of bacon and y represent the cost of a carton of eggs. To find each item's price, we solve the system of equations

$$\begin{cases} 3x + 2y = 7.45 & (1) \\ 2x + 3y = 6.45 & (2) \end{cases}$$

We solve this system using the method of elimination.

$$\begin{array}{llll} 3x + 2y = 7.45 \quad (1) & \text{Multiply by 2} & 6x + 4y = 14.90 & (1) \\ 2x + 3y = 6.45 \quad (2) & \text{Multiply by 3} & \underline{6x + 9y = 19.35} & (2) \\ & & -5y = -4.45 & \text{(Subtract) (2)} \\ & & y = 0.89 & \end{array}$$

Back-substitute 0.89 for y in equation (1) to find x.

$$\begin{aligned} 3x + 2(0.89) &= 7.45 \quad (1) \\ 3x &= 5.67 \\ x &= 1.89 \end{aligned}$$

Bacon costs \$1.89 per pound and eggs cost \$0.89 per dozen. So the refund when we return 2 pounds of bacon and 2 cartons of eggs will be $2(1.89) + 2(0.89) = \$5.56$.

63. Let x represent the milligrams of liquid 1 and y represent the milligrams of liquid 2 necessary to obtain the desired mixture. To learn the amount of each liquid to use, we solve the system of equations

$$\begin{cases} 0.20x+0.40y=40 & (1) \\ 0.30x+0.20y=30 & (2) \end{cases} \quad \text{or} \quad \begin{cases} 2x+4y=400 & (1) \\ 3x+2y=300 & (2) \end{cases}$$

We will solve this system of equations by the method of elimination.

$$\begin{array}{lll} 2x+4y=400 \quad (1) & & 2x+4y=\ \ 400 \quad (1) \\ 3x+2y=300 \quad (2) & \text{Multiply by 2} & \underline{6x+4y=\ \ 600} \quad (2) \\ & & -4x \qquad =-200 \quad \text{(Subtract) (2)} \\ & & \qquad\quad x=50 \end{array}$$

Back-substitute 50 for x in equation (1) and solve for y.

$$2(50)+4y=400 \quad (1)$$
$$4y=300 \quad \text{or} \quad y=75$$

The pharmacist should mix 50 mg of liquid 1 and 75 mg of liquid 2 to fill the prescription.

65. Let x represent the pounds of rolled oats and y represent the pounds of molasses in the horse's diet. To determine the amount of oats and molasses to feed the horse, we solve the system of equations

$$\begin{cases} 0.41x+3.35y=33 & (1) \\ 1.95x+0.36y=21 & (2) \end{cases}$$

We will solve the system using the method of substitution. First we solve equation (1) for x:

$$x=\frac{33-3.35y}{0.41} \quad (1)$$

Then we substitute this value for x in equation (2) and solve for y.

$$1.95\left(\frac{33-3.35y}{0.41}\right)+0.36y=21$$

$$64.35-6.5325y+0.1476y=8.61 \quad \text{(Multiply both sides by 0.41)}$$
$$64.35-6.3849y=8.61$$
$$6.3849y=55.74$$
$$y=8.730$$

Back-substituting into (1) we get $x=\dfrac{33-3.35(8.730)}{0.41}=9.157$

The farmer should feed the horse 9.157 pounds of rolled oats and 8.730 pounds of molasses each day.

67. Let x represent the number of orchestra seats, y represent the number of main seats, and z represent the number of balcony seats in the theater. To find the number of each kind of seat solve the system of equations

$$\begin{cases} x+\ \ y+\ \ z=500 & (1) \\ 50x+35y+25z=17{,}100 & (2) \\ 50\left(\frac{1}{2}x\right)+35y+25z=14{,}600 & (3) \end{cases}$$

$$x+\ \ y+\ \ z=\ \ 500 \quad (1) \quad \text{Multiply by 50} \quad 50x+50y+50z=25{,}000 \quad (1)$$

$$50x + 35y + 25z = 17{,}100 \quad (2) \qquad\qquad 50x + 35y + 25z = 17{,}100 \quad (2)$$

$$\text{Subtract} \qquad 15y + 25z = 7900 \quad (2)$$

$$x + y + z = 500 \quad (1) \quad \text{Multiply by 25} \quad 25x + 25y + 25z = 12{,}500 \quad (1)$$
$$25x + 35y + 25z = 14{,}600 \quad (3) \qquad\qquad 25x + 35y + 25z = 14{,}600 \quad (3)$$

$$\text{Subtract} \qquad -10y = -2{,}100 \quad (3)$$
$$y = 210$$

Back-substitute 210 for y in (2) and solve for z: $15(210) + 25z = 7900$

$$3150 + 25z = 7900$$
$$25z = 4750 \text{ or } z = 190$$

Finally substitute for both y and z in equation (1) and solve for x: $x + 190 + 210 = 500$ or $x = 100$.

There are 100 orchestra seats, 210 main seats, and 190 balcony seats in the theater.

69. Let x represent the price of a hamburger, y represent the price of fries, and z represent the price of cola. To find the price of each item, we need to solve the system of equations.

$$\begin{cases} 8x + 6y + 6z = 26.10 & (1) \\ 10x + 6y + 8z = 31.60 & (2) \end{cases}$$

By subtracting equation (1) from (2), we eliminate y. Then we will solve for x in terms of z.

$$2x + 2z = 5.50 \qquad (2)$$
$$x = 2.75 - z$$

Back-substitute into equation (2) and solve for y.

$$10(2.75 - z) + 6y + 8z = 31.60$$
$$27.5 - 10z + 6y + 8z = 31.60$$
$$6y - 2z = 31.60 - 27.5$$
$$6y = 4.10 + 2z$$
$$y = \frac{41}{60} + \frac{1}{3}z$$

There is not enough information to determine the price of each food item. We have $x = 2.75 - z$ and $y = \frac{1}{3}z + \frac{41}{60}$ where z is the parameter. Using these equations and the restrictions $\$1.75 \le x \le \$2.25$, $\$0.75 \le y \le \$1.00$, and $\$0.60 \le z \le \$0.90$, possible prices are

Price of a Hamburger	Price of Fries	Price of Cola
$2.15	$0.88	$0.60
$2.05	$0.92	$0.70
$1.95	$0.95	$0.80
$1.85	$0.98	$0.90

71. (a) To find the equilibrium price, solve the system of equations.

$$\begin{cases} S = -200 + 50p & (1) \\ D = 1000 - 25p & (2) \end{cases}$$

Setting the equations equal, since at equilibrium supply equals demand, we find

$$-200 + 50p = 1000 - 25p$$
$$50p + 25p = 1000 + 200$$
$$75p = 1200$$
$$p = 16$$

The equilibrium price of the T-shirts is \$16.00.

(b) Using the equilibrium price in either equation (1) or (2) gives the equilibrium quantity.

$$S = -200 + 50 \cdot 16 = -200 + 800 = 600$$

The equilibrium quantity is 600 T-shirts.

(c)

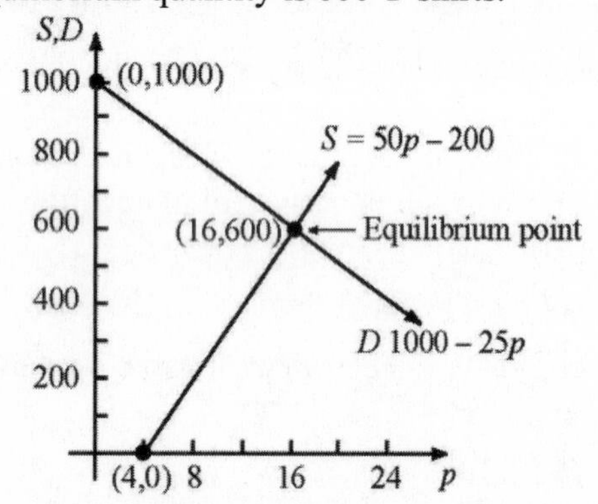

73. Using elimination to solve the system, we subtract equation (1) from equation (2).

$$\begin{array}{rl} 0.06Y + 6000r = 900 & (2) \\ 0.06Y - 5000r = 240 & (1) \\ \hline 11{,}000r = 660 & \\ r = 0.06 & \end{array}$$

Back-substituting r into equation (1), we get

$$0.06Y - 5000(0.06) = 240$$
$$0.06Y = 540$$
$$Y = 9000$$

At equilibrium, income is \$9000 million or \$9 billion, and the interest rate is 6%.

75. (a) First, set up a system of equations relating the income from the three possible investments.

Let x denote the amount invested in Treasury Bills earning 4% per year,
y denote the amount invested in CDs earning 6% per year.
z denote the amount invested in corporate paper earning 8% per year.

There are two equations: one that distributes the \$70,000 among the investments and the other that generates the \$5000.00 in earnings.

$$\begin{cases} x + y + z = 70{,}000 & (1) \\ 0.04x + 0.06y + 0.08z = 5{,}000 & (2) \end{cases}$$

We will use elimination, but first we will subtract the z-term from each side. Doing this allows us to eliminate one of the two variables remaining on the left.

$$\begin{cases} x + y = 70{,}000 - z & (1) \\ 0.04x + 0.06y = 5{,}000 - 0.08z & (2) \end{cases}$$

Now multiply equation (1) by 6 and equation (2) by 100 and then subtract.

$$\begin{array}{rll} 6x+6y &= 420{,}000-6z & (1) \\ 4x+6y &= 500{,}000-8z & (2) \\ \hline 2x \quad &= -80{,}000+2z & (1) \end{array}$$

$$x = -40{,}000 + z$$

Back-substituting for x in equation (1), we can solve for y in terms of z.

$$(-40{,}000+z)+y+z=70{,}000 \quad (1)$$
$$-40{,}000 + 2z + y = 70{,}000$$
$$2y = 110{,}000 - 2z$$
$$y = 55{,}000 - z$$

where z is a real number between 40,000 and 55,000. (Any other value of z will cause either x or y to be negative.) Using the three restrictions form a table such as the one below.

$x = z - 40{,}000$	$y = 55{,}000 - z$	z	Income
0	\$30,000	\$40,000	\$5000
\$ 5,000	\$20,000	\$45,000	\$5000
\$10,000	\$10,000	\$50,000	\$5000
\$15,000	0	\$55,000	\$5000

2.2 Systems of Linear Equations: Matrix Method

1. matrix

3. True. An augmented matrix has as many rows as there are equations and as many columns as there are variables plus 1.

5. $\left[\begin{array}{cc|c} 2 & -3 & 5 \\ 1 & -1 & 3 \end{array}\right]$

7. $\left[\begin{array}{cc|c} 2 & 1 & -6 \\ 3 & 1 & -1 \end{array}\right]$

9. $\left[\begin{array}{ccc|c} 2 & -1 & -1 & 0 \\ 1 & -1 & 1 & 1 \\ 3 & -1 & 0 & 2 \end{array}\right]$

11. $\left[\begin{array}{ccc|c} 2 & -3 & 1 & 7 \\ 1 & 1 & -1 & 1 \\ 2 & 2 & -3 & -4 \end{array}\right]$

13. $\left[\begin{array}{cccc|c} 4 & -1 & 2 & -1 & 4 \\ 1 & 1 & 0 & 0 & -6 \\ 0 & 2 & -1 & 1 & 5 \end{array}\right]$

15. $\left[\begin{array}{cccc|c} 1 & -1 & 1 & -1 & 0 \\ 2 & 3 & -1 & 4 & 5 \end{array}\right]$

17. $\left[\begin{array}{cc|c} 1 & -3 & -2 \\ 2 & -5 & 5 \end{array}\right] \xrightarrow{R_2=-2r_1+r_2} \left[\begin{array}{cc|c} 1 & -3 & -2 \\ -2(1)+2 & -2(-3)+(-5) & -2(-2)+5 \end{array}\right] = \left[\begin{array}{cc|c} 1 & -3 & -2 \\ 0 & 1 & 9 \end{array}\right]$

19. (a)

$$\left[\begin{array}{ccc|c} 1 & -3 & 4 & 3 \\ 2 & -5 & 6 & 6 \\ -3 & 3 & 4 & 6 \end{array}\right] \xrightarrow{R_2 = -2r_1 + r_2} \left[\begin{array}{ccc|c} 1 & -3 & 4 & 3 \\ -2(1)+2 & -2(-3)-5 & -2(4)+6 & -2(3)+6 \\ -3 & 3 & 4 & 6 \end{array}\right]$$

$$= \left[\begin{array}{ccc|c} 1 & -3 & 4 & 3 \\ 0 & 1 & -2 & 0 \\ -3 & 3 & 4 & 6 \end{array}\right]$$

(b)

$$\left[\begin{array}{ccc|c} 1 & -3 & 4 & 3 \\ 2 & -5 & 6 & 6 \\ -3 & 3 & 4 & 6 \end{array}\right] \xrightarrow{R_3 = 3r_1 + r_3} \left[\begin{array}{ccc|c} 1 & -3 & 4 & 3 \\ 2 & -5 & 6 & 6 \\ 3(1)-3 & 3(-3)+3 & 3(4)+4 & 3(3)+6 \end{array}\right] = \left[\begin{array}{ccc|c} 1 & -3 & 4 & 3 \\ 2 & -5 & 6 & 6 \\ 0 & -6 & 16 & 15 \end{array}\right]$$

21. (a)

$$\left[\begin{array}{ccc|c} 1 & -3 & 2 & -6 \\ 2 & -5 & 3 & -4 \\ -3 & -6 & 2 & 6 \end{array}\right] \xrightarrow{R_2 = -2r_1 + r_2} \left[\begin{array}{ccc|c} 1 & -3 & 2 & -6 \\ -2(1)+2 & -2(-3)-5 & -2(2)+3 & -2(-6)-4 \\ -3 & -6 & 2 & 6 \end{array}\right]$$

$$= \left[\begin{array}{ccc|c} 1 & -3 & 2 & -6 \\ 0 & 1 & -1 & 8 \\ -3 & -6 & 2 & 6 \end{array}\right]$$

(b)

$$\left[\begin{array}{ccc|c} 1 & -3 & 2 & -6 \\ 2 & -5 & 3 & -4 \\ -3 & -6 & 2 & 6 \end{array}\right] \xrightarrow{R_3 = 3r_1 + r_3} \left[\begin{array}{ccc|c} 1 & -3 & 2 & -6 \\ 2 & -5 & 3 & -4 \\ 3(1)-3 & 3(-3)-6 & 3(2)+2 & 3(-6)+6 \end{array}\right]$$

$$= \left[\begin{array}{ccc|c} 1 & -3 & 2 & -6 \\ 2 & -5 & 3 & -4 \\ 0 & -15 & 8 & -12 \end{array}\right]$$

23. (a)

$$\left[\begin{array}{ccc|c} 1 & -3 & 1 & -2 \\ 2 & -5 & 6 & -2 \\ -3 & 1 & 4 & 6 \end{array}\right] \xrightarrow{R_2 = -2r_1 + r_2} \left[\begin{array}{ccc|c} 1 & -3 & 1 & -2 \\ -2(1)+2 & -2(-3)-5 & -2(1)+6 & -2(-2)-2 \\ -3 & 1 & 4 & 6 \end{array}\right]$$

$$= \left[\begin{array}{ccc|c} 1 & -3 & 1 & -2 \\ 0 & 1 & 4 & 2 \\ -3 & 1 & 4 & 6 \end{array}\right]$$

(b)

$$\left[\begin{array}{ccc|c} 1 & -3 & 1 & -2 \\ 2 & -5 & 6 & -2 \\ -3 & 1 & 4 & 6 \end{array}\right] \xrightarrow{R_3 = 3r_1 + r_3} \left[\begin{array}{ccc|c} 1 & -3 & 1 & -2 \\ 2 & -5 & 6 & -2 \\ 3(1)-3 & 3(-3)+1 & 3(1)+4 & 3(-2)+6 \end{array}\right]$$

$$= \left[\begin{array}{ccc|c} 1 & -3 & 1 & -2 \\ 2 & -5 & 6 & -2 \\ 0 & -8 & 7 & 0 \end{array}\right]$$

25. (a) $\begin{cases} x + 2y = 5 \\ y = -1 \end{cases}$

(b) The system is consistent.
The solution is $x = 7$, $y = -1$ or, written as an ordered pair, $(7, -1)$.

27. (a) $\begin{cases} x + 2y + 3z = 1 \\ y + 4z = 2 \\ 0 = 3 \end{cases}$

(b) The system is inconsistent. There is no solution.

29. (a) $\begin{cases} x + 2z = -1 \\ y - 4z = -2 \\ 0 = 0 \end{cases}$

(b) The system is consistent and has an infinite number of solutions. The solutions are:
$x = -2z - 1$, $y = 4z - 2$ where z is any real number.

31. (a) $\begin{cases} x_1 + 2x_2 - x_3 + x_4 = 1 \\ x_2 + 4x_3 + x_4 = 2 \\ x_3 + 2x_4 = 3 \\ x_4 = 4 \end{cases}$

(b) The system is consistent.
To find the solution start with $x_4 = 4$ and back-substitute.

$$x_3 = 3 - 2x_4 = 3 - 2(4) = -5$$
$$x_2 = 2 - 4x_3 - x_4 = 2 - 4(-5) - (4) = 18$$
$$x_1 = 1 - 2x_2 + x_3 - x_4$$
$$= 1 - 2(18) + (-5) - 4 = -44$$

or $(-44, 18, -5, 4)$

33. (a) $\begin{cases} x_1 + 2x_2 + 4x_4 = 2 \\ x_2 + x_3 + 3x_4 = 3 \\ x_3 = 0 \\ 0 = 0 \end{cases}$

(b) The system is consistent.
To find the solutions, express x_1 and x_2 in terms of x_4.

$$x_3 = 0$$
$$x_2 = -3x_4 - x_3 + 3 = -3x_4 + 3$$
$$x_1 = -4x_4 - 2x_2 + 2$$
$$= -4x_4 - 2(-3x_4 + 3) + 2$$
$$= 2x_4 - 4$$

where x_4 is any real number.

35. (a) $\begin{cases} x_1 - 2x_2 + x_4 = -2 \\ x_2 - 3x_3 + 2x_4 = 2 \\ x_3 - x_4 = 0 \end{cases}$

(b) The system is consistent.
To find the solutions, express x_1, x_2, and x_3 in terms of x_4.

$$x_3 = x_4$$
$$x_2 = -2x_4 + 3x_3 + 2 = -2x_4 + 3x_4 + 2$$
$$= x_4 + 2$$
$$x_1 = -x_4 + 2x_2 - 2 = -x_4 + 2(x_4 + 2) - 2$$
$$= x_4 + 2$$

where x_4 is any real number.

37. Write the system as:

$$\left[\begin{array}{cc|c} 1 & 1 & 6 \\ 2 & -1 & 0 \end{array}\right] \xrightarrow{R_2=-2r_1+r_2} \left[\begin{array}{cc|c} 1 & 1 & 6 \\ 0 & -3 & -12 \end{array}\right] \xrightarrow{R_2=-\frac{1}{3}r_2} \left[\begin{array}{cc|c} 1 & 1 & 6 \\ 0 & 1 & 4 \end{array}\right]$$

The row-echelon form of the system is $\begin{cases} x+y=6 \\ \quad y=4 \end{cases}$

Back-substitute 4 for y in the first equation: $x+4=6$ or $x=2$
The solution of the system of equations is $x = 2$ and $y = 4$ or (2, 4).

39. Write the system as

$$\left[\begin{array}{cc|c} 2 & 1 & 5 \\ 1 & -1 & 1 \end{array}\right] \xrightarrow[\text{rows 1 and 2}]{\text{Interchange}} \left[\begin{array}{cc|c} 1 & -1 & 1 \\ 2 & 1 & 5 \end{array}\right] \xrightarrow{R_2=-2r_1+r_2} \left[\begin{array}{cc|c} 1 & -1 & 1 \\ 0 & 3 & 3 \end{array}\right] \xrightarrow{R_2=\frac{1}{3}r_2} \left[\begin{array}{cc|c} 1 & -1 & 1 \\ 0 & 1 & 1 \end{array}\right]$$

The row-echelon form of the system is $\begin{cases} x-y=1 \\ \quad y=1 \end{cases}$

Back-substitute 1 for y in the first equation, giving $x - 1 = 1$ or $x = 2$.
The solution of the system of equations is $x = 2$ and $y = 1$ or (2, 1).

41. Write the system as

$$\left[\begin{array}{cc|c} 2 & 3 & 7 \\ 3 & -1 & 5 \end{array}\right] \xrightarrow{R_1=\frac{1}{2}r_1} \left[\begin{array}{cc|c} 1 & \frac{3}{2} & \frac{7}{2} \\ 3 & -1 & 5 \end{array}\right] \xrightarrow{R_2=-3r_1+r_2r_2} \left[\begin{array}{cc|c} 1 & \frac{3}{2} & \frac{7}{2} \\ 0 & -\frac{11}{2} & -\frac{11}{2} \end{array}\right] \xrightarrow{R_2=-\frac{2}{11}r_2} \left[\begin{array}{cc|c} 1 & \frac{3}{2} & \frac{7}{2} \\ 0 & 1 & 1 \end{array}\right]$$

The row-echelon form of the system is $\begin{cases} x+\frac{3}{2}y=\frac{7}{2} \\ \quad y=1 \end{cases}$. Back-substitute 1 for y in the first equation: $2x + 3(1) = 7$ giving $x = 2$. The solution of the system is $x = 2$ and $y = 1$ or (2, 1).

43. Write the system as

$$\left[\begin{array}{cc|c} 2 & -3 & 6 \\ 6 & -9 & 10 \end{array}\right] \xrightarrow{R_1=\frac{1}{2}r_1} \left[\begin{array}{cc|c} 1 & -\frac{3}{2} & 3 \\ 6 & -9 & 10 \end{array}\right] \xrightarrow{R_2=-6r_1+r_2} \left[\begin{array}{cc|c} 1 & -\frac{3}{2} & 3 \\ 0 & 0 & -8 \end{array}\right]$$

The system is inconsistent.

45. Write the system as

$$\left[\begin{array}{cc|c} 2 & -3 & 0 \\ 4 & 9 & 5 \end{array}\right] \xrightarrow{R_1=\frac{1}{2}r_1} \left[\begin{array}{cc|c} 1 & -\frac{3}{2} & 0 \\ 4 & 9 & 5 \end{array}\right] \xrightarrow{R_2=-4r_1+r_2} \left[\begin{array}{cc|c} 1 & -\frac{3}{2} & 0 \\ 0 & 15 & 5 \end{array}\right] \xrightarrow{R_2=\frac{1}{15}r_2} \left[\begin{array}{cc|c} 1 & -\frac{3}{2} & 0 \\ 0 & 1 & \frac{1}{3} \end{array}\right]$$

The row-echelon form of the system is $\begin{cases} x-\frac{3}{2}y=0 \\ \quad y=\frac{1}{3} \end{cases}$

Back-substitute $\frac{1}{3}$ for y in the first equation, giving $x-\frac{3}{2}\cdot\frac{1}{3}=0$ or $x=\frac{1}{2}$.

The solution of the system of equations is $x = \frac{1}{2}$ and $y = \frac{1}{3}$ or $\left(\frac{1}{2}, \frac{1}{3}\right)$.

47. Write the system as

$$\left[\begin{array}{cc|c} 2 & 6 & 4 \\ 5 & 15 & 10 \end{array}\right] \xrightarrow{R_1 = \frac{1}{2}r_1} \left[\begin{array}{cc|c} 1 & 3 & 2 \\ 5 & 15 & 10 \end{array}\right] \xrightarrow{R_2 = -5r_1 + r_2} \left[\begin{array}{cc|c} 1 & 3 & 2 \\ 0 & 0 & 0 \end{array}\right]$$

This system has an infinite number of solutions. They are $x = 2 - 3y$, where y is any real number, or written as ordered pairs $\{(x, y) | x = -3y + 2,\ y \text{ any real number}\}$.

49. Write the system as

$$\left[\begin{array}{cc|c} \frac{1}{2} & \frac{1}{3} & 2 \\ 1 & 1 & 5 \end{array}\right] \xrightarrow[\text{rows 1 and 2}]{\text{Interchange}} \left[\begin{array}{cc|c} 1 & 1 & 5 \\ \frac{1}{2} & \frac{1}{3} & 2 \end{array}\right] \xrightarrow{R_2 = 6r_2} \left[\begin{array}{cc|c} 1 & 1 & 5 \\ 3 & 2 & 12 \end{array}\right]$$

$$\xrightarrow{R_2 = -3r_1 + r_2} \left[\begin{array}{cc|c} 1 & 1 & 5 \\ 0 & -1 & -3 \end{array}\right] \xrightarrow{R_2 = -r_2} \left[\begin{array}{cc|c} 1 & 1 & 5 \\ 0 & 1 & 3 \end{array}\right]$$

The row echelon form of the system of equations is

$$\begin{cases} x = -y + 5 & (1) \\ y = 3 & (2) \end{cases}$$

Back-substitute 3 for y in equation (1), to get $x = -3 + 5 = 2$.

The solution of the system of equations is $x = 2$ and $y = 3$ or, written as an ordered pair, (2, 3).

51. Write the system as

$$\left[\begin{array}{cc|c} 1 & 1 & 1 \\ 3 & -2 & \frac{4}{3} \end{array}\right] \xrightarrow{R_2 = -3r_1 + r_2} \left[\begin{array}{cc|c} 1 & 1 & 1 \\ 0 & -5 & -\frac{5}{3} \end{array}\right] \xrightarrow{R_2 = -\frac{1}{5}r_2} \left[\begin{array}{cc|c} 1 & 1 & 1 \\ 0 & 1 & \frac{1}{3} \end{array}\right]$$

The row echelon form of the system of equations is

$$\begin{cases} x = -y + 1 & (1) \\ y = \frac{1}{3} & (2) \end{cases}$$

Back-substitute $\frac{1}{3}$ for y in equation (1), to get $x = -\frac{1}{3} + 1 = \frac{2}{3}$.

The solution of the system of equations is $x = \frac{2}{3}$ and $y = \frac{1}{3}$ or written as an ordered pair, $\left(\frac{2}{3}, \frac{1}{3}\right)$.

53. Write the system as

$$\left[\begin{array}{ccc|c} 2 & 1 & 1 & 6 \\ 1 & -1 & -1 & -3 \\ 3 & 1 & 2 & 7 \end{array}\right] \xrightarrow[\text{rows 1 and 2}]{\text{Interchange}} \left[\begin{array}{ccc|c} 1 & -1 & -1 & -3 \\ 2 & 1 & 1 & 6 \\ 3 & 1 & 2 & 7 \end{array}\right] \xrightarrow[R_3 = -3r_1 + r_3]{R_2 = -2r_1 + r_2} \left[\begin{array}{ccc|c} 1 & -1 & -1 & -3 \\ 0 & 3 & 3 & 12 \\ 0 & 4 & 5 & 16 \end{array}\right]$$

$$\xrightarrow{R_2 = \frac{1}{3}r_2} \left[\begin{array}{ccc|c} 1 & -1 & -1 & -3 \\ 0 & 1 & 1 & 4 \\ 0 & 4 & 5 & 16 \end{array}\right] \xrightarrow{R_3 = -4r_2 + r_3} \left[\begin{array}{ccc|c} 1 & -1 & -1 & -3 \\ 0 & 1 & 1 & 4 \\ 0 & 0 & 1 & 0 \end{array}\right]$$

The row echelon form of the system of equations is

$$\begin{cases} x = y + z - 3 & (1) \\ y = -z + 4 & (2) \\ z = 0 & (3) \end{cases}$$

Back-substitute 0 for z in (2) to get $y = 4$.
Then back-substitute $z = 0$ and $y = 4$ in equation (1), to get $x = 4 + 0 - 3 = 1$.

The solution of the system of equations is $x = 1$, $y = 4$, and $z = 0$ or written as an ordered triple, (1, 4, 0).

55. Write the system as

$$\left[\begin{array}{ccc|c} 2 & -2 & -1 & 2 \\ 2 & 3 & 1 & 2 \\ 3 & 2 & 0 & 0 \end{array}\right] \xrightarrow{R_1 = \frac{1}{2}r_1} \left[\begin{array}{ccc|c} 1 & -1 & -\frac{1}{2} & 1 \\ 2 & 3 & 1 & 2 \\ 3 & 2 & 0 & 0 \end{array}\right] \xrightarrow[R_3 = -3r_1 + r_3]{R_2 = -2r_1 + r_2} \left[\begin{array}{ccc|c} 1 & -1 & -\frac{1}{2} & 1 \\ 0 & 5 & 2 & 0 \\ 0 & 5 & \frac{3}{2} & -3 \end{array}\right]$$

$$\xrightarrow{R_2 = \frac{1}{5}r_2} \left[\begin{array}{ccc|c} 1 & -1 & -\frac{1}{2} & 1 \\ 0 & 1 & \frac{2}{5} & 0 \\ 0 & 5 & \frac{3}{2} & -3 \end{array}\right] \xrightarrow{R_3 = -5r_2 + r_3} \left[\begin{array}{ccc|c} 1 & -1 & -\frac{1}{2} & 1 \\ 0 & 1 & \frac{2}{5} & 0 \\ 0 & 0 & -\frac{1}{2} & -3 \end{array}\right] \xrightarrow{R_3 = -2r_3} \left[\begin{array}{ccc|c} 1 & -1 & -\frac{1}{2} & 1 \\ 0 & 1 & \frac{2}{5} & 0 \\ 0 & 0 & 1 & 6 \end{array}\right]$$

The row echelon form of the system of equations is

$$\begin{cases} x = y + \frac{1}{2}z + 1 & (1) \\ y = -\frac{2}{5}z & (2) \\ z = 6 & (3) \end{cases}$$

Back-substitute 6 for z in (2) to get $y = -\frac{12}{5}$.

Then back-substitute $z = 6$ and $y = -\frac{12}{5}$ in equation (1), to get $x = -\frac{12}{5} + \frac{1}{2}(6) + 1 = \frac{8}{5}$.

The solution of the system of equations is $x = \frac{8}{5}$, $y = -\frac{12}{5}$, and $z = 6$ or $\left(\frac{8}{5}, -\frac{12}{5}, 6\right)$.

57. Write the system as

$$\left[\begin{array}{ccc|c} 2 & 1 & -1 & 2 \\ 1 & 3 & 2 & 1 \\ 1 & 1 & 1 & 2 \end{array}\right] \xrightarrow[\text{rows 1 and 2}]{\text{Interchange}} \left[\begin{array}{ccc|c} 1 & 3 & 2 & 1 \\ 2 & 1 & -1 & 2 \\ 1 & 1 & 1 & 2 \end{array}\right] \xrightarrow[R_3 = -r_1 + r_3]{R_2 = -2r_1 + r_2} \left[\begin{array}{ccc|c} 1 & 3 & 2 & 1 \\ 0 & -5 & -5 & 0 \\ 0 & -2 & -1 & 1 \end{array}\right]$$

$$\xrightarrow{R_2 = -\frac{1}{5}r_2} \left[\begin{array}{ccc|c} 1 & 3 & 2 & 1 \\ 0 & 1 & 1 & 0 \\ 0 & -2 & -1 & 1 \end{array}\right] \xrightarrow{R_3 = 2r_2 + r_3} \left[\begin{array}{ccc|c} 1 & 3 & 2 & 1 \\ 0 & 1 & 1 & 0 \\ 0 & 0 & 1 & 1 \end{array}\right]$$

The row echelon form of the system of equations is

$$\begin{cases} x = -3y - 2z + 1 & (1) \\ y = -z & (2) \\ z = 1 & (3) \end{cases}$$

Back-substitute $z = 1$ in equation (2) to get $y = -1$, and then back-substitute $y = -1$ and $z = 1$ in equation (1), to get $x = -3(-1) - 2(1) + 1 = 2$.

The solution of the system of equations is $x = 2$, $y = -1$, and $z = 1$ or $(2, -1, 1)$.

59. Write the system as

$$\left[\begin{array}{ccc|c} 1 & 1 & -1 & 0 \\ 4 & 4 & -4 & -1 \\ 2 & 1 & 1 & 2 \end{array}\right] \xrightarrow[R_3=-2r_1+r_3]{R_2=-4r_1+r_2} \left[\begin{array}{ccc|c} 1 & 1 & -1 & 0 \\ 0 & 0 & 0 & -1 \\ 0 & -1 & 3 & 2 \end{array}\right]$$

The system is inconsistent; the second row of the matrix has $0 = -1$ which is contradictory.

61. Write the system as

$$\left[\begin{array}{ccc|c} 3 & 1 & -1 & \frac{2}{3} \\ 2 & -1 & 1 & 1 \\ 4 & 2 & 0 & \frac{8}{3} \end{array}\right] \xrightarrow{R_1=r_1-r_2} \left[\begin{array}{ccc|c} 1 & 2 & -2 & -\frac{1}{3} \\ 2 & -1 & 1 & 1 \\ 4 & 2 & 0 & \frac{8}{3} \end{array}\right] \xrightarrow[R_3=-4r_1+r_3]{R_2=-2r_1+r_2} \left[\begin{array}{ccc|c} 1 & 2 & -2 & -\frac{1}{3} \\ 0 & -5 & 5 & \frac{5}{3} \\ 0 & -6 & 8 & 4 \end{array}\right]$$

$$\xrightarrow{R_2=-\frac{1}{5}r_2} \left[\begin{array}{ccc|c} 1 & 2 & -2 & -\frac{1}{3} \\ 0 & 1 & -1 & -\frac{1}{3} \\ 0 & -6 & 8 & 4 \end{array}\right] \xrightarrow{R_3=6r_2+r_3} \left[\begin{array}{ccc|c} 1 & 2 & -2 & -\frac{1}{3} \\ 0 & 1 & -1 & -\frac{1}{3} \\ 0 & 0 & 2 & 2 \end{array}\right] \xrightarrow{R_3=\frac{1}{2}r_3} \left[\begin{array}{ccc|c} 1 & 2 & -2 & -\frac{1}{3} \\ 0 & 1 & -1 & -\frac{1}{3} \\ 0 & 0 & 1 & 1 \end{array}\right]$$

The row echelon form of the system of equations is

$$\begin{cases} x = -2y + 2z - \frac{1}{3} & (1) \\ y = z - \frac{1}{3} & (2) \\ z = 1 & (3) \end{cases}$$

Back-substitute $z = 1$ in equation (2) to get $y = 1 - \frac{1}{3} = \frac{2}{3}$, and then back-substitute $y = \frac{2}{3}$ and $z = 1$ in equation (1), to get $x = -2\left(\frac{2}{3}\right) + 2(1) - \frac{1}{3} = \frac{1}{3}$.

The solution of the system of equations is $x = \frac{1}{3}$, $y = \frac{2}{3}$, and $z = 1$, or $\left(\frac{1}{3}, \frac{2}{3}, 1\right)$.

63. Using a TI-84 graphing utility, the row-echelon form is

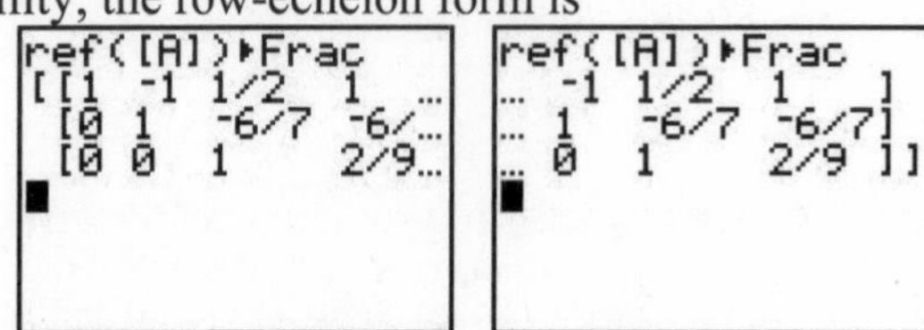

The reduced row-echelon form is

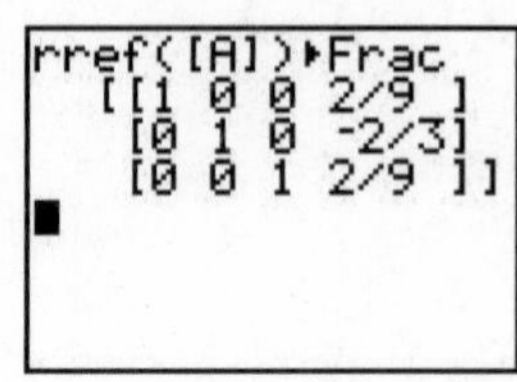

The solution to the system is $x=\frac{2}{9}$, $y=-\frac{2}{3}$, and $z=\frac{2}{9}$ or $\left(\frac{2}{9},-\frac{2}{3},\frac{2}{9}\right)$.

65. Using a TI-84 graphing utility, the row-echelon form is

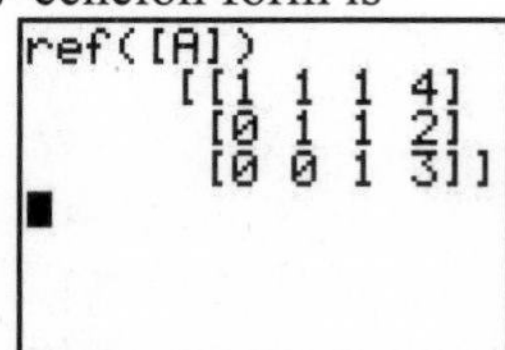

The reduced row-echelon form is

```
rref([A])
 [[1 0 0 2 ]
  [0 1 0 -1]
  [0 0 1 3 ]]
```

The solution to the system is $x = 2$, $y = -1$, and $z = 3$ or $(2, -1, 3)$.

67. Using a TI-84 graphing utility, the row-echelon form is

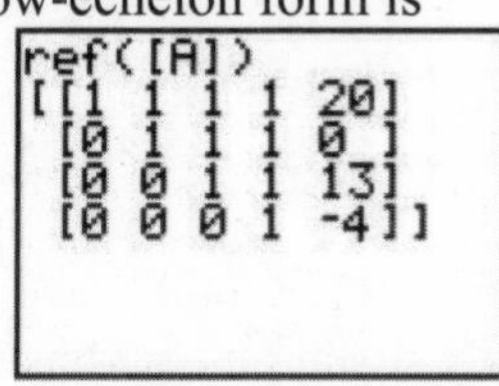

The reduced row-echelon form is

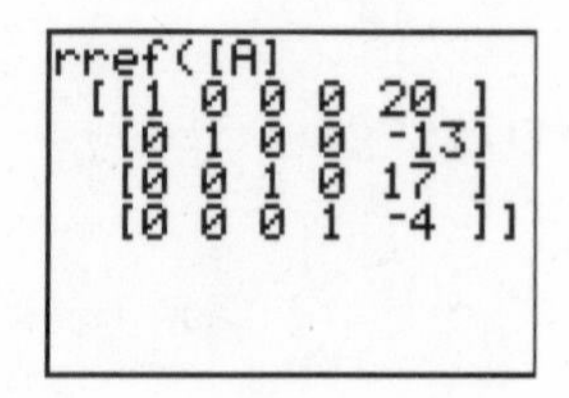

The solution to the system is $x_1 = 20$, $x_2 = -13$, $x_3 = 17$ and $x_4 = -4$ or $(20, -13, 17, -4)$.

69. Let x = the price of a mezzanine ticket,
y = the price of a lower balcony ticket, and
z = the price of a middle balcony ticket.

The system of equations which models the problem, can be written:

$$\begin{cases} 4x+6y=585 & (1) \\ 2x+7y+8z=842.50 & (2) \\ 3y+12z=667.50 & (3) \end{cases}$$

We can then represent the system as:

$$\left[\begin{array}{ccc|c} 4 & 6 & 0 & 585 \\ 2 & 7 & 8 & 842.5 \\ 0 & 3 & 12 & 667.5 \end{array}\right] \xrightarrow{R_1=0.25r_1} \left[\begin{array}{ccc|c} 1 & 1.5 & 0 & 146.25 \\ 2 & 7 & 8 & 842.5 \\ 0 & 3 & 12 & 667.5 \end{array}\right] \xrightarrow{R_2=-2r_1+r_2} \left[\begin{array}{ccc|c} 1 & \frac{3}{2} & 0 & 146.25 \\ 0 & 4 & 8 & 550 \\ 0 & 3 & 12 & 667.5 \end{array}\right]$$

$$\xrightarrow{R_2=0.25r_2} \left[\begin{array}{ccc|c} 1 & 1.5 & 0 & 146.25 \\ 0 & 1 & 2 & 137.5 \\ 0 & 3 & 12 & 667.5 \end{array}\right] \xrightarrow{R_3=-3r_2+r_3} \left[\begin{array}{ccc|c} 1 & 1.5 & 0 & 146.25 \\ 0 & 1 & 2 & 137.5 \\ 0 & 0 & 6 & 255 \end{array}\right] \xrightarrow{R_3=\frac{1}{6}r_3} \left[\begin{array}{ccc|c} 1 & 1.5 & 0 & 146.25 \\ 0 & 1 & 2 & 137.5 \\ 0 & 0 & 1 & 42.5 \end{array}\right]$$

The row-echelon form of the system is

$$\begin{cases} x=-1.5y+146.25 & (1) \\ y=-2z+137.5 & (2) \\ z=42.5 & (3) \end{cases}$$

Back-substitute 42.5 for z in (2) to get $y = -2(42.5) + 137.5 = 52.5$.
Then back-substitute $y = 52.5$ in (1) to get $x = (-1.5)(52.5) + 146.25 = 67.5$.
So, mezzanine tickets cost \$67.50, lower balcony tickets cost \$52.50, and middle balcony tickets cost \$42.50.

71. Let x = number of workstations set up for 2 students, and
y = number of workstations set up for 3 students.

The problem can be modeled with the system of equations,

$$\begin{cases} x+y=16 & (1) \\ 2x+3y=38 & (2) \end{cases}$$

The system can be represented by the augmented matrices

$$\left[\begin{array}{cc|c} 1 & 1 & 16 \\ 2 & 3 & 38 \end{array}\right] \xrightarrow{R_2=-2r_1+r_2} \left[\begin{array}{cc|c} 1 & 1 & 16 \\ 0 & 1 & 6 \end{array}\right]$$

The row-echelon form of the system is

$$\begin{cases} x=-y+16 & (1) \\ y=6 & (2) \end{cases}$$

There are 10 workstations set up for 2 students and 6 workstations set up for 3 students.

73. Let x = amount Carletta invests in Treasury Bills yielding 0.06,
y = amount Carletta invests in Treasury Bonds yielding 0.07, and
z = amount Carletta invests in corporate bonds yielding 0.08.

The problem can be modeled with the system of equations,

$$\begin{cases} x+y+x=10{,}000 & (1) \\ 0.06x+0.07y+0.08z = 680 & (2) \\ z=0.5x & (3) \end{cases}$$

If equation (2) is multiplied by 100, and equation (3) is multiplied by 10 and set equal to zero, the system can be represented by the augmented matrices

$$\left[\begin{array}{ccc|c} 1 & 1 & 1 & 10{,}000 \\ 6 & 7 & 8 & 68{,}000 \\ -5 & 0 & 10 & 0 \end{array}\right] \xrightarrow[R_3=5r_2+r_3]{R_2=-6r_1+r_2} \left[\begin{array}{ccc|c} 1 & 1 & 1 & 10{,}000 \\ 0 & 1 & 2 & 8{,}000 \\ 0 & 5 & 15 & 50{,}000 \end{array}\right] \xrightarrow{R_3=-5r_2+r_3} \left[\begin{array}{ccc|c} 1 & 1 & 1 & 10{,}000 \\ 0 & 1 & 2 & 8{,}000 \\ 0 & 0 & 5 & 10{,}000 \end{array}\right]$$

$$\xrightarrow{R_3=\frac{1}{5}r_3} \left[\begin{array}{ccc|c} 1 & 1 & 1 & 10{,}000 \\ 0 & 1 & 2 & 8{,}000 \\ 0 & 0 & 1 & 2{,}000 \end{array}\right]$$

The row-echelon form of the system is

$$\begin{cases} x=-y-z+10{,}000 & (1) \\ y=-2z+8000 & (2) \\ z=2000 & (3) \end{cases}$$

Carletta should invest \$2000 in corporate bonds, \$4000 in Treasury Bonds, and \$4000 in Treasury Bills to attain \$680 annual income.

75. Let x = number of servings of chicken breast needed,
y = number of potatoes needed, and
z = number of cups of spinach needed in the diet.

The problem can be modeled with the system of equations,

$$\begin{cases} 24x+4y+5z=38 & (1) \\ 26y+7z=40 & (2) \\ 1.5x+0.5z= 2.5 & (3) \end{cases}$$

The system can be represented by the augmented matrices

$$\left[\begin{array}{ccc|c} 24 & 4 & 5 & 38 \\ 0 & 26 & 7 & 40 \\ 1.5 & 0 & 0.5 & 2.5 \end{array}\right] \xrightarrow{R_3=2r_3} \left[\begin{array}{ccc|c} 24 & 4 & 5 & 38 \\ 0 & 26 & 7 & 40 \\ 3 & 0 & 1 & 5 \end{array}\right] \xrightarrow{\text{Interchange } R_1 \text{ and } R_3} \left[\begin{array}{ccc|c} 3 & 0 & 1 & 5 \\ 0 & 26 & 7 & 40 \\ 24 & 4 & 5 & 38 \end{array}\right]$$

$$\xrightarrow{R_1=\frac{1}{3}r_1} \left[\begin{array}{ccc|c} 1 & 0 & \frac{1}{3} & \frac{5}{3} \\ 0 & 26 & 7 & 40 \\ 24 & 4 & 5 & 38 \end{array}\right] \xrightarrow{R_3=-24r_1+r_3} \left[\begin{array}{ccc|c} 1 & 0 & \frac{1}{3} & \frac{5}{3} \\ 0 & 26 & 7 & 40 \\ 0 & 4 & -3 & -2 \end{array}\right] \xrightarrow{\text{Interchange } R_2 \text{ and } R_3} \left[\begin{array}{ccc|c} 1 & 0 & \frac{1}{3} & \frac{5}{3} \\ 0 & 4 & -3 & -2 \\ 0 & 26 & 7 & 40 \end{array}\right]$$

$$\xrightarrow{R_2=\frac{1}{4}r_2} \left[\begin{array}{ccc|c} 1 & 0 & \frac{1}{3} & \frac{5}{3} \\ 0 & 1 & -\frac{3}{4} & -\frac{1}{2} \\ 0 & 26 & 7 & 40 \end{array}\right] \xrightarrow{R_3=-26r_2+r_3} \left[\begin{array}{ccc|c} 1 & 0 & \frac{1}{3} & \frac{5}{3} \\ 0 & 1 & -\frac{3}{4} & -\frac{1}{2} \\ 0 & 0 & \frac{53}{2} & 53 \end{array}\right] \xrightarrow{R_3=\frac{2}{53}r_3} \left[\begin{array}{ccc|c} 1 & 0 & \frac{1}{3} & \frac{5}{3} \\ 0 & 1 & -\frac{3}{4} & -\frac{1}{2} \\ 0 & 0 & 1 & 2 \end{array}\right]$$

The row-echelon form of the system is

$$\begin{cases} x = -\frac{1}{3}z + \frac{5}{3} & (1) \\ y = \frac{3}{4}z - \frac{1}{2} & (2) \\ z = 2 & (3) \end{cases}$$

Back-substituting $z = 2$ into (2) gives $y = 1$. Then back-substituting $z = 2$ and $y = 1$ into (1) gives $x = 1$.

The dietician should serve 1 serving of chicken, 1 potato, and 2 cups of spinach.

77. Let x = number of cases of orange juice to be prepared,
y = number of cases of tomato juice to be prepared, and
z = number of cases of pineapple juice to be prepared.

The problem can be modeled with the system of equations,

$$\begin{cases} 10x + 12y + 9z = 398 & (1) \\ 4x + 4y + 6z = 164 & (2) \\ 2x + y + z = 58 & (3) \end{cases}$$

The system can be represented by the augmented matrices

$$\left[\begin{array}{ccc|c} 10 & 12 & 9 & 398 \\ 4 & 4 & 6 & 164 \\ 2 & 1 & 1 & 58 \end{array}\right] \xrightarrow[R_1 = \frac{1}{10}r_1]{} \left[\begin{array}{ccc|c} 1 & 1.2 & 0.9 & 39.8 \\ 4 & 4 & 6 & 164 \\ 2 & 1 & 1 & 58 \end{array}\right] \xrightarrow[\substack{R_2 = -4r_1 + r_2 \\ R_3 = -2r_1 + r_3}]{} \left[\begin{array}{ccc|c} 1 & 1.2 & 0.9 & 39.8 \\ 0 & -0.8 & 2.4 & 4.8 \\ 0 & -1.4 & -0.8 & -21.6 \end{array}\right]$$

$$\xrightarrow[R_2 = -\frac{10}{8}r_2]{} \left[\begin{array}{ccc|c} 1 & 1.2 & 0.9 & 39.8 \\ 0 & 1 & -3 & -6 \\ 0 & -1.4 & -0.8 & -21.6 \end{array}\right] \xrightarrow[R_3 = 1.4r_2 + r_3]{} \left[\begin{array}{ccc|c} 1 & 1.2 & 0.9 & 39.8 \\ 0 & 1 & -3 & -6 \\ 0 & 0 & -5 & -30 \end{array}\right]$$

$$\xrightarrow[R_3 = -\frac{1}{5}r_3]{} \left[\begin{array}{ccc|c} 1 & 1.2 & 0.9 & 39.8 \\ 0 & 1 & -3 & -6 \\ 0 & 0 & 1 & 6 \end{array}\right]$$

The row-echelon form of the system is

$$\begin{cases} x = -1.2y - 0.9z + 39.8 & (1) \\ y = 3z - 6 & (2) \\ z = 6 & (3) \end{cases}$$

The company prepared 6 cases of pineapple juice, 12 cases of tomato juice, and 20 cases of orange juice.

79. Let x = number of packages of package 1 to be ordered,
y = number of packages of package 2 to be ordered, and
z = number of packages of package 3 to be ordered.

The problem can be modeled with the system of equations,

$$\begin{cases} 20x + 40z = 200 & (1) \\ 15x + 3y + 30z = 180 & (2) \\ x + y = 12 & (3) \end{cases}$$

The system can be represented by the augmented matrices

$$\left[\begin{array}{ccc|c} 20 & 0 & 40 & 200 \\ 15 & 3 & 30 & 180 \\ 1 & 1 & 0 & 12 \end{array}\right] \xrightarrow[\text{rows 1 and 3}]{\text{Interchange}} \left[\begin{array}{ccc|c} 1 & 1 & 0 & 12 \\ 15 & 3 & 30 & 180 \\ 20 & 0 & 40 & 200 \end{array}\right] \xrightarrow[\substack{R_2 = \frac{1}{3}r_2 \\ R_3 = \frac{1}{20}r_3}]{} \left[\begin{array}{ccc|c} 1 & 1 & 0 & 12 \\ 5 & 1 & 10 & 60 \\ 1 & 0 & 2 & 10 \end{array}\right]$$

$$\xrightarrow[\substack{R_2 = -5r_1 + r_2 \\ R_3 = -r_1 + r_3}]{} \left[\begin{array}{ccc|c} 1 & 1 & 0 & 12 \\ 0 & -4 & 10 & 0 \\ 0 & -1 & 2 & -2 \end{array}\right] \xrightarrow[\text{rows 2 and 3}]{\text{Interchange}} \left[\begin{array}{ccc|c} 1 & 1 & 0 & 12 \\ 0 & -1 & 2 & -2 \\ 0 & -4 & 10 & 0 \end{array}\right] \xrightarrow[R_2 = -r_2]{} \left[\begin{array}{ccc|c} 1 & 1 & 0 & 12 \\ 0 & 1 & -2 & 2 \\ 0 & -4 & 10 & 0 \end{array}\right]$$

$$\xrightarrow[R_3 = 4r_2 + r_3]{} \left[\begin{array}{ccc|c} 1 & 1 & 0 & 12 \\ 0 & 1 & -2 & 2 \\ 0 & 0 & 2 & 8 \end{array}\right] \xrightarrow[R_3 = \frac{1}{2}r_3]{} \left[\begin{array}{ccc|c} 1 & 1 & 0 & 12 \\ 0 & 1 & -2 & 2 \\ 0 & 0 & 1 & 4 \end{array}\right]$$

The row-echelon form of the system is

$$\begin{cases} x = -y + 12 & (1) \\ y = 2z + 2 & (2) \\ z = 4 & (3) \end{cases}$$

The teacher should order 4 packages of package 3, 10 packages of package 2, and 2 packages of package 1 paper.

81. Let x = number of assorted cartons needed,
y = number of mixed cartons needed, and
z = number of single cartons needed to fill the recreation center's order.

The problem can be modeled with the system of equations,

$$\begin{cases} 2x + 4y = 40 & (1) \\ 4x + 2y = 32 & (2) \\ x + 2z = 14 & (3) \end{cases}$$

The system can be represented by the augmented matrices

$$\left[\begin{array}{ccc|c} 2 & 4 & 0 & 40 \\ 4 & 2 & 0 & 32 \\ 1 & 0 & 2 & 14 \end{array}\right] \xrightarrow[R_1 = \frac{1}{2}r_1]{} \left[\begin{array}{ccc|c} 1 & 2 & 0 & 20 \\ 4 & 2 & 0 & 32 \\ 1 & 0 & 2 & 14 \end{array}\right] \xrightarrow[\substack{R_2 = -4r_1 + r_2 \\ R_3 = -r_1 + r_3}]{} \left[\begin{array}{ccc|c} 1 & 2 & 0 & 20 \\ 0 & -6 & 0 & -48 \\ 0 & -2 & 2 & -6 \end{array}\right]$$

$$\xrightarrow{R_2=-\frac{1}{6}r_2}\left[\begin{array}{ccc|c}1&2&0&20\\0&1&0&8\\0&-2&2&-6\end{array}\right]\xrightarrow{R_3=2r_2+r_3}\left[\begin{array}{ccc|c}1&2&0&20\\0&1&0&8\\0&0&2&10\end{array}\right]\xrightarrow{R_3=\frac{1}{2}r_3}\left[\begin{array}{ccc|c}1&2&0&20\\0&1&0&8\\0&0&1&5\end{array}\right]$$

The row-echelon form of the system is

$$\begin{cases}x=-2y+20 & (1)\\ y=8 & (2)\\ z=5 & (3)\end{cases}$$

The supplier should send the recreation center 5 single cartons, 8 mixed cartons, and 4 assorted cartons.

83. Let x = number of large cans needed,
y = number of mammoth cans needed, and
z = number of giant cans needed to fulfill the order.

The problem can be modeled with the system of equations,

$$\begin{cases}y+z=5 & (1)\\ 2x+6y+4z=26 & (2)\\ x+2y+2z=12 & (3)\end{cases}$$

The system can be represented by the augmented matrices

$$\left[\begin{array}{ccc|c}0&1&1&5\\2&6&4&26\\1&2&2&12\end{array}\right]\xrightarrow[\text{rows 1and 3}]{\text{Interchange}}\left[\begin{array}{ccc|c}1&2&2&12\\2&6&4&26\\0&1&1&5\end{array}\right]\xrightarrow{R_2=-2r_1+r_2}\left[\begin{array}{ccc|c}1&2&2&12\\0&2&0&2\\0&1&1&5\end{array}\right]$$

$$\xrightarrow{R_2=\frac{1}{2}r_2}\left[\begin{array}{ccc|c}1&2&2&12\\0&1&0&1\\0&1&1&5\end{array}\right]\xrightarrow{R_3=-r_2+r_3}\left[\begin{array}{ccc|c}1&2&2&12\\0&1&0&1\\0&0&1&4\end{array}\right]$$

The row-echelon form of the system is

$$\begin{cases}x=-2y-2z+12 & (1)\\ y=1 & (2)\\ z=4 & (3)\end{cases}$$

The store should use 4 giant size cans, 1 mammoth size can, and 2 large size cans to fill the order.

85. Let x denote the number of shares of Apple Computer purchased, y denote the number of shares of Dell Inc. purchased, and z denote the number of shares of Microsoft Corporation purchased. The problem is modeled by the system of equations:

$$\begin{cases}55.4x+23.87y+23.3z=27{,}288.1 & (1)\\ x+y+z=890 & (2)\\ z=2y & (3)\end{cases}$$

Reordering the equations to make the augmented matrix easier to work with and rewriting equation (3) as $-2y+z=0$, we get the revised system and its corresponding augmented matrix.

$$\begin{cases}x+y+z=890 & (2)\\ -2y+z=0 & (3)\\ 55.4x+23.87y+23.3z=27{,}881.1 & (1)\end{cases}\quad\text{and}\quad\left[\begin{array}{ccc|c}1&1&1&890\\0&-2&1&0\\55.4&23.87&23.3&27{,}881.1\end{array}\right]$$

The augmented matrix is then row reduced using the following row operations

$$\xrightarrow{R_3=-55.4r_1+r_3}\left[\begin{array}{ccc|c}1&1&1&890\\0&-2&1&0\\0&-31.53&-32.1&-22{,}017.9\end{array}\right]\xrightarrow{R_2=-\frac{1}{2}r_2}\left[\begin{array}{ccc|c}1&1&1&890\\0&1&-0.5&0\\0&-31.53&-32.1&-22{,}017.9\end{array}\right]$$

$$\xrightarrow{R_3=31.53r_2+r_3}\left[\begin{array}{ccc|c}1&1&1&890\\0&1&-0.5&0\\0&0&-47.865&-22{,}017.9\end{array}\right]\xrightarrow{R_3=\frac{r_3}{-47.825}}\left[\begin{array}{ccc|c}1&1&1&890\\0&1&-0.5&0\\0&0&1&460\end{array}\right]$$

Back-substitute to solve the system. $z = 460$

$$y = 0.5z = 230$$
$$x = 890 - y - z = 890 - 460 - 230 = 200$$

The investor purchased 200 shares of Apple Computer, 230 shares of Dell Inc., and 460 shares of Microsoft Corporation.

87. Let x_1 denote the number of shares of Bank of America purchased,
x_2 denote the number of shares of Citigroup, Inc. purchased,
x_3 denote the number of shares of JP Morgan Chase & Co. purchased,
x_4 denote the number of shares of Wells Fargo & Co. purchased.

To determine the number of shares of each company bought, set up a system of equations. We must first subtract the total brokerage fee of \$40.00 from the total cost of purchasing the stocks.

$$\begin{cases}48.8x_1 + 49.08x_2 + 42.41x_3 + 68.11x_4 = 63{,}487.7 & (1)\\ x_1 + x_2 + x_3 + x_4 = 1150 & (2)\\ x_4 = 2x_1 + 90 & (3)\\ x_4 = x_1 + x_2 - 30 & (4)\end{cases}$$

To use matrix methods to solve this system, first rewrite equations (3) and (4) with the variables on the left, then form the augmented matrix that represents the system.

$$\begin{cases}48.8x_1 + 49.08x_2 + 42.41x_3 + 68.11x_4 = 63{,}487.7 & (1)\\ x_1 + x_2 + x_3 + x_4 = 1{,}150 & (2)\\ -2x_1 + x_4 = 90 & (3)\\ -x_1 - x_2 + x_4 = -30 & (4)\end{cases}$$

$$\left[\begin{array}{cccc|c}48.8&49.08&42.41&68.11&63{,}487.7\\1&1&1&1&1{,}150\\-2&0&0&1&90\\-1&-1&0&1&-30\end{array}\right]$$

We now use row operations to write the augmented matrix in row echelon form.

$$\xrightarrow{\text{Interchange rows 1 and 2}}\left[\begin{array}{cccc|c}1&1&1&1&1{,}150\\48.8&49.08&42.41&68.11&63{,}487.7\\-2&0&0&1&90\\-1&-1&0&1&-30\end{array}\right]\xrightarrow{\substack{R_2=-48.8r_1+r_2\\ R_3=2r_1+r_3\\ R_4=r_1+r_4}}\left[\begin{array}{cccc|c}1&1&1&1&1150\\0&0.28&-6.39&19.31&7367.7\\0&2&2&3&2390\\0&0&1&2&1120\end{array}\right]$$

$$\xrightarrow[\text{rows 2 and 3}]{\text{Interchange}} \left[\begin{array}{cccc|c} 1 & 1 & 1 & 1 & 1150 \\ 0 & 2 & 2 & 3 & 2390 \\ 0 & 0.28 & -6.39 & 19.31 & 7367.7 \\ 0 & 0 & 1 & 2 & 1120 \end{array}\right] \xrightarrow{R_2 = \frac{1}{2}r_2} \left[\begin{array}{cccc|c} 1 & 1 & 1 & 1 & 1150 \\ 0 & 1 & 1 & 1.5 & 1195 \\ 0 & 0.28 & -6.39 & 19.31 & 7367.7 \\ 0 & 0 & 1 & 2 & 1120 \end{array}\right]$$

$$\xrightarrow[\substack{R_1 = -r_2 + r_1 \\ R_3 = -0.28r_2 + r_3}]{} \left[\begin{array}{cccc|c} 1 & 0 & 0 & -0.5 & -45 \\ 0 & 1 & 1 & 1.5 & 1195 \\ 0 & 0 & -6.67 & 18.89 & 7033.1 \\ 0 & 0 & 1 & 2 & 1120 \end{array}\right] \xrightarrow[\text{rows 3 and 4}]{\text{Interchange}} \left[\begin{array}{cccc|c} 1 & 0 & 0 & -0.5 & -45 \\ 0 & 1 & 1 & 1.5 & 1195 \\ 0 & 0 & 1 & 2 & 1120 \\ 0 & 0 & -6.67 & 18.89 & 7033.1 \end{array}\right]$$

$$\xrightarrow[\substack{R_2 = -r_3 + r_2 \\ R_4 = 6.67r_3 + r_4}]{} \left[\begin{array}{cccc|c} 1 & 0 & 0 & -0.5 & -45 \\ 0 & 1 & 0 & -0.5 & 75 \\ 0 & 0 & 1 & 2 & 1120 \\ 0 & 0 & 0 & 32.23 & 14{,}503.5 \end{array}\right] \xrightarrow{R_4 = \frac{1}{32.23}r_4} \left[\begin{array}{cccc|c} 1 & 0 & 0 & -0.5 & -45 \\ 0 & 1 & 0 & -0.5 & 75 \\ 0 & 0 & 1 & 2 & 1120 \\ 0 & 0 & 0 & 1 & 450 \end{array}\right]$$

We see from the final matrix that $x_4 = 450$ shares of Wells Fargo were purchased. Now either back-substitute to find the values of the remaining variables, or use row operations to write the augmented matrix in reduced row echelon form. We back-substitute here.

$x_1 = 0.5x_4 - 45 = 0.5(450) - 45 = 225 - 45 = 180$; 180 shares of Bank of America were bought

$x_2 = 0.5x_4 + 75 = 0.5(450) + 75 = 225 + 75 = 300$; 300 shares of Citigroup were purchased

$x_3 = -2x_4 + 1120 = (-2)(450) + 1120 = -900 + 1120 = 220$; 220 shares of JP Morgan Chase were bought.

2.3 Systems of *m* Linear Equations Containing *n* Variables

1. False

3. Yes.

5. No, the leftmost 1 in the second row is not to the right of the leftmost 1 in the first row.

7. Yes.

9. There are infinitely many solutions represented by the system

$$\begin{cases} x = 2z + 6 \\ y = -3z + 1 \end{cases}$$

where z is the parameter, can be assigned any value, and used to compute values of x and y.

11. The system has one solution. It is $x = -1$, $y = 3$, and $z = 4$ or, written as an ordered triple, (–1, 3, 4).

13. The system has infinitely many solutions. They are represented by the system

$$\begin{cases} x = z + 1 \\ y = -2z + 1 \end{cases}$$

where z is the parameter and can be assigned any real number.

15. The system has infinitely many solutions. They are represented by the system

$$\begin{cases} x_1 = x_4 + 4 \\ x_2 = -2x_3 - 3x_4 \end{cases}$$

where x_3 and x_4 are the parameters and can be assigned any real numbers.

17. Write the system as the augmented matrix, $\left[\begin{array}{rr|r} 3 & -3 & 12 \\ 3 & 2 & -3 \\ 2 & 1 & 4 \end{array}\right]$.

Then use row operations to find the reduced row-echelon form.

$$\left[\begin{array}{rr|r} 3 & -3 & 12 \\ 3 & 2 & -3 \\ 2 & 1 & 4 \end{array}\right] \xrightarrow{R_1 = \frac{1}{3}r_1} \left[\begin{array}{rr|r} 1 & -1 & 4 \\ 3 & 2 & -3 \\ 2 & 1 & 4 \end{array}\right] \xrightarrow[R_3 = -2r_1 + r_3]{R_2 = -3r_1 + r_2} \left[\begin{array}{rr|r} 1 & -1 & 4 \\ 0 & 5 & -15 \\ 0 & 3 & -4 \end{array}\right]$$

$$\xrightarrow{R_2 = \frac{1}{5}r_2} \left[\begin{array}{rr|r} 1 & -1 & 4 \\ 0 & 1 & -3 \\ 0 & 3 & -4 \end{array}\right] \xrightarrow[R_3 = -3r_2 + r_3]{R_1 = r_2 + r_1} \left[\begin{array}{rr|r} 1 & 0 & 1 \\ 0 & 1 & -3 \\ 0 & 0 & 5 \end{array}\right] \xrightarrow{R_3 = \frac{1}{5}r_3} \left[\begin{array}{rr|r} 1 & 0 & 1 \\ 0 & 1 & -3 \\ 0 & 0 & 1 \end{array}\right]$$

There is no solution. The system is inconsistent.

19. Write the system as the augmented matrix, $\left[\begin{array}{rr|r} 2 & -4 & 8 \\ 1 & -2 & 4 \\ -1 & 2 & -4 \end{array}\right]$.

Then use row operations to find the reduced row-echelon form.

$$\left[\begin{array}{rr|r} 2 & -4 & 8 \\ 1 & -2 & 4 \\ -1 & 2 & -4 \end{array}\right] \xrightarrow[\text{rows 1 and 2}]{\text{Interchange}} \left[\begin{array}{rr|r} 1 & -2 & 4 \\ 2 & -4 & 8 \\ -1 & 2 & -4 \end{array}\right] \xrightarrow[R_3 = r_1 + r_3]{R_2 = -2r_1 + r_2} \left[\begin{array}{rr|r} 1 & -2 & 4 \\ 0 & 0 & 0 \\ 0 & 0 & 0 \end{array}\right]$$

The system has an infinite number of solutions. They are $x = 2y + 4$, where y is the parameter and can be assigned any real number.

21. Write the system as the augmented matrix, $\left[\begin{array}{rrr|r} 2 & 1 & 3 & -1 \\ -1 & 1 & 3 & 8 \end{array}\right]$.

Then use row operations to find the reduced row-echelon form.

$$\left[\begin{array}{rrr|r} 2 & 1 & 3 & -1 \\ -1 & 1 & 3 & 8 \end{array}\right] \xrightarrow[\text{rows 1 and 2}]{\text{Interchange}} \left[\begin{array}{rrr|r} -1 & 1 & 3 & 8 \\ 2 & 1 & 3 & -1 \end{array}\right] \xrightarrow{R_1 = -r_1} \left[\begin{array}{rrr|r} 1 & -1 & -3 & -8 \\ 2 & 1 & 3 & -1 \end{array}\right]$$

$$\xrightarrow{R_2 = -2r_1 + r_2} \left[\begin{array}{rrr|r} 1 & -1 & -3 & -8 \\ 0 & 3 & 9 & 15 \end{array}\right] \xrightarrow{R_2 = \frac{1}{3}r_2} \left[\begin{array}{rrr|r} 1 & -1 & -3 & -8 \\ 0 & 1 & 3 & 5 \end{array}\right] \xrightarrow{R_1 = r_2 + r_1} \left[\begin{array}{rrr|r} 1 & 0 & 0 & -3 \\ 0 & 1 & 3 & 5 \end{array}\right]$$

There are an infinite number of solutions. They are represented by the system of equations

$$\begin{cases} x = -3 \\ y = -3z + 5 \end{cases}$$

where z is the parameter and can be assigned any real number.

23. Write the system as the augmented matrix, $\left[\begin{array}{cccc|c} 1 & 1 & 0 & 0 & 7 \\ 0 & 1 & -1 & 1 & 5 \\ 1 & -1 & 1 & 1 & 6 \\ 0 & 1 & 0 & -1 & 10 \end{array}\right]$.

Then use row operations to find the reduced row-echelon form.

$$\left[\begin{array}{cccc|c} 1 & 1 & 0 & 0 & 7 \\ 0 & 1 & -1 & 1 & 5 \\ 1 & -1 & 1 & 1 & 6 \\ 0 & 1 & 0 & -1 & 10 \end{array}\right] \xrightarrow{R_3 = -r_1 + r_3} \left[\begin{array}{cccc|c} 1 & 1 & 0 & 0 & 7 \\ 0 & 1 & -1 & 1 & 5 \\ 0 & -2 & 1 & 1 & -1 \\ 0 & 1 & 0 & -1 & 10 \end{array}\right] \xrightarrow[\substack{R_3 = 2r_2 + r_3 \\ R_4 = -r_2 + r_4}]{R_1 = -r_2 + r_1} \left[\begin{array}{cccc|c} 1 & 0 & 1 & -1 & 2 \\ 0 & 1 & -1 & 1 & 5 \\ 0 & 0 & -1 & 3 & 9 \\ 0 & 0 & 1 & -2 & 5 \end{array}\right]$$

$$\xrightarrow{R_3 = -r_3} \left[\begin{array}{cccc|c} 1 & 0 & 1 & -1 & 2 \\ 0 & 1 & -1 & 1 & 5 \\ 0 & 0 & 1 & -3 & -9 \\ 0 & 0 & 1 & -2 & 5 \end{array}\right] \xrightarrow[\substack{R_2 = r_3 + r_2 \\ R_4 = -r_3 + r_4}]{R_1 = -r_3 + r_1} \left[\begin{array}{cccc|c} 1 & 0 & 0 & 2 & 11 \\ 0 & 1 & 0 & -2 & -4 \\ 0 & 0 & 1 & -3 & -9 \\ 0 & 0 & 0 & 1 & 14 \end{array}\right] \xrightarrow[\substack{R_2 = 2r_4 + r_2 \\ R_3 = 3r_4 + r_3}]{R_1 = -2r_4 + r_1} \left[\begin{array}{cccc|c} 1 & 0 & 0 & 0 & -17 \\ 0 & 1 & 0 & 0 & 24 \\ 0 & 0 & 1 & 0 & 33 \\ 0 & 0 & 0 & 1 & 14 \end{array}\right]$$

There is one solution, $x_1 = -17$, $x_2 = 24$, $x_3 = 33$, and $x_4 = 14$ or $(-17, 24, 33, 14)$.

25. Write the system as the augmented matrix, $\left[\begin{array}{ccc|c} 2 & -3 & 4 & 7 \\ 1 & -2 & 3 & 2 \end{array}\right]$.

Then use row operations to find the reduced row-echelon form.

$$\left[\begin{array}{ccc|c} 2 & -3 & 4 & 7 \\ 1 & -2 & 3 & 2 \end{array}\right] \xrightarrow[\text{rows 1 and 2}]{\text{Interchange}} \left[\begin{array}{ccc|c} 1 & -2 & 3 & 2 \\ 2 & -3 & 4 & 7 \end{array}\right] \xrightarrow{R_2 = -2r_1 + r_2} \left[\begin{array}{ccc|c} 1 & -2 & 3 & 2 \\ 0 & 1 & -2 & 3 \end{array}\right]$$

$$\xrightarrow{R_1 = 2r_2 + r_1} \left[\begin{array}{ccc|c} 1 & 0 & -1 & 8 \\ 0 & 1 & -2 & 3 \end{array}\right]$$

There are an infinite number of solutions. They can be represented by the system of equations

$$\begin{cases} x = z + 8 \\ y = 2z + 3 \end{cases}$$

where z is the parameter and can be assigned any real number.

27. Write the system as the augmented matrix, $\left[\begin{array}{rrrr|r} 1 & 1 & 1 & 1 & 4 \\ 2 & -1 & 1 & 0 & 0 \\ 3 & 2 & 1 & -1 & 6 \\ 1 & -2 & -2 & 2 & -1 \end{array}\right]$.

Then use row operations to find the reduced row-echelon form.

$$\left[\begin{array}{rrrr|r} 1 & 1 & 1 & 1 & 4 \\ 2 & -1 & 1 & 0 & 0 \\ 3 & 2 & 1 & -1 & 6 \\ 1 & -2 & -2 & 2 & -1 \end{array}\right] \xrightarrow[\substack{R_2 = -2r_1 + r_2 \\ R_3 = -3r_1 + r_3 \\ R_4 = -r_1 + r_4}]{} \left[\begin{array}{rrrr|r} 1 & 1 & 1 & 1 & 4 \\ 0 & -3 & -1 & -2 & -8 \\ 0 & -1 & -2 & -4 & -6 \\ 0 & -3 & -3 & 1 & -5 \end{array}\right]$$

$$\xrightarrow[\substack{R_2 = -r_3 \\ R_3 = -r_2}]{} \left[\begin{array}{rrrr|r} 1 & 1 & 1 & 1 & 4 \\ 0 & 1 & 2 & 4 & 6 \\ 0 & 3 & 1 & 2 & 8 \\ 0 & -3 & -3 & 1 & -5 \end{array}\right] \xrightarrow[\substack{R_1 = -r_2 + r_1 \\ R_3 = -3r_2 + r_3 \\ R_4 = 3r_2 + r_4}]{} \left[\begin{array}{rrrr|r} 1 & 0 & -1 & -3 & -2 \\ 0 & 1 & 2 & 4 & 6 \\ 0 & 0 & -5 & -10 & -10 \\ 0 & 0 & 3 & 13 & 13 \end{array}\right]$$

$$\xrightarrow[R_3 = -\frac{1}{5}r_3]{} \left[\begin{array}{rrrr|r} 1 & 0 & -1 & -3 & -2 \\ 0 & 1 & 2 & 4 & 6 \\ 0 & 0 & 1 & 2 & 2 \\ 0 & 0 & 3 & 13 & 13 \end{array}\right] \xrightarrow[\substack{R_1 = r_3 + r_1 \\ R_2 = -2r_3 + r_2 \\ R_4 = -3r_3 + r_4}]{} \left[\begin{array}{rrrr|r} 1 & 0 & 0 & -1 & 0 \\ 0 & 1 & 0 & 0 & 2 \\ 0 & 0 & 1 & 2 & 2 \\ 0 & 0 & 0 & 7 & 7 \end{array}\right]$$

$$\xrightarrow[R_4 = \frac{1}{7}r_4]{} \left[\begin{array}{rrrr|r} 1 & 0 & 0 & -1 & 0 \\ 0 & 1 & 0 & 0 & 2 \\ 0 & 0 & 1 & 2 & 2 \\ 0 & 0 & 0 & 1 & 1 \end{array}\right] \xrightarrow[\substack{R_1 = r_4 + r_1 \\ R_3 = -2r_4 + r_3}]{} \left[\begin{array}{rrrr|r} 1 & 0 & 0 & 0 & 1 \\ 0 & 1 & 0 & 0 & 2 \\ 0 & 0 & 1 & 0 & 0 \\ 0 & 0 & 0 & 1 & 1 \end{array}\right]$$

There is one solution to the system. It is written $x_1 = 1, x_2 = 2, x_3 = 0$ and $x_4 = 1$ or (1, 2, 0, 1).

29. Let x be the amount of supplement 1 used,
y be the amount of supplement 2 used, and
z be the amount of supplement 3 used.

The problem is modeled by the system of equations

$$\begin{cases} 0.2x + 0.4y + 0.3z = 40 & (1) \\ 0.3x + 0.2y + 0.5z = 30 & (2) \end{cases}$$

and can be represented by the augmented matrix

$$\left[\begin{array}{ccc|c} 0.2 & 0.4 & 0.3 & 40 \\ 0.3 & 0.2 & 0.5 & 30 \end{array}\right] \text{ or multiplying by 10 (to remove decimals) } \left[\begin{array}{ccc|c} 2 & 4 & 3 & 400 \\ 3 & 2 & 5 & 300 \end{array}\right]$$

Use the row operations to write the matrix in row reduced form.

$$\left[\begin{array}{ccc|c} 2 & 4 & 3 & 400 \\ 3 & 2 & 5 & 300 \end{array}\right] \xrightarrow[R_1 = \frac{1}{2}r_1]{} \left[\begin{array}{ccc|c} 1 & 2 & 1.5 & 200 \\ 3 & 2 & 5 & 300 \end{array}\right] \xrightarrow[R_2 = -3r_1 + r_2]{} \left[\begin{array}{ccc|c} 1 & 2 & 1.5 & 200 \\ 0 & -4 & 0.5 & -300 \end{array}\right]$$

$$\xrightarrow[R_2 = -\frac{1}{4}r_2]{} \left[\begin{array}{ccc|c} 1 & 2 & 1.5 & 200 \\ 0 & 1 & -0.125 & 75 \end{array}\right]$$

There are an infinite number of solutions. They are $y = 0.125z + 75$ and

$$x = -2y - 1.5z + 200$$
$$x = -2(0.125z + 75) - 1.5z + 200$$
$$x = -1.75z + 50, \text{ where } z \text{ is a nonnegative real number.}$$

Since x and y also must be nonnegative, we find that

$$x = -1.75z + 50 \geq 0$$
$$50 \geq 1.75z$$
$$z \leq 28.57$$

So we create a table,

$x = -1.75z + 50$	$y = 0.125z + 75$	z
50	75	0
41.25	75.625	5
32.5	76.25	10
23.75	76.875	15
15	77.5	20
0	78.571	28.57

31. (a) Let x_1 denote the servings of turkey bologna he consumes,
x_2 denote the number of bananas he eats,
x_3 denote the servings of cottage cheese he eats, and
x_4 denote the servings of chocolate milk he drinks.

$$\begin{cases} 184x_1 + 92x_2 + 90x_3 + 72x_4 = 700 & (1) \\ 13.2x_1 + 0.48x_2 + 1.93x_3 + 2x_4 = 20 & (2) \\ 4.85x_1 + 23.43x_2 + 0.01x_3 + 10.4x_4 = 100 & (3) \end{cases}$$

(b) Write the augmented matrix, then because of the complexity of the arithmetic, we suggest using a graphing utility to solve the system. Below are the results using a TI-84 graphing calculator.

The augmented matrix:

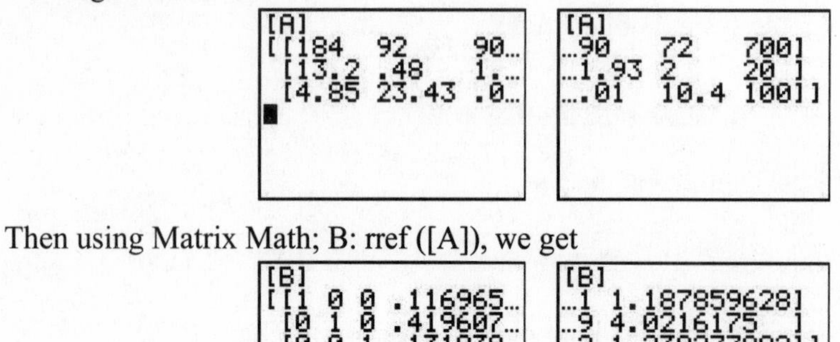

Then using Matrix Math; B: rref ([A]), we get

For those who need to do it by hand, below are the steps (rounded to 5 decimal places).

$$\left[\begin{array}{cccc|c} 184 & 92 & 90 & 72 & 700 \\ 13.2 & 0.48 & 1.93 & 2 & 20 \\ 4.85 & 23.43 & 0.01 & 10.4 & 100 \end{array}\right] \xrightarrow[R_1=\frac{1}{184}r_1]{} \left[\begin{array}{cccc|c} 1 & 0.5 & 0.48313 & 0.39130 & 3.80435 \\ 13.2 & 0.48 & 1.93 & 2 & 20 \\ 4.85 & 23.43 & 0.01 & 10.4 & 100 \end{array}\right]$$

$$\xrightarrow[\substack{R_2=-13.2r_1+r_2 \\ R_3=-4.85r_1+r_3}]{} \left[\begin{array}{cccc|c} 1 & 0.5 & 0.48313 & 0.39130 & 3.80435 \\ 0 & -6.12 & -4.52652 & -3.16522 & -30.21739 \\ 0 & 21.005 & -2.36228 & 8.50217 & 81.54891 \end{array}\right]$$

$$\xrightarrow[R_2=-\frac{1}{6.12}r_2]{} \left[\begin{array}{cccc|c} 1 & 0.5 & 0.48313 & 0.39130 & 3.80435 \\ 0 & 1 & 0.73963 & 0.51719 & 4.93748 \\ 0 & 21.005 & -2.36228 & 8.50217 & 81.54891 \end{array}\right]$$

$$\xrightarrow[\substack{R_1=-0.5r_2+r_1 \\ R_3=-21.005r_2+r_3}]{} \left[\begin{array}{cccc|c} 1 & 0 & 0.11932 & 0.13271 & 1.33561 \\ 0 & 1 & 0.73963 & 0.51719 & 4.93748 \\ 0 & 0 & -17.89816 & -2.36145 & -22.16290 \end{array}\right]$$

$$\xrightarrow[R_3=-\frac{1}{17.89816}r_3]{} \left[\begin{array}{cccc|c} 1 & 0 & 0.11932 & 0.13271 & 1.33561 \\ 0 & 1 & 0.73963 & 0.51719 & 4.93748 \\ 0 & 0 & 1 & 0.13194 & 1.23828 \end{array}\right]$$

$$\xrightarrow[\substack{R_1=-0.11932r_3+r_1 \\ R_2=-0.73963r_3+r_2}]{} \left[\begin{array}{cccc|c} 1 & 0 & 0 & 0.11697 & 1.18786 \\ 0 & 1 & 0 & 0.41961 & 4.02161 \\ 0 & 0 & 1 & 0.13194 & 1.23827 \end{array}\right]$$

There are multiple solutions, which can be represented by the system of equations:

$$\begin{cases} x_1 = -0.117x_4 + 1.188 \\ x_2 = -0.420x_4 + 4.022 \\ x_3 = -0.132x_4 + 1.238 \end{cases}$$

where x_4 is the parameter. Since each of the variables must be nonnegative (you cannot eat a negative amount), we find that $0 \le x_4 \le 9.385$. If $x_4 > 9.385$, x_3 would become negative. We then can make a table.

Turkey Bologna x_1	Bananas x_2	Cottage Cheese x_3	Chocolate Milk x_4
1.2	4.0	1.2	0
1.1	3.6	1.1	1
1.0	3.2	1.0	2
0.8	2.8	0.8	3
0.7	2.3	0.7	4
0.6	1.9	0.6	5
0.5	1.5	0.4	6
0.4	1.1	0.3	7
0.3	0.7	0.2	8
0.1	0.2	0.1	9

33. (a) Let x_1 represent the amount invested in Janus Large Cap Growth (at 9%),
x_2 represent the amount invested in FI Mid Cap (at 13%),
x_3 represent the amount invested in Alliance Bernstein Value (at 14%), and
x_4 represent the amount invested in Calvert Socially Responsible (at 7%).

$$\begin{cases} x_1 + x_2 + x_3 + x_4 = 50{,}000 & (1) \\ 0.09x_1 + 0.13x_2 + 0.14x_3 + 0.07x_4 = 6{,}000 & (2) \end{cases}$$

(b) The system is represented by the following augmented matrices.

$$\left[\begin{array}{cccc|c} 1 & 1 & 1 & 1 & 50{,}000 \\ 0.09 & 0.13 & 0.14 & 0.07 & 6{,}000 \end{array}\right] \xrightarrow[R_2=-0.09r_1+r_2]{} \left[\begin{array}{cccc|c} 1 & 1 & 1 & 1 & 50{,}000 \\ 0 & 0.04 & 0.05 & -0.02 & 1{,}500 \end{array}\right]$$

$$\xrightarrow[R_2=25r_2]{} \left[\begin{array}{cccc|c} 1 & 1 & 1 & 1 & 50{,}000 \\ 0 & 1 & 1.25 & -0.5 & 37{,}500 \end{array}\right] \xrightarrow[R_1=-r_2+r_1]{} \left[\begin{array}{cccc|c} 1 & 0 & -0.25 & 0.5 & 12{,}500 \\ 0 & 1 & 1.25 & -0.5 & 37{,}500 \end{array}\right]$$

The reduced row echelon form of the augmented matrix indicates that there are an infinite number of solutions. They can be represented by the system of equations:

$$\begin{cases} x_1 = 0.25x_3 - 0.5x_4 + 12{,}500 \\ x_2 = -1.25x_3 + 0.5x_4 + 37{,}500 \end{cases}$$

where x_3 and x_4 are parameters. All the variables must be nonnegative, so we begin with both parameters at zero and increase each one individually until x_1 or x_2 equal zero.

(c) A sample table follows. Your table might be different.

Amount Invested			
Janus Large Cap	FI Mid Cap	Alliance Bernstein Value	Calvert Socially Responsible
$12,500	$37,500	$ 0	$ 0
$15,000	$25,000	$10,000	$ 0
$17,500	$12,500	$20,000	$ 0
$20,000	$ 0	$30,000	$ 0
$ 5,000	$40,000	$ 0	$ 5,000
$ 5,000	$ 5,000	$30,000	$10,000

35. Let x_1 represent the amount invested in Treasury Bills yielding 5%,
x_2 represent the amount invested in corporate paper yielding 6%,
x_3 represent the amount invested in corporate bonds yielding 8%, and
x_4 represent the amount invested in junk bonds which yielding 12% in interest.

(a) The first couple has $20,000 to invest, and require $2000 in income.
To determine how they should allot their funds we need to solve the system of equations:

$$\begin{cases} x_1 + x_2 + x_3 + x_4 = 20{,}000 & (1) \\ 0.05x_1 + 0.06x_2 + 0.08x_3 + 0.12x_4 = 2{,}000 & (2) \end{cases}$$

We write the system as an augmented matrix and use row operations to put it in reduced row echelon form.

$$\left[\begin{array}{cccc|c} 1 & 1 & 1 & 1 & 20{,}000 \\ 0.05 & 0.06 & 0.08 & 0.12 & 2{,}000 \end{array}\right] \xrightarrow{R_2=-0.05r_1+r_2} \left[\begin{array}{cccc|c} 1 & 1 & 1 & 1 & 20{,}000 \\ 0 & 0.01 & 0.03 & 0.07 & 1{,}000 \end{array}\right]$$

$$\xrightarrow{R_2=100r_2} \left[\begin{array}{cccc|c} 1 & 1 & 1 & 1 & 20{,}000 \\ 0 & 1 & 3 & 7 & 100{,}000 \end{array}\right] \xrightarrow{R_1=-r_2+r_1} \left[\begin{array}{cccc|c} 1 & 0 & -2 & -6 & -80{,}000 \\ 0 & 1 & 3 & 7 & 100{,}000 \end{array}\right]$$

The solutions to the system can be expressed as the system of equations,

$$\begin{cases} x_1 = 2x_3 + 6x_4 - 80{,}000 \\ x_2 = -3x_3 - 7x_4 + 100{,}000 \end{cases}$$

where x_3 and x_4 are parameters. In this system all variables must be nonnegative, no variable can be greater than $16,666.67.

A possible investment strategy for a couple with $20,000.00 to invest is

Amount Invested			
Treasury Bills	Corporate Paper	Corporate Bonds	Junk Bonds
$0.00	$0.00	$10,000.00	$10,000.00
$0.00	$6666.67	$0.00	$13,333.33
$5714.29	$0.00	$0.00	$14,285.71

(b) The second couple has \$25,000 to invest, and require \$2000 in income.
To determine how they should allot their funds we need to solve the system of equations:

$$\begin{cases} x_1 + x_2 + x_3 + x_4 = 25{,}000 & (1) \\ 0.05x_1 + 0.06x_2 + 0.08x_3 + 0.12x_4 = 2{,}000 & (2) \end{cases}$$

We write the system as an augmented matrix and use row operations to put it in reduced row echelon form.

$$\left[\begin{array}{cccc|c} 1 & 1 & 1 & 1 & 25{,}000 \\ 0.05 & 0.06 & 0.08 & 0.12 & 2{,}000 \end{array}\right] \xrightarrow{R_2=-0.05r_1+r_2} \left[\begin{array}{cccc|c} 1 & 1 & 1 & 1 & 25{,}000 \\ 0 & 0.01 & 0.03 & 0.07 & 750 \end{array}\right]$$

$$\xrightarrow{R_2=100r_2} \left[\begin{array}{cccc|c} 1 & 1 & 1 & 1 & 25{,}000 \\ 0 & 1 & 3 & 7 & 75{,}000 \end{array}\right] \xrightarrow{R_1=-r_2+r_1} \left[\begin{array}{cccc|c} 1 & 0 & -2 & -6 & -50{,}000 \\ 0 & 1 & 3 & 7 & 75{,}000 \end{array}\right]$$

The solutions to the system can be written as the system

$$\begin{cases} x_1 = 2x_3 + 6x_4 - 50{,}000 \\ x_2 = -3x_3 - 7x_4 + 75{,}000 \end{cases}$$

where x_3 and x_4 are parameters with the restrictions all variables must be nonnegative.

The couple with \$25,000.00 could achieve their goal of earning \$2000 in investment income by investing in the following ways.

Amount Invested			
Treasury Bills	Corporate Paper	Corporate Bonds	Junk Bonds
\$0.00	\$0.00	\$25,000.00	\$0.00
\$10,000.00	\$5,000.00	\$0.00	\$10,000.00
\$13,333.33	\$0.00	\$1,666.67	\$10,000.00

(c) The third couple has \$30,000 to invest, and requires \$2000 in income
To determine how they should allot their funds we need to solve the system of equations:

$$\begin{cases} x_1 + x_2 + x_3 + x_4 = 30{,}000 & (1) \\ 0.05x_1 + 0.06x_2 + 0.08x_3 + 0.12x_4 = 2{,}000 & (2) \end{cases}$$

We write the system as an augmented matrix and use row operations to put it in reduced row echelon form.

$$\left[\begin{array}{cccc|c} 1 & 1 & 1 & 1 & 30{,}000 \\ 0.05 & 0.06 & 0.08 & 0.12 & 2{,}000 \end{array}\right] \xrightarrow{R_2=-0.05r_1+r_2} \left[\begin{array}{cccc|c} 1 & 1 & 1 & 1 & 30{,}000 \\ 0 & 0.01 & 0.03 & 0.07 & 500 \end{array}\right]$$

$$\xrightarrow{R_2=100r_2} \left[\begin{array}{cccc|c} 1 & 1 & 1 & 1 & 30{,}000 \\ 0 & 1 & 3 & 7 & 50{,}000 \end{array}\right] \xrightarrow{R_1=-r_2+r_1} \left[\begin{array}{cccc|c} 1 & 0 & -2 & -6 & -20{,}000 \\ 0 & 1 & 3 & 7 & 50{,}000 \end{array}\right]$$

The solutions to the system can be written as the system:

$$\begin{cases} x_1 = 2x_3 + 6x_4 - 20,000 \\ x_2 = -3x_3 - 7x_4 + 50,000 \end{cases}$$

where x_3 and x_4 are parameters with the restrictions all variables must be nonnegative.

The couple with \$30,000.00 could achieve their goal of earning \$2000 in investment income by investing in the following ways.

Amount Invested			
Treasury Bills	Corporate Paper	Corporate Bonds	Junk Bonds
\$0.00	\$26,666.67	\$0.00	\$3333.33
\$0.00	\$20,000.00	\$10,000.00	\$0.00
\$22,857.14	\$0.00	\$0.00	\$7142.86
\$13,333.33	\$0.00	\$16,666.67	\$0.00

37. Let x represent the number of species 1 bacteria, y represent the number of species 2 bacteria, and z represent the number of species 3 in the test tube. To find the number of each species that can coexist in the test tube solve the system of equations.

$$\begin{cases} 3x + y + 2z = 12000 \\ 2y + 4z = 12000 \\ x + 2y + 4z = 14000 \end{cases}$$

We represent the system with the following augmented matrix, and then use row operations to write the matrix in reduced row-echelon form.

$$\left[\begin{array}{ccc|c} 3 & 1 & 2 & 12000 \\ 0 & 2 & 4 & 12000 \\ 1 & 2 & 4 & 14000 \end{array}\right] \xrightarrow[\text{rows 1 and 3}]{\text{Interchange}} \left[\begin{array}{ccc|c} 1 & 2 & 4 & 14000 \\ 0 & 2 & 4 & 12000 \\ 3 & 1 & 2 & 12000 \end{array}\right] \xrightarrow{R_3 = -3r_1 + r_3} \left[\begin{array}{ccc|c} 1 & 2 & 4 & 14000 \\ 0 & 2 & 4 & 12000 \\ 0 & -5 & -10 & -30000 \end{array}\right]$$

$$\xrightarrow{R_2 = \frac{1}{2}r_3} \left[\begin{array}{ccc|c} 1 & 2 & 4 & 14000 \\ 0 & 1 & 2 & 6000 \\ 0 & -5 & -10 & -30000 \end{array}\right] \xrightarrow[R_3 = 5r_2 + r_3]{R_1 = -2r_2 + r_1} \left[\begin{array}{ccc|c} 1 & 0 & 0 & 2000 \\ 0 & 1 & 2 & 6000 \\ 0 & 0 & 0 & 0 \end{array}\right]$$

This system has an infinite number of solutions. They can be expressed as a system of equations.

$$\begin{cases} x = 2000 \\ y = -2z + 6000 \end{cases}$$

where z is the parameter, and all variables must be nonnegative. The nonnegativity constraint requires $z \le 3000$. Possible combinations of the population are shown in the table. Your table might look different.

Number of Species 1	Number of Species 2	Number of Species 3
2000	6000	0
2000	4000	1000
2000	2000	2000
2000	0	3000

2.4 Matrix Algebra

1. True

3. 2×3 (number of rows × number of columns)

5. True. For any two matrices A and B with the same dimensions, $A + B = B + A$.

7. $\begin{bmatrix} 3 & 2 \\ -1 & 3 \end{bmatrix}$ is a 2×2 matrix. It is square.

9. $\begin{bmatrix} 2 & 1 & -3 \\ 1 & 0 & -1 \end{bmatrix}$ is a 2×3 matrix.

11. $\begin{bmatrix} 4 & 0 \\ -1 & 2 \\ 5 & 8 \end{bmatrix}$ is a 3×2 matrix.

13. $\begin{bmatrix} 1 & 4 \\ -2 & 8 \\ 0 & 0 \end{bmatrix}$ is 3×2 matrix.

15. $\begin{bmatrix} 4 \\ 1 \end{bmatrix}$ is a 2×1 column matrix.

17. [2] is a 1×1 matrix. It is a column matrix, a row matrix, and a square matrix.

19. False, to be equal 2 matrices must have the same dimension.

21. True

23. True

25. True

27. True

29. False, to be equal 2 matrices must have the same dimensions. The sum of

$$\begin{bmatrix} 8 \\ 1 \end{bmatrix} + \begin{bmatrix} 2 \\ 9 \end{bmatrix} = \begin{bmatrix} 10 \\ 10 \end{bmatrix}$$

31. $$\begin{bmatrix} 3 & -1 \\ 4 & 2 \end{bmatrix} + \begin{bmatrix} -2 & 2 \\ 2 & 5 \end{bmatrix} = \begin{bmatrix} 3+(-2) & (-1)+2 \\ 4+\ 2 & 2+5 \end{bmatrix} = \begin{bmatrix} 1 & 1 \\ 6 & 7 \end{bmatrix}$$

33. $$3\begin{bmatrix} 2 & 6 & 0 \\ 4 & -2 & 1 \end{bmatrix} = \begin{bmatrix} 3\cdot 2 & 3\cdot 6 & 3\cdot 0 \\ 3\cdot 4 & 3\cdot(-2) & 3\cdot 1 \end{bmatrix} = \begin{bmatrix} 6 & 18 & 0 \\ 12 & -6 & 3 \end{bmatrix}$$

35. $$2\begin{bmatrix} 1 & -1 & 8 \\ 2 & 4 & 1 \end{bmatrix} - 3\begin{bmatrix} 0 & -2 & 8 \\ 1 & 4 & 1 \end{bmatrix} = \begin{bmatrix} 2 & -2 & 16 \\ 4 & 8 & 2 \end{bmatrix} - \begin{bmatrix} 0 & -6 & 24 \\ 3 & 12 & 3 \end{bmatrix}$$
$$= \begin{bmatrix} 2-0 & (-2)-(-6) & 16-24 \\ 4-3 & 8-12 & 2-3 \end{bmatrix}$$
$$= \begin{bmatrix} 2 & 4 & -8 \\ 1 & -4 & -1 \end{bmatrix}$$

37. $$3\begin{bmatrix} a & 8 \\ b & 1 \\ c & -2 \end{bmatrix} + 5\begin{bmatrix} 2a & 6 \\ -b & -2 \\ -c & 0 \end{bmatrix} = \begin{bmatrix} 3a & 24 \\ 3b & 3 \\ 3c & -6 \end{bmatrix} + \begin{bmatrix} 10a & 30 \\ -5b & -10 \\ -5c & 0 \end{bmatrix} = \begin{bmatrix} 3a+10a & 24+30 \\ 3b-5b & 3-10 \\ 3c-5c & -6+0 \end{bmatrix} = \begin{bmatrix} 13a & 54 \\ -2b & -7 \\ -2c & -6 \end{bmatrix}$$

39. $A-B=\begin{bmatrix}2&-3&4\\0&2&1\end{bmatrix}-\begin{bmatrix}1&-2&0\\5&1&2\end{bmatrix}=\begin{bmatrix}2-1&-3-(-2)&4-0\\0-5&2-1&1-2\end{bmatrix}=\begin{bmatrix}1&-1&4\\-5&1&-1\end{bmatrix}$

41. $2A-3C=2\begin{bmatrix}2&-3&4\\0&2&1\end{bmatrix}-3\begin{bmatrix}-3&0&5\\2&1&3\end{bmatrix}=\begin{bmatrix}4&-6&8\\0&4&2\end{bmatrix}-\begin{bmatrix}-9&0&15\\6&3&9\end{bmatrix}$

$=\begin{bmatrix}4+9&-6+0&8-15\\0-6&4-3&2-9\end{bmatrix}=\begin{bmatrix}13&-6&-7\\-6&1&-7\end{bmatrix}$

43. $(A+B)-2C=\left(\begin{bmatrix}2&-3&4\\0&2&1\end{bmatrix}+\begin{bmatrix}1&-2&0\\5&1&2\end{bmatrix}\right)-2\begin{bmatrix}-3&0&5\\2&1&3\end{bmatrix}$

$=\begin{bmatrix}3&-5&4\\5&3&3\end{bmatrix}-\begin{bmatrix}-6&0&10\\4&2&6\end{bmatrix}=\begin{bmatrix}3+6&-5-0&4-10\\5-4&3-2&3-6\end{bmatrix}=\begin{bmatrix}9&-5&-6\\1&1&-3\end{bmatrix}$

45. $3A+4(B+C)=3\begin{bmatrix}2&-3&4\\0&2&1\end{bmatrix}+4\left(\begin{bmatrix}1&-2&0\\5&1&2\end{bmatrix}+\begin{bmatrix}-3&0&5\\2&1&3\end{bmatrix}\right)$

$=\begin{bmatrix}6&-9&12\\0&6&3\end{bmatrix}+4\begin{bmatrix}-2&-2&5\\7&2&5\end{bmatrix}=\begin{bmatrix}6&-9&12\\0&6&3\end{bmatrix}+\begin{bmatrix}-8&-8&20\\28&8&20\end{bmatrix}$

$=\begin{bmatrix}6-8&-9-8&12+20\\0+28&6+8&3+20\end{bmatrix}=\begin{bmatrix}-2&-17&32\\28&14&23\end{bmatrix}$

47. $2(A-B)-C=2\left(\begin{bmatrix}2&-3&4\\0&2&1\end{bmatrix}-\begin{bmatrix}1&-2&0\\5&1&2\end{bmatrix}\right)-\begin{bmatrix}-3&0&5\\2&1&3\end{bmatrix}$

$=2\begin{bmatrix}2-1&-3-(-2)&4-0\\0-5&2-1&1-2\end{bmatrix}-\begin{bmatrix}-3&0&5\\2&1&3\end{bmatrix}$

$=2\begin{bmatrix}1&-1&4\\-5&1&-1\end{bmatrix}+\begin{bmatrix}3&0&-5\\-2&-1&-3\end{bmatrix}=\begin{bmatrix}2+3&-2-0&8-5\\-10-2&2-1&-2-3\end{bmatrix}$

$=\begin{bmatrix}5&-2&3\\-12&1&-5\end{bmatrix}$

49. $3A-B-6C=3\begin{bmatrix}2&-3&4\\0&2&1\end{bmatrix}-\begin{bmatrix}1&-2&0\\5&1&2\end{bmatrix}-6\begin{bmatrix}-3&0&5\\2&1&3\end{bmatrix}$

$=\begin{bmatrix}6&-9&12\\0&6&3\end{bmatrix}+\begin{bmatrix}-1&2&0\\-5&-1&-2\end{bmatrix}+\begin{bmatrix}18&0&-30\\-12&-6&-18\end{bmatrix}$

$=\begin{bmatrix}6-1+18&-9+2-0&12-0-30\\0-5-12&6-1-6&3-2-18\end{bmatrix}=\begin{bmatrix}23&-7&-18\\-17&-1&-17\end{bmatrix}$

51. The commutative property for addition: $A + B = B + A$.

$$A + B = \begin{bmatrix} 2 & -3 & 4 \\ 0 & 2 & 1 \end{bmatrix} + \begin{bmatrix} 1 & -2 & 0 \\ 5 & 1 & 2 \end{bmatrix} = \begin{bmatrix} 2+1 & -3-2 & 4+0 \\ 0+5 & 2+1 & 1+2 \end{bmatrix} = \begin{bmatrix} 3 & -5 & 4 \\ 5 & 3 & 3 \end{bmatrix}$$

$$B + A = \begin{bmatrix} 1 & -2 & 0 \\ 5 & 1 & 2 \end{bmatrix} + \begin{bmatrix} 2 & -3 & 4 \\ 0 & 2 & 1 \end{bmatrix} = \begin{bmatrix} 1+2 & -2-3 & 0+4 \\ 5+0 & 1+2 & 2+1 \end{bmatrix} = \begin{bmatrix} 3 & -5 & 4 \\ 5 & 3 & 3 \end{bmatrix} = A + B$$

53. The additive inverse property: $A + (-A) = 0$

$$A+(-A) = \begin{bmatrix} 2 & -3 & 4 \\ 0 & 2 & 1 \end{bmatrix} + \begin{bmatrix} -2 & 3 & -4 \\ 0 & -2 & -1 \end{bmatrix} = \begin{bmatrix} 2+(-2) & (-3)+3 & 4+(-4) \\ 0+0 & 2+(-2) & 1+(-1) \end{bmatrix} = \begin{bmatrix} 0 & 0 & 0 \\ 0 & 0 & 0 \end{bmatrix} = \mathbf{0}$$

55. Property of scalar multiplication: $(k + h)A = kA + hA$.

$$(2+3)B = 5B = 5\begin{bmatrix} 1 & -2 & 0 \\ 5 & 1 & 2 \end{bmatrix} = \begin{bmatrix} 5 & -10 & 0 \\ 25 & 5 & 10 \end{bmatrix}$$

$$2B + 3B = 2\begin{bmatrix} 1 & -2 & 0 \\ 5 & 1 & 2 \end{bmatrix} + 3\begin{bmatrix} 1 & -2 & 0 \\ 5 & 1 & 2 \end{bmatrix} = \begin{bmatrix} 2 & -4 & 0 \\ 10 & 2 & 4 \end{bmatrix} + \begin{bmatrix} 3 & -6 & 0 \\ 15 & 3 & 6 \end{bmatrix}$$

$$= \begin{bmatrix} 2+3 & -4-6 & 0+0 \\ 10+15 & 2+3 & 4+6 \end{bmatrix} = \begin{bmatrix} 5 & -10 & 0 \\ 25 & 5 & 10 \end{bmatrix} = 5B$$

57. Two matrices are equal if they have the same dimension and if corresponding entries are equal.

$\begin{bmatrix} x \\ 4 \end{bmatrix}$ and $\begin{bmatrix} -4 \\ z \end{bmatrix}$ have the same dimension so the matrices are equal if $x = -4$ and $z = 4$.

59. Two matrices are equal if they have the same dimension and if corresponding entries are equal.

$\begin{bmatrix} x-2y & 0 \\ -2 & 6 \end{bmatrix}$ and $\begin{bmatrix} 3 & 0 \\ -2 & x+y \end{bmatrix}$ have the same dimension so the matrices are equal if

$x - 2y = 3$ and $x + y = 6$. To find the values of x and y, solve the system $\begin{cases} x - 2y = 3 \\ x + \ \ y = 6 \end{cases}$.

Writing the system as an augmented matrix and then using row operations to put it in reduced row-echelon form, we get

$$\left[\begin{array}{cc|c} 1 & -2 & 3 \\ 1 & 1 & 6 \end{array}\right] \xrightarrow{R_2 = -r_1 + r_2} \left[\begin{array}{cc|c} 1 & -2 & 3 \\ 0 & 3 & 3 \end{array}\right] \xrightarrow{R_2 = \frac{1}{3}r_2} \left[\begin{array}{cc|c} 1 & -2 & 3 \\ 0 & 1 & 1 \end{array}\right] \xrightarrow{R_1 = 2r_2 + r_1} \left[\begin{array}{cc|c} 1 & 0 & 5 \\ 0 & 1 & 1 \end{array}\right]$$

The matrices are equal when $x = 5$ and $y = 1$.

61. $[2 \quad 3 \quad -4] + [x \quad 2y \quad z] = [x+2 \quad 2y+3 \quad z-4] = [6 \quad -9 \quad 2]$

Two matrices are equal if they have the same dimension and if corresponding entries are equal. $[x+2 \quad 2y+3 \quad z-4]$ and $[6 \quad -9 \quad 2]$ have the same dimension so the matrices are equal if the corresponding entries are equal.

The values of x, y, and z, solve the system $\begin{cases} x + 2 = 6 \\ 2y + 3 = -9 \\ z - 4 = 2 \end{cases}$.

So we find that the sum of the two matrices equals $[6 \quad -9 \quad 2]$ when $x = 4$, $y = -6$, and $z = 6$.

63.

```
[A]+[B]
[[-2   1  7  5 ...
 [4    6  7  5 ...
 [-3.5 8  -4 13...
 [12   -1 7  6 ...
```

$$A+B=\begin{bmatrix} -2 & 1 & 7 & 5 \\ 4 & 6 & 7 & 5 \\ -3.5 & 8 & -4 & 13 \\ 12 & -1 & 7 & 6 \end{bmatrix}$$

65.

```
[C]-3*([A]+[B])
[[19   -11 -14 ...
 [-12  -13 -21 ...
 [15.5 -24 19  ...
 [-29  10  -14 ...
```

```
[C]-3*([A]+[B])
...  -11 -14 -15]
...  -13 -21 -17]
...5 -24 19  -39]
...  10  -14 -11]]
```

$$C-3(A+B)=\begin{bmatrix} 19 & -11 & -14 & -15 \\ -12 & -13 & -21 & -17 \\ 15.5 & -24 & 19 & -39 \\ -29 & 10 & -14 & -11 \end{bmatrix}$$

67.

```
3([B]+[C])-[A]
[[37   -17 30 1...
 [4    9   13 1...
 [20.5 16  -3 3...
 [29   18  22 4...
```

```
3([B]+[C])-[A]
...7  -17 30 15]
...   9   13 1 ]
...0.5 16 -3 31]
...9  18  22 43]]
```

$$3(B+C)-A=\begin{bmatrix} 37 & -17 & 30 & 15 \\ 4 & 9 & 13 & 1 \\ 20.5 & 16 & -3 & 31 \\ 29 & 18 & 22 & 43 \end{bmatrix}$$

69. The matrix will have dimension 2×3.

	DEMOCRATS	REPUBLICANS	INDEPENDENTS
UNDER $25,000	351	271	73
OVER $25,000	203	215	55

71. Before constructing the matrix, we need to determine how many students of each gender majored in each of the three areas.

LAS: females: (0.50)(500) = 250; males: (0.50)(500) = 250
ENG: males: (0.75)(300) = 225; females: 300 – 225 = 75
EDUC Majors: 1000 – (500 + 300) = 200
females: (0.60)(200) = 120; males: 200 – 120 = 80

	LAS	ENG	EDUC
MALE	250	225	80
FEMALE	250	75	120

73. Before constructing any matrix, we need to find the number of degrees earned by each gender.

Associate: women: 689,000 – 262,000 = 427,000 degrees
Bachelor: men: 1,475,000 – 877,000 = 598,000 degrees
Master: men: 615,000 – 373,000 = 242,000 degrees
Doctoral: women: 50,200 – 25,700 = 24,500 degrees

(a) The data can be displayed in a matrix with dimension 2×4.

	Associate	Bachelor	Master's	Doctoral
Male	262,000	598,000	242,000	25,700
Female	427,000	877,000	373,000	24,500

(b) The rows represent the distribution of projected post-secondary degrees for each gender. Row 1 represents the degrees projected to be earned by males, and row 2 represents the projections for females. The columns represent the distribution of the projected number of each post secondary degree by gender. Column 1 represents associate degrees, column 2 bachelor's degrees, column 3 master's degrees, and column 4 doctoral degrees.

(c) The data can also be displayed in a matrix with dimension 4 × 2.

$$\begin{array}{l} \\ \text{Associate's} \\ \text{Bachelor's} \\ \text{Master's} \\ \text{Doctoral} \end{array} \begin{array}{c} \begin{array}{cc} \text{Male} & \text{Female} \end{array} \\ \begin{bmatrix} 262{,}000 & 427{,}000 \\ 598{,}000 & 877{,}000 \\ 242{,}000 & 373{,}000 \\ 25{,}700 & 24{,}500 \end{bmatrix} \end{array}$$

(d) The rows of this matrix represent the numbers of various post-secondary degrees projected to be earned distributed between the two genders: male (column 1) and female (column 2). Column 1 represents the numbers of the various degrees projected to be earned by males, and column 2 represents the numbers of the various degrees projected to be earned by females.

75. Before constructing the matrix, we need to find the numbers of males and females in each type of prison.

Local: female: (0.127)(747,529) = 94,936; male: 747,529 – 94,936 = 652,593
State: male: (0.936)(1,328,339) = 1,243,325; female: 1,328,339 – 1,243,325 = 85,014
Federal: female: (0.071)(184,484) = 13,098; male: 184,484 – 13,098 = 171,386

$$\begin{array}{l} \\ \text{MALE} \\ \text{FEMALE} \end{array} \begin{array}{c} \begin{array}{ccc} \text{LOCAL} & \text{STATE} & \text{FED} \end{array} \\ \begin{bmatrix} 652{,}593 & 1{,}243{,}325 & 171{,}386 \\ 94{,}936 & 85{,}014 & 13{,}098 \end{bmatrix} \end{array}$$

77. (a) A 2 × 3 matrix will have rows representing the location of the dealer and columns representing the style of the cars sold.

January Sales	Subcompacts	Intermediate	SUV
City	350	225	80
Suburban	375	200	75

February Sales	Subcompacts	Intermediate	SUV
City	300	175	40
Suburban	325	150	50

(b) Combined sales from January and February are found by adding the two matrices in part (a).

$$\begin{bmatrix} 350 & 225 & 80 \\ 375 & 200 & 75 \end{bmatrix} + \begin{bmatrix} 300 & 175 & 40 \\ 325 & 150 & 50 \end{bmatrix} = \begin{bmatrix} 650 & 400 & 120 \\ 700 & 350 & 125 \end{bmatrix}$$

(c) A 3 × 2 will have the rows representing the style of car sold and the columns representing the location of the dealer.

January Sales	City	Suburban
Subcompact	350	375
Intermediate	225	200
SUV	80	75

February Sales	City	Suburban
Subcompact	300	325
Intermediate	175	150
SUV	40	50

(d) Combined sales from January and February are found by adding the two matrices in part (c).

$$\begin{bmatrix} 350 & 375 \\ 225 & 200 \\ 80 & 75 \end{bmatrix} + \begin{bmatrix} 300 & 325 \\ 175 & 150 \\ 40 & 50 \end{bmatrix} = \begin{bmatrix} 650 & 700 \\ 400 & 350 \\ 120 & 125 \end{bmatrix}$$

79. Before writing any matrices, we determine the distribution of the degrees between the genders.

2004:
Social Science: 6782 degrees; 3741 to females; 6782 – 3741 = 3041 to males,
Humanities: 5461 degrees; 2627 to males; 5461 – 2627 = 2834 to females,
Education: 6627 degrees; 4361 to females; 6627 – 4361 = 2266 to males.

2003:
Social Science: 6763 degrees; 3745 to females; 6763 – 3745 = 3018 to males,
Humanities: 5401 degrees; 2656 to males; 5401 – 2656 = 2745 to females,
Education: 6602 degrees; 4363 to females; 6602 – 4363 = 2239 to males.

(a) In 2×3 matrices the rows represent the genders and the columns represent the area of study.

2004	Social Sciences	Humanities	Education
Male	3041	2627	2266
Female	3741	2834	4361

2003	Social Sciences	Humanities	Education
Male	3018	2656	2239
Female	3745	2745	4363

(b) The number of doctoral degrees awarded over the two year period is:

$$\begin{bmatrix} 3041 & 2627 & 2266 \\ 3741 & 2834 & 4361 \end{bmatrix} + \begin{bmatrix} 3018 & 2656 & 2239 \\ 3745 & 2745 & 4363 \end{bmatrix} = \begin{bmatrix} 6059 & 5283 & 4505 \\ 7486 & 5579 & 8724 \end{bmatrix}$$

(c) The difference in doctoral degrees awarded between 2003 and 2004 is:

$$\begin{bmatrix} 3041 & 2627 & 2266 \\ 3741 & 2834 & 4361 \end{bmatrix} - \begin{bmatrix} 3018 & 2656 & 2239 \\ 3745 & 2745 & 4363 \end{bmatrix} = \begin{bmatrix} 23 & -29 & 27 \\ -4 & 89 & -2 \end{bmatrix}$$

81. (a) The rows of the matrix represent the existence of internet access; the columns represent community type.

$$A = \begin{array}{r} \\ \text{Access} \\ \text{No Access} \end{array} \begin{array}{c} \begin{array}{ccc} \text{Rural} & \text{Suburban} & \text{Urban} \end{array} \\ \begin{bmatrix} 0.13 & 0.35 & 0.19 \\ 0.08 & 0.16 & 0.09 \end{bmatrix} \end{array}$$

(b) The matrix $5262A$ is the number of persons surveyed who were in each category.

$$5262A = 5262 \cdot \begin{bmatrix} 0.13 & 0.35 & 0.19 \\ 0.08 & 0.16 & 0.09 \end{bmatrix} = \begin{bmatrix} 684 & 1842 & 1000 \\ 421 & 842 & 474 \end{bmatrix}$$

2.5 Multiplication of Matrices

1. True

3. 3×7

5. True. For example, $\begin{bmatrix} 3 & 2 \\ 4 & 2 \end{bmatrix}\begin{bmatrix} -1 & 1 \\ 2 & -\frac{3}{2} \end{bmatrix} = \begin{bmatrix} -1 & 1 \\ 2 & -\frac{3}{2} \end{bmatrix}\begin{bmatrix} 3 & 2 \\ 4 & 2 \end{bmatrix}$.

7. $\begin{bmatrix} 1 & 3 \end{bmatrix}\begin{bmatrix} 2 \\ 4 \end{bmatrix} = [(1)(2) + (3)(4)] = [14]$

9. $\begin{bmatrix} 1 & -2 & 3 \end{bmatrix}\begin{bmatrix} 0 \\ 1 \\ 2 \end{bmatrix} = \begin{bmatrix} 1 \cdot 0 + (-2) \cdot 1 + 3 \cdot 2 \end{bmatrix} = \begin{bmatrix} 4 \end{bmatrix}$

11. $\begin{bmatrix} 1 & 4 \end{bmatrix}\begin{bmatrix} 2 & 0 \\ 4 & -2 \end{bmatrix} = \begin{bmatrix} 1 \cdot 2 + 4 \cdot 4 & 1 \cdot 0 + 4 \cdot (-2) \end{bmatrix} = \begin{bmatrix} 18 & -8 \end{bmatrix}$

13. $\begin{bmatrix} 2 & 0 \\ 4 & -2 \end{bmatrix}\begin{bmatrix} 2 & 1 \\ 3 & -2 \end{bmatrix} = \begin{bmatrix} 2 \cdot 2 + 0 \cdot 3 & 2 \cdot 1 + 0 \cdot (-2) \\ 4 \cdot 2 + (-2) \cdot 3 & 4 \cdot 1 + (-2)(-2) \end{bmatrix} = \begin{bmatrix} 4 & 2 \\ 2 & 8 \end{bmatrix}$

15. $\begin{bmatrix} 1 & -2 & 3 \end{bmatrix}\begin{bmatrix} 0 & 1 \\ 1 & 2 \\ 2 & 3 \end{bmatrix} = \begin{bmatrix} 1 \cdot 0 + (-2) \cdot 1 + 3 \cdot 2 & 1 \cdot 1 + (-2) \cdot 2 + 3 \cdot 3 \end{bmatrix} = \begin{bmatrix} 4 & 6 \end{bmatrix}$

17. $\begin{bmatrix} 1 & -2 & 3 \\ 4 & 0 & 6 \end{bmatrix}\begin{bmatrix} 0 & -2 \\ 1 & 0 \\ 2 & -4 \end{bmatrix} = \begin{bmatrix} 1 \cdot 0 + (-2) \cdot 1 + 3 \cdot 2 & 1 \cdot (-2) + (-2) \cdot 0 + 3 \cdot (-4) \\ 4 \cdot 0 + 0 \cdot 1 + 6 \cdot 2 & 4 \cdot (-2) + 0 \cdot 0 + 6 \cdot (-4) \end{bmatrix} = \begin{bmatrix} 4 & -14 \\ 12 & -32 \end{bmatrix}$

19. $\begin{bmatrix} 2 & 0 \\ 4 & -2 \\ 6 & -1 \end{bmatrix}\begin{bmatrix} 2 & 1 \\ 3 & -2 \end{bmatrix} = \begin{bmatrix} 2 \cdot 2 + 0 \cdot 3 & 2 \cdot 1 + 0 \cdot (-2) \\ 4 \cdot 2 + (-2) \cdot 3 & 4 \cdot 1 + (-2)(-2) \\ 6 \cdot 2 + (-1) \cdot 3 & 6 \cdot 1 + (-1)(-2) \end{bmatrix} = \begin{bmatrix} 4 & 2 \\ 2 & 8 \\ 9 & 8 \end{bmatrix}$

21. $\begin{bmatrix} 1 & -1 & 6 \\ 2 & 0 & -1 \\ 3 & 1 & 2 \end{bmatrix}\begin{bmatrix} 3 & 2 \\ 0 & 1 \\ 1 & 0 \end{bmatrix} = \begin{bmatrix} 1 \cdot 3 + (-1) \cdot 0 + 6 \cdot 1 & 1 \cdot 2 + (-1) \cdot 1 + 6 \cdot 0 \\ 2 \cdot 3 + 0 \cdot 0 + (-1) \cdot 1 & 2 \cdot 2 + 0 \cdot 1 + (-1) \cdot 0 \\ 3 \cdot 3 + 1 \cdot 0 + 2 \cdot 1 & 3 \cdot 2 + 1 \cdot 1 + 2 \cdot 0 \end{bmatrix} = \begin{bmatrix} 9 & 1 \\ 5 & 4 \\ 11 & 7 \end{bmatrix}$

23. BA is defined; the dimension of the product is a 3×4 matrix.

25. AB is not defined.

27. $(BA)C$ is not defined. (BA is 3×4 and C is 2×3.)

29. $BA + A$ is defined and is a 3×4 matrix.

31. $CB - A$ is not defined. (The dimension of CB is 2×3, and A is a 3×4 matrix. To find the difference between 2 matrices, they must have the same dimension.)

33. $AB = \begin{bmatrix} 1-2 & 2+8 & 3-4 \\ 0-4 & 0+16 & 0-8 \end{bmatrix} = \begin{bmatrix} -1 & 10 & -1 \\ -4 & 16 & -8 \end{bmatrix}$

35. $BC = \begin{bmatrix} 3+8+0 & 1-2+6 \\ -3+16+0 & -1-4-4 \end{bmatrix} = \begin{bmatrix} 11 & 5 \\ 13 & -9 \end{bmatrix}$

37. $(D+I_3)C=\begin{bmatrix}2 & 0 & 4\\ 0 & 2 & 2\\ 0 & -1 & 2\end{bmatrix}\begin{bmatrix}3 & 1\\ 4 & -1\\ 0 & 2\end{bmatrix}=\begin{bmatrix}6+0+0 & 2+0+8\\ 0+8+0 & 0-2+4\\ 0-4+0 & 0+1+4\end{bmatrix}=\begin{bmatrix}6 & 10\\ 8 & 2\\ -4 & 5\end{bmatrix}$

39. $EI_2=\begin{bmatrix}3\cdot1+-1\cdot0 & 3\cdot0+-1\cdot1\\ 4\cdot1+2\cdot0 & 4\cdot0+2\cdot1\end{bmatrix}=\begin{bmatrix}3 & -1\\ 4 & 2\end{bmatrix}=E$

41. $(2E)B=\begin{bmatrix}6 & -2\\ 8 & 4\end{bmatrix}\cdot B=\begin{bmatrix}6\cdot1+-2\cdot-1 & 6\cdot2+-2\cdot4 & 6\cdot3+-2\cdot-2\\ 8\cdot1+4\cdot-1 & 8\cdot2+4\cdot4 & 8\cdot3+4\cdot-2\end{bmatrix}=\begin{bmatrix}8 & 4 & 22\\ 4 & 32 & 16\end{bmatrix}$

43 $-5E+A=\begin{bmatrix}-15 & 5\\ -20 & -10\end{bmatrix}+\begin{bmatrix}1 & 2\\ 0 & 4\end{bmatrix}=\begin{bmatrix}-14 & 7\\ -20 & -6\end{bmatrix}$

45. $3CB+4D=3(CB)+4D$

$$=3\begin{bmatrix}2 & 10 & 7\\ 5 & 4 & 14\\ -2 & 8 & -4\end{bmatrix}+4\begin{bmatrix}1 & 0 & 4\\ 0 & 1 & 2\\ 0 & -1 & 1\end{bmatrix}=\begin{bmatrix}6 & 30 & 21\\ 15 & 12 & 42\\ -6 & 24 & -12\end{bmatrix}+\begin{bmatrix}4 & 0 & 16\\ 0 & 4 & 8\\ 0 & -4 & 4\end{bmatrix}=\begin{bmatrix}10 & 30 & 37\\ 15 & 16 & 50\\ -6 & 20 & -8\end{bmatrix}$$

47. First we will find $D(CB)$ and then $(DC)B$ and compare the results.

$$CB=\begin{bmatrix}3\cdot1+1\cdot-1 & 3\cdot2+1\cdot4 & 3\cdot3+1\cdot-2\\ 4\cdot1+-1\cdot-1 & 4\cdot2-1\cdot4 & 4\cdot3+-1\cdot-2\\ 0\cdot1+2\cdot-1 & 0\cdot2+2\cdot4 & 0\cdot3+2\cdot-2\end{bmatrix}=\begin{bmatrix}2 & 10 & 7\\ 5 & 4 & 14\\ -2 & 8 & -4\end{bmatrix}$$

$$D(CB)=\begin{bmatrix}1 & 0 & 4\\ 0 & 1 & 2\\ 0 & -1 & 1\end{bmatrix}\begin{bmatrix}2 & 10 & 7\\ 5 & 4 & 14\\ -2 & 8 & -4\end{bmatrix}$$

$$=\begin{bmatrix}1\cdot2+0\cdot5+4\cdot-2 & 1\cdot10+0\cdot4+4\cdot8 & 1\cdot7+0\cdot14+4\cdot-4\\ 0\cdot2+1\cdot5+2\cdot-2 & 0\cdot10+1\cdot4+2\cdot8 & 0\cdot7+1\cdot14+2\cdot-4\\ 0\cdot2+-1\cdot5+1\cdot-2 & 0\cdot10+-1\cdot4+1\cdot8 & 0\cdot7+-1\cdot14+1\cdot-4\end{bmatrix}=\begin{bmatrix}-6 & 42 & -9\\ 1 & 20 & 6\\ -7 & 4 & -18\end{bmatrix}$$

$$DC=\begin{bmatrix}1\cdot3+0\cdot4+4\cdot0 & 1\cdot1+0\cdot-1+4\cdot2\\ 0\cdot3+1\cdot4+2\cdot0 & 0\cdot1+1\cdot-1+2\cdot2\\ 0\cdot3+-1\cdot4+1\cdot0 & 0\cdot1+-1\cdot-1+1\cdot2\end{bmatrix}=\begin{bmatrix}3 & 9\\ 4 & 3\\ -4 & 3\end{bmatrix}$$

$$(DC)B=\begin{bmatrix}3 & 9\\ 4 & 3\\ -4 & 3\end{bmatrix}\begin{bmatrix}1 & 2 & 3\\ -1 & 4 & -2\end{bmatrix}$$

$$=\begin{bmatrix}3\cdot1+9\cdot-1 & 3\cdot2+9\cdot4 & 3\cdot3+9\cdot-2\\ 4\cdot1+3\cdot-1 & 4\cdot2+3\cdot4 & 4\cdot3+3\cdot-2\\ -4\cdot1+3\cdot-1 & -4\cdot2+3\cdot4 & -4\cdot3+3\cdot-2\end{bmatrix}=\begin{bmatrix}-6 & 42 & -9\\ 1 & 20 & 6\\ -7 & 4 & -18\end{bmatrix}=D(CB)$$

49. $AB=\begin{bmatrix}1 & -1\\ 2 & 0\end{bmatrix}\begin{bmatrix}3 & 2\\ -2 & 4\end{bmatrix}=\begin{bmatrix}1\cdot 3+(-1)\cdot(-2) & 1\cdot 2+(-1)\cdot 4\\ 2\cdot 3+0\cdot(-2) & 2\cdot 2+0\cdot 4\end{bmatrix}=\begin{bmatrix}5 & -2\\ 6 & 4\end{bmatrix}$

$BA=\begin{bmatrix}3 & 2\\ -2 & 4\end{bmatrix}\begin{bmatrix}1 & -1\\ 2 & 0\end{bmatrix}=\begin{bmatrix}3\cdot 1+2\cdot 2 & 3\cdot(-1)+2\cdot 0\\ (-2)\cdot 1+4\cdot 2 & (-2)\cdot(-1)+4\cdot 0\end{bmatrix}=\begin{bmatrix}7 & -3\\ 6 & 2\end{bmatrix}$

$AB\neq BA$

51.

```
[A]*[B]
[[.5   16 -30 2...
 [21   14 28  6...
 [19.5 8  -23 3...
 [-9.5 45 -9  8...
```

```
[A]*[B]
...5   16 -30 25]
...1   14 28  64]
...9.5 8  -23 33]
...9.5 45 -9  83]]
```

$$AB=\begin{bmatrix}0.5 & 16 & -30 & 25\\ 21 & 14 & 28 & 64\\ 19.5 & 8 & -23 & 33\\ -9.5 & 45 & -9 & 83\end{bmatrix}$$

53.

```
([A]*[B])[C]
[[31.5  251 -31...
 [861   350 791...
 [369.5 115 206...
 [412.5 882 451...
```

```
([A]*[B])[C]
...251 -31.5 143]
...350 791   420]
...115 206.5 215]
...882 451.5 491]]
```

$$(AB)C=\begin{bmatrix}31.5 & 251 & -31.5 & 143\\ 861 & 350 & 791 & 420\\ 369.5 & 115 & 206.5 & 215\\ 412.5 & 882 & 451.5 & 491\end{bmatrix}$$

55.

```
[B]*([A]+[C])
[[66  74    94 ...
 [71  13    106...
 [165 124.5 79 ...
 [158 -3    152...
```

```
[B]*([A]+[C])
... 74    94  38]
... 13    106 28]
... 124.5 79  52]
... -3    152 46]]
```

$$B(A+C)=\begin{bmatrix}66 & 74 & 94 & 38\\ 71 & 13 & 106 & 28\\ 165 & 124.5 & 79 & 52\\ 158 & -3 & 152 & 46\end{bmatrix}$$

57.

```
[A]*(2[B]-3[C])
[[-5   23   -10...
 [-108 -56  -70...
 [108  -152 -67...
 [-346 279  -24...
```

```
[A]*(2[B]-3[C])
...23   -102 44 ]
...-56  -70  122]
...-152 -67  36 ]
...279  -249 187]]
```

$$A(2B-3C)=\begin{bmatrix}-5 & 23 & -102 & 44\\ -108 & -56 & -70 & 122\\ 108 & -152 & -67 & 36\\ -346 & 279 & -249 & 187\end{bmatrix}$$

59. (a) Matrix A represents the price at the close of trading on July 7, 2006 of a chare of Chevron Corporation, ConocoPhillips, and Exxon Mobil Corporation respectively. The dimension of A is 1×3.

(b) $AB=\begin{bmatrix}63.68 & 67.19 & 62.84\end{bmatrix}\begin{bmatrix}100 & 415 & 175\\ 250 & 90 & 200\\ 325 & 225 & 130\end{bmatrix}=\begin{bmatrix}43{,}588.50 & 46{,}613.30 & 32{,}751.20\end{bmatrix}$

Matrix AB represents the total amount each of the three funds spent on the three stocks on July 7, 2006 at their closing price.

(c) $C=\begin{bmatrix}1\\ 1\\ 1\end{bmatrix}$ $\quad BC=\begin{bmatrix}100 & 415 & 175\\ 250 & 90 & 200\\ 325 & 225 & 130\end{bmatrix}\begin{bmatrix}1\\ 1\\ 1\end{bmatrix}=\begin{bmatrix}100+415+175\\ 250+90+200\\ 325+225+130\end{bmatrix}=\begin{bmatrix}690\\ 540\\ 680\end{bmatrix}$

Matrix BC represents the total number of shares of Chevron Corporation, Conoco Phillips, and Exxon Mobil Corporation, respectively, purchased by the three funds at the close of trading on July 7, 2006.

61. (a) Since the columns of A must have the same names as the rows of B,

$$A = \begin{array}{c} \\ \text{John} \\ \text{Marsha} \end{array} \overset{\begin{array}{c}\text{Credit Hours} \\ \text{JJC} \quad \text{CSU}\end{array}}{\begin{bmatrix} 12 & 4 \\ 8 & 8 \end{bmatrix}} \qquad B = \begin{array}{c} \text{JJC} \\ \text{CSU} \end{array} \overset{\begin{array}{c}\text{Cost per} \\ \text{Credit}\end{array}}{\begin{bmatrix} 73 \\ 189 \end{bmatrix}}$$

(b) A is 2×2 (c) B is 2×1

(d) $AB = \begin{bmatrix} 12 & 4 \\ 8 & 8 \end{bmatrix} \begin{bmatrix} 73 \\ 189 \end{bmatrix} = \begin{bmatrix} 1632 \\ 2096 \end{bmatrix}$

(e) AB has dimension 2×1. Its rows are called John and Marsha, and its column is called total cost of tuition.

(f) The first entry in AB is the total cost, \$1632.00, John spends on tuition at the two schools, and the second entry is the total cost, \$2096.00, Marsha spends on tuition.

63. (a) Since the columns of A must have the same names as the rows of B,

$$A = \begin{array}{c} \text{Feb. 2006} \\ \text{Aug. 2006} \\ \text{Feb. 2007} \end{array} \overset{\text{GE} \quad \text{GM} \quad \text{JNJ}}{\begin{bmatrix} 33 & 20 & 57 \\ 32.5 & 31 & 64 \\ 35 & 35 & 64 \end{bmatrix}} \qquad B = \begin{array}{c} \text{GE} \\ \text{GM} \\ \text{JNJ} \end{array} \overset{\text{Bill} \quad \text{Dan}}{\begin{bmatrix} 50 & 40 \\ 30 & 64 \\ 20 & 30 \end{bmatrix}}$$

(b) A has dimension 3×3. (c) B has dimension 3×2.

(d) $AB = \begin{bmatrix} 33 & 20 & 57 \\ 32.5 & 31 & 64 \\ 35 & 35 & 64 \end{bmatrix} \begin{bmatrix} 50 & 40 \\ 30 & 64 \\ 20 & 30 \end{bmatrix} = \begin{bmatrix} 3390 & 4230 \\ 3835 & 5080 \\ 4080 & 5420 \end{bmatrix}$

(e) AB has dimension 3×2. The rows are the dates. The columns are named Bill and Dan.

(f) The entries in AB represent the value of each man's total investment on each of the particular dates. For example, 3390 tells us that Bill's total investment in February 2006 was worth \$3390.00.

65.

	Number of pants bought	Number of shirts bought	Number of jackets bought
Lee	6	8	2
Chan	2	5	3

	Cost per item
Pants	\$25.00
Shirts	\$18.00
Jackets	\$39.00

$$\begin{bmatrix} 6 & 8 & 2 \\ 2 & 5 & 3 \end{bmatrix} \begin{bmatrix} 25 \\ 18 \\ 39 \end{bmatrix} = \begin{bmatrix} (6)(25)+(8)(18)+(2)(39) \\ (2)(25)+(5)(18)+(3)(39) \end{bmatrix} = \begin{bmatrix} 372 \\ 257 \end{bmatrix}$$

Lee spent \$372 and Chan spent \$257.

67. To find the value of the variable, x, first multiply the matrices.

$$\begin{bmatrix} x & 4 & 1 \end{bmatrix}\begin{bmatrix} 2 & 1 & 0 \\ 1 & 0 & 2 \\ 0 & 2 & 4 \end{bmatrix}\begin{bmatrix} x \\ -7 \\ 5/4 \end{bmatrix} = \begin{bmatrix} 2x+4 & x+2 & 12 \end{bmatrix}\begin{bmatrix} x \\ -7 \\ 5/4 \end{bmatrix}$$
$$= (2x+4)x + (x+2)(-7) + 15$$
$$= 2x^2 - 3x + 1$$

Since the product of the three matrices is **0**, set $2x^2 - 3x + 1$ equal to 0 and solve the quadratic equation.

$$2x^2 - 3x + 1 = 0$$
$$(2x-1)(x-1) = 0$$
$$2x - 1 = 0 \quad \text{or} \quad x - 1 = 0$$
$$x = \frac{1}{2} \quad \text{or} \quad x = 1$$

69. When $A = \begin{bmatrix} a & b \\ c & d \end{bmatrix}$ and $B = \begin{bmatrix} 1 & 1 \\ -1 & 1 \end{bmatrix}$, then

$$AB = \begin{bmatrix} a-b & a+b \\ c-d & c+d \end{bmatrix} \text{ and } BA = \begin{bmatrix} a+c & b+d \\ -a+c & -b+d \end{bmatrix}.$$

If $AB = BA$ then corresponding entries must be equal, or

$$\begin{cases} a-b = a+c & \text{or} \quad -b = c \\ a+b = b+d & \text{or} \quad a = d \\ c-d = -a+c & \text{or} \quad d = a \\ c+d = -b+d & \text{or} \quad c = -b \end{cases}$$

So, if A and B are both 2×2 matrices, and A is not the identity matrix, then $AB = BA$ whenever $c = -b$ and $d = a$.

71.
$$A^2 = A \cdot A = \begin{bmatrix} a & 1-a \\ 1+a & -a \end{bmatrix}\begin{bmatrix} a & 1-a \\ 1+a & -a \end{bmatrix} = \begin{bmatrix} a^2 + (1-a)(1+a) & a(1-a) + (1-a)(-a) \\ (1+a)a - a(1+a) & (1+a)(1-a) - a(-a) \end{bmatrix}$$
$$= \begin{bmatrix} a^2 + 1 - a^2 & a - a^2 - a + a^2 \\ a + a^2 - a - a^2 & 1 - a^2 + a^2 \end{bmatrix} = \begin{bmatrix} 1 & 0 \\ 0 & 1 \end{bmatrix} = I_2$$

73.
$$AB = \begin{bmatrix} a & b \\ -b & a \end{bmatrix}\begin{bmatrix} c & d \\ -d & c \end{bmatrix} = \begin{bmatrix} ac - bd & ad + bc \\ -bc - ad & -bd + ac \end{bmatrix}$$

$$BA = \begin{bmatrix} c & d \\ -d & c \end{bmatrix}\begin{bmatrix} a & b \\ -b & a \end{bmatrix} = \begin{bmatrix} ca - db & cb + da \\ -da - cb & -db + ca \end{bmatrix} = \begin{bmatrix} ac - bd & ad + bc \\ -bc - ad & ac - bd \end{bmatrix} = AB$$

Since the entries a, b, c, and d represent numbers, the commutative properties of arithmetic can be used in the last matrix of product BA.

75. $A^2 = A \cdot A = \begin{bmatrix} 3 & 1 \\ -2 & -1 \end{bmatrix} \begin{bmatrix} 3 & 1 \\ -2 & -1 \end{bmatrix} = \begin{bmatrix} 7 & 2 \\ -4 & -1 \end{bmatrix}$

$A^3 = A \cdot A^2 = \begin{bmatrix} 3 & 1 \\ -2 & -1 \end{bmatrix} \begin{bmatrix} 7 & 2 \\ -4 & -1 \end{bmatrix} = \begin{bmatrix} 17 & 5 \\ -10 & -3 \end{bmatrix}$

$A^4 = A \cdot A^3 = \begin{bmatrix} 3 & 1 \\ -2 & -1 \end{bmatrix} \begin{bmatrix} 17 & 5 \\ -10 & -3 \end{bmatrix} = \begin{bmatrix} 41 & 12 \\ -24 & -7 \end{bmatrix}$

77. $A^2 = A \cdot A = \begin{bmatrix} \frac{1}{2} & \frac{1}{2} \\ \frac{1}{4} & \frac{3}{4} \end{bmatrix} \begin{bmatrix} \frac{1}{2} & \frac{1}{2} \\ \frac{1}{4} & \frac{3}{4} \end{bmatrix} = \begin{bmatrix} \frac{3}{8} & \frac{5}{8} \\ \frac{5}{16} & \frac{11}{16} \end{bmatrix}$

$A^3 = A \cdot A^2 = \begin{bmatrix} \frac{1}{2} & \frac{1}{2} \\ \frac{1}{4} & \frac{3}{4} \end{bmatrix} \begin{bmatrix} \frac{3}{8} & \frac{5}{8} \\ \frac{5}{16} & \frac{11}{16} \end{bmatrix} = \begin{bmatrix} \frac{11}{32} & \frac{21}{32} \\ \frac{21}{64} & \frac{43}{64} \end{bmatrix}$

$A^4 = A \cdot A^3 = \begin{bmatrix} \frac{1}{2} & \frac{1}{2} \\ \frac{1}{4} & \frac{3}{4} \end{bmatrix} \begin{bmatrix} \frac{11}{32} & \frac{21}{32} \\ \frac{21}{64} & \frac{43}{64} \end{bmatrix} = \begin{bmatrix} \frac{43}{128} & \frac{85}{128} \\ \frac{85}{256} & \frac{171}{256} \end{bmatrix}$

79. Answers will vary. However, $A^n = \begin{bmatrix} \frac{1}{3} & \frac{2}{3} \\ \frac{1}{3} & \frac{2}{3} \end{bmatrix}$.

81.

```
[B]²
[[.674 .043 .23…
 [.382 .096 .21…
 [.500 .070 .22…
 [.130 .110 .20…
```

```
[B]²
….043 .230 .054]
….096 .212 .310]
….070 .220 .210]
….110 .200 .560]]
```

```
[B]^10
[[.514 .065 .22…
 [.491 .068 .22…
 [.500 .067 .22…
 [.470 .070 .21…
```

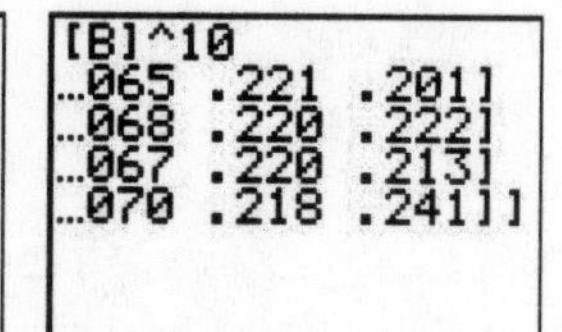

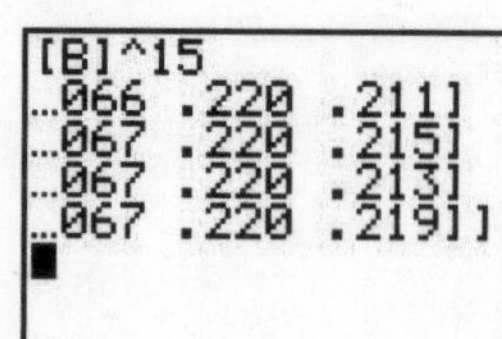

2.6 The Inverse of a Matrix

1. True.

3. inverse

5. Since the product $\begin{bmatrix}1&2\\2&3\end{bmatrix}\begin{bmatrix}-3&2\\2&-1\end{bmatrix}=\begin{bmatrix}-3+4&2-2\\-6+6&4-3\end{bmatrix}=\begin{bmatrix}1&0\\0&1\end{bmatrix}=I_2$

and the product $\begin{bmatrix}-3&2\\2&-1\end{bmatrix}\begin{bmatrix}1&2\\2&3\end{bmatrix}=\begin{bmatrix}-3+4&-6+6\\2-2&4-3\end{bmatrix}=\begin{bmatrix}1&0\\0&1\end{bmatrix}=I_2$,

the matrices are inverses of each other.

7. Since the product $\begin{bmatrix}-1&-2\\3&4\end{bmatrix}\begin{bmatrix}2&1\\-\frac{3}{2}&-\frac{1}{2}\end{bmatrix}=\begin{bmatrix}-2+3&-1+1\\6-6&3-2\end{bmatrix}=\begin{bmatrix}1&0\\0&1\end{bmatrix}=I_2$ and the product

$\begin{bmatrix}2&1\\-\frac{3}{2}&-\frac{1}{2}\end{bmatrix}\begin{bmatrix}-1&-2\\3&4\end{bmatrix}=\begin{bmatrix}1&0\\0&1\end{bmatrix}=I_2$, the matrices are inverses of each other.

9. Since the product

$\begin{bmatrix}1&2&3\\2&3&4\\1&2&1\end{bmatrix}\begin{bmatrix}-\frac{5}{2}&2&-\frac{1}{2}\\1&-1&1\\\frac{1}{2}&0&-\frac{1}{2}\end{bmatrix}=\begin{bmatrix}-\frac{5}{2}+\frac{4}{2}+\frac{3}{2}&2-2+0&-\frac{1}{2}+\frac{4}{2}-\frac{3}{2}\\-5+3+2&4-3+0&-1+3-2\\-\frac{5}{2}+\frac{4}{2}+\frac{1}{2}&2-2+0&-\frac{1}{2}+\frac{4}{2}-\frac{1}{2}\end{bmatrix}=\begin{bmatrix}1&0&0\\0&1&0\\0&0&1\end{bmatrix}=I_3$ and the

product $\begin{bmatrix}-\frac{5}{2}&2&-\frac{1}{2}\\1&-1&1\\\frac{1}{2}&0&-\frac{1}{2}\end{bmatrix}\begin{bmatrix}1&2&3\\2&3&4\\1&2&1\end{bmatrix}=\begin{bmatrix}1&0&0\\0&1&0\\0&0&1\end{bmatrix}=I_3$, the matrices are inverses of each other.

11. First, we augment the matrix with I_2; then we use row operations to obtain the reduced row echelon form of the matrix.

$$\left[\begin{array}{cc|cc}3&7&1&0\\2&5&0&1\end{array}\right]\xrightarrow[R_1=r_1-r_2]{}\left[\begin{array}{cc|cc}1&2&1&-1\\2&5&0&1\end{array}\right]\xrightarrow[R_2=-2r_1+r_2]{}\left[\begin{array}{cc|cc}1&2&1&-1\\0&1&-2&3\end{array}\right]$$

$$\xrightarrow[R_1=-2r_2+r_1]{}\left[\begin{array}{cc|cc}1&0&5&-7\\0&1&-2&3\end{array}\right]$$

Since the identity matrix I_2 is on the left side, the matrix on the right is the inverse.

$$\begin{bmatrix}3&7\\2&5\end{bmatrix}^{-1}=\begin{bmatrix}5&-7\\-2&3\end{bmatrix}$$

13. First, we augment the matrix with I_2, and then we use row operations to obtain the reduced row echelon form of the matrix.

$$\left[\begin{array}{cc|cc} 1 & -1 & 1 & 0 \\ 3 & -4 & 0 & 1 \end{array}\right] \xrightarrow[R_2=-3r_1+r_2]{} \left[\begin{array}{cc|cc} 1 & -1 & 1 & 0 \\ 0 & -1 & -3 & 1 \end{array}\right] \xrightarrow[R_2=-r_2]{} \left[\begin{array}{cc|cc} 1 & -1 & 1 & 0 \\ 0 & 1 & 3 & -1 \end{array}\right]$$

$$\xrightarrow[R_1=r_2+r_1]{} \left[\begin{array}{cc|cc} 1 & 0 & 4 & -1 \\ 0 & 1 & 3 & -1 \end{array}\right]$$

Since the identity matrix I_2 is on the left side, the matrix on the right is the inverse.

$$\begin{bmatrix} 1 & -1 \\ 3 & -4 \end{bmatrix}^{-1} = \begin{bmatrix} 4 & -1 \\ 3 & -1 \end{bmatrix}$$

15. First, we augment the matrix with I_2; then we use row operations to obtain the reduced row echelon form of the matrix.

$$\left[\begin{array}{cc|cc} 2 & 1 & 1 & 0 \\ 4 & 3 & 0 & 1 \end{array}\right] \xrightarrow[R_1=\frac{1}{2}r_1]{} \left[\begin{array}{cc|cc} 1 & \frac{1}{2} & \frac{1}{2} & 0 \\ 4 & 3 & 0 & 1 \end{array}\right] \xrightarrow[R_2=-4r_1+r_2]{} \left[\begin{array}{cc|cc} 1 & \frac{1}{2} & \frac{1}{2} & 0 \\ 0 & 1 & -2 & 1 \end{array}\right]$$

$$\xrightarrow[R_1=-\frac{1}{2}r_2+r_1]{} \left[\begin{array}{cc|cc} 1 & 0 & \frac{3}{2} & -\frac{1}{2} \\ 0 & 1 & -2 & 1 \end{array}\right]$$

Since the identity matrix I_2 is on the left side, the matrix on the right is the inverse.

$$\begin{bmatrix} 2 & 1 \\ 4 & 3 \end{bmatrix}^{-1} = \begin{bmatrix} \frac{3}{2} & -\frac{1}{2} \\ -2 & 1 \end{bmatrix}$$

17. First, we augment the matrix with I_3; then we use row operations to obtain the reduced row echelon form of the matrix.

$$\left[\begin{array}{ccc|ccc} 0 & 0 & 1 & 1 & 0 & 0 \\ 0 & 1 & 0 & 0 & 1 & 0 \\ 1 & 0 & 0 & 0 & 0 & 1 \end{array}\right] \xrightarrow[\text{rows 1 and 3}]{\text{Interchange}} \left[\begin{array}{ccc|ccc} 1 & 0 & 0 & 0 & 0 & 1 \\ 0 & 1 & 0 & 0 & 1 & 0 \\ 0 & 0 & 1 & 1 & 0 & 0 \end{array}\right]$$

Since the identity matrix I_3 is on the left side, the matrix on the right is the inverse.

$$\begin{bmatrix} 0 & 0 & 1 \\ 0 & 1 & 0 \\ 1 & 0 & 0 \end{bmatrix}^{-1} = \begin{bmatrix} 0 & 0 & 1 \\ 0 & 1 & 0 \\ 1 & 0 & 0 \end{bmatrix}$$

19. First, we augment the matrix with I_3, and then we use row operations to obtain the reduced row echelon form of the matrix.

$$\left[\begin{array}{ccc|ccc} 1 & 1 & -1 & 1 & 0 & 0 \\ 3 & -1 & 0 & 0 & 1 & 0 \\ 2 & -3 & 4 & 0 & 0 & 1 \end{array}\right] \xrightarrow[\substack{R_2=-3r_1+r_2 \\ R_3=-2r_1+r_3}]{} \left[\begin{array}{ccc|ccc} 1 & 1 & -1 & 1 & 0 & 0 \\ 0 & -4 & 3 & -3 & 1 & 0 \\ 0 & -5 & 6 & -2 & 0 & 1 \end{array}\right]$$

$$\xrightarrow[R_2=-\frac{1}{4}r_2]{} \left[\begin{array}{ccc|ccc} 1 & 1 & -1 & 1 & 0 & 0 \\ 0 & 1 & -\frac{3}{4} & \frac{3}{4} & -\frac{1}{4} & 0 \\ 0 & -5 & 6 & -2 & 0 & 1 \end{array}\right] \xrightarrow[\substack{R_1=-r_2+r_1 \\ R_3=5r_2+r_3}]{} \left[\begin{array}{ccc|ccc} 1 & 0 & -\frac{1}{4} & \frac{1}{4} & \frac{1}{4} & 0 \\ 0 & 1 & -\frac{3}{4} & \frac{3}{4} & -\frac{1}{4} & 0 \\ 0 & 0 & \frac{9}{4} & \frac{7}{4} & -\frac{5}{4} & 1 \end{array}\right]$$

$$\xrightarrow[R_3=\frac{4}{9}r_3]{} \left[\begin{array}{ccc|ccc} 1 & 0 & -\frac{1}{4} & \frac{1}{4} & \frac{1}{4} & 0 \\ 0 & 1 & -\frac{3}{4} & \frac{3}{4} & -\frac{1}{4} & 0 \\ 0 & 0 & 1 & \frac{7}{9} & -\frac{5}{9} & \frac{4}{9} \end{array}\right] \xrightarrow[\substack{R_1=\frac{1}{4}r_3+r_1 \\ R_2=\frac{3}{4}r_3+r_2}]{} \left[\begin{array}{ccc|ccc} 1 & 0 & 0 & \frac{4}{9} & \frac{1}{9} & \frac{1}{9} \\ 0 & 1 & 0 & \frac{4}{3} & -\frac{2}{3} & \frac{1}{3} \\ 0 & 0 & 1 & \frac{7}{9} & -\frac{5}{9} & \frac{4}{9} \end{array}\right]$$

Since the identity matrix I_3 is on the left side, the matrix on the right is the inverse.

$$\begin{bmatrix} 1 & 1 & -1 \\ 3 & -1 & 0 \\ 2 & -3 & 4 \end{bmatrix}^{-1} = \begin{bmatrix} \frac{4}{9} & \frac{1}{9} & \frac{1}{9} \\ \frac{4}{3} & -\frac{2}{3} & \frac{1}{3} \\ \frac{7}{9} & -\frac{5}{9} & \frac{4}{9} \end{bmatrix}$$

21. First, we augment the matrix with I_3; then we use row operations to obtain the reduced row echelon form of the matrix.

$$\left[\begin{array}{ccc|ccc} 1 & 1 & -1 & 1 & 0 & 0 \\ 2 & 1 & 1 & 0 & 1 & 0 \\ 1 & 0 & 1 & 0 & 0 & 1 \end{array}\right] \xrightarrow[\substack{R_2=-2r_1+r_2 \\ R_3=-r_1+r_3}]{} \left[\begin{array}{ccc|ccc} 1 & 1 & -1 & 1 & 0 & 0 \\ 0 & -1 & 3 & -2 & 1 & 0 \\ 0 & -1 & 2 & -1 & 0 & 1 \end{array}\right]$$

$$\xrightarrow[R_2=-r_2]{} \left[\begin{array}{ccc|ccc} 1 & 1 & -1 & 1 & 0 & 0 \\ 0 & 1 & -3 & 2 & -1 & 0 \\ 0 & -1 & 2 & -1 & 0 & 1 \end{array}\right] \xrightarrow[\substack{R_1=-r_2+r_1 \\ R_3=r_2+r_3}]{} \left[\begin{array}{ccc|ccc} 1 & 0 & 2 & -1 & 1 & 0 \\ 0 & 1 & -3 & 2 & -1 & 0 \\ 0 & 0 & -1 & 1 & -1 & 1 \end{array}\right]$$

$$\xrightarrow[R_3=-r_3]{} \left[\begin{array}{ccc|ccc} 1 & 0 & 2 & -1 & 1 & 0 \\ 0 & 1 & -3 & 2 & -1 & 0 \\ 0 & 0 & 1 & -1 & 1 & -1 \end{array}\right] \xrightarrow[\substack{R_1=-2r_3+r_1 \\ R_2=3r_3+r_2}]{} \left[\begin{array}{ccc|ccc} 1 & 0 & 0 & 1 & -1 & 2 \\ 0 & 1 & 0 & -1 & 2 & -3 \\ 0 & 0 & 1 & -1 & 1 & -1 \end{array}\right]$$

Since the identity matrix I_3 is on the left side, the matrix on the right is the inverse.

$$\begin{bmatrix} 1 & 1 & -1 \\ 2 & 1 & 1 \\ 1 & 0 & 1 \end{bmatrix}^{-1} = \begin{bmatrix} 1 & -1 & 2 \\ -1 & 2 & -3 \\ -1 & 1 & -1 \end{bmatrix}$$

23. First, we augment the matrix with I_4, and then we use row operations to obtain the reduced row echelon form of the matrix.

$$\left[\begin{array}{cccc|cccc} 1 & 1 & 0 & 0 & 1 & 0 & 0 & 0 \\ 0 & 1 & -1 & 1 & 0 & 1 & 0 & 0 \\ 1 & -1 & 1 & 1 & 0 & 0 & 1 & 0 \\ 0 & 1 & 0 & -1 & 0 & 0 & 0 & 1 \end{array}\right] \xrightarrow{R_3=-r_1+r_3} \left[\begin{array}{cccc|cccc} 1 & 1 & 0 & 0 & 1 & 0 & 0 & 0 \\ 0 & 1 & -1 & 1 & 0 & 1 & 0 & 0 \\ 0 & -2 & 1 & 1 & -1 & 0 & 1 & 0 \\ 0 & 1 & 0 & -1 & 0 & 0 & 0 & 1 \end{array}\right]$$

$$\xrightarrow[\substack{R_3=2r_2+r_3 \\ R_4=-r_2+r_4}]{R_1=-r_2+r_1} \left[\begin{array}{cccc|cccc} 1 & 0 & 1 & -1 & 1 & -1 & 0 & 0 \\ 0 & 1 & -1 & 1 & 0 & 1 & 0 & 0 \\ 0 & 0 & -1 & 3 & -1 & 2 & 1 & 0 \\ 0 & 0 & 1 & -2 & 0 & -1 & 0 & 1 \end{array}\right] \xrightarrow{R_3=-r_3} \left[\begin{array}{cccc|cccc} 1 & 0 & 1 & -1 & 1 & -1 & 0 & 0 \\ 0 & 1 & -1 & 1 & 0 & 1 & 0 & 0 \\ 0 & 0 & 1 & -3 & 1 & -2 & -1 & 0 \\ 0 & 0 & 1 & -2 & 0 & -1 & 0 & 1 \end{array}\right]$$

$$\xrightarrow[\substack{R_2=r_3+r_2 \\ R_4=-r_3+r_4}]{R_1=-r_3+r_1} \left[\begin{array}{cccc|cccc} 1 & 0 & 0 & 2 & 0 & 1 & 1 & 0 \\ 0 & 1 & 0 & -2 & 1 & -1 & -1 & 0 \\ 0 & 0 & 1 & -3 & 1 & -2 & -1 & 0 \\ 0 & 0 & 0 & 1 & -1 & 1 & 1 & 1 \end{array}\right] \xrightarrow[\substack{R_2=2r_4+r_2 \\ R_3=3r_4+r_3}]{R_1=-2r_4+r_1} \left[\begin{array}{cccc|cccc} 1 & 0 & 0 & 0 & 2 & -1 & -1 & -2 \\ 0 & 1 & 0 & 0 & -1 & 1 & 1 & 2 \\ 0 & 0 & 1 & 0 & -2 & 1 & 2 & 3 \\ 0 & 0 & 0 & 1 & -1 & 1 & 1 & 1 \end{array}\right]$$

Since the identity matrix I_4 is on the left side, the matrix on the right is the inverse.

$$\begin{bmatrix} 1 & 1 & 0 & 0 \\ 0 & 1 & -1 & 1 \\ 1 & -1 & 1 & 1 \\ 0 & 1 & 0 & -1 \end{bmatrix}^{-1} = \begin{bmatrix} 2 & -1 & -1 & -2 \\ -1 & 1 & 1 & 2 \\ -2 & 1 & 2 & 3 \\ -1 & 1 & 1 & 1 \end{bmatrix}$$

25. First, we augment the matrix with I_2; then we use row operations to obtain the reduced row echelon form of the matrix.

$$\left[\begin{array}{cc|cc} 4 & 6 & 1 & 0 \\ 2 & 3 & 0 & 1 \end{array}\right] \xrightarrow{R_1=\frac{1}{4}r_1} \left[\begin{array}{cc|cc} 1 & \frac{3}{2} & \frac{1}{4} & 0 \\ 2 & 3 & 0 & 1 \end{array}\right] \xrightarrow{R_2=-2r_1+r_2} \left[\begin{array}{cc|cc} 1 & \frac{3}{2} & \frac{1}{4} & 0 \\ 0 & 0 & -\frac{1}{2} & 1 \end{array}\right]$$

The 0s in row 2 indicate that we cannot get the identity matrix. This tells us that the original matrix has no inverse.

27. First, we augment the matrix with I_2; then we use row operations to obtain the reduced row echelon form of the matrix.

$$\left[\begin{array}{cc|cc} -8 & 4 & 1 & 0 \\ -4 & 2 & 0 & 1 \end{array}\right] \xrightarrow{R_1=-\frac{1}{8}r_1} \left[\begin{array}{cc|cc} 1 & -\frac{1}{2} & -\frac{1}{8} & 0 \\ -4 & 2 & 0 & 1 \end{array}\right] \xrightarrow{R_2=4r_1+r_2} \left[\begin{array}{cc|cc} 1 & -\frac{1}{2} & -\frac{1}{8} & 0 \\ 0 & 0 & -\frac{1}{2} & 1 \end{array}\right]$$

The 0s in row 2 indicate that we cannot get the identity matrix. This tells us that the original matrix has no inverse.

29. Since the matrix has a row of zeros, it has no inverse.

31. To find the inverse we augment the matrix with I_2, and then use row operations to obtain the reduced row echelon form.

$$\left[\begin{array}{cc|cc} 1 & 1 & 1 & 0 \\ 1 & 2 & 0 & 1 \end{array}\right] \xrightarrow{R_2=-r_1+r_2} \left[\begin{array}{cc|cc} 1 & 1 & 1 & 0 \\ 0 & 1 & -1 & 1 \end{array}\right] \xrightarrow{R_1=-r_2+r_1} \left[\begin{array}{cc|cc} 1 & 0 & 2 & -1 \\ 0 & 1 & -1 & 1 \end{array}\right]$$

Since the identity matrix I_2 is on the left side, $\begin{bmatrix} 1 & 1 \\ 1 & 2 \end{bmatrix}^{-1} = \begin{bmatrix} 2 & -1 \\ -1 & 1 \end{bmatrix}$.

33. To find the inverse we augment the matrix with I_2, and then use row operations to obtain the reduced row echelon form.

$$\left[\begin{array}{cc|cc} 3 & -2 & 1 & 0 \\ 0 & 4 & 0 & 1 \end{array}\right] \xrightarrow{R_1=\frac{1}{3}r_1} \left[\begin{array}{cc|cc} 1 & -\frac{2}{3} & \frac{1}{3} & 0 \\ 0 & 4 & 0 & 1 \end{array}\right] \xrightarrow{R_2=\frac{1}{4}r_2} \left[\begin{array}{cc|cc} 1 & -\frac{2}{3} & \frac{1}{3} & 0 \\ 0 & 1 & 0 & \frac{1}{4} \end{array}\right]$$

$$\xrightarrow{R_1=\frac{2}{3}r_2+r_1} \left[\begin{array}{cc|cc} 1 & 0 & \frac{1}{3} & \frac{1}{6} \\ 0 & 1 & 0 & \frac{1}{4} \end{array}\right]$$

Since the identity matrix I_2 is on the left side, $\begin{bmatrix} 3 & -2 \\ 0 & 4 \end{bmatrix}^{-1} = \begin{bmatrix} \frac{1}{3} & \frac{1}{6} \\ 0 & \frac{1}{4} \end{bmatrix}$.

35. To find the inverse we augment the matrix with I_2, and then use row operations to obtain the reduced row echelon form.

$$\left[\begin{array}{cc|cc} 3 & 2 & 1 & 0 \\ 6 & 4 & 0 & 1 \end{array}\right] \xrightarrow{R_1=\frac{1}{3}r_1} \left[\begin{array}{cc|cc} 1 & \frac{2}{3} & \frac{1}{3} & 0 \\ 6 & 4 & 0 & 1 \end{array}\right] \xrightarrow{R_2=-6r_1+r_2} \left[\begin{array}{cc|cc} 1 & \frac{2}{3} & \frac{1}{3} & 0 \\ 0 & 0 & -2 & 1 \end{array}\right]$$

The 0s in the row 2 tell us we cannot get the identity matrix. This indicates that the original matrix has no inverse.

37. To find the inverse we augment the matrix with I_3, and then use row operations to obtain the reduced row echelon form.

$$\left[\begin{array}{ccc|ccc} 1 & -2 & -1 & 1 & 0 & 0 \\ -2 & 5 & 4 & 0 & 1 & 0 \\ 3 & -8 & -5 & 0 & 0 & 1 \end{array}\right] \xrightarrow[R_3=-3r_1+r_3]{R_2=2r_1+r_2} \left[\begin{array}{ccc|ccc} 1 & -2 & -1 & 1 & 0 & 0 \\ 0 & 1 & 2 & 2 & 1 & 0 \\ 0 & -2 & -2 & -3 & 0 & 1 \end{array}\right]$$

$$\xrightarrow[R_3=2r_2+r_3]{R_1=2r_2+r_1} \left[\begin{array}{ccc|ccc} 1 & 0 & 3 & 5 & 2 & 0 \\ 0 & 1 & 2 & 2 & 1 & 0 \\ 0 & 0 & 2 & 1 & 2 & 1 \end{array}\right] \xrightarrow{R=\frac{1}{2}r_3} \left[\begin{array}{ccc|ccc} 1 & 0 & 3 & 5 & 2 & 0 \\ 0 & 1 & 2 & 2 & 1 & 0 \\ 0 & 0 & 1 & \frac{1}{2} & 1 & \frac{1}{2} \end{array}\right]$$

$$\xrightarrow[R_2=-2r_3+r_2]{R_1=-3r_3+r_1} \left[\begin{array}{ccc|ccc} 1 & 0 & 0 & \frac{7}{2} & -1 & -\frac{3}{2} \\ 0 & 1 & 0 & 1 & -1 & -1 \\ 0 & 0 & 1 & \frac{1}{2} & 1 & \frac{1}{2} \end{array}\right]$$

Since the identity matrix I_3 is on the left side, $\begin{bmatrix} 1 & -2 & -1 \\ -2 & 5 & 4 \\ 3 & -8 & -5 \end{bmatrix}^{-1} = \begin{bmatrix} \frac{7}{2} & -1 & -\frac{3}{2} \\ 1 & -1 & -1 \\ \frac{1}{2} & 1 & \frac{1}{2} \end{bmatrix}$.

39. To find A^{-1} we use row operations to obtain the reduced row echelon form of the matrix $[A \mid I_2]$.

$$\left[\begin{array}{cc|cc} 1 & 2 & 1 & 0 \\ 2 & -1 & 0 & 1 \end{array}\right] \xrightarrow{R_2=-2r_1+r_2} \left[\begin{array}{cc|cc} 1 & 2 & 1 & 0 \\ 0 & -5 & -2 & 1 \end{array}\right] \xrightarrow{R_2=-\frac{1}{5}r_2} \left[\begin{array}{cc|cc} 1 & 2 & 1 & 0 \\ 0 & 1 & \frac{2}{5} & -\frac{1}{5} \end{array}\right]$$

$$\xrightarrow{R_1=-2r_2+r_1} \left[\begin{array}{cc|cc} 1 & 0 & \frac{1}{5} & \frac{2}{5} \\ 0 & 1 & \frac{2}{5} & -\frac{1}{5} \end{array}\right] \quad \text{So, } A^{-1} = \begin{bmatrix} \frac{1}{5} & \frac{2}{5} \\ \frac{2}{5} & -\frac{1}{5} \end{bmatrix}.$$

To find B^{-1} we use row operations to obtain the reduced row-echelon form of the matrix $[B \mid I_2]$.

$$\left[\begin{array}{cc|cc} 1 & 3 & 1 & 0 \\ 2 & 1 & 0 & 1 \end{array}\right] \xrightarrow{R_2=-2r_1+r_2} \left[\begin{array}{cc|cc} 1 & 3 & 1 & 0 \\ 0 & -5 & -2 & 1 \end{array}\right] \xrightarrow{R_2=-\frac{1}{5}r_2} \left[\begin{array}{cc|cc} 1 & 3 & 1 & 0 \\ 0 & 1 & \frac{2}{5} & -\frac{1}{5} \end{array}\right]$$

$$\xrightarrow{R_1=-3r_2+r_1} \left[\begin{array}{cc|cc} 1 & 0 & -\frac{1}{5} & \frac{3}{5} \\ 0 & 1 & \frac{2}{5} & -\frac{1}{5} \end{array}\right] \quad \text{So, } B^{-1} = \begin{bmatrix} -\frac{1}{5} & \frac{3}{5} \\ \frac{2}{5} & -\frac{1}{5} \end{bmatrix} \text{ and}$$

$$A^{-1} - B^{-1} = \begin{bmatrix} \frac{1}{5} & \frac{2}{5} \\ \frac{2}{5} & -\frac{1}{5} \end{bmatrix} - \begin{bmatrix} -\frac{1}{5} & \frac{3}{5} \\ \frac{2}{5} & -\frac{1}{5} \end{bmatrix} = \begin{bmatrix} \frac{1}{5}+\frac{1}{5} & \frac{2}{5}-\frac{3}{5} \\ \frac{2}{5}-\frac{2}{5} & -\frac{1}{5}+\frac{1}{5} \end{bmatrix} = \begin{bmatrix} \frac{2}{5} & -\frac{1}{5} \\ 0 & 0 \end{bmatrix}$$

41. To solve the system $\begin{cases} x+3y+2z=2 \\ 2x+7y+3z=1 \\ x \qquad +6z=3 \end{cases}$ we define $A = \begin{bmatrix} 1 & 3 & 2 \\ 2 & 7 & 3 \\ 1 & 0 & 6 \end{bmatrix}$, $X = \begin{bmatrix} x \\ y \\ z \end{bmatrix}$, and $B = \begin{bmatrix} 2 \\ 1 \\ 3 \end{bmatrix}$.

The solution to the system is $X = A^{-1}B$

$$\begin{bmatrix} x \\ y \\ z \end{bmatrix} = \begin{bmatrix} 1 & 3 & 2 \\ 2 & 7 & 3 \\ 1 & 0 & 6 \end{bmatrix}^{-1} \begin{bmatrix} 2 \\ 1 \\ 3 \end{bmatrix} = \begin{bmatrix} 42 & -18 & -5 \\ -9 & 4 & 1 \\ -7 & 3 & 1 \end{bmatrix} \begin{bmatrix} 2 \\ 1 \\ 3 \end{bmatrix} = \begin{bmatrix} 51 \\ -11 \\ -8 \end{bmatrix}$$

or $x = 51, y = -11$, and $z = -8$ or $(51, -11, -8)$.

43. To solve the system $\begin{cases} 3x+7y=10 \\ 2x+5y=\ \ 2 \end{cases}$ we define $A=\begin{bmatrix} 3 & 7 \\ 2 & 5 \end{bmatrix}$, $X=\begin{bmatrix} x \\ y \end{bmatrix}$, and $B=\begin{bmatrix} 10 \\ 2 \end{bmatrix}$.

The solution to the system is $X=A^{-1}B$

$$\begin{bmatrix} x \\ y \end{bmatrix}=\begin{bmatrix} 3 & 7 \\ 2 & 5 \end{bmatrix}^{-1}\begin{bmatrix} 10 \\ 2 \end{bmatrix}=\begin{bmatrix} 5 & -7 \\ -2 & 3 \end{bmatrix}\begin{bmatrix} 10 \\ 2 \end{bmatrix}=\begin{bmatrix} 36 \\ -14 \end{bmatrix}$$

So, $x=36$ and $y=-14$ or $(36, -14)$.

45. To solve the system $\begin{cases} 3x+7y=13 \\ 2x+5y=\ \ 9 \end{cases}$ we define $A=\begin{bmatrix} 3 & 7 \\ 2 & 5 \end{bmatrix}$, $X=\begin{bmatrix} x \\ y \end{bmatrix}$, and $B=\begin{bmatrix} 13 \\ 9 \end{bmatrix}$.

The solution to the system is $X=A^{-1}B$

$$\begin{bmatrix} x \\ y \end{bmatrix}=\begin{bmatrix} 3 & 7 \\ 2 & 5 \end{bmatrix}^{-1}\begin{bmatrix} 13 \\ 9 \end{bmatrix}=\begin{bmatrix} 5 & -7 \\ -2 & 3 \end{bmatrix}\begin{bmatrix} 13 \\ 9 \end{bmatrix}=\begin{bmatrix} 2 \\ 1 \end{bmatrix}$$

So, $x=2$ and $y=1$ or $(2, 1)$.

47. To solve the system $\begin{cases} 3x+7y=\ \ 12 \\ 2x+5y=-4 \end{cases}$ we define $A=\begin{bmatrix} 3 & 7 \\ 2 & 5 \end{bmatrix}$, $X=\begin{bmatrix} x \\ y \end{bmatrix}$, and $B=\begin{bmatrix} 12 \\ -4 \end{bmatrix}$.

The solution to the system is $X=A^{-1}B$

$$\begin{bmatrix} x \\ y \end{bmatrix}=\begin{bmatrix} 3 & 7 \\ 2 & 5 \end{bmatrix}^{-1}\begin{bmatrix} 12 \\ -4 \end{bmatrix}=\begin{bmatrix} 5 & -7 \\ -2 & 3 \end{bmatrix}\begin{bmatrix} 12 \\ -4 \end{bmatrix}=\begin{bmatrix} 88 \\ -36 \end{bmatrix}$$

So, $x=88$ and $y=-36$ or $(88, -36)$.

49. To solve $\begin{cases} x+\ y-\ z=\ 3 \\ 3x-\ y\qquad =-4 \\ 2x-3y+4z=\ 6 \end{cases}$ we define $A=\begin{bmatrix} 1 & 1 & -1 \\ 3 & -1 & 0 \\ 2 & -3 & 4 \end{bmatrix}$, $X=\begin{bmatrix} x \\ y \\ z \end{bmatrix}$, and $B=\begin{bmatrix} 3 \\ -4 \\ 6 \end{bmatrix}$.

The solution to the system is $X=A^{-1}B$

$$\begin{bmatrix} x \\ y \\ z \end{bmatrix}=\begin{bmatrix} 1 & 1 & -1 \\ 3 & -1 & 0 \\ 2 & -3 & 4 \end{bmatrix}^{-1}\begin{bmatrix} 3 \\ -4 \\ 6 \end{bmatrix}=\begin{bmatrix} \frac{4}{9} & \frac{1}{9} & \frac{1}{9} \\ \frac{4}{3} & -\frac{2}{3} & \frac{1}{3} \\ \frac{7}{9} & -\frac{5}{9} & \frac{4}{9} \end{bmatrix}\begin{bmatrix} 3 \\ -4 \\ 6 \end{bmatrix}=\begin{bmatrix} \frac{14}{9} \\ \frac{26}{3} \\ \frac{65}{9} \end{bmatrix}$$

So, $x=\frac{14}{9}$, $y=\frac{26}{3}$, and $z=\frac{65}{9}$ or $\left(\frac{14}{9}, \frac{26}{3}, \frac{65}{9}\right)$.

51. To solve $\begin{cases} x+\ y-\ z=\ 12 \\ 3x-\ y\qquad =-4 \\ 2x-3y+4z=\ 16 \end{cases}$ we define $A=\begin{bmatrix} 1 & 1 & -1 \\ 3 & -1 & 0 \\ 2 & -3 & 4 \end{bmatrix}$, $X=\begin{bmatrix} x \\ y \\ z \end{bmatrix}$, and $B=\begin{bmatrix} 12 \\ -4 \\ 16 \end{bmatrix}$.

The solution to the system is $X = A^{-1}B$

$$\begin{bmatrix} x \\ y \\ z \end{bmatrix} = \begin{bmatrix} 1 & 1 & -1 \\ 3 & -1 & 0 \\ 2 & -3 & 4 \end{bmatrix}^{-1} \begin{bmatrix} 12 \\ -4 \\ 16 \end{bmatrix} = \begin{bmatrix} \frac{4}{9} & \frac{1}{9} & \frac{1}{9} \\ \frac{4}{3} & -\frac{2}{3} & \frac{1}{3} \\ \frac{7}{9} & -\frac{5}{9} & \frac{4}{9} \end{bmatrix} \begin{bmatrix} 12 \\ -4 \\ 16 \end{bmatrix} = \begin{bmatrix} \frac{60}{9} \\ \frac{72}{3} \\ \frac{168}{9} \end{bmatrix} = \begin{bmatrix} \frac{20}{3} \\ 24 \\ \frac{56}{3} \end{bmatrix}$$

So, $x = \frac{20}{3}$, $y = 24$, and $z = \frac{56}{3}$ or $\left(\frac{20}{3}, 24, \frac{56}{3}\right)$.

53. To solve $\begin{cases} x + y - z = 0 \\ 3x - y = -8 \\ 2x - 3y + 4z = -6 \end{cases}$ we define $A = \begin{bmatrix} 1 & 1 & -1 \\ 3 & -1 & 0 \\ 2 & -3 & 4 \end{bmatrix}$, $X = \begin{bmatrix} x \\ y \\ z \end{bmatrix}$, and $B = \begin{bmatrix} 0 \\ -8 \\ -6 \end{bmatrix}$.

The solution to the system is $X = A^{-1}B$

$$\begin{bmatrix} x \\ y \\ z \end{bmatrix} = \begin{bmatrix} 1 & 1 & -1 \\ 3 & -1 & 0 \\ 2 & -3 & 4 \end{bmatrix}^{-1} \begin{bmatrix} 0 \\ -8 \\ -6 \end{bmatrix} = \begin{bmatrix} \frac{4}{9} & \frac{1}{9} & \frac{1}{9} \\ \frac{4}{3} & -\frac{2}{3} & \frac{1}{3} \\ \frac{7}{9} & -\frac{5}{9} & \frac{4}{9} \end{bmatrix} \begin{bmatrix} 0 \\ -8 \\ -6 \end{bmatrix} = \begin{bmatrix} -\frac{14}{9} \\ \frac{10}{3} \\ \frac{16}{9} \end{bmatrix}$$

So, $x = -\frac{14}{9}$, $y = \frac{10}{3}$, and $z = \frac{16}{9}$ or $\left(-\frac{14}{9}, \frac{10}{3}, \frac{16}{9}\right)$.

55. To find the inverse of an $n \times n$ matrix using a graphing utility, enter the matrix augmented with I_n, and compute the reduced row echelon (RREF) form.

```
rref([A])
[[1 0 0 .005447...
 [0 1 0 .010356...
 [0 0 1 -.01933...
```

```
rref([A])
....0509   -.0066]
...-.0186 .0095 ]
....0116   .0344 ]]
```

The inverse is $\begin{bmatrix} 0.00545 & 0.0509 & -0.0066 \\ 0.01036 & -0.0186 & 0.0095 \\ -0.0193 & 0.0116 & 0.0344 \end{bmatrix}$.

57. To find the inverse of an $n \times n$ matrix using a graphing utility, enter the matrix augmented with I_n, and compute the reduced row echelon (RREF) form.

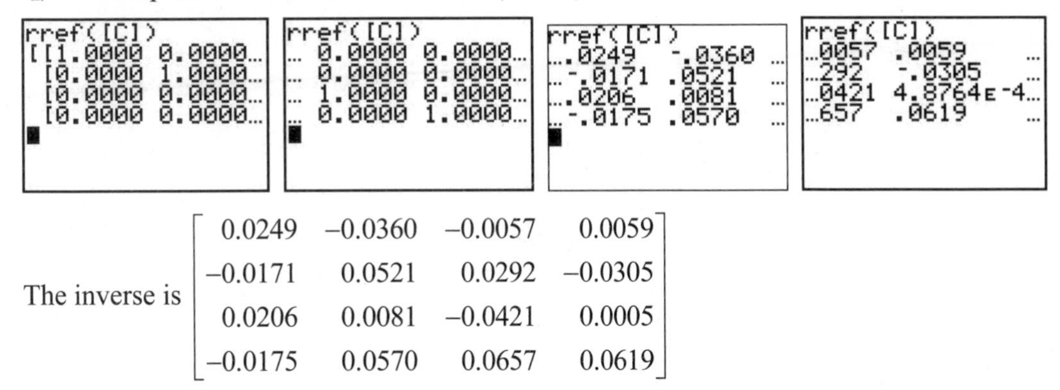

The inverse is $\begin{bmatrix} 0.0249 & -0.0360 & -0.0057 & 0.0059 \\ -0.0171 & 0.0521 & 0.0292 & -0.0305 \\ 0.0206 & 0.0081 & -0.0421 & 0.0005 \\ -0.0175 & 0.0570 & 0.0657 & 0.0619 \end{bmatrix}$

59. To find the inverse of an $n \times n$ matrix using a graphing utility, enter the matrix augmented with I_n, and compute the reduced row echelon (RREF) form.

```
rref([E])
[[1 0 0 0 0 .25...
 [0 1 0 0 0 -1....
 [0 0 1 0 0 1.7...
 [0 0 0 1 0 -.5...
 [0 0 0 0 1 -1....
```

```
rref([E])
... .25   -.0625 ...
... -1.5  .125   ...
... 1.75  -.1875 ...
... -.5   .375   ...
... -1.25 .3125  ...
```

```
rref([E])
...-.28125  .3125...
...3.0625   -2.62...
...-2.84375 2.937...
....6875    -.875...
...2.40625  -2.56...
```

```
rref([E])
...125   .09375 ]
...625   .3125  ]
...9375  -.71875]
...875   .4375  ]
....5625 .53125 ]]
```

The inverse is $\begin{bmatrix} 0.25 & -0.0625 & -0.28125 & 0.3125 & 0.09375 \\ -1.5 & 0.125 & 3.0625 & -2.625 & 0.3125 \\ 1.75 & -0.1875 & -2.84375 & 2.9375 & -0.71875 \\ -0.5 & 0.375 & 0.6875 & -0.875 & 0.4375 \\ -1.25 & 0.3125 & 2.40625 & -2.5625 & 0.53125 \end{bmatrix}$ or written as fractions

$$\begin{bmatrix} \frac{1}{4} & -\frac{1}{16} & -\frac{9}{32} & \frac{5}{16} & \frac{3}{32} \\ -\frac{3}{2} & \frac{1}{8} & \frac{49}{16} & -\frac{21}{8} & \frac{5}{16} \\ \frac{7}{4} & -\frac{3}{16} & -\frac{91}{32} & \frac{47}{16} & -\frac{25}{32} \\ -\frac{1}{2} & \frac{3}{8} & \frac{11}{16} & -\frac{7}{8} & \frac{7}{16} \\ -\frac{5}{4} & \frac{5}{16} & \frac{77}{32} & -\frac{41}{16} & \frac{17}{32} \end{bmatrix}$$

61. To solve the system of equations using a graphing utility (or Excel) first we write the system in the form $AX = B$, where

$$A = \begin{bmatrix} 25 & 61 & -12 \\ 18 & -12 & 7 \\ 3 & 4 & -1 \end{bmatrix}, X = \begin{bmatrix} x \\ y \\ z \end{bmatrix} \text{ and } B = \begin{bmatrix} 10 \\ -9 \\ 12 \end{bmatrix}$$

Enter matrices A and B into the graphing utility's matrix editor. Then calculate $A^{-1}B$.

```
[A]-1*[B]
[[4.566617862 ]
 [-6.44363104 ]
 [-24.07467057]]
```

So $x = 4.567$, $y = -6.4436$, and $z = -24.0747$ or $(4.567, -6.444, -24.075)$.

63. To solve the system of equations using a graphing utility (or Excel) first we write the system in the form $AX = D$, where

$$A = \begin{bmatrix} 25 & 61 & -12 \\ 18 & -12 & 7 \\ 3 & 4 & -1 \end{bmatrix}, X = \begin{bmatrix} x \\ y \\ z \end{bmatrix} \text{ and } D = \begin{bmatrix} 21 \\ 7 \\ -2 \end{bmatrix}$$

Enter matrices A and D into the graphing utility's matrix editor. Then calculate $A^{-1}D$.

So $x = -1.187, y = 2.457$, and $z = 8.265$ or $(-1.187, 2.457, 8.265)$.

65. For each part of the problem, let x denote the amount invested in the Mid Cap Growth Fund and let y denote the amount invested in the bond fund. Then solve the system of equations

$$\begin{cases} x + y = 3000 \\ 0.1118x + 0.055y = 3000i \end{cases}$$

where i is the average return desired. Therefore, we can write the system as

$$AX = B$$

$$\begin{bmatrix} 1 & 1 \\ 0.1118 & 0.055 \end{bmatrix}\begin{bmatrix} x \\ y \end{bmatrix} = \begin{bmatrix} 3000 \\ 3000i \end{bmatrix}$$

$$X = A^{-1}B$$

$$\begin{bmatrix} x \\ y \end{bmatrix} = \begin{bmatrix} 1 & 1 \\ 0.1118 & 0.055 \end{bmatrix}^{-1}\begin{bmatrix} 3000 \\ 3000i \end{bmatrix}$$

We first find A^{-1} and then find the needed products.

$$\left[\begin{array}{cc|cc} 1 & 1 & 1 & 0 \\ 0.1118 & 0.055 & 0 & 1 \end{array}\right] \xrightarrow{R_2 = -0.1118r_1 + r_2} \left[\begin{array}{cc|cc} 1 & 1 & 1 & 0 \\ 0 & -0.0568 & -0.1118 & 1 \end{array}\right]$$

$$\xrightarrow{R_2 = -\frac{1}{0.0568}r_2} \left[\begin{array}{cc|cc} 1 & 1 & 1 & 0 \\ 0 & 1 & 1.9683 & -17.6056 \end{array}\right]$$

$$\xrightarrow{R_1 = -r_2 + r_1} \left[\begin{array}{cc|cc} 1 & 0 & -0.9683 & 17.6056 \\ 0 & 1 & 1.9683 & -17.6056 \end{array}\right]$$

(a) When the average rate of return is 6%, (\$3000)(0.06) = \$180.00 will be earned, and

$$X = \begin{bmatrix} x \\ y \end{bmatrix} = \begin{bmatrix} -0.9683 & 17.6056 \\ 1.9683 & -17.6056 \end{bmatrix}\begin{bmatrix} 3000 \\ 180 \end{bmatrix} = \begin{bmatrix} 264.08 \\ 2735.92 \end{bmatrix}$$

Invest \$264.08 in the Mid Cap Growth Fund and \$2735.92 in the Bond Fund.

(b) When the average rate of return is 8%, (\$3000)(0.08) = \$240.00 will be earned, and

$$X = \begin{bmatrix} x \\ y \end{bmatrix} = \begin{bmatrix} -0.9683 & 17.6056 \\ 1.9683 & -17.6056 \end{bmatrix}\begin{bmatrix} 3000 \\ 240 \end{bmatrix} = \begin{bmatrix} 1320.42 \\ 1679.58 \end{bmatrix}$$

Invest \$1320.42 in the Mid Cap Growth Fund and \$1679.58 in the Bond Fund.

(c) When the average rate of return is 10%, ($3000)(0.10) = $300.00 will be earned, and

$$X=\begin{bmatrix}x\\y\end{bmatrix}=\begin{bmatrix}-0.9683 & 17.6056\\1.9683 & -17.6056\end{bmatrix}\begin{bmatrix}3000\\300\end{bmatrix}=\begin{bmatrix}2376.76\\623.24\end{bmatrix}$$

Invest $2376.76 in the Mid Cap Growth Fund and $623.24 in the Bond Fund.

67. Let x denote the amount invested in the Small Cap Growth Fund,
y denote the amount invested in the Large Cap Value Fund, and
z denote the amount invested in the International Fund.
For each section we will need to solve the following system of equations.

$$\begin{cases} x+y+z=8000 \\ x=y+1000 \\ 0.1x+0.12y+0.15z=8000(\text{average return rate}) \end{cases}$$

We will write the system as

$$AX=B$$

where $A=\begin{bmatrix}1 & 1 & 1\\1 & -1 & 0\\0.1 & 0.12 & 0.15\end{bmatrix}$; $X=\begin{bmatrix}x\\y\\z\end{bmatrix}$; and $B=\begin{bmatrix}8000\\1000\\8000(\text{average rate of return})\end{bmatrix}$.

The solution will be given as $X=A^{-1}B$. First, we need to find A^{-1}.

$$A^{-1}=\begin{bmatrix}0.950 & 0.550 & -4.916\\0.950 & -0.450 & -4.916\\-0.901 & -0.100 & 9.833\end{bmatrix}$$

(a) For an expected average return of 10%,

$$X=\begin{bmatrix}0.950 & 0.550 & -4.916\\0.950 & -0.450 & -4.916\\-0.901 & -0.100 & 9.833\end{bmatrix}\begin{bmatrix}8000\\1000\\800\end{bmatrix}=\begin{bmatrix}4219.76\\3219.76\\560.47\end{bmatrix}$$

You should invest $4219.76 in the Small Cap Growth Fund, $3219.76 in the Large Cap Value Fund, and $560.47 in the International Fund to get an expected annual return of 10% ($800.00).

(b) For an expected average return of 12%,

$$X=\begin{bmatrix}0.950 & 0.550 & -4.916\\0.950 & -0.450 & -4.916\\-0.901 & -0.100 & 9.833\end{bmatrix}\begin{bmatrix}8000\\1000\\960\end{bmatrix}=\begin{bmatrix}3433.14\\2433.14\\2133.73\end{bmatrix}$$

You should invest $3433.14 in the Small Cap Growth Fund, $2433.14 in the Large Cap Value Fund, and $2133.73 in the International Fund to get an expected annual return of 12% ($960.00).

(c) For an expected average return of 15%,

$$X=\begin{bmatrix}0.950 & 0.550 & -4.916\\0.950 & -0.450 & -4.916\\-0.901 & -0.100 & 9.833\end{bmatrix}\begin{bmatrix}8000\\1000\\1200\end{bmatrix}=\begin{bmatrix}2253.20\\1253.20\\4493.61\end{bmatrix}$$

You should invest $2253.20 in the Small Cap Growth Fund, $1253.20 in the Large Cap Value Fund, and $1253.20 in the International Fund to get an expected annual return of 15% ($1200.00).

69. Let x represent the President's bonus, y represent the CFO's bonus, and z represent the vice-president's bonus.
To set up this problem, let's call the profit P. Then we find that the bonuses can be determined as follows.

President: $x = 0.08\left[P - (x + y + z)\right]$

$$x = 0.08P - 0.08x - 0.08y - 0.08z$$

$$1.08x + 0.08y + 0.08z = 0.08P \qquad (1)$$

CFO: $y = 0.05\left[P - (x + y + z)\right]$

$$y = 0.05P - 0.05x - 0.05y - 0.05z$$

$$0.05x + 1.05y + 0.05z = 0.05P \qquad (2)$$

V-Pres. $z = 0.04\left[P - (x + y + z)\right]$

$$z = 0.04P - 0.04x - 0.04y - 0.04z$$

$$0.04x + 0.04y + 1.04z = 0.04P \qquad (3)$$

For each section we need to solve the system of equations:

$$\begin{cases} 1.08x + 0.08y + 0.08z = 0.08P \\ 0.05x + 1.05y + 0.05z = 0.05P \\ 0.04x + 0.04y + 1.04z = 0.04P \end{cases} \text{ or } \begin{cases} 13.5x + y + z = P \\ x + 21y + z = P \\ x + y + 26z = P \end{cases}$$

We will write the system as

$$AX = B$$

where $A = \begin{bmatrix} 13.5 & 1 & 1 \\ 1 & 21 & 1 \\ 1 & 1 & 46 \end{bmatrix}$; $X = \begin{bmatrix} x \\ y \\ z \end{bmatrix}$; and $\begin{bmatrix} P \\ P \\ P \end{bmatrix}$ for each value of P.

The solution is given by $X = A^{-1}B$. First, we need to find A^{-1}.

$$A^{-1} = \begin{bmatrix} 0.0745 & -0.0034 & -0.0027 \\ -0.0034 & 0.0479 & -0.0017 \\ -0.0027 & -0.0017 & 0.0386 \end{bmatrix}$$

(a) If the profits are \$300,000, then $P = 300{,}000$, and the bonuses are given by:

$$X = \begin{bmatrix} 0.0745 & -0.0034 & -0.0027 \\ -0.0034 & 0.0479 & -0.0017 \\ -0.0027 & -0.0017 & 0.0386 \end{bmatrix}\begin{bmatrix} 300{,}000 \\ 300{,}000 \\ 300{,}000 \end{bmatrix} = \begin{bmatrix} 20{,}512.82 \\ 12{,}820.51 \\ 10{,}256.41 \end{bmatrix}$$

So when profits equal \$300.000, the president's bonus is \$20,512.82, the CFO's bonus is \$12,820.51, and the vice-president's bonus is \$10,256.41.

(b) If the profits are \$500,000, then $P = 500{,}000$, and the bonuses are given by:

$$X = \begin{bmatrix} 0.0745 & -0.0034 & -0.0027 \\ -0.0034 & 0.0479 & -0.0017 \\ -0.0027 & -0.0017 & 0.0386 \end{bmatrix}\begin{bmatrix} 500{,}000 \\ 500{,}000 \\ 500{,}000 \end{bmatrix} = \begin{bmatrix} 34{,}188.03 \\ 21{,}367.52 \\ 17{,}097.02 \end{bmatrix}$$

So when profits equal \$500.000, the president's bonus is \$34,188.03, the CFO's bonus is \$21,367.52, and the vice-president's bonus is \$17,097.02.

(c) If the profits are \$750,000, then $P = 750{,}000$, and the bonuses are given by:

$$X = \begin{bmatrix} 0.0745 & -0.0034 & -0.0027 \\ -0.0034 & 0.0479 & -0.0017 \\ -0.0027 & -0.0017 & 0.0386 \end{bmatrix} \begin{bmatrix} 750{,}000 \\ 750{,}000 \\ 750{,}000 \end{bmatrix} = \begin{bmatrix} 51{,}282.05 \\ 32{,}051.28 \\ 25{,}641.03 \end{bmatrix}$$

So when profits equal \$750.000, the president's bonus is \$51,282.05, the CFO's bonus is \$32,051.28, and the vice-president's bonus is \$25,641.03.

71. First we find A^{-1}.

$$\left[\begin{array}{ccc|ccc} 1 & 0 & 0 & 1 & 0 & 0 \\ 3 & 1 & 5 & 0 & 1 & 0 \\ -2 & 0 & 1 & 0 & 0 & 1 \end{array}\right] \to \left[\begin{array}{ccc|ccc} 1 & 0 & 0 & 1 & 0 & 0 \\ 0 & 1 & 5 & -3 & 1 & 0 \\ 0 & 0 & 1 & 2 & 0 & 1 \end{array}\right] \to \left[\begin{array}{ccc|ccc} 1 & 0 & 0 & 1 & 0 & 0 \\ 0 & 1 & 0 & -13 & 1 & -5 \\ 0 & 0 & 1 & 2 & 0 & 1 \end{array}\right]$$

Since the left side of the augmented matrix is I_3, the right side is A^{-1}.
To decode the message we form 3×1 column matrices from the code and multiply each column matrix on the left with A^{-1}.

$$\begin{bmatrix} 1 & 0 & 0 \\ -13 & 1 & -5 \\ 2 & 0 & 1 \end{bmatrix} \begin{bmatrix} 4 \\ 152 \\ 17 \end{bmatrix} = \begin{bmatrix} 4 \\ 15 \\ 25 \end{bmatrix} = \begin{matrix} \text{D} \\ \text{O} \\ \text{Y} \end{matrix}; \quad \begin{bmatrix} 1 & 0 & 0 \\ -13 & 1 & -5 \\ 2 & 0 & 1 \end{bmatrix} \begin{bmatrix} 15 \\ 106 \\ -22 \end{bmatrix} = \begin{bmatrix} 15 \\ 21 \\ 8 \end{bmatrix} = \begin{matrix} \text{O} \\ \text{U} \\ \text{H} \end{matrix};$$

$$\begin{bmatrix} 1 & 0 & 0 \\ -13 & 1 & -5 \\ 2 & 0 & 1 \end{bmatrix} \begin{bmatrix} 1 \\ 50 \\ 3 \end{bmatrix} = \begin{bmatrix} 1 \\ 22 \\ 5 \end{bmatrix} = \begin{matrix} \text{A} \\ \text{V} \\ \text{E} \end{matrix}; \quad \begin{bmatrix} 1 & 0 & 0 \\ -13 & 1 & -5 \\ 2 & 0 & 1 \end{bmatrix} \begin{bmatrix} 1 \\ 52 \\ 7 \end{bmatrix} = \begin{bmatrix} 1 \\ 4 \\ 9 \end{bmatrix} = \begin{matrix} \text{A} \\ \text{D} \\ \text{I} \end{matrix};$$

$$\begin{bmatrix} 1 & 0 & 0 \\ -13 & 1 & -5 \\ 2 & 0 & 1 \end{bmatrix} \begin{bmatrix} 19 \\ 195 \\ -12 \end{bmatrix} = \begin{bmatrix} 19 \\ 8 \\ 26 \end{bmatrix} = \begin{matrix} \text{S} \\ \text{H} \\ \text{Z} \end{matrix}$$

The message is: Do you have a dish

73. To encode a message using A, assign each letter in the message a number according to the scheme, then group the numbers into 3×1 column matrices, and multiply each on the left by matrix A.

I H A V E T I C K E T S T O T H E F I N A L F O U R
9 8 1 22 5 20 9 3 11 5 20 19 20 15 20 8 5 6 9 14 1 12 6 15 21 18 26

$$\begin{bmatrix}1&0&0\\3&1&5\\-2&0&1\end{bmatrix}\begin{bmatrix}9\\8\\1\end{bmatrix}=\begin{bmatrix}9\\40\\-17\end{bmatrix};\quad \begin{bmatrix}1&0&0\\3&1&5\\-2&0&1\end{bmatrix}\begin{bmatrix}22\\5\\20\end{bmatrix}=\begin{bmatrix}22\\171\\-24\end{bmatrix};\quad \begin{bmatrix}1&0&0\\3&1&5\\-2&0&1\end{bmatrix}\begin{bmatrix}9\\3\\11\end{bmatrix}=\begin{bmatrix}9\\85\\-7\end{bmatrix};$$

$$\begin{bmatrix}1&0&0\\3&1&5\\-2&0&1\end{bmatrix}\begin{bmatrix}5\\20\\19\end{bmatrix}=\begin{bmatrix}5\\130\\9\end{bmatrix};\quad \begin{bmatrix}1&0&0\\3&1&5\\-2&0&1\end{bmatrix}\begin{bmatrix}20\\15\\20\end{bmatrix}=\begin{bmatrix}20\\175\\-20\end{bmatrix};\quad \begin{bmatrix}1&0&0\\3&1&5\\-2&0&1\end{bmatrix}\begin{bmatrix}8\\5\\6\end{bmatrix}=\begin{bmatrix}8\\59\\-10\end{bmatrix};$$

$$\begin{bmatrix}1&0&0\\3&1&5\\-2&0&1\end{bmatrix}\begin{bmatrix}9\\14\\1\end{bmatrix}=\begin{bmatrix}9\\46\\-17\end{bmatrix};\quad \begin{bmatrix}1&0&0\\3&1&5\\-2&0&1\end{bmatrix}\begin{bmatrix}12\\6\\15\end{bmatrix}=\begin{bmatrix}12\\117\\-9\end{bmatrix};\quad \begin{bmatrix}1&0&0\\3&1&5\\-2&0&1\end{bmatrix}\begin{bmatrix}21\\18\\26\end{bmatrix}=\begin{bmatrix}21\\211\\-16\end{bmatrix}$$

The encoded message is; 9 40 −17 22 171 −24 9 85 −7 5 130 9 20 175 −20 8 59 −10 9 46 −17 12 117 −9 21 211 −16

75. (a) To encode a message using A, assign each letter in the message a number according to the scheme, then group the numbers into 3×1 column matrices, and multiply each on the left by A.

I A M G O I N G T O D I S N E Y
9 1 13 7 15 9 14 7 20 15 4 9 19 14 5 25

$$\begin{bmatrix}1&0&0\\3&1&5\\-2&0&1\end{bmatrix}\begin{bmatrix}9\\1\\13\end{bmatrix}=\begin{bmatrix}9\\93\\-5\end{bmatrix};\quad \begin{bmatrix}1&0&0\\3&1&5\\-2&0&1\end{bmatrix}\begin{bmatrix}7\\15\\9\end{bmatrix}=\begin{bmatrix}7\\81\\-5\end{bmatrix};\quad \begin{bmatrix}1&0&0\\3&1&5\\-2&0&1\end{bmatrix}\begin{bmatrix}14\\7\\20\end{bmatrix}=\begin{bmatrix}14\\149\\-8\end{bmatrix};$$

$$\begin{bmatrix}1&0&0\\3&1&5\\-2&0&1\end{bmatrix}\begin{bmatrix}15\\4\\9\end{bmatrix}=\begin{bmatrix}15\\94\\-21\end{bmatrix};\quad \begin{bmatrix}1&0&0\\3&1&5\\-2&0&1\end{bmatrix}\begin{bmatrix}19\\14\\5\end{bmatrix}=\begin{bmatrix}19\\96\\-33\end{bmatrix};$$

$$\begin{bmatrix}1&0&0\\3&1&5\\-2&0&1\end{bmatrix}\begin{bmatrix}25\\25\\26\end{bmatrix}=\begin{bmatrix}25\\230\\-24\end{bmatrix}$$

The encoded message is: 9 93 −5 7 81 −5 14 149 −8 15 94 −21 19 96 −33 25 230 −24.

(b)

W	E	S	U	R	F	T	H	E	N	E	T
23	5	19	21	18	6	20	8	5	14	5	20

$$\begin{bmatrix} 1 & 0 & 0 \\ 3 & 1 & 5 \\ -2 & 0 & 1 \end{bmatrix}\begin{bmatrix} 23 \\ 5 \\ 19 \end{bmatrix} = \begin{bmatrix} 23 \\ 169 \\ -27 \end{bmatrix}; \begin{bmatrix} 1 & 0 & 0 \\ 3 & 1 & 5 \\ -2 & 0 & 1 \end{bmatrix}\begin{bmatrix} 21 \\ 18 \\ 6 \end{bmatrix} = \begin{bmatrix} 21 \\ 111 \\ -36 \end{bmatrix}; \begin{bmatrix} 1 & 0 & 0 \\ 3 & 1 & 5 \\ -2 & 0 & 1 \end{bmatrix}\begin{bmatrix} 20 \\ 8 \\ 5 \end{bmatrix} = \begin{bmatrix} 20 \\ 93 \\ -35 \end{bmatrix};$$

$$\begin{bmatrix} 1 & 0 & 0 \\ 3 & 1 & 5 \\ -2 & 0 & 1 \end{bmatrix}\begin{bmatrix} 14 \\ 5 \\ 20 \end{bmatrix} = \begin{bmatrix} 14 \\ 147 \\ -8 \end{bmatrix}$$

The encoded message is: 23 169 –27 21 111 –36 20 93 –35 14 147 –8.

(c)

L	E	T	S	G	O	C	L	U	B	B	I	N	G
12	5	20	19	7	15	3	12	21	2	2	9	14	7

$$\begin{bmatrix} 1 & 0 & 0 \\ 3 & 1 & 5 \\ -2 & 0 & 1 \end{bmatrix}\begin{bmatrix} 12 \\ 5 \\ 20 \end{bmatrix} = \begin{bmatrix} 12 \\ 141 \\ -4 \end{bmatrix}; \begin{bmatrix} 1 & 0 & 0 \\ 3 & 1 & 5 \\ -2 & 0 & 1 \end{bmatrix}\begin{bmatrix} 19 \\ 7 \\ 15 \end{bmatrix} = \begin{bmatrix} 19 \\ 139 \\ -23 \end{bmatrix};$$

$$\begin{bmatrix} 1 & 0 & 0 \\ 3 & 1 & 5 \\ -2 & 0 & 1 \end{bmatrix}\begin{bmatrix} 3 \\ 12 \\ 21 \end{bmatrix} = \begin{bmatrix} 3 \\ 126 \\ 15 \end{bmatrix};$$

$$\begin{bmatrix} 1 & 0 & 0 \\ 3 & 1 & 5 \\ -2 & 0 & 1 \end{bmatrix}\begin{bmatrix} 2 \\ 2 \\ 9 \end{bmatrix} = \begin{bmatrix} 2 \\ 53 \\ 5 \end{bmatrix}; \begin{bmatrix} 1 & 0 & 0 \\ 3 & 1 & 5 \\ -2 & 0 & 1 \end{bmatrix}\begin{bmatrix} 14 \\ 7 \\ 26 \end{bmatrix} = \begin{bmatrix} 14 \\ 179 \\ -2 \end{bmatrix}$$

The encoded message is: 12 141 –4 19 139 –23 3 126 15 2 53 5 14 179 –2.

77. (a) First, we write the student ID number as a 4 × 3 matrix $D = \begin{bmatrix} 8 & 3 & 0 \\ 1 & 9 & 6 \\ 8 & 7 & 0 \\ 2 & 4 & 9 \end{bmatrix}$, then we use K to encrypt D.

$$E = D \cdot K = \begin{bmatrix} 8 & 3 & 0 \\ 1 & 9 & 6 \\ 8 & 7 & 0 \\ 2 & 4 & 9 \end{bmatrix}\begin{bmatrix} 0 & 0 & 1 \\ 1 & 1 & 1 \\ 0 & 1 & 0 \end{bmatrix} = \begin{bmatrix} 0+3+0 & 0+3+0 & 8+3+0 \\ 0+9+0 & 0+9+6 & 1+9+0 \\ 0+7+0 & 0+7+0 & 8+7+0 \\ 0+4+0 & 0+4+9 & 2+4+0 \end{bmatrix} = \begin{bmatrix} 3 & 3 & 11 \\ 9 & 15 & 10 \\ 7 & 7 & 15 \\ 4 & 13 & 6 \end{bmatrix}$$

or 3, 3, 11, 9, 15, 10, 7, 7, 15, 4, 13, 6.

(b) The encrypted ID, written as a 4 × 3 matrix $E = \begin{bmatrix} 6 & 14 & 10 \\ 2 & 6 & 2 \\ 6 & 7 & 15 \\ 9 & 14 & 17 \end{bmatrix}$, then the original ID number is found by multiplying E and $K^{-1} = \begin{bmatrix} -1 & 1 & -1 \\ 0 & 0 & 1 \\ 1 & 0 & 0 \end{bmatrix}$.

$$D = E \cdot K^{-1} = \begin{bmatrix} 6 & 14 & 10 \\ 2 & 6 & 2 \\ 6 & 7 & 15 \\ 9 & 14 & 17 \end{bmatrix} \begin{bmatrix} -1 & 1 & -1 \\ 0 & 0 & 1 \\ 1 & 0 & 0 \end{bmatrix} = \begin{bmatrix} 4 & 6 & 8 \\ 0 & 2 & 4 \\ 9 & 6 & 1 \\ 8 & 9 & 5 \end{bmatrix}$$

or 4, 6, 8, 0, 2, 4, 9, 6, 1, 8, 9, 5

79. To find the inverse of $\begin{bmatrix} 1 & 2 \\ 2 & 3 \end{bmatrix}$ we first find $\Delta = ad - bc = 1 \cdot 3 - 2 \cdot 2 = -1$, and then write the inverse $\begin{bmatrix} 1 & 2 \\ 2 & 3 \end{bmatrix}^{-1} = \begin{bmatrix} \frac{3}{-1} & \frac{-2}{-1} \\ \frac{-2}{-1} & \frac{1}{-1} \end{bmatrix} = \begin{bmatrix} -3 & 2 \\ 2 & -1 \end{bmatrix}$

81. To find the inverse of $\begin{bmatrix} -1 & -2 \\ 3 & 4 \end{bmatrix}$ we first find $\Delta = ad - bc = -1 \cdot 4 - (-2) \cdot 3 = 2$, and then write the inverse $\begin{bmatrix} -1 & -2 \\ 3 & 4 \end{bmatrix}^{-1} = \begin{bmatrix} \frac{4}{2} & \frac{2}{2} \\ \frac{-3}{2} & \frac{-1}{2} \end{bmatrix} = \begin{bmatrix} 2 & 1 \\ -\frac{3}{2} & -\frac{1}{2} \end{bmatrix}$

2.7 Applications

Application to Economics: Leontief Models

1. We let x represent A's wages, y represent B's wages, and z represent C's wages, which we are told equal $30,000. The amount paid out by each A, B, and C equals the amount each receives, so

$$\begin{bmatrix} x \\ y \\ z \end{bmatrix} = \begin{bmatrix} \frac{1}{2} & \frac{1}{3} & \frac{1}{4} \\ \frac{1}{4} & \frac{1}{3} & \frac{1}{4} \\ \frac{1}{4} & \frac{1}{3} & \frac{1}{2} \end{bmatrix} \begin{bmatrix} x \\ y \\ z \end{bmatrix} \text{ giving the system } \begin{cases} x = \frac{1}{2}x + \frac{1}{3}y + \frac{1}{4}z \\ y = \frac{1}{4}x + \frac{1}{3}y + \frac{1}{4}z \\ z = \frac{1}{4}x + \frac{1}{3}y + \frac{1}{2}z \end{cases} \text{ or } \begin{cases} \frac{1}{2}x - \frac{1}{3}y - \frac{1}{4}z = 0 \\ -\frac{1}{4}x + \frac{2}{3}y - \frac{1}{4}z = 0 \\ -\frac{1}{4}x - \frac{1}{3}y + \frac{1}{2}z = 0 \end{cases}$$

Solving the equations for x, y, and z, we find

$$\left[\begin{array}{ccc|c} \frac{1}{2} & -\frac{1}{3} & -\frac{1}{4} & 0 \\ -\frac{1}{4} & \frac{2}{3} & -\frac{1}{4} & 0 \\ -\frac{1}{4} & -\frac{1}{3} & \frac{1}{2} & 0 \end{array}\right] \xrightarrow[\substack{R_1=2r_1 \\ R_2=\frac{1}{4}r_1+r_2 \\ R_3=\frac{1}{4}r_1+r_3}]{} \left[\begin{array}{ccc|c} 1 & -\frac{2}{3} & -\frac{1}{2} & 0 \\ 0 & \frac{1}{2} & -\frac{3}{8} & 0 \\ 0 & -\frac{1}{2} & \frac{3}{8} & 0 \end{array}\right] \xrightarrow[\substack{R_2=2r_2 \\ R_1=\frac{2}{3}r_2+r_1 \\ R_3=\frac{1}{2}r_2+r_3}]{} \left[\begin{array}{ccc|c} 1 & 0 & -1 & 0 \\ 0 & 1 & -\frac{3}{4} & 0 \\ 0 & 0 & 0 & 0 \end{array}\right]$$

or $x = z$ and $y = \frac{3}{4}z$ where z is the parameter. Since we are told C's wages are \$30,000, we know A's wages are also \$30,000, and B's wages are $\frac{3}{4}(30{,}000) = \$22{,}500$.

3. We let x represent A's wages, y represent B's wages, and z represent C's wages, which we are told equal \$30,000. Since the amount paid out by each A, B, and C equals the amount each receives, we get $\begin{bmatrix} x \\ y \\ z \end{bmatrix} = \begin{bmatrix} 0.2 & 0.3 & 0.1 \\ 0.6 & 0.4 & 0.2 \\ 0.2 & 0.3 & 0.7 \end{bmatrix} \begin{bmatrix} x \\ y \\ z \end{bmatrix}$ which gives the system

$$\begin{cases} x = 0.2x + 0.3y + 0.1z \\ y = 0.6x + 0.4y + 0.2z \\ z = 0.2x + 0.3y + 0.7z \end{cases} \text{ or } \begin{cases} 0.8x - 0.3y - 0.1z = 0 \\ -0.6x + 0.6y - 0.2z = 0 \\ -0.2x - 0.3y + 0.3z = 0 \end{cases} \text{ or } \begin{cases} 8x - 3y - 1z = 0 \\ -6x + 6y - 2z = 0 \\ -2x - 3y + 3z = 0 \end{cases}$$

Solving the equations for x, y, and z, we find

$$\left[\begin{array}{ccc|c} 8 & -3 & -1 & 0 \\ -6 & 6 & -2 & 0 \\ -2 & -3 & 3 & 0 \end{array}\right] \xrightarrow[\substack{R_1=\frac{1}{8}r_1 \\ R_2=6r_1+r_2 \\ R_3=2r_1+r_3}]{} \left[\begin{array}{ccc|c} 1 & -\frac{3}{8} & -\frac{1}{8} & 0 \\ 0 & \frac{15}{4} & -\frac{11}{4} & 0 \\ 0 & -\frac{15}{4} & \frac{11}{4} & 0 \end{array}\right] \xrightarrow[\substack{R_2=\frac{4}{15}r_2 \\ R_1=\frac{3}{8}r_2+r_1 \\ R_3=\frac{15}{4}r_2+r_3}]{} \left[\begin{array}{ccc|c} 1 & 0 & -\frac{2}{5} & 0 \\ 0 & 1 & -\frac{11}{15} & 0 \\ 0 & 0 & 0 & 0 \end{array}\right]$$

or $x = \frac{2}{5}z$ and $y = \frac{11}{15}z$ where z is the parameter. Since we are told C's wages are \$30,000, we find A's wages are $\frac{2}{5}(30{,}000) = \$12{,}000$, and B's wages are $\frac{11}{15}(30{,}000) = \$22{,}000$.

5. The total output vector X for an open Leontief model is from the system $X = [I_3 - A]^{-1} \cdot D$ where D represents future demand for the goods produced in the system. If $D_2 = \begin{bmatrix} 80 \\ 90 \\ 60 \end{bmatrix}$, then using the information from Table 5, we get

$$X = \begin{bmatrix} 1.6048 & 0.3568 & 0.7131 \\ 0.2946 & 1.3363 & 0.3857 \\ 0.3660 & 0.2721 & 1.4013 \end{bmatrix} \begin{bmatrix} 80 \\ 90 \\ 60 \end{bmatrix} = \begin{bmatrix} 203.28 \\ 166.98 \\ 137.85 \end{bmatrix}$$

The total output of R, S, and T required for the forecast demand D_2 is to produce 203.28 units of product R, 166.98 units of product S, and 137.85 units of product T.

7. We place the information in the 4 × 4 input/output matrix, A,

$$A:\quad \begin{array}{r} \\ \text{farmer} \\ \text{builder} \\ \text{tailor} \\ \text{rancher} \end{array} \begin{array}{c} \begin{array}{cccc} \text{farmer} & \text{builder} & \text{tailor} & \text{rancher} \end{array} \\ \begin{bmatrix} 0.3 & 0.3 & 0.3 & 0.2 \\ 0.2 & 0.3 & 0.3 & 0.2 \\ 0.2 & 0.1 & 0.1 & 0.2 \\ 0.3 & 0.3 & 0.3 & 0.4 \end{bmatrix} \end{array}$$

and define the variables x_1: the farmer's income; x_2; the builder's income; x_3: the tailor's income; and x_4: the rancher's income (which we are told is \$25,000). Since in a closed Leontief model the amount paid equals the amount received for each member, we form the system of equations $X = AX$ or $(I_4 - A)X = 0$.

$$\begin{cases} 0.3x_1 + 0.3x_2 + 0.3x_3 + 0.2x_4 = x_1 \\ 0.2x_1 + 0.3x_2 + 0.3x_3 + 0.2x_4 = x_2 \\ 0.2x_1 + 0.1x_2 + 0.1x_3 + 0.2x_4 = x_3 \\ 0.3x_1 + 0.3x_2 + 0.3x_3 + 0.4x_4 = x_4 \end{cases} \quad \text{or} \quad \begin{cases} 0.7x_1 - 0.3x_2 - 0.3x_3 - 0.2x_4 = 0 \\ -0.2x_1 + 0.7x_2 - 0.3x_3 - 0.2x_4 = 0 \\ -0.2x_1 - 0.1x_2 + 0.9x_3 - 0.2x_4 = 0 \\ -0.3x_1 - 0.3x_2 - 0.3x_3 + 0.6x_4 = 0 \end{cases}$$

We solve the system of equations for X.

$$\left[\begin{array}{cccc|c} 7 & -3 & -3 & -2 & 0 \\ -2 & 7 & -3 & -2 & 0 \\ -2 & -1 & 9 & -2 & 0 \\ -3 & -3 & -3 & 6 & 0 \end{array}\right] \xrightarrow[\substack{R_2=2r_1+r_2 \\ R_3=2r_1+r_3 \\ R_4=3r_1+r_4}]{R_1=\frac{1}{7}r_1} \left[\begin{array}{cccc|c} 1 & -\frac{3}{7} & -\frac{3}{7} & -\frac{2}{7} & 0 \\ 0 & \frac{43}{7} & -\frac{27}{7} & -\frac{18}{7} & 0 \\ 0 & -\frac{13}{7} & \frac{57}{7} & -\frac{18}{7} & 0 \\ 0 & -\frac{30}{7} & -\frac{30}{7} & \frac{36}{7} & 0 \end{array}\right]$$

$$\xrightarrow[\substack{R_1=\frac{3}{7}r_2+r_1 \\ R_3=\frac{13}{7}r_2+r_3 \\ R_4=\frac{30}{7}r_2+r_4}]{R_2=\frac{7}{43}r_2} \left[\begin{array}{cccc|c} 1 & 0 & -\frac{30}{43} & -\frac{20}{43} & 0 \\ 0 & 1 & -\frac{27}{43} & -\frac{18}{43} & 0 \\ 0 & 0 & \frac{300}{43} & -\frac{144}{43} & 0 \\ 0 & 0 & -\frac{300}{43} & \frac{144}{0} & 0 \end{array}\right] \xrightarrow[\substack{R_1=\frac{30}{43}r_3+r_1 \\ R_2=\frac{27}{43}r_3+r_2 \\ R_4=\frac{300}{43}r_3+r_4}]{R_3=\frac{43}{300}r_3} \left[\begin{array}{cccc|c} 1 & 0 & 0 & -\frac{4}{5} & 0 \\ 0 & 1 & 0 & -\frac{18}{25} & 0 \\ 0 & 0 & 1 & -\frac{12}{25} & 0 \\ 0 & 0 & 0 & 0 & 0 \end{array}\right]$$

The solutions to the system are $x_1 = \frac{4}{5}x_4$, $x_2 = \frac{18}{25}x_4$, and $x_3 = \frac{12}{25}x_4$ where x_4 is the parameter. Since the rancher's income was \$25,000, we get the farmer's income was $\frac{4}{5}(25{,}000) = \$20{,}000$; the builder's income was $\frac{18}{25}(25{,}000) = \$18{,}000$; and the tailor's income was $\frac{12}{25}(25{,}000) = \$12{,}000$.

9. In an open Leontief model total output must satisfy producer needs and consumer demand. First we construct matrix A which specifies the material needed per unit of production.

$$A: \begin{array}{c} \\ R \\ S \end{array} \begin{array}{c} \begin{array}{cc} R & S \end{array} \\ \begin{bmatrix} \frac{3}{13} & \frac{4}{7} \\ \frac{2}{13} & \frac{1}{7} \end{bmatrix} \end{array}$$

We define X as the total output of R and S needed to meet future demand. $AX + D = X$ or

$$X = [I - A]^{-1} \cdot D = \begin{bmatrix} \frac{10}{13} & -\frac{4}{7} \\ -\frac{2}{13} & \frac{6}{7} \end{bmatrix}^{-1} \begin{bmatrix} 80 \\ 40 \end{bmatrix} = \begin{bmatrix} \frac{3}{2} & 1 \\ \frac{7}{26} & \frac{35}{26} \end{bmatrix} \begin{bmatrix} 80 \\ 40 \end{bmatrix} = \begin{bmatrix} 160 \\ \frac{980}{13} \end{bmatrix}$$

To meet future demand, produce 160 units of product R and 75.385 units of product S. (We get $[I - A]^{-1}$ using the formula from Section 2.6 Problem 78.)

$$\Delta = \frac{10}{13} \cdot \frac{6}{7} - \left(-\frac{4}{7}\right)\left(-\frac{2}{13}\right) = \frac{60-8}{7 \cdot 13} = \frac{4}{7} \text{ and}$$

$$(I_2 - A)^{-1} = \begin{bmatrix} \frac{6}{7} \cdot \frac{7}{4} & \frac{4}{7} \cdot \frac{7}{4} \\ \frac{2}{13} \cdot \frac{7}{4} & \frac{10}{13} \cdot \frac{7}{4} \end{bmatrix} = \begin{bmatrix} \frac{3}{2} & 1 \\ \frac{7}{26} & \frac{35}{26} \end{bmatrix}$$

11. We place the information in the 3 × 3 input/output matrix, A, in which the columns represent the work done by each of the members and the rows represent the work done for each of the members.

A:	Physician	Attorney	Financial Planner
Physician	0.3	0.5	0.2
Attorney	0.4	0.2	0.2
Financial Planner	0.3	0.3	0.6

and define the variables x: the physician's income; y: the attorney's income; and z: the financial planner's income. Since in a closed Leontief model the amount paid equals the amount received by each member, we form the system of equations

$$\begin{cases} 0.3x + 0.5y + 0.2z = x \\ 0.4x + 0.2y + 0.2z = y \\ 0.3x + 0.3y + 0.6z = z \end{cases}$$

and represent it by $X = AX$ or $(I_3 - A)X = 0$ where $X = \begin{bmatrix} x \\ y \\ z \end{bmatrix}$.

We solve the system of equations for X as we did before, and get

$$\left[\begin{array}{ccc|c} 0.7 & -0.5 & -0.2 & 0 \\ -0.4 & 0.8 & -0.2 & 0 \\ -0.3 & -0.3 & 0.4 & 0 \end{array}\right] \xrightarrow{A=10A} \left[\begin{array}{ccc|c} 7 & -5 & -2 & 0 \\ -4 & 8 & -2 & 0 \\ -3 & -3 & 4 & 0 \end{array}\right] \xrightarrow[\text{rows 1 and 3}]{\text{Interchange}} \left[\begin{array}{ccc|c} -3 & -3 & 4 & 0 \\ -4 & 8 & -2 & 0 \\ 7 & -5 & -2 & 0 \end{array}\right]$$

$$\xrightarrow{R_1=r_1-r_2} \left[\begin{array}{ccc|c} 1 & -11 & 6 & 0 \\ -4 & 8 & -2 & 0 \\ 7 & -5 & -2 & 0 \end{array}\right] \xrightarrow[R_3=-7r_1+r_3]{R_2=4r_1+r_2} \left[\begin{array}{ccc|c} 1 & -11 & 6 & 0 \\ 0 & -36 & 22 & 0 \\ 0 & 72 & -44 & 0 \end{array}\right]$$

$$\xrightarrow{R_2=-\frac{1}{36}r_2} \left[\begin{array}{ccc|c} 1 & -11 & 6 & 0 \\ 0 & 1 & -\frac{11}{18} & 0 \\ 0 & 72 & -44 & 0 \end{array}\right] \xrightarrow[R_3=-72r_2+r_3]{R_1=11r_2+r_1} \left[\begin{array}{ccc|c} 1 & 0 & -\frac{13}{18} & 0 \\ 0 & 1 & -\frac{11}{18} & 0 \\ 0 & 0 & 0 & 0 \end{array}\right]$$

The solutions to the system are $x = \frac{13}{18}z$ and $y = \frac{11}{18}z$ where z is the parameter. Since it was decided that each person's work was valued at approximately \$20,000, if we let the financial planner earn \$25,000, then the physician will earn \$18,055.56, and the attorney will earn \$15,277.78. (You can choose a different amount for the financial planner's income and recalculate the incomes of the other two members of the model.)

Application to Accounting

1. We first find the total costs for the 2 service departments by solving the system

$$\begin{cases} x_1 = 2000 + \frac{1}{9}x_1 + \frac{3}{9}x_2 \\ x_2 = 1000 + \frac{3}{9}x_1 + \frac{1}{9}x_2 \end{cases}$$

We will denote

$$X = \begin{bmatrix} x_1 \\ x_2 \end{bmatrix},\ C = \begin{bmatrix} \frac{1}{9} & \frac{3}{9} \\ \frac{3}{9} & \frac{1}{9} \end{bmatrix},\ \text{and } D = \begin{bmatrix} 2000 \\ 1000 \end{bmatrix}$$

The system can now be represented by $X = D + CX$, which means the total costs of the 2 service departments can be obtained by solving $X = [I_2 - C]^{-1}D$.

$$[I_2 - C]^{-1} = \begin{bmatrix} \frac{8}{9} & -\frac{3}{9} \\ -\frac{3}{9} & \frac{8}{9} \end{bmatrix}^{-1} = \begin{bmatrix} \frac{72}{55} & \frac{27}{55} \\ \frac{27}{55} & \frac{72}{55} \end{bmatrix},\ \text{so } X = \begin{bmatrix} \frac{72}{55} & \frac{27}{55} \\ \frac{27}{55} & \frac{72}{55} \end{bmatrix} \begin{bmatrix} 2000 \\ 1000 \end{bmatrix} = \begin{bmatrix} 3109.09 \\ 2290.91 \end{bmatrix}$$

The problem has a solution because $[I_2 - C]^{-1}$ exists and because both $[I_2 - C]^{-1}$ and D have only nonnegative entries.

By substituting $x_1 = \$3109.09$ and $x_2 = \$2290.91$ into the table we find all direct and indirect costs:

Dept.	Total Costs	Direct Costs	Indirect Costs S_1	Indirect Costs S_2
S_1	\$3109.09	\$2000	\$345.45	\$763.64
S_2	\$2290.91	\$1000	\$1036.36	\$254.55
P_1	\$3354.54	\$2500	\$345.45	\$509.09
P_2	\$2790.91	\$1500	\$1036.36	\$254.55
P_3	\$3854.54	\$3000	\$345.45	\$509.09
Totals:	\$15,399.99	\$10,000	\$3109.07	\$2290.92

Finally, we show that the total of service charges allocated to production departments P_1, P_2, and P_3 equals the sum of the direct costs of the service departments S_1 and S_2.
Service charges allocated to P_1, P_2, P_3:

$$\$345.45 + \$509.09 + \$1036.36 + \$254.55 + \$345.45 + \$509.09 = \$2999.99$$

Direct costs of S_1 & S_2: $\$2000 + \$1000 = \$3000$

3. The total costs for the 2 service departments come from the two first lines of the table. They are given by the system of equations

$$\begin{cases} x_1 = 800 + 0.2x_1 + 0.1x_2 \\ x_2 = 4000 + 0.1x_1 + 0.3x_2 \end{cases}$$

We define the matrices

$$X = \begin{bmatrix} x_1 \\ x_2 \end{bmatrix}, C = \begin{bmatrix} 0.2 & 0.1 \\ 0.1 & 0.3 \end{bmatrix}, \text{ and } D = \begin{bmatrix} 800 \\ 4000 \end{bmatrix}$$

We can use the matrices to express the system of equations as $X = D + CX$ and solve for X.

$$X = [I_2 - C]^{-1}D = \begin{bmatrix} 0.8 & -0.1 \\ -0.1 & 0.7 \end{bmatrix}^{-1} \begin{bmatrix} 800 \\ 4000 \end{bmatrix} = \begin{bmatrix} \frac{70}{55} & \frac{10}{55} \\ \frac{10}{55} & \frac{80}{55} \end{bmatrix} \begin{bmatrix} 800 \\ 4000 \end{bmatrix} = \begin{bmatrix} 1745.4545 \\ 5963.6363 \end{bmatrix}$$

So we get $x_1 = \$1745.45$ and $x_2 = \$5963.64$. The table gives all direct and indirect costs.

Dept.	Total Costs	Direct Costs	Indirect Costs S_1	Indirect Costs S_2
S_1	\$1745.45	\$800	\$349.09	\$596.36
S_2	\$5963.64	\$4000	\$174.55	\$1789.09
P_1	\$2445.45	\$1500	\$349.09	\$596.36
P_2	\$2216.37	\$500	\$523.64	\$1192.73
P_3	\$3338.18	\$1200	\$349.09	\$1789.09
Totals:	\$15709.09	\$8000	\$1745.46	\$5963.63

Finally, we show that the total of service charges allocated to production departments P_1, P_2, and P_3 equal the sum of the direct costs of the service departments S_1 and S_2.
Service charges: $\$349.09 + \$596.36 + \$523.64 + \$1192.73 + \$349.09 + \$1789.09 = \$4800$
Direct costs of S_1 and S_2: $\$800 + \$4000 = \$4800$

Application to Statistics: The Method of Least Squares

1. To find the transpose A^{T} of matrix A we interchange the rows and the columns of A.

$$A^T = \begin{bmatrix} 4 & 3 \\ 1 & 1 \\ 2 & 0 \end{bmatrix}$$

3. To find the transpose A^{T} of matrix A we interchange the rows and the columns of A.

$$A^T = \begin{bmatrix} 1 & 0 & 1 \\ 11 & 12 & 4 \end{bmatrix}$$

5. To find the transpose A^{T} of matrix A we interchange the rows and the columns of A.

$$A^T = \begin{bmatrix} 8 & 6 & 3 \end{bmatrix}$$

7. (a) $\begin{bmatrix} 1 & 1 & 2 \\ 1 & 0 & 1 \\ 3 & 2 & 3 \end{bmatrix}^{\mathrm{T}} = \begin{bmatrix} 1 & 1 & 3 \\ 1 & 0 & 2 \\ 2 & 1 & 3 \end{bmatrix}$ Since the matrix does not equal its transpose, the matrix is not symmetric.

(b) $\begin{bmatrix} 0 & 1 & 3 \\ 1 & 4 & 7 \\ 3 & 7 & 5 \end{bmatrix}^{\mathrm{T}} = \begin{bmatrix} 0 & 1 & 3 \\ 1 & 4 & 7 \\ 3 & 7 & 5 \end{bmatrix}$ Since the matrix equals its transpose, the matrix is symmetric.

(c) $\begin{bmatrix} 1 & 2 & 3 & 0 \\ 2 & 4 & 5 & 0 \\ 3 & 5 & 1 & 0 \end{bmatrix}^{\mathrm{T}} = \begin{bmatrix} 1 & 2 & 3 \\ 2 & 4 & 5 \\ 3 & 5 & 1 \\ 0 & 0 & 0 \end{bmatrix}$ Since the matrix does not equal its transpose, the matrix is not symmetric.

A symmetric matrix must be square. If A has dimension $n \times m$, then A^{T} has dimension $m \times n$. If A is symmetric, then $A = A^{\mathrm{T}}$. For two matrices to be equal they must have the same dimensions, so $n = m$.

9. (a) To find the least squares line, we define $A = \begin{bmatrix} 3 & 1 \\ 5 & 1 \\ 6 & 1 \\ 7 & 1 \end{bmatrix}$, $Y = \begin{bmatrix} 10 \\ 13 \\ 15 \\ 16 \end{bmatrix}$ and $X = \begin{bmatrix} m \\ b \end{bmatrix}$ and use these matrices to form the equation $A^{\mathrm{T}}AX = A^{\mathrm{T}}Y$.

$$\begin{bmatrix} 3 & 5 & 6 & 7 \\ 1 & 1 & 1 & 1 \end{bmatrix} \begin{bmatrix} 3 & 1 \\ 5 & 1 \\ 6 & 1 \\ 7 & 1 \end{bmatrix} \begin{bmatrix} m \\ b \end{bmatrix} = \begin{bmatrix} 3 & 5 & 6 & 7 \\ 1 & 1 & 1 & 1 \end{bmatrix} \begin{bmatrix} 10 \\ 13 \\ 15 \\ 16 \end{bmatrix}$$

which simplifies to

$$\begin{bmatrix} 119 & 21 \\ 21 & 4 \end{bmatrix}\begin{bmatrix} m \\ b \end{bmatrix} = \begin{bmatrix} 297 \\ 54 \end{bmatrix} \text{ and represents the system } \begin{cases} 119m + 21b = 297 \\ 21m + 4b = 54 \end{cases}.$$

We solve the system of equations for m and b.

$$\left[\begin{array}{cc|c} 119 & 21 & 297 \\ 21 & 4 & 54 \end{array}\right] \xrightarrow[R_2=-21r_1+r_2]{R_1=\frac{1}{119}r} \left[\begin{array}{cc|c} 1 & \frac{3}{17} & \frac{297}{119} \\ 0 & \frac{5}{17} & \frac{27}{17} \end{array}\right] \xrightarrow[R_1=-\frac{3}{17}r_2+r_1]{R_2=\frac{17}{5}r_2} \left[\begin{array}{cc|c} 1 & 0 & \frac{54}{35} \\ 0 & 1 & \frac{27}{5} \end{array}\right]$$

and we get $m = \dfrac{54}{35}$ and $b = \dfrac{27}{5}$. So the least squares line of best fit is $y = \dfrac{54}{35}x + \dfrac{27}{5}$.

(b) The predicted supply when the price \$8.00 per item is $y = \dfrac{54}{35}(8) + \dfrac{27}{5} = 17.7429$. That is, when the price is \$8, the supply is predicted to be about 17,743 units.

11. To find the least squares line, we define $A = \begin{bmatrix} 10 & 1 \\ 17 & 1 \\ 11 & 1 \\ 18 & 1 \\ 21 & 1 \end{bmatrix}$, $Y = \begin{bmatrix} 50 \\ 61 \\ 55 \\ 60 \\ 70 \end{bmatrix}$, and $X = \begin{bmatrix} m \\ b \end{bmatrix}$ and use these matrices to form the equation $A^{\mathrm{T}}AX = A^{\mathrm{T}}Y$.

$$\begin{bmatrix} 10 & 17 & 11 & 18 & 21 \\ 1 & 1 & 1 & 1 & 1 \end{bmatrix}\begin{bmatrix} 10 & 1 \\ 17 & 1 \\ 11 & 1 \\ 18 & 1 \\ 21 & 1 \end{bmatrix}\begin{bmatrix} m \\ b \end{bmatrix} = \begin{bmatrix} 10 & 17 & 11 & 18 & 21 \\ 1 & 1 & 1 & 1 & 1 \end{bmatrix}\begin{bmatrix} 50 \\ 61 \\ 55 \\ 60 \\ 70 \end{bmatrix}$$

which simplifies to

$$\begin{bmatrix} 1275 & 77 \\ 77 & 5 \end{bmatrix}\begin{bmatrix} m \\ b \end{bmatrix} = \begin{bmatrix} 4692 \\ 296 \end{bmatrix} \text{ and represents the system } \begin{cases} 1275m + 77b = 4692 \\ 77m + 5b = 296 \end{cases}.$$

We solve the system of equations for m and b.

$$\left[\begin{array}{cc|c} 1275 & 77 & 4692 \\ 77 & 5 & 296 \end{array}\right] \xrightarrow{R_1=\frac{1}{1275}r_1} \left[\begin{array}{cc|c} 1 & \frac{77}{1275} & \frac{92}{25} \\ 77 & 5 & 296 \end{array}\right] \xrightarrow{R_2=-77r_1+r_2} \left[\begin{array}{cc|c} 1 & \frac{77}{1275} & \frac{92}{25} \\ 0 & \frac{446}{1275} & \frac{316}{25} \end{array}\right]$$

$$\xrightarrow{R_2=\frac{1275}{446}r_2} \left[\begin{array}{cc|c} 1 & \frac{77}{1275} & \frac{92}{25} \\ 0 & 1 & \frac{8058}{223} \end{array}\right] \xrightarrow{R_1=-\frac{77}{1275}r_2+r_1} \left[\begin{array}{cc|c} 1 & 0 & \frac{334}{223} \\ 0 & 1 & \frac{8058}{223} \end{array}\right]$$

We get $m = \dfrac{334}{223}$ and $b = \dfrac{8058}{223}$. So the least squares line of best fit for these data is

$$y = \frac{334}{223}x + \frac{8058}{223}.$$

13. In this problem we let

$$A=\begin{bmatrix} 0 & 1 \\ 1 & 1 \\ 2 & 1 \\ 3 & 1 \\ 4 & 1 \\ 5 & 1 \\ 6 & 1 \\ 7 & 1 \\ 8 & 1 \\ 9 & 1 \\ 10 & 1 \\ 11 & 1 \\ 12 & 1 \\ 13 & 1 \\ 14 & 1 \end{bmatrix}, \quad X=\begin{bmatrix} m \\ b \end{bmatrix}, \quad \text{and} \quad Y=\begin{bmatrix} 6.7 \\ 7.0 \\ 7.6 \\ 7.8 \\ 8.3 \\ 8.2 \\ 8.5 \\ 10.3 \\ 10.5 \\ 11.1 \\ 12.0 \\ 12.9 \\ 13.6 \\ 14.3 \\ 14.7 \end{bmatrix}$$

The line of best fit is found from solving the system of equations formed by

$$A^T AX = A^T Y$$

$$\begin{bmatrix} 1015 & 105 \\ 105 & 15 \end{bmatrix}\begin{bmatrix} m \\ b \end{bmatrix}=\begin{bmatrix} 1243.6 \\ 153.5 \end{bmatrix}$$

This reduces to a system of two equations in two unknowns:

$$\begin{cases} 1015m+105b=1243.6 \\ 105m+15b=\ 153.5 \end{cases}$$

Solving the system using the method of substitution, we find

$$b=\frac{153.5-105m}{15}=10.233-7m$$

Back-substituting into equation (1) gives

$$\begin{aligned} 1015m+105(10.2333-7m)&=1243.6 \\ 1015m+1074.5-735m&=1243.6 \\ 280m&=169.1 \\ m&=0.6039 \end{aligned}$$

We then compute $b = 10.2333 - 7 \cdot (0.6039) = 6.006$. The least squares solution to the problem is given by the line

$$N = 0.604t + 6.006$$

Using $t = 20$ to represent the year 2010, we find

$$N = 0.604 \cdot 20 + 6.006 = 18.086$$

18.1 million persons are predicted to have diagnosed diabetes in the year 2010.

Chapter 2 Review Exercises

1. We will solve the system by elimination.

$$\begin{array}{lll} 2x - y = 5 \quad (1) & \text{Multiply by 2:} & 4x - 2y = 10 \quad (1) \\ 5x + 2y = 8 \quad (2) & & 5x + 2y = 8 \quad (2) \\ \hline & & 9x \quad = 18 \quad \text{(Add) (2)} \end{array}$$

Solving (2) for x we get $x = 2$, and back-substituting 2 for x in equation (1) we get $2(2) - y = 5$ or $y = -1$

The solution to the system is $x = 2, y = -1$ or $(2, -1)$.

3. We will solve the system by elimination.

$$\begin{array}{ll} x - 2y - 4 = 0 & (1) \\ 3x + 2y - 4 = 0 & (2) \\ \hline 4x \quad - 8 = 0 & \text{(Add)} \quad \text{or } x = 2 \end{array}$$

Back-substituting $x = 2$ in equation (1) we get $2 - 2y - 4 = 0$ or $-2y = 2$ or $y = -1$.

The solution to the system is $x = 2$ and $y = -1$ or $(2, -1)$.

5. We will solve the system by elimination.

$$\begin{array}{lll} 3x - 2y = 8 \quad (1) & & 3x - 2y = 8 \quad (1) \\ x - \frac{2}{3}y = 12 \quad (2) & \text{Multiply by 3:} & 3x - 2y = 36 \quad (2) \\ & & \overline{\quad 0 = -28} \quad \text{(Subtract) (2)} \end{array}$$

Since equation (2) has no solution, the system is inconsistent.

7.

$$\begin{array}{lll} x + 2y - z = 6 \quad (1) & \text{Multiply by 2:} & 2x + 4y - 2z = 12 \quad (1) \\ 2x - y + 3z = -13 \quad (2) & & 2x - y + 3z = -13 \quad (2) \\ \hline & & 5y - 5z = 25 \quad \text{(Subtract) (2)} \end{array}$$

$$\begin{array}{lll} x + 2y - z = 6 \quad (1) & \text{Multiply by 3:} & 3x + 6y - 3z = 18 \quad (1) \\ 3x - 2y + 3z = -16 \quad (3) & & 3x - 2y - 3z = -16 \quad (3) \\ \hline & & 8y - 6z = 34 \quad \text{(Subtract) (3)} \end{array}$$

Now we work with the revised (2) and (3) and solve for z.

$$\begin{array}{lllll} 5y - 5z = 25 \quad (2) & \text{Divide by 5:} & y - z = 5 \quad (2) & \text{Multiply by 4:} & 4y - 4z = 20 \quad (2) \\ 8y - 6z = 34 \quad (3) & \text{Divide by 2:} & 4y - 3z = 17 \quad (3) & & 4y - 3z = 17 \quad (3) \\ \hline & & & & -z = 3 \quad \text{Subtract (3)} \end{array}$$

Solving (3) for z, we get $z = -3$. Back-substituting into (2) we find $y - (-3) = 5$ or $y = 2$. Finally substituting both y and z into equation (1), we solve for x:

$$x + 2(2) - (-3) = 6 \text{ or } x + 7 = 6 \quad \text{or } x = -1$$

The solution to the system is $x = -1, y = 2$, and $z = -3$ or $(-1, 2, -3)$.

9.

$$\begin{array}{rll} 2x-4y+\ z=-15 & (1) & \\ x+2y-4z=\ 27 & (2) & \text{Multiply by 2:} \end{array} \qquad \begin{array}{rl} 2x-4y+\ z=-15 & (1) \\ 2x+4y-8z=\ 54 & (2) \\ \hline -8y+9z=-69 & \text{(Subtract) (2)} \end{array}$$

$$\begin{array}{rll} 2x-4y+\ z=-15 & (1) & \text{Multiply by 5:} \\ 5x-6y-2z=\ -3 & (3) & \text{Multiply by 2:} \end{array} \qquad \begin{array}{rl} 10x-20y+5z=-75 & (1) \\ 10x-12y-4z=\ -6 & (3) \\ \hline -8y+9z=-69 & \text{(Subtract) (3)} \end{array}$$

Now we work with revised (2) and (3).

$$\begin{array}{rl} -8y+9z=-69 & (2) \\ -8y+9z=-69 & (3) \\ \hline 0=0 & \text{(Subtract)} \end{array}$$

The system is equivalent to a system with only two equations, so the equations are dependent and the system has infinitely many solutions. We let z be the parameter and solve (2) for y:

$$-8y+9z=-69 \quad (2)$$

$$y=\frac{9}{8}z+\frac{69}{8}$$

Substitute this expression into (1) to determine x in terms of z.

$$2x-4y+\ z=-15 \quad (1)$$

$$2x-4\left(\frac{9}{8}z+\frac{69}{8}\right)+z=-15$$

$$2x-\frac{9}{2}z-\frac{69}{2}+z=-15$$

$$2x=\frac{7}{2}z+\frac{39}{2} \quad \text{or } x=\frac{7}{4}z+\frac{39}{4}$$

The solutions to the system can be written as the system

$$\begin{cases} x=\dfrac{7}{4}z+\dfrac{39}{4} \\ y=\dfrac{9}{8}z+\dfrac{69}{8} \end{cases} \quad \text{where } z \text{ is the parameter.}$$

11. The system of equations is:

$$\begin{cases} 3x+2y=8 \\ x+4y=-1 \end{cases}$$

13. The system of equations is

$$\begin{cases} x=\ 4 \\ y=\ 6 \\ z=-1 \end{cases}$$

This system has one solution. It is $x=4$, $y=6$, and $z=-1$ or $(4, 6, -1)$.

15. We write the system as an augmented matrix and then use row operations to write it in reduced row echelon form.

$$\left[\begin{array}{cc|c} -5 & 2 & -2 \\ -3 & 3 & 4 \end{array}\right] \xrightarrow[R_1=-\frac{1}{5}r_1]{} \left[\begin{array}{cc|c} 1 & -\frac{2}{5} & \frac{2}{5} \\ -3 & 3 & 4 \end{array}\right] \xrightarrow[R_2=3r_1+r_2]{} \left[\begin{array}{cc|c} 1 & -\frac{2}{5} & \frac{2}{5} \\ 0 & \frac{9}{5} & \frac{26}{5} \end{array}\right]$$

$$\xrightarrow[R_2=\frac{5}{9}r_2]{} \left[\begin{array}{cc|c} 1 & -\frac{2}{5} & \frac{2}{5} \\ 0 & 1 & \frac{26}{9} \end{array}\right] \xrightarrow[R_1=\frac{2}{5}r_2+r_1]{} \left[\begin{array}{cc|c} 1 & 0 & \frac{14}{9} \\ 0 & 1 & \frac{26}{9} \end{array}\right]$$

The solution to the system is $x = \frac{14}{9}$ and $y = \frac{26}{9}$ or $\left(\frac{14}{9}, \frac{26}{9}\right)$

17. We write the system as an augmented matrix and then use row operations to write it in reduced row echelon form.

$$\left[\begin{array}{ccc|c} 1 & 2 & 5 & 6 \\ 3 & 7 & 12 & 23 \\ 1 & 4 & 0 & 25 \end{array}\right] \xrightarrow[\substack{R_2=-3r_1+r_2 \\ R_3=-r_1+r_3}]{} \left[\begin{array}{ccc|c} 1 & 2 & 5 & 6 \\ 0 & 1 & -3 & 5 \\ 0 & 2 & -5 & 19 \end{array}\right] \xrightarrow[\substack{R_1=-2r_2+r_1 \\ R_3=-2r_2+r_3}]{} \left[\begin{array}{ccc|c} 1 & 0 & 11 & -4 \\ 0 & 1 & -3 & 5 \\ 0 & 0 & 1 & 9 \end{array}\right]$$

$$\xrightarrow[\substack{R_1=-11r_3+r_1 \\ R_2=3r_3+r_2}]{} \left[\begin{array}{ccc|c} 1 & 0 & 0 & -103 \\ 0 & 1 & 0 & 32 \\ 0 & 0 & 1 & 9 \end{array}\right]$$

The solution to the system is $x = -103$, $y = 32$, and $z = 9$ or $(-103, 32, 9)$.

19. We write the system as an augmented matrix and then use row operations to write it in reduced row echelon form.

$$\left[\begin{array}{ccc|c} 1 & 2 & 7 & 2 \\ 3 & 7 & 18 & -1 \\ 1 & 4 & 2 & -13 \end{array}\right] \xrightarrow[\substack{R_2=-3r_1+r_2 \\ R_3=-r_1+r_3}]{} \left[\begin{array}{ccc|c} 1 & 2 & 7 & 2 \\ 0 & 1 & -3 & -7 \\ 0 & 2 & -5 & -15 \end{array}\right] \xrightarrow[\substack{R_1=-2r_2+r_1 \\ R_3=-2r_2+r_3}]{} \left[\begin{array}{ccc|c} 1 & 0 & 13 & 16 \\ 0 & 1 & -3 & -7 \\ 0 & 0 & 1 & -1 \end{array}\right]$$

$$\xrightarrow[\substack{R_1=-13r_3+r_1 \\ R_2=3r_3+r_2}]{} \left[\begin{array}{ccc|c} 1 & 0 & 0 & 29 \\ 0 & 1 & 0 & -10 \\ 0 & 0 & 1 & -1 \end{array}\right]$$

The solution to the system is $x = 29$, $y = -10$, and $z = -1$ or $(29, -10, -1)$.

21. We write the system as an augmented matrix and then use row operations to write it in reduced row echelon form.

$$\left[\begin{array}{ccc|c} 2 & -1 & 1 & 1 \\ 1 & 1 & -1 & 2 \\ 3 & -1 & 1 & 0 \end{array}\right] \xrightarrow{\text{Interchange rows 1 and 2}} \left[\begin{array}{ccc|c} 1 & 1 & -1 & 2 \\ 2 & -1 & 1 & 1 \\ 3 & -1 & 1 & 0 \end{array}\right] \xrightarrow[R_3=-3r_1+r_3]{R_2=-2r_1+r_2} \left[\begin{array}{ccc|c} 1 & 1 & -1 & 2 \\ 0 & -3 & 3 & -3 \\ 0 & -4 & 4 & -6 \end{array}\right]$$

$$\xrightarrow{R_2=-\frac{1}{3}r_2} \left[\begin{array}{ccc|c} 1 & 1 & -1 & 2 \\ 0 & 1 & -1 & 1 \\ 0 & -4 & 4 & -6 \end{array}\right] \xrightarrow{R_3=4r_2+r_3} \left[\begin{array}{ccc|c} 1 & 1 & -1 & 2 \\ 0 & 1 & -1 & 1 \\ 0 & 0 & 0 & \mathbf{-2} \end{array}\right]$$

Row 3 of the final matrix indicates that the third equation has no solution. The system is inconsistent.

23. We write the system as an augmented matrix and then use row operations to write it in reduced row echelon form.

$$\left[\begin{array}{ccc|c} 0 & 1 & -2 & 6 \\ 3 & 2 & -1 & 2 \\ 4 & 0 & 3 & -1 \end{array}\right] \xrightarrow{\text{Interchange rows 1 and 3}} \left[\begin{array}{ccc|c} 4 & 0 & 3 & -1 \\ 3 & 2 & -1 & 2 \\ 0 & 1 & -2 & 6 \end{array}\right] \xrightarrow{R_1=r_1-r_2} \left[\begin{array}{ccc|c} 1 & -2 & 4 & -3 \\ 3 & 2 & -1 & 2 \\ 0 & 1 & -2 & 6 \end{array}\right]$$

$$\xrightarrow{R_2=-3r_1+r_2} \left[\begin{array}{ccc|c} 1 & -2 & 4 & -3 \\ 0 & 8 & -13 & 11 \\ 0 & 1 & -2 & 6 \end{array}\right] \xrightarrow{\text{Interchange rows 2 and 3}} \left[\begin{array}{ccc|c} 1 & -2 & 4 & -3 \\ 0 & 1 & -2 & 6 \\ 0 & 8 & -13 & 11 \end{array}\right]$$

$$\xrightarrow[R_3=-8r_2+r_3]{R_1=2r_2+r_1} \left[\begin{array}{ccc|c} 1 & 0 & 0 & 9 \\ 0 & 1 & -2 & 6 \\ 0 & 0 & 3 & -37 \end{array}\right] \xrightarrow{R_3=\frac{1}{3}r_3} \left[\begin{array}{ccc|c} 1 & 0 & 0 & 9 \\ 0 & 1 & -2 & 6 \\ 0 & 0 & 1 & -\frac{37}{3} \end{array}\right]$$

$$\xrightarrow{R_2=2r_3+r_2} \left[\begin{array}{ccc|c} 1 & 0 & 0 & 9 \\ 0 & 1 & 0 & -\frac{56}{3} \\ 0 & 0 & 1 & -\frac{37}{3} \end{array}\right]$$

The solution to the system is $x=9$, $y=-\frac{56}{3}$, and $z=-\frac{37}{3}$ or $\left(9, -\frac{56}{3}, -\frac{37}{3}\right)$.

25. We write the system as an augmented matrix and then use row operations to write it in reduced row echelon form.

$$\left[\begin{array}{ccc|c} 1 & -3 & 0 & 5 \\ 0 & 3 & 1 & 0 \\ 2 & -1 & 2 & 2 \end{array}\right] \xrightarrow{R_3=-2r_1+r_3} \left[\begin{array}{ccc|c} 1 & -3 & 0 & 5 \\ 0 & 3 & 1 & 0 \\ 0 & 5 & 2 & -8 \end{array}\right] \xrightarrow{R_2=\frac{1}{3}r_2} \left[\begin{array}{ccc|c} 1 & -3 & 0 & 5 \\ 0 & 1 & \frac{1}{3} & 0 \\ 0 & 5 & 2 & -8 \end{array}\right]$$

$$\xrightarrow[R_3=-5r_2+r_3]{R_1=3r_2+r_1} \left[\begin{array}{ccc|c} 1 & 0 & 1 & 5 \\ 0 & 1 & \frac{1}{3} & 0 \\ 0 & 0 & \frac{1}{3} & -8 \end{array}\right] \xrightarrow{R_3=3r_3} \left[\begin{array}{ccc|c} 1 & 0 & 1 & 5 \\ 0 & 1 & \frac{1}{3} & 0 \\ 0 & 0 & 1 & -24 \end{array}\right] \xrightarrow[R_2=-\frac{1}{3}r_3+r_2]{R_1=-r_3+r_1} \left[\begin{array}{ccc|c} 1 & 0 & 0 & 29 \\ 0 & 1 & 0 & 8 \\ 0 & 0 & 1 & -24 \end{array}\right]$$

The solution to the system is $x = 29$, $y = 8$ and $z = -24$ or (29, 8, –24).

27. We write the system as an augmented matrix and then use row operations to write it in reduced row echelon form.

$$\left[\begin{array}{ccc|c} 3 & 1 & -2 & 3 \\ 1 & -2 & 1 & 4 \end{array}\right] \xrightarrow[\text{rows 1 and 2}]{\text{Interchange}} \left[\begin{array}{ccc|c} 1 & -2 & 1 & 4 \\ 3 & 1 & -2 & 3 \end{array}\right] \xrightarrow{R_2=-3r_1+r_2} \left[\begin{array}{ccc|c} 1 & -2 & 1 & 4 \\ 0 & 7 & -5 & -9 \end{array}\right]$$

$$\xrightarrow{R_2=\frac{1}{7}r_2} \left[\begin{array}{ccc|c} 1 & -2 & 1 & 4 \\ 0 & 1 & -\frac{5}{7} & -\frac{9}{7} \end{array}\right] \xrightarrow{R_1=2r_2+r_1} \left[\begin{array}{ccc|c} 1 & 0 & -\frac{3}{7} & \frac{10}{7} \\ 0 & 1 & -\frac{5}{7} & -\frac{9}{7} \end{array}\right]$$

The system has an infinite number of solutions. If we let z be the parameter, the solutions can be written as the system $\begin{cases} x = \frac{3}{7}z + \frac{10}{7} \\ y = \frac{5}{7}z - \frac{9}{7} \end{cases}$. Sample solutions will vary and are found by choosing a z-value and substituting it into the equations for x and y. Possible solutions are $\left(\frac{13}{7}, -\frac{4}{7}, 1\right)$, $\left(\frac{10}{7}, -\frac{9}{7}, 0\right)$, and $(1, -2, -1)$.

29. We write the system as an augmented matrix and then use row operations to write it in reduced row echelon form.

$$\left[\begin{array}{rrr|r} 1 & 2 & -1 & 5 \\ 2 & -1 & 2 & 0 \end{array}\right] \xrightarrow{R_2=-2r_1+r_2} \left[\begin{array}{rrr|r} 1 & 2 & -1 & 5 \\ 0 & -5 & 4 & -10 \end{array}\right] \xrightarrow{R_2=-\frac{1}{5}r_2} \left[\begin{array}{rrr|r} 1 & 2 & -1 & 5 \\ 0 & 1 & -\frac{4}{5} & 2 \end{array}\right]$$

$$\xrightarrow{R_1=-2r_2+r_1} \left[\begin{array}{rrr|r} 1 & 0 & \frac{3}{5} & 1 \\ 0 & 1 & -\frac{4}{5} & 2 \end{array}\right]$$

The system has an infinite number of solutions. If we let z be the parameter, the solutions are expressed as the system $\begin{cases} x = -\frac{3}{5}z + 1 \\ y = \frac{4}{5}z + 2 \end{cases}$. Sample solutions will vary and are found by choosing a z-value and substituting it into the equations for x and y. Sample solutions are (–2, 6, 5), (1, 2, 0), and (–5, 10, 10).

31. We write the system as an augmented matrix and then use row operations to write it in reduced row echelon form.

$$\left[\begin{array}{rr|r} 2 & -1 & 6 \\ 1 & -2 & 0 \\ 3 & -1 & 6 \end{array}\right] \xrightarrow{\text{Interchange rows 1 and 2}} \left[\begin{array}{rr|r} 1 & -2 & 0 \\ 2 & -1 & 6 \\ 3 & -1 & 6 \end{array}\right] \xrightarrow[R_3=-3r_1+r_3]{R_2=-2r_1+r_2} \left[\begin{array}{rr|r} 1 & -2 & 0 \\ 0 & 3 & 6 \\ 0 & 5 & 6 \end{array}\right] \xrightarrow{R_2=\frac{1}{3}r_2} \left[\begin{array}{rr|r} 1 & -2 & 0 \\ 0 & 1 & 2 \\ 0 & 5 & 6 \end{array}\right]$$

$$\xrightarrow{R_3=-5r_2+r_3} \left[\begin{array}{rr|r} 1 & -2 & 0 \\ 0 & 1 & 2 \\ 0 & 0 & -4 \end{array}\right]$$

Since equation (3) has no solution, the system is inconsistent.

33. The augmented matrix is in row echelon form and represents the system of equations:

$$\begin{cases} x + 4y + 3z = 4 & (1) \\ y = -1 & (2) \\ z = 1 & (3) \end{cases}$$

We can read two of the solutions, $y = -1$ and $z = 1$ directly from the system. We obtain the third variable, x, by back-substituting y and z in equation (1) and solving for x.

$$\begin{aligned} x + 4(-1) + 3(1) &= 4 \\ x - 1 &= 4 \\ x &= 5 \end{aligned}$$

The system has one solution; it is $x = 5$, $y = -1$, and $z = 1$ or (5, –1, 1).

35. The augmented matrix represents the system of equations

$$\begin{cases} x_1 \qquad\qquad + 2x_4 = 1 & (1) \\ x_2 + x_3 + 2x_4 = 2 & (2) \\ x_3 \qquad = 3 & (3) \end{cases}$$

The system has an infinite number of solutions. If we let x_4 be the parameter, and using $x_3 = 3$, we can express the solutions by the system:

$$\begin{cases} x_1 = -2x_4 + 1 \\ x_2 = -2x_4 - 1 \\ x_3 = 3 \end{cases}$$

37. $A + C = \begin{bmatrix} 1 & 0 \\ 2 & 4 \\ -1 & 2 \end{bmatrix} + \begin{bmatrix} 3 & -4 \\ 1 & 5 \\ 5 & -2 \end{bmatrix} = \begin{bmatrix} 4 & -4 \\ 3 & 9 \\ 4 & 0 \end{bmatrix}$ The sum has dimension 3 × 2.

39. $6A = 6 \cdot \begin{bmatrix} 1 & 0 \\ 2 & 4 \\ -1 & 2 \end{bmatrix} = \begin{bmatrix} 6 & 0 \\ 12 & 24 \\ -6 & 12 \end{bmatrix}$ The scalar product has dimension 3 × 2.

41. $AB = \begin{bmatrix} 1 & 0 \\ 2 & 4 \\ -1 & 2 \end{bmatrix} \begin{bmatrix} 4 & -3 & 0 \\ 1 & 1 & -2 \end{bmatrix} = \begin{bmatrix} 4+0 & -3+0 & 0+0 \\ 8+4 & -6+4 & 0-8 \\ -4+2 & 3+2 & 0-4 \end{bmatrix} = \begin{bmatrix} 4 & -3 & 0 \\ 12 & -2 & -8 \\ -2 & 5 & -4 \end{bmatrix}$

The matrix product has dimension 3 × 3.

43. $CB = \begin{bmatrix} 3 & -4 \\ 1 & 5 \\ 5 & -2 \end{bmatrix} \begin{bmatrix} 4 & -3 & 0 \\ 1 & 1 & -2 \end{bmatrix} = \begin{bmatrix} 12-4 & -9-4 & 0+8 \\ 4+5 & -3+5 & 0-10 \\ 20-2 & -15-2 & 0+4 \end{bmatrix} = \begin{bmatrix} 8 & -13 & 8 \\ 9 & 2 & -10 \\ 18 & -17 & 4 \end{bmatrix}$

The matrix product has dimension 3 × 3.

45. $(A + C)B = \left(\begin{bmatrix} 1 & 0 \\ 2 & 4 \\ -1 & 2 \end{bmatrix} + \begin{bmatrix} 3 & -4 \\ 1 & 5 \\ 5 & -2 \end{bmatrix} \right) \cdot \begin{bmatrix} 4 & -3 & 0 \\ 1 & 1 & -2 \end{bmatrix} = \begin{bmatrix} 4 & -4 \\ 3 & 9 \\ 4 & 0 \end{bmatrix} \begin{bmatrix} 4 & -3 & 0 \\ 1 & 1 & -2 \end{bmatrix}$

$= \begin{bmatrix} 16-4 & -12-4 & 0+8 \\ 12+9 & -9+9 & 0-18 \\ 16+0 & -12+0 & 0+0 \end{bmatrix} = \begin{bmatrix} 12 & -16 & 8 \\ 21 & 0 & -18 \\ 16 & -12 & 0 \end{bmatrix}$

The resulting matrix has dimension 3 × 3.

47. $3A + 2C = 3 \begin{bmatrix} 1 & 0 \\ 2 & 4 \\ -1 & 2 \end{bmatrix} + 2 \begin{bmatrix} 3 & -4 \\ 1 & 5 \\ 5 & -2 \end{bmatrix} = \begin{bmatrix} 3 & 0 \\ 6 & 12 \\ -3 & 6 \end{bmatrix} + \begin{bmatrix} 6 & -8 \\ 2 & 10 \\ 10 & -4 \end{bmatrix} = \begin{bmatrix} 9 & -8 \\ 8 & 22 \\ 7 & 2 \end{bmatrix}$ The dimension is 3 × 2.

49. $C+\mathbf{0}=\begin{bmatrix}3 & -4\\ 1 & 5\\ 5 & -2\end{bmatrix}+\begin{bmatrix}0 & 0\\ 0 & 0\\ 0 & 0\end{bmatrix}=\begin{bmatrix}3 & -4\\ 1 & 5\\ 5 & -2\end{bmatrix}=C$. The sum has dimension 3×2.

51. $\begin{bmatrix}a & b\\ c & d\end{bmatrix}^{-1}=\begin{bmatrix}\frac{d}{ad-bc} & -\frac{b}{ad-bc}\\ -\frac{c}{ad-bc} & \frac{a}{ad-bc}\end{bmatrix}$ provided $ad-bc\neq 0$.

Here $ad-bc=(3)(1)-(0)(-2)=3$. So $\begin{bmatrix}3 & 0\\ -2 & 1\end{bmatrix}^{-1}=\begin{bmatrix}\frac{1}{3} & 0\\ \frac{2}{3} & 1\end{bmatrix}$.

53. $\begin{bmatrix}a & b\\ c & d\end{bmatrix}^{-1}=\begin{bmatrix}\frac{d}{ad-bc} & -\frac{b}{ad-bc}\\ -\frac{c}{ad-bc} & \frac{a}{ad-bc}\end{bmatrix}$ provided $ad-bc\neq 0$.

Here $ad-bc=(4)(3)-(2)(6)=0$. So the inverse does not exist.

55. To find the inverse of a 3×3 matrix we augment it with I_3 and use row operations to write the matrix in reduced row-echelon form.

$$\left[\begin{array}{ccc|ccc}4 & 3 & -1 & 1 & 0 & 0\\ 0 & 2 & 2 & 0 & 1 & 0\\ 3 & -1 & 0 & 0 & 0 & 1\end{array}\right]\xrightarrow{R_1=r_1-r_3}\left[\begin{array}{ccc|ccc}1 & 4 & -1 & 1 & 0 & -1\\ 0 & 2 & 2 & 0 & 1 & 0\\ 3 & -1 & 0 & 0 & 0 & 1\end{array}\right]$$

$$\xrightarrow{R_3=-3r_1+r_3}\left[\begin{array}{ccc|ccc}1 & 4 & -1 & 1 & 0 & -1\\ 0 & 2 & 2 & 0 & 1 & 0\\ 0 & -13 & 3 & -3 & 0 & 4\end{array}\right]\xrightarrow[\substack{R_1=-4r_2+r_1\\ R_3=13r_2+r_3}]{R_2=\frac{1}{2}r_2}\left[\begin{array}{ccc|ccc}1 & 0 & -5 & 1 & -2 & -1\\ 0 & 1 & 1 & 0 & \frac{1}{2} & 0\\ 0 & 0 & 16 & -3 & \frac{13}{2} & 4\end{array}\right]$$

$$\xrightarrow{R_3=\frac{1}{16}r_3}\left[\begin{array}{ccc|ccc}1 & 0 & -5 & 1 & -2 & -1\\ 0 & 1 & 1 & 0 & \frac{1}{2} & 0\\ 0 & 0 & 1 & -\frac{3}{16} & \frac{13}{32} & \frac{1}{4}\end{array}\right]\xrightarrow[R_2=-r_3+r_2]{R_1=5r_3+r_1}\left[\begin{array}{ccc|ccc}1 & 0 & 0 & \frac{1}{16} & \frac{1}{32} & \frac{1}{4}\\ 0 & 1 & 0 & \frac{3}{16} & \frac{3}{32} & -\frac{1}{4}\\ 0 & 0 & 1 & -\frac{3}{16} & \frac{13}{32} & \frac{1}{4}\end{array}\right]$$

Since the left side is now I_3, the right side of the matrix is the inverse.

$$\begin{bmatrix}4 & 3 & -1\\ 0 & 2 & 2\\ 3 & -1 & 0\end{bmatrix}^{-1}=\begin{bmatrix}\frac{1}{16} & \frac{1}{32} & \frac{1}{4}\\ \frac{3}{16} & \frac{3}{32} & -\frac{1}{4}\\ -\frac{3}{16} & \frac{13}{32} & \frac{1}{4}\end{bmatrix}$$

57. We want $AB = BA$ when $A = \begin{bmatrix} x & y \\ z & w \end{bmatrix}$ and $B = \begin{bmatrix} 1 & 1 \\ -1 & 1 \end{bmatrix}$.

$$AB = \begin{bmatrix} x & y \\ z & w \end{bmatrix} \begin{bmatrix} 1 & 1 \\ -1 & 1 \end{bmatrix} = \begin{bmatrix} x-y & x+y \\ z-w & z+w \end{bmatrix} \text{ and } BA = \begin{bmatrix} 1 & 1 \\ -1 & 1 \end{bmatrix} \begin{bmatrix} x & y \\ z & w \end{bmatrix} = \begin{bmatrix} x+z & y+w \\ -x+z & -y+w \end{bmatrix}$$

If the two products are equal then $\begin{bmatrix} x-y & x+y \\ z-w & z+w \end{bmatrix}$ and $\begin{bmatrix} x+z & y+w \\ -x+z & -y+w \end{bmatrix}$ must be equal.

These 2 matrices are equal when their corresponding entries are equal. So

$$\begin{array}{cccc} x-y=x+z & x+y=y+w & z-w=-x+z & \text{and} \quad z+w=-y+w \\ -y=z & x=w & w=x & z=-y \end{array}$$

So for $AB = BA$, $x = w$ and $y = -z$.

59. Let x represent the number of caramels in the box and y represent the number of creams in the box. To find out how many of each kind of candy to package, we solve the system of equations

$$\begin{cases} x+y=50 & (1) \\ 0.05x+0.10y=4.00 & (2) \end{cases}$$

We will solve the system using substitution. First we solve (1) for y, and substitute its value into equation (2).

$$\begin{aligned} y &= 50-x & (1) \\ 0.05x+0.10(50-x) &= 4.00 & (2) \\ 0.05x+5-0.10x &= 4.00 \\ -0.05x &= -1 \\ x &= 20 \end{aligned}$$

Sweet Delight Candies, Inc should pack 20 caramels and 30 creams in each box if they want no profit and no loss.

If they want to increase profit while keeping 50 pieces of candy in the box, Sweet Delight Candies Inc. should increase the number of caramels (and decrease the number of creams) in each box.

61. Let x represent the amount of almonds needed, y represent the amount of cashews needed, and z represent the amount of peanuts needed to make the 100 bags of nuts.
To determine how many pounds of each type of nut the store needs, we solve the system

$$\begin{cases} x+y+z=100 & (1) \\ 6x+5y+2z=4.00(100) & (2) \end{cases}$$

We form an augmented matrix and use row operations to write it in reduced row echelon form.

$$\left[\begin{array}{ccc|c} 1 & 1 & 1 & 100 \\ 6 & 5 & 2 & 400 \end{array}\right] \xrightarrow{R_2=-6r_1+r_2} \left[\begin{array}{ccc|c} 1 & 1 & 1 & 100 \\ 0 & -1 & -4 & -200 \end{array}\right] \xrightarrow{R_2=-r_2} \left[\begin{array}{ccc|c} 1 & 1 & 1 & 100 \\ 0 & 1 & 4 & 200 \end{array}\right]$$

$$\xrightarrow{R_1=-r_2+r_1} \left[\begin{array}{ccc|c} 1 & 0 & -3 & -100 \\ 0 & 1 & 4 & 200 \end{array}\right]$$

There are an infinite number of solutions to this system. If z is the parameter, then the solutions are represented by the system

$$\begin{cases} x = 3z - 100 \\ y = -4z + 200 \end{cases}$$

The physical constraints of this problem limit z. The three variables need to be nonnegative.

$$y = -4z + 200 \geq 0 \text{ or } z \leq 50$$

Possible combinations of the nuts that can be packaged include:

Almonds (pounds)	Cashews (pounds)	Peanuts (pounds)
5	60	35
20	40	40
35	20	45
50	0	50

63. Let x represent the money invested in Treasury Bills, y represent the money invested in corporate bonds, and z represent the money invested in junk bonds.
The couple invests \$40,000. We find how they should allocate their funds to meet their investment goal by solving a system of equations. We solve the system by writing an augmented matrix which we put into reduced row-echelon form. Before writing the matrix we multiplied equation (2) by 100 to remove decimal points.

(a) The couple requires \$2500 in investment income.

$$\begin{cases} x + y + z = 40{,}000 & (1) \\ 0.06x + 0.08y + 0.10z = 2{,}500 & (2) \end{cases}$$

$$\left[\begin{array}{ccc|c} 1 & 1 & 1 & 40{,}000 \\ 6 & 8 & 10 & 250{,}000 \end{array}\right] \xrightarrow[R_2 = -6r_1 + r_2]{} \left[\begin{array}{ccc|c} 1 & 1 & 1 & 40{,}000 \\ 0 & 2 & 4 & 10{,}000 \end{array}\right] \xrightarrow[R_2 = \frac{1}{2}r_2]{} \left[\begin{array}{ccc|c} 1 & 1 & 1 & 40{,}000 \\ 0 & 1 & 2 & 5{,}000 \end{array}\right]$$

$$\xrightarrow[R_1 = -r_2 + r_1]{} \left[\begin{array}{ccc|c} 1 & 0 & -1 & 35{,}000 \\ 0 & 1 & 2 & 5{,}000 \end{array}\right]$$

The system of equations has infinite solutions. If we let z be the parameter, we can express the solutions as the system

$$\begin{cases} x = z + 35{,}000 \\ y = -2z + 5000 \end{cases}$$

Since the variables must be nonnegative, we find

$$y \geq 0$$
$$-2z + 5000 \geq 0$$
$$z \leq 2500.$$

Possible investments allocations available to the couple include

Treasury Bills	Corporate Bonds	Junk Bonds
$35,000	$5000	0
$36,000	$3000	$1000
$36,500	$2000	$1500
$37,000	$1000	$2000
$37,500	0	$2500

(b) The couple requires $3000 in investment income.

$$\begin{cases} x + y + z = 40{,}000 & (1) \\ 0.06x + 0.08y + 0.10z = 3{,}000 & (2) \end{cases}$$

$$\left[\begin{array}{ccc|c} 1 & 1 & 1 & 40{,}000 \\ 6 & 8 & 10 & 300{,}000 \end{array}\right] \xrightarrow[R_2=-6r_1+r_2]{} \left[\begin{array}{ccc|c} 1 & 1 & 1 & 40{,}000 \\ 0 & 2 & 4 & 60{,}000 \end{array}\right] \xrightarrow[R_2=\frac{1}{2}r_2]{} \left[\begin{array}{ccc|c} 1 & 1 & 1 & 40{,}000 \\ 0 & 1 & 2 & 30{,}000 \end{array}\right]$$

$$\xrightarrow[R_1=-r_2+r_1]{} \left[\begin{array}{ccc|c} 1 & 0 & -1 & 10{,}000 \\ 0 & 1 & 2 & 30{,}000 \end{array}\right]$$

The system of equations has infinite solutions. If we let z be the parameter, we can express the solutions as the system

$$\begin{cases} x = z + 10{,}000 \\ y = -2z + 30{,}000 \end{cases}$$

Since the variables must be nonnegative,

$$-2z + 30{,}000 \geq 0 \text{ or } z \leq 15{,}000.$$

Possible investments allocations available to the couple include

Treasury Bills	Corporate Bonds	Junk Bonds
$10,000	$30,000	0
$15,000	$20,000	$5000
$20,000	$10,000	$10,000
$22,000	$6,000	$12,000
$25,000	0	$15,000

(c) The couple requires $3500 in investment income.

$$\begin{cases} x + y + z = 40{,}000 & (1) \\ 0.06x + 0.08y + 0.10z = 3{,}500 & (2) \end{cases}$$

$$\left[\begin{array}{ccc|c} 1 & 1 & 1 & 40{,}000 \\ 6 & 8 & 10 & 350{,}000 \end{array}\right] \xrightarrow[R_2=-6r_1+r_2]{} \left[\begin{array}{ccc|c} 1 & 1 & 1 & 40{,}000 \\ 0 & 2 & 4 & 110{,}000 \end{array}\right] \xrightarrow[R_2=\frac{1}{2}r_2]{} \left[\begin{array}{ccc|c} 1 & 1 & 1 & 40{,}000 \\ 0 & 1 & 2 & 55{,}000 \end{array}\right]$$

$$\xrightarrow[R_1=-r_2+r_1]{} \left[\begin{array}{ccc|c} 1 & 0 & -1 & -15{,}000 \\ 0 & 1 & 2 & 55{,}000 \end{array}\right]$$

The system of equations has infinite solutions. If we let z be the parameter, we can express the solutions as the system

$$\begin{cases} x = z - 15{,}000 \\ y = -2z + 55{,}000 \end{cases}$$

Since the variables must be nonnegative,

$$z - 15{,}000 \geq 0 \text{ or } z \geq 15{,}000$$

and

$$-2z + 55{,}000 \geq 0 \text{ or } z \leq 27{,}500.$$

Possible investments allocations available to the couple include

Treasury Bills	Corporate Bonds	Junk Bonds
0	$25,000	$15,000
$5,000	$15,000	$20,000
$7,500	$10,000	$22,500
$10,000	$5,000	$25,000
$12,500	0	$27,500

65. (a) To encode the message first we assign a number to each letter.

H E Y D U D E W H A T S U P
19 22 2 23 6 23 22 4 19 26 7 8 6 11

(I) To do (I) we group the numbers into 2×1 column matrices and multiply each column matrix on the left by A.

$$\begin{bmatrix} 3 & 1 \\ 2 & 2 \end{bmatrix}\begin{bmatrix} 19 \\ 22 \end{bmatrix} = \begin{bmatrix} 79 \\ 82 \end{bmatrix}; \quad \begin{bmatrix} 3 & 1 \\ 2 & 2 \end{bmatrix}\begin{bmatrix} 2 \\ 23 \end{bmatrix} = \begin{bmatrix} 29 \\ 50 \end{bmatrix}; \quad \begin{bmatrix} 3 & 1 \\ 2 & 2 \end{bmatrix}\begin{bmatrix} 6 \\ 23 \end{bmatrix} = \begin{bmatrix} 41 \\ 58 \end{bmatrix}; \quad \begin{bmatrix} 3 & 1 \\ 2 & 2 \end{bmatrix}\begin{bmatrix} 22 \\ 4 \end{bmatrix} = \begin{bmatrix} 70 \\ 52 \end{bmatrix};$$

$$\begin{bmatrix} 3 & 1 \\ 2 & 2 \end{bmatrix}\begin{bmatrix} 19 \\ 26 \end{bmatrix} = \begin{bmatrix} 83 \\ 90 \end{bmatrix}; \quad \begin{bmatrix} 3 & 1 \\ 2 & 2 \end{bmatrix}\begin{bmatrix} 7 \\ 8 \end{bmatrix} = \begin{bmatrix} 29 \\ 30 \end{bmatrix}; \quad \begin{bmatrix} 3 & 1 \\ 2 & 2 \end{bmatrix}\begin{bmatrix} 6 \\ 11 \end{bmatrix} = \begin{bmatrix} 29 \\ 34 \end{bmatrix}$$

The encoded message is 79 82 29 50 41 58 70 52 83 90 29 30 29 34.

(II) To do (II) we group the numbers into 3×1 column matrices and multiply each column matrix on the left by B.

$$\begin{bmatrix} 1 & 4 & 2 \\ 2 & 0 & 2 \\ 0 & 0 & 4 \end{bmatrix}\begin{bmatrix} 19 \\ 22 \\ 2 \end{bmatrix} = \begin{bmatrix} 111 \\ 42 \\ 8 \end{bmatrix}; \quad \begin{bmatrix} 1 & 4 & 2 \\ 2 & 0 & 2 \\ 0 & 0 & 4 \end{bmatrix}\begin{bmatrix} 23 \\ 6 \\ 23 \end{bmatrix} = \begin{bmatrix} 93 \\ 92 \\ 92 \end{bmatrix}; \quad \begin{bmatrix} 1 & 4 & 2 \\ 2 & 0 & 2 \\ 0 & 0 & 4 \end{bmatrix}\begin{bmatrix} 22 \\ 4 \\ 19 \end{bmatrix} = \begin{bmatrix} 76 \\ 82 \\ 76 \end{bmatrix};$$

$$\begin{bmatrix} 1 & 4 & 2 \\ 2 & 0 & 2 \\ 0 & 0 & 4 \end{bmatrix}\begin{bmatrix} 26 \\ 7 \\ 8 \end{bmatrix} = \begin{bmatrix} 70 \\ 68 \\ 32 \end{bmatrix}; \quad \begin{bmatrix} 1 & 4 & 2 \\ 2 & 0 & 2 \\ 0 & 0 & 4 \end{bmatrix}\begin{bmatrix} 6 \\ 11 \\ 1 \end{bmatrix} = \begin{bmatrix} 52 \\ 14 \\ 4 \end{bmatrix}$$

The encoded message is 111 42 8 93 92 92 76 82 76 70 68 32 52 14 4.

(b) To encode the message we first assign a number to each letter.

C	A	L	L	M	E	O	N	M	Y	C	E	L	L
24	26	15	15	14	22	12	13	14	2	24	22	15	15

(I) To do (I) we group the numbers into 2 × 1 column matrices and multiply each column matrix on the left by A.

$$\begin{bmatrix}3&1\\2&2\end{bmatrix}\begin{bmatrix}24\\26\end{bmatrix}=\begin{bmatrix}98\\100\end{bmatrix};\quad \begin{bmatrix}3&1\\2&2\end{bmatrix}\begin{bmatrix}15\\15\end{bmatrix}=\begin{bmatrix}60\\60\end{bmatrix};\quad \begin{bmatrix}3&1\\2&2\end{bmatrix}\begin{bmatrix}14\\22\end{bmatrix}=\begin{bmatrix}64\\72\end{bmatrix};$$

$$\begin{bmatrix}3&1\\2&2\end{bmatrix}\begin{bmatrix}12\\13\end{bmatrix}=\begin{bmatrix}49\\50\end{bmatrix};\quad \begin{bmatrix}3&1\\2&2\end{bmatrix}\begin{bmatrix}14\\2\end{bmatrix}=\begin{bmatrix}44\\32\end{bmatrix};\quad \begin{bmatrix}3&1\\2&2\end{bmatrix}\begin{bmatrix}24\\22\end{bmatrix}=\begin{bmatrix}94\\92\end{bmatrix};\quad \begin{bmatrix}3&1\\2&2\end{bmatrix}\begin{bmatrix}15\\15\end{bmatrix}=\begin{bmatrix}60\\60\end{bmatrix}$$

The encoded message is 98 100 60 60 64 72 49 50 44 32 94 92 60 60.

(II) To do (II) we group the numbers into 3 × 1 column matrices and multiply each column matrix on the left by B.

$$\begin{bmatrix}1&4&2\\2&0&2\\0&0&4\end{bmatrix}\begin{bmatrix}24\\26\\15\end{bmatrix}=\begin{bmatrix}158\\78\\60\end{bmatrix};\quad \begin{bmatrix}1&4&2\\2&0&2\\0&0&4\end{bmatrix}\begin{bmatrix}15\\14\\22\end{bmatrix}=\begin{bmatrix}115\\74\\88\end{bmatrix};\quad \begin{bmatrix}1&4&2\\2&0&2\\0&0&4\end{bmatrix}\begin{bmatrix}12\\13\\14\end{bmatrix}=\begin{bmatrix}92\\52\\56\end{bmatrix};$$

$$\begin{bmatrix}1&4&2\\2&0&2\\0&0&4\end{bmatrix}\begin{bmatrix}2\\24\\22\end{bmatrix}=\begin{bmatrix}142\\48\\88\end{bmatrix};\quad \begin{bmatrix}1&4&2\\2&0&2\\0&0&4\end{bmatrix}\begin{bmatrix}15\\15\\1\end{bmatrix}=\begin{bmatrix}77\\32\\4\end{bmatrix}$$

The encoded message is 158 78 60 115 74 88 92 52 56 142 48 88 77 32 4.

(c) To decode the message: 75 78 45 42 72 64 35 42 17 26 67 62 using A we group the numbers into 2 × 1 column matrices and multiply each column matrix on the left by A^{-1}.

$$A^{-1}=\begin{bmatrix}3&1\\2&2\end{bmatrix}^{-1}=\begin{bmatrix}\frac{1}{2}&-\frac{1}{4}\\-\frac{1}{2}&\frac{3}{4}\end{bmatrix}$$

$$\begin{bmatrix}\frac{1}{2}&-\frac{1}{4}\\-\frac{1}{2}&\frac{3}{4}\end{bmatrix}\begin{bmatrix}75\\78\end{bmatrix}=\begin{bmatrix}18\\21\end{bmatrix};\quad \begin{bmatrix}\frac{1}{2}&-\frac{1}{4}\\-\frac{1}{2}&\frac{3}{4}\end{bmatrix}\begin{bmatrix}45\\42\end{bmatrix}=\begin{bmatrix}12\\9\end{bmatrix};\quad \begin{bmatrix}\frac{1}{2}&-\frac{1}{4}\\-\frac{1}{2}&\frac{3}{4}\end{bmatrix}\begin{bmatrix}72\\64\end{bmatrix}=\begin{bmatrix}20\\12\end{bmatrix};$$

$$\begin{bmatrix}\frac{1}{2}&-\frac{1}{4}\\-\frac{1}{2}&\frac{3}{4}\end{bmatrix}\begin{bmatrix}35\\42\end{bmatrix}=\begin{bmatrix}7\\14\end{bmatrix};\quad \begin{bmatrix}\frac{1}{2}&-\frac{1}{4}\\-\frac{1}{2}&\frac{3}{4}\end{bmatrix}\begin{bmatrix}17\\26\end{bmatrix}=\begin{bmatrix}2\\11\end{bmatrix};\quad \begin{bmatrix}\frac{1}{2}&-\frac{1}{4}\\-\frac{1}{2}&\frac{3}{4}\end{bmatrix}\begin{bmatrix}67\\62\end{bmatrix}=\begin{bmatrix}18\\13\end{bmatrix}$$

The decoded numbers are 18 21 12 9 20 12 7 14 2 11 18 13. Replacing the numbers with the appropriate letters gives the message: I F O R G O T M Y P I N
which translates to "I forgot my PIN"

67. (a) On December 1, 2006 the three stocks cost

$$\begin{array}{cc} & \begin{array}{ccc}\text{Microsoft} & \text{Intel} & \text{Dell}\end{array} \\ A = 12/1/2006 & \begin{bmatrix} 29.50 & 21.40 & 27.50 \end{bmatrix} \end{array}$$

(b) To answer part (c), the rows of matrix B must be the same as the columns of matrix A.

$$\begin{array}{cc} & \begin{array}{ccc}\text{Juanita} & \text{Debbie} & \text{Dawn}\end{array} \\ B = \begin{array}{r}\text{Microsoft} \\ \text{Intel} \\ \text{Dell}\end{array} & \begin{bmatrix} 20 & 15 & 10 \\ 30 & 25 & 20 \\ 10 & 20 & 25 \end{bmatrix} \end{array}$$

(c) The entries of matrix AB is represents each woman's cost of purchasing the three stocks.

$$AB = \begin{bmatrix} 29.50 & 21.40 & 27.50 \end{bmatrix} \begin{bmatrix} 20 & 15 & 10 \\ 30 & 25 & 20 \\ 10 & 20 & 25 \end{bmatrix}$$
$$= \begin{bmatrix} 590+642+275 & 442.5+535+550 & 295+428+687.5 \end{bmatrix}$$
$$= \begin{bmatrix} 1507 & 1527.5 & 1410.5 \end{bmatrix}$$

$$\begin{array}{cc} & \begin{array}{ccc}\text{Juanita} & \text{Debbie} & \text{Dawn}\end{array} \\ AB = 12/1/2006 & \begin{bmatrix} \$1507 & \$1527.50 & \$1410.50 \end{bmatrix} \end{array}$$

Chapter 2 Project

1. Matrix A represents the amounts of materials needed to produce 1 unit of a product. The entries in the matrix are quotients. Column 1 is found by dividing each entry in the agriculture column of the table by total gross output of agriculture. Column 2 entries are the quotients found by dividing each entry in the manufacturing column by total gross output of manufacturing, and column 3 entries are the quotients found by dividing entries by total gross outcome of services. D is the consumer demand. It is labeled Open Sector in Table 1.

$$A = \begin{bmatrix} 0.410 & 0.030 & 0.026 \\ 0.062 & 0.378 & 0.105 \\ 0.124 & 0.159 & 0.192 \end{bmatrix} \qquad D = \begin{bmatrix} 39.24 \\ 60.02 \\ 130.65 \end{bmatrix}$$

3. To calculate the new X_1, we need to solve the equation $X_1 = [I_3 - A]^{-1} D_1$

$$X_1 = \begin{bmatrix} 1.720 & 0.100 & 0.068 \\ 0.223 & 1.676 & 0.225 \\ 0.308 & 0.345 & 1.292 \end{bmatrix} \begin{bmatrix} 40.24 \\ 60.02 \\ 130.65 \end{bmatrix} = \begin{bmatrix} 84.099 \\ 138.963 \\ 201.901 \end{bmatrix}$$

5. Using the interpretation from Problem 4, we find that service production would need to increase by 0.308 unit for a one unit increase in agriculture.

7.

$$A = \begin{bmatrix} 0.2424 & 0.0005 & 0.0058 & 0.0366 & 0.0001 & 0.0012 & 0.0045 & 0.0354 & 0.0005 \\ 0.0013 & 0.2131 & 0.0073 & 0.0207 & 0.0419 & 0.0000 & 0.0000 & 0.0000 & 0.0025 \\ 0.049 & 0.0318 & .0009 & 0.0073 & 0.0379 & 0.0082 & 0.0261 & 0.0083 & 0.0214 \\ 0.1744 & 0.0982 & 0.2976 & 0.3492 & 0.0564 & 0.0438 & 0.0076 & 0.0980 & 0.0145 \\ 0.0446 & 0.0856 & 0.0247 & 0.0455 & 0.1607 & 0.0439 & 0.0207 & 0.0347 & 0.0189 \\ 0.0492 & 0.0237 & 0.0812 & 0.0583 & 0.0121 & 0.0211 & 0.0019 & 0.0196 & 0.0022 \\ 0.0729 & 0.2251 & 0.0164 & 0.0180 & 0.0322 & 0.0698 & 0.1749 & 0.0701 & 0.0066 \\ 0.0318 & 0.0396 & 0.1031 & 0.0607 & 0.1156 & 0.1412 & 0.0751 & 0.1526 & 0.0112 \\ 0.0006 & 0.0002 & 0.0011 & 0.0035 & 0.0026 & 0.0072 & 0.0111 & 0.0071 & 0.0025 \end{bmatrix}$$

$$D_0 = \begin{bmatrix} 34940 \\ -39241 \\ 787208 \\ 1611520 \\ 586248 \\ 1103110 \\ 1520718 \\ 2214382 \\ 1032052 \end{bmatrix}$$

9.

$$D_1 = \begin{bmatrix} 34940 \\ -39241 \\ 787209 \\ 1611520 \\ 586248 \\ 1103110 \\ 1520718 \\ 2214382 \\ 1032052 \end{bmatrix} \quad X_1 = [I_9 - A]^{-1} D_1 = \begin{bmatrix} 434735.5291 \\ 135766.6225 \\ 1009114.196 \\ 3906797.921 \\ 1300135.743 \\ 1566308.417 \\ 2531704.148 \\ 3715940.758 \\ 119064.343 \end{bmatrix}$$

If demand for construction increases by 1 million dollars, demand for transportation, communication, and utilities increases by \$80,000.00.

$$1300135.74 - 1300135.66 = 0.08$$

11. An increase of \$1 million in demand for construction, produces an composite increase of \$2.1 million in the economy.

Mathematical Questions From Professional Exams

1. (b)

3. (d)

Chapter 3

Linear Programming with Two Variables

3.1 Systems of Linear Inequalities

7. True

9. $x \geq 0$

The corresponding linear equation is $x = 0$. We graph a solid line, since the inequality is nonstrict. The test point (1, 1) satisfies the inequality, so we shade the region containing it.

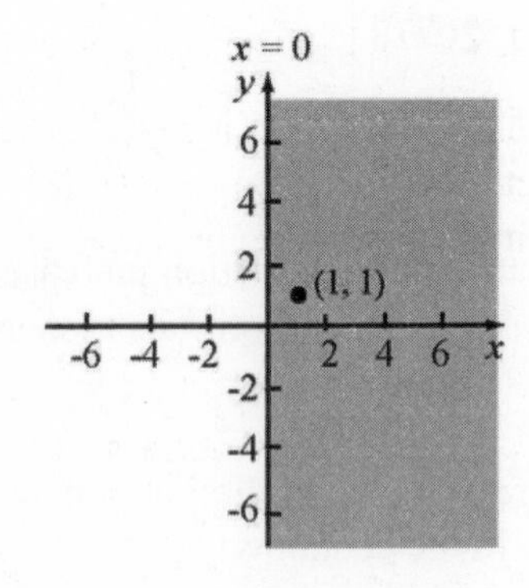

11. $x < 4$

The corresponding linear equation is $x = 4$. We graph a dashed line, since the inequality is strict. The test point (0, 0) satisfies the inequality, so we shade the region containing it.

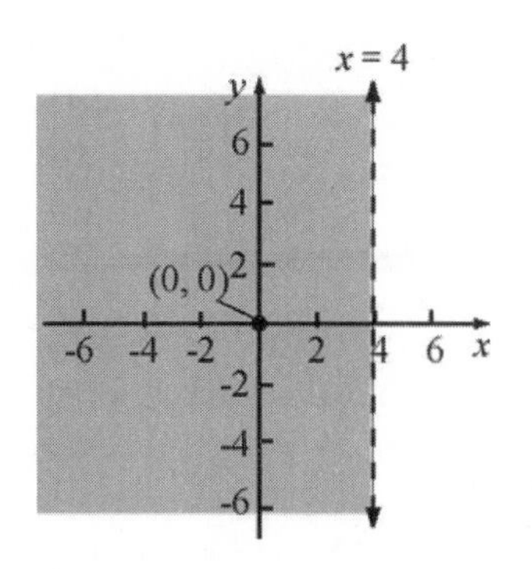

13. $y \geq 1$

The corresponding linear equation is $y = 1$. We graph a solid line, since the inequality is non-strict. The test point (0, 0) does not satisfy the inequality, so we shade the region opposite it.

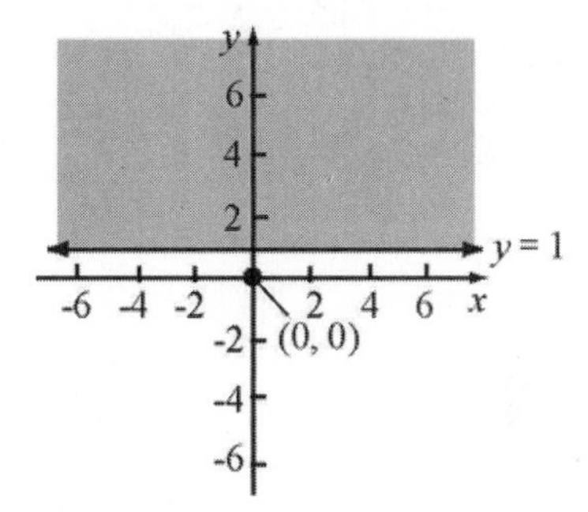

15. The corresponding linear equation is $2x + y = 4$. We graph a solid line, since the inequality is nonstrict. The test point (0, 0) we chose satisfies the inequality, so we shade the region containing it.

$$\begin{aligned} 2x + y &\leq 4 \\ 2 \cdot 0 + 0 &= 0 \\ 0 &< 4 \end{aligned}$$

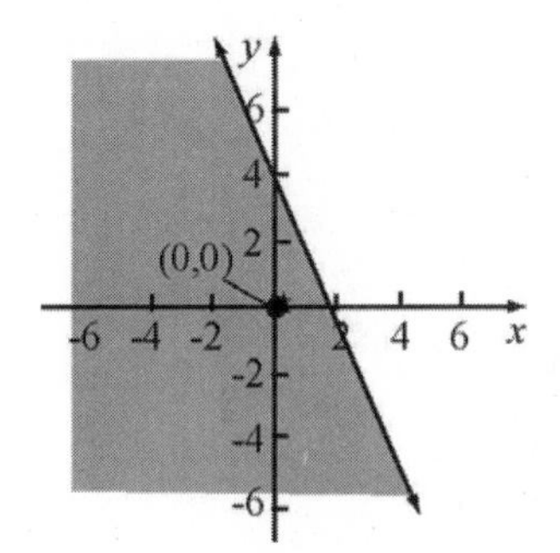

17. The corresponding linear equation is $5x + y = 10$. We graph a solid line, since the inequality is nonstrict. The test point (0, 0) does not satisfy the inequality, so we shade the region opposite it.

$$\begin{aligned} 5x + y &\geq 10 \\ 5 \cdot 0 + 0 &= 0 \\ 0 &< 10 \end{aligned}$$

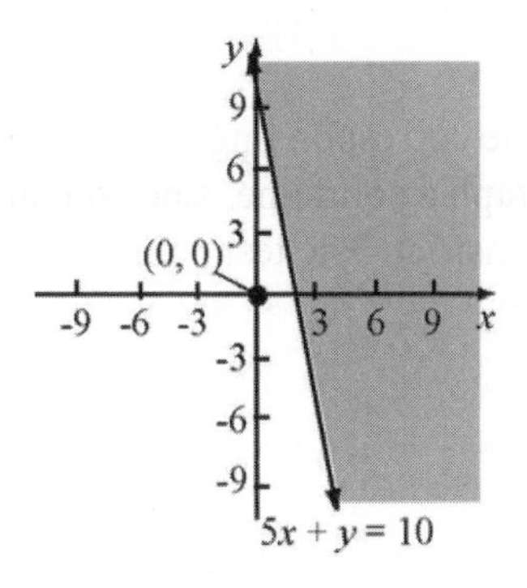

19. The corresponding linear equation is $x + 5y = 5$. We graph a dashed line, since the inequality is strict. The test point (0, 0) satisfies the inequality, so we shade the region containing it.

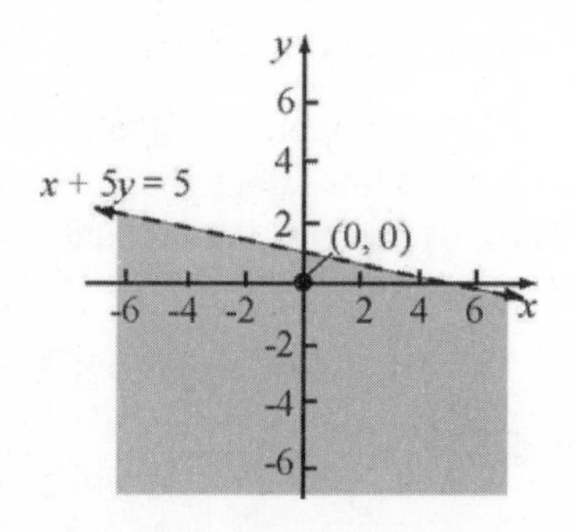

$$\begin{aligned} x+5y&<5 \\ 0+5\cdot 0&=0 \\ 0&<5 \end{aligned}$$

21. To determine which points are part of the graph of the system, check each point to see if it satisfies both of the inequalities:

Point	Inequality 1	Inequality 2	Conclusion
$P_1 = (3, 8)$	Is $3+3\cdot 8 \ge 0$? $27 > 0$	Is $-3\cdot 3+2\cdot 8 \ge 0$? $7 > 0$	Both inequalities are satisfied; (3, 8) is part of the graph.
$P_2 = (12, 9)$	Is $12+3\cdot 9 \ge 0$? $39 > 0$	Is $-3\cdot 12+2\cdot 9 \ge 0$? $-18 < 0$	Since the second inequality is not satisfied, (12, 9) is not part of the graph.
$P_3 = (5, 1)$	Is $5+3\cdot 1 \ge 0$? $8 > 0$	Is $-3\cdot 5+2\cdot 1 \ge 0$? $-13 < 0$	Since the second inequality is not satisfied, (5, 1) is not part of the graph.

23. To determine which points are part of the graph of the system, check each point to see if it satisfies all of the inequalities:

Point	Inequality 1	Inequality	Conclusion
$P_1 = (2, 3)$	Is $3\cdot 2+2\cdot 3 \ge 0$? $12 > 0$	Is $2+3 \le 15$? $5 < 15$	Both inequalities are satisfied; (2, 3) is part of the graph.
$P_2 = (10, 10)$	Is $3\cdot 10+2\cdot 10 \ge 0$? $50 > 0$	Is $10+10 \le 15$? $20 > 15$	The second inequality is not satisfied; (10, 10) is not part of the graph.
$P_3 = (5, 1)$	Is $3\cdot 5+2\cdot 1 \ge 0$? $17 > 0$	Is $5+1 \le 15$? $6 < 15$	Both inequalities are satisfied; (2, 3) is part of the graph.

25. The region that represents the graph of the system is the set of points common to the solutions of each individual inequality. We use the test point (0, 0).

$$\begin{aligned} 5x-4y&\le 8 \\ 5\cdot 0-4\cdot 0&=0 \\ 0&<8 \end{aligned} \qquad \begin{aligned} 2x+5y&\le 23 \\ 2\cdot 0+5\cdot 0&=0 \\ 0&<23 \end{aligned}$$

The test point (0, 0) satisfies both inequalities, so the region (***b***) represents the graph of the system.

27. The region that represents the graph of the system is the set of points common to the solutions of each individual inequality. We use the test point (0, 0).

$$\begin{array}{rr} 2x-3y\ge -3 & 2x+3y\le 16 \\ 2\cdot 0-3\cdot 0=0 & 2\cdot 0+6\cdot 0=0 \\ 0>-3 & 0<16 \end{array}$$

The test point (0, 0) satisfies both inequalities, so the region (***c***) represents the graph of the system.

29. The region that represents the graph of the system is the set of points common to the solutions of each individual inequality. We use the test point (0, 0).

$$\begin{array}{rr} 5x-3y\ge 3 & 2x+6y\ge 30 \\ 5\cdot 0-3\cdot 0=0 & 2\cdot 0+6\cdot 0=0 \\ 0<3 & 0<30 \end{array}$$

The test point (0, 0) satisfies neither inequality, so the region representing each inequality is on the opposite side of the line from (0, 0). This means (***d***) represents the graph of the system.

31. The region that represents the graph of the system is the set of points common to the solutions of each individual inequality. We use the test point (0, 2).

$$\begin{array}{rr} 5x-4y\le 0 & 2x+4y\le 28 \\ 5\cdot 0-5\cdot 2=-10 & 2\cdot 0+4\cdot 2=8 \\ -10<0 & 8<28 \end{array}$$

The test point (0, 2) satisfies both inequalities, so the region (***c***) represents the graph of the system.

33. First graph each inequality separately. Then graph the lines with the points the separate graphs have in common. The solution to the system is the set of all points common to both graphs.

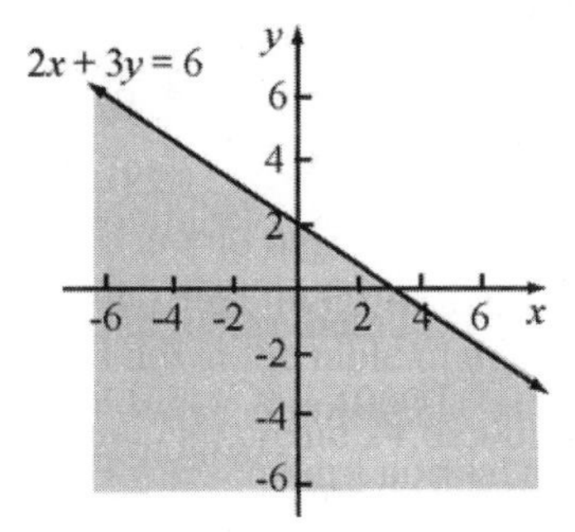

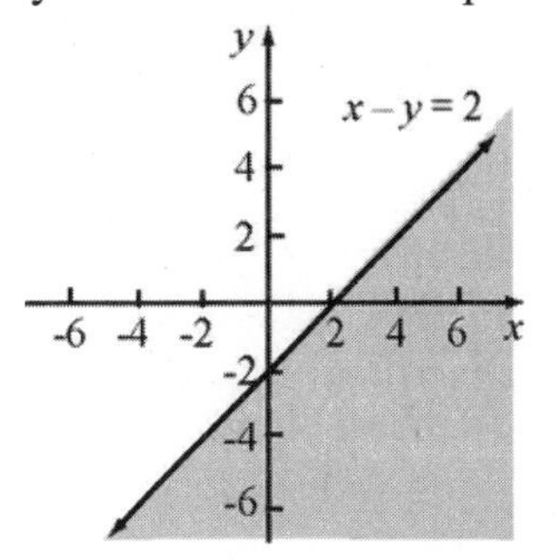

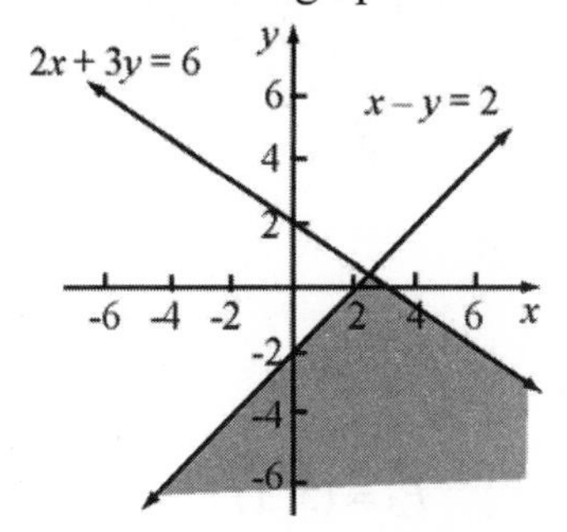

35. First graph each inequality separately. Then graph the lines with the points the separate graphs have in common. The solution to the system is the set of all points common to both graphs.

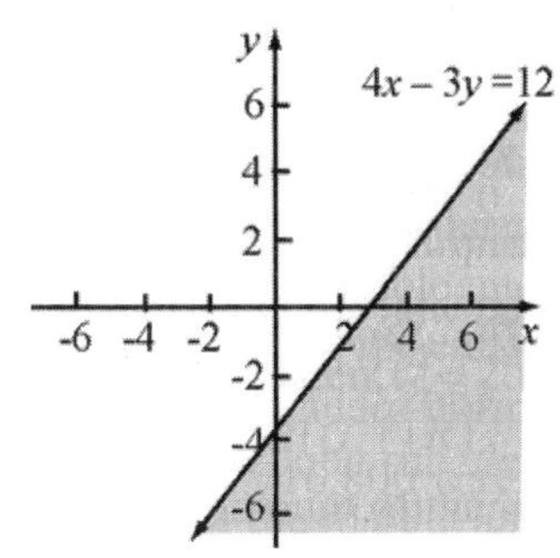

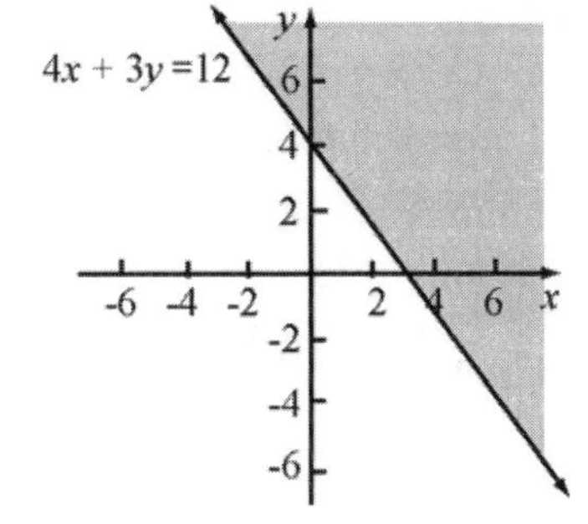

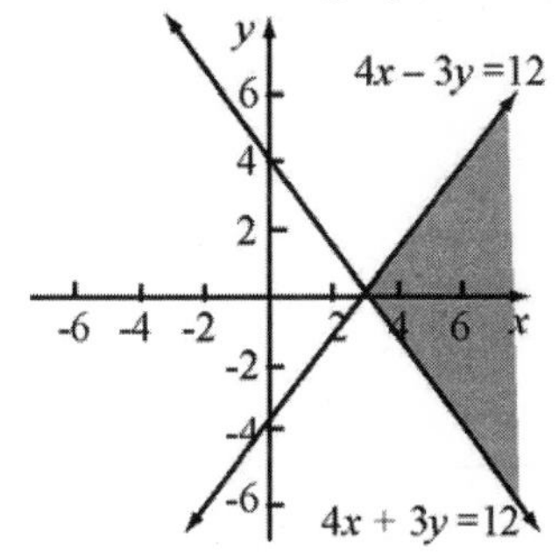

37. First graph each inequality separately. Then graph the lines with the points the separate graphs have in common. The solution to the system is the set of all points common to both graphs.

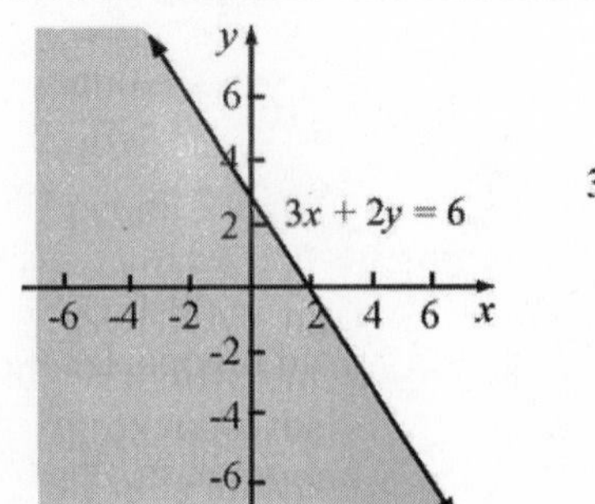

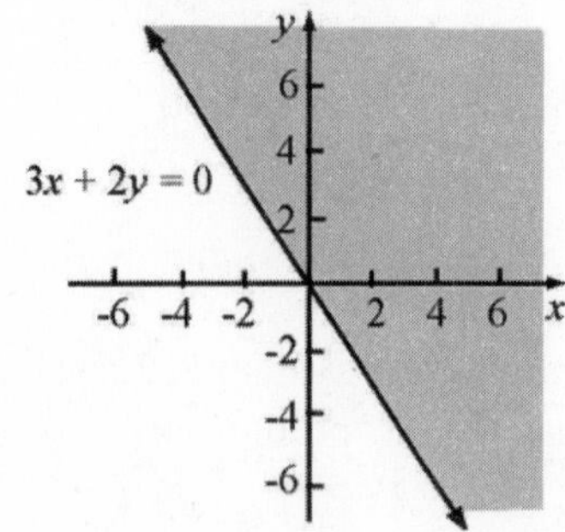

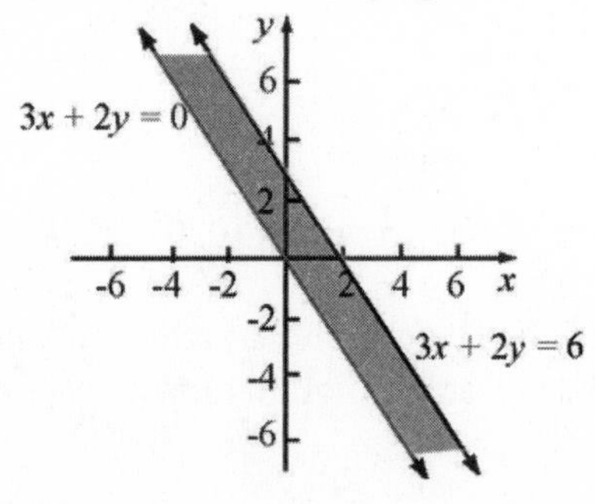

39. First graph each inequality separately. Then graph the lines with the points the separate graphs have in common. The solution to the system is the set of all points common to both graphs.

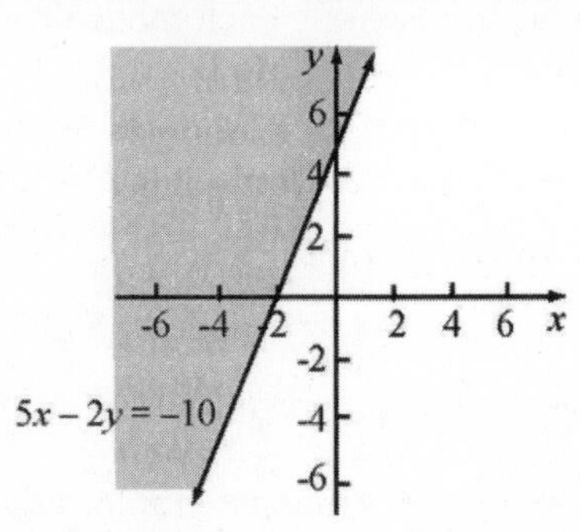

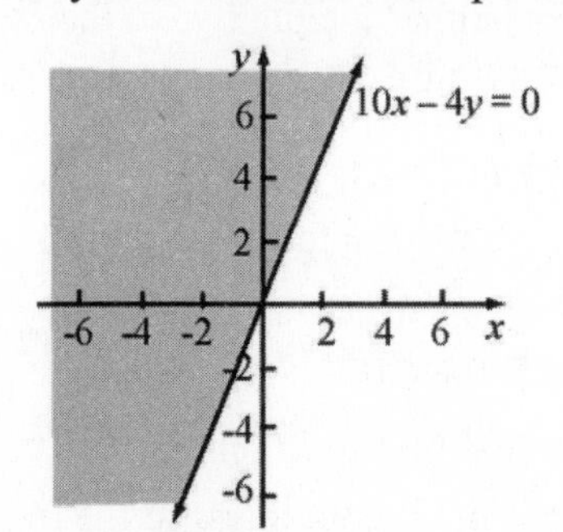

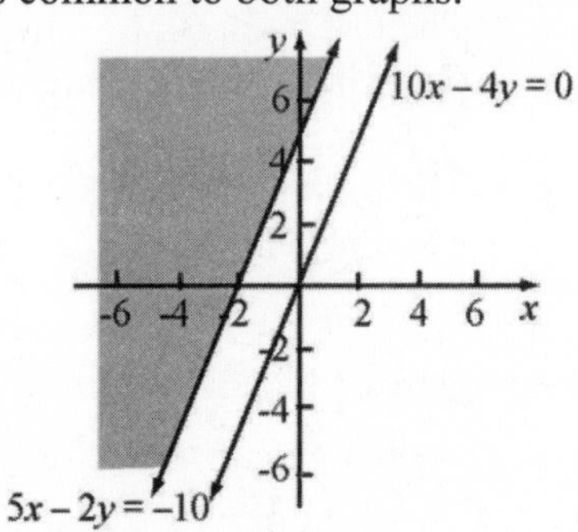

41. (a) $\begin{cases} x+y\geq 2 \\ x\geq 0 \\ y\geq 0 \end{cases}$

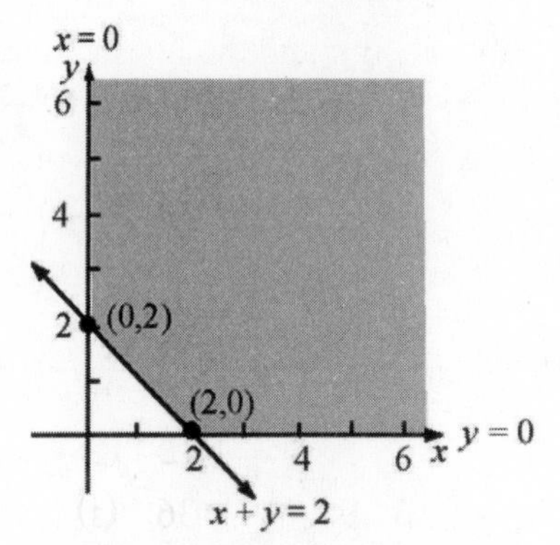

(b) The graph is unbounded.
The corner points are:
(0, 2)
(2, 0)

43. (a) $\begin{cases} x+y\geq 2 \\ 2x+3y\leq 6 \\ x\geq 0 \\ y\geq 0 \end{cases}$

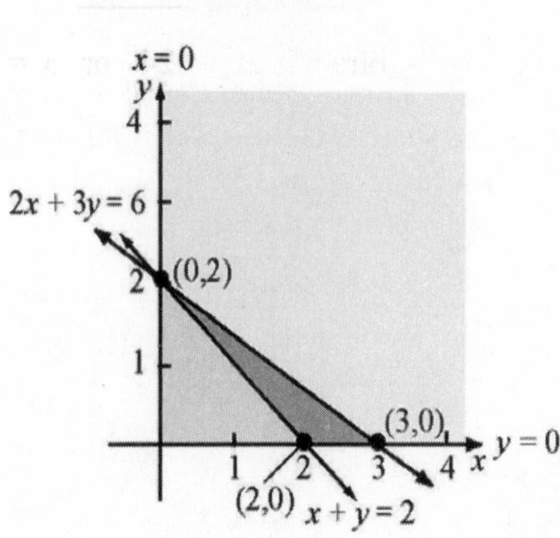

(b) The graph is bounded.
The corner points are:
(0, 2)
(2, 0)
(3, 0)

45. (a)
$$\begin{cases} x+y\geq 2 \\ x+y\leq 8 \\ 2x+y\leq 10 \\ x\geq 0 \\ y\geq 0 \end{cases}$$

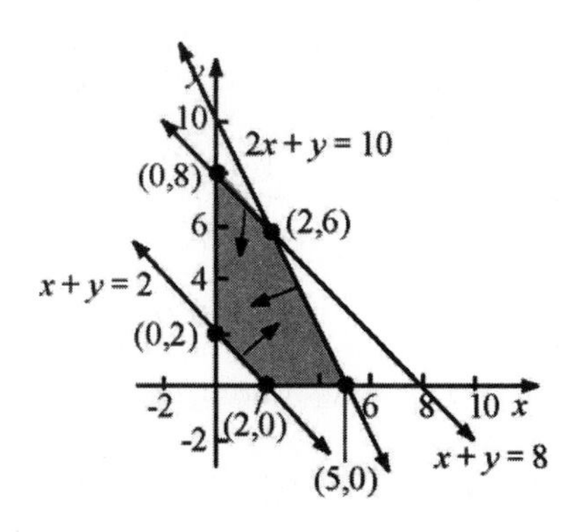

We find the point (2, 6) by solving the system formed by the 2nd and 3rd equations.
$$\begin{cases} x+y=8 & (2) \\ 2x+y=10 & (3) \end{cases}$$
Solving (2) for y, we get
$$y=8-x \quad (2)$$
Back-substituting y into (3), gives
$$2x+(8-x)=10$$
$$x=2$$
Then using x in (2), we solve for y.
$$y=8-2=6$$

(b) The graph is bounded. The corner points are: (0, 8), (0, 2), (2, 0), (5, 0), (2, 6).

47. (a)
$$\begin{cases} x+y\geq 2 \\ 2x+3y\leq 12 \\ 3x+y\leq 12 \\ x\geq 0 \\ y\geq 0 \end{cases}$$

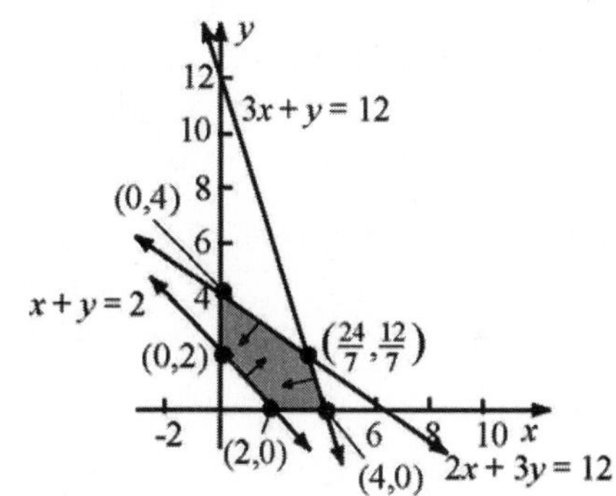

(b) The graph is bounded. To find the point $\left(\frac{24}{7}, \frac{12}{4}\right)$ solve the system:

$$\begin{cases} 2x+3y=12 & (2) \\ 3x+y=12 & (3) \end{cases} \quad \text{multiply by 3} \quad \begin{array}{ll} 2x+3y=12 & (2) \\ \underline{9x+3y=36} & (3) \end{array}$$

subtract $-7x=-24$ or $x=\dfrac{24}{7}$ and $y=12-3x=\dfrac{12}{7}$

The corner points are: (0, 4), (0, 2), (2, 0), (4, 0) and $\left(\frac{24}{7}, \frac{12}{7}\right)$.

49. (a)

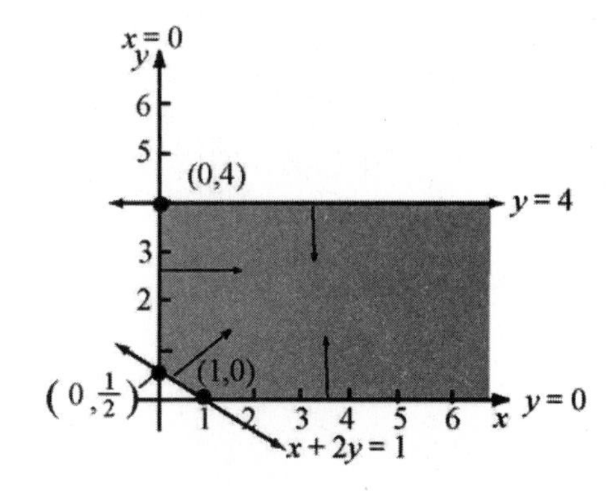

(b) The graph is unbounded. The corner points are: (1, 0), $\left(0, \frac{1}{2}\right)$, and (0, 4).

51. (a)

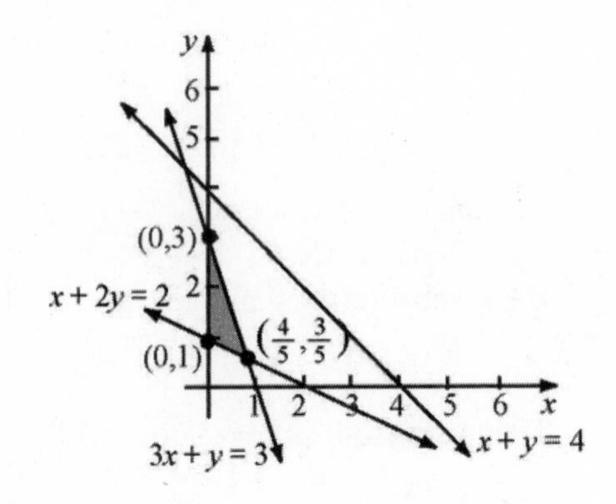

(b) The graph is bounded. To find the corner point $\left(\frac{4}{5}, \frac{3}{5}\right)$, we find the intersection of the lines

$$x + 2y = 2 \text{ and } 3x + y = 3$$

$$\begin{cases} x + 2y = 2 & (1) \\ 3x + y = 3 & (3) \end{cases}$$

Using the substitution method, we solve (1) for x and substitute the result in (3).

$$x = 2 - 2y \quad (1)$$

$$3(2 - 2y) + y = 3 \quad (3)$$

$$-5y = -3 \text{ or } y = \frac{3}{5}$$

Back-substituting into (1), we find

$$x = 2 - 2\left(\frac{3}{5}\right) = 2 - \frac{6}{5} = \frac{4}{5}$$

The corner points are (0, 3), (0, 1), and $\left(\frac{4}{5}, \frac{3}{5}\right)$.

53. We let x = the number of pound packages of low-grade mix, and y = the number of packages of high-grade mix to be prepared. There are 60 pounds (960 ounces) of almonds and 90 pounds (1440 ounces) of peanuts available.

(a) The system of inequalities representing the system is:

$$\begin{cases} \frac{1}{4}x + \frac{1}{2}y \le 60 & (1) \\ \frac{3}{4}x + \frac{1}{2}y \le 90 & (2) \\ x \ge 0 & (3) \\ y \ge 0 & (4) \end{cases} \quad \text{or} \quad \begin{cases} x + 2y \le 240 & (1) \\ 3x + 2y \le 360 & (2) \\ x \ge 0 & (3) \\ y \ge 0 & (4) \end{cases}$$

(b)

x = 0
y
$\frac{3}{4}x + \frac{1}{2}y = 90$
180
(0, 120)
(60, 90)
x
y = 0
(0, 0)
(120, 0)
240
$\frac{1}{4}x + \frac{1}{2}y = 60$

From the graph we see that three of the corner points are (0, 0), (120, 0), and (0, 120). The fourth corner (60, 90), is found by solving the system of equations

$$\begin{cases} x + 2y = 240 & (1) \\ 3x + 2y = 360 & (2) \end{cases}$$

Using elimination, we subtract (1) from (2) and get $2x = 120$ or $x = 60$. Back-substituting into (1) we find

$$60 + 2y = 240 \quad (1)$$

$$2y = 180$$

$$y = 90$$

55. (a) First we note that meaningful values for x and y are nonnegative values. Next we note that there are only limited numbers of grinding and finishing hours. There are:

$80 = 2 \cdot 40$ grinding hours available, and $120 = 3 \cdot 40$ finishing hours available

Using these constraints, the system is:

$$\begin{cases} 3x+2y \leq 80 & (1) \\ 4x+3y \leq 120 & (2) \\ x \geq 0 & (3) \\ y \geq 0 & (4) \end{cases}$$

(b)

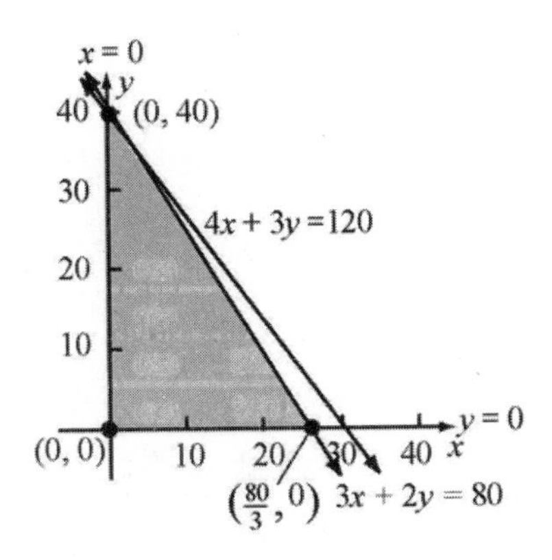

The corner points are:

$(0, 40)$, $(0, 0)$, and $\left(\frac{80}{3}, 0\right)$.

57. (a) First we note that meaningful values of x and y are nonnegative. Then we see that there is at most \$25,000 to be invested. Using these constraints, the system is

$$\begin{cases} x+y \leq 25{,}000 \\ x \geq 15{,}000 \\ y \leq 10{,}000 \\ x \geq 0 \\ y \geq 0 \end{cases}$$

(b)

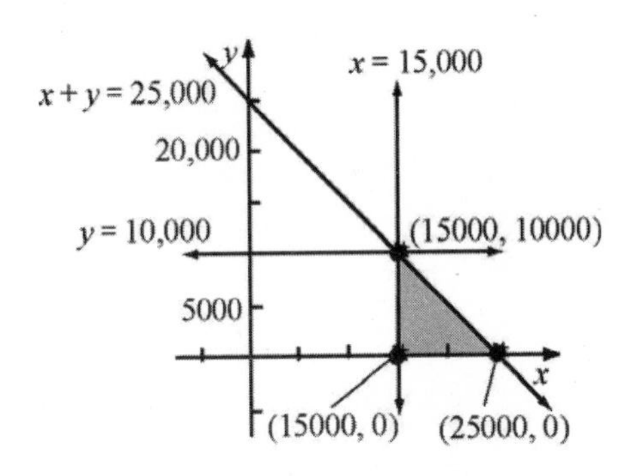

Corner points: $(15000, 10000)$, $(15000, 0)$, and $(25000, 0)$

(c) The x-value of each corner point represents the amount in dollars to be invested in Treasury Bills, the y-value is the amount in dollars to be invested in corporate bonds.

59. (a) Let x = the units of grain 1 used, and

y = the units of grain 2 used.

$$\begin{cases} x + 2y \geq 5 \\ 5x + y \geq 16 \\ x \geq 0 \\ y \geq 0 \end{cases}$$

(b) Corner Points: (5, 0), (3, 1), and (0, 16).

(b)

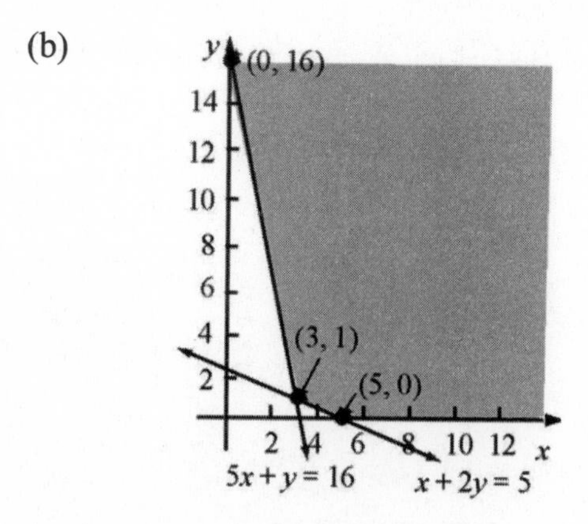

61. (a) Let x = amount of food A to be eaten,
Let y = amount of food B to be eaten.

$$\begin{cases} 5x + 4y \geq 85 \\ 3x + 3y \geq 70 \\ 2x + 3y \geq 50 \\ x \geq 0 \\ y \geq 0 \end{cases}$$

(b) Corner Points: $\left(0, \frac{70}{3}\right), \left(20, \frac{10}{3}\right)$, (25, 0).

(b)

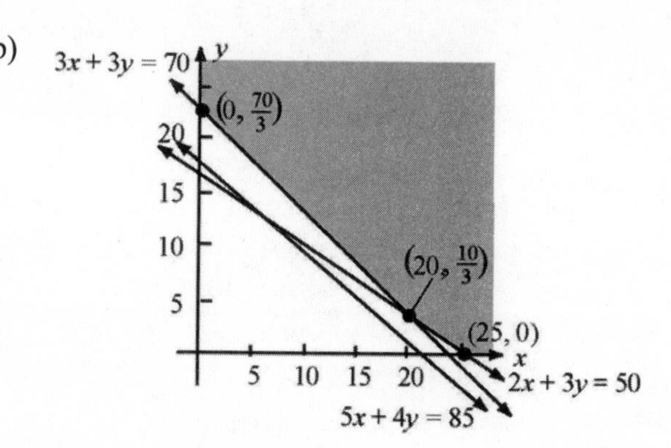

63. (a)

$$\begin{cases} x + y \leq 30{,}000 & (1) \\ x \geq 3{,}000 & (2) \\ y \geq 5{,}000 & (3) \\ 0.0435x + 0.0505y \geq 500 & (4) \end{cases}$$

Corner points:
(5690, 5000), (25000, 5000),
(3000, 7317), (3000, 27000)

(b)

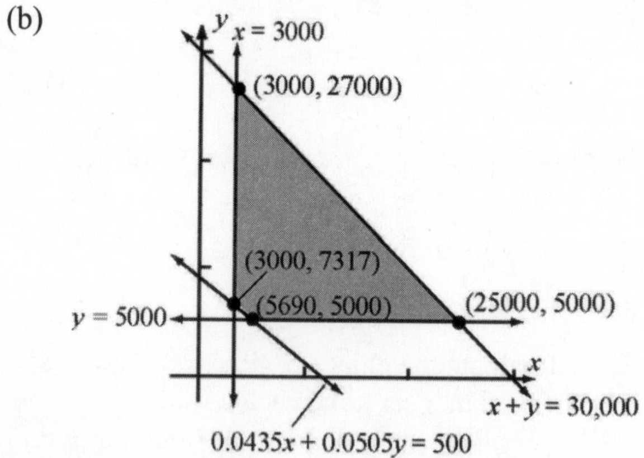

(c) Each corner point represents an amount x deposited in INGdirect and an amount y deposited in HSBCdirect that satisfy the conditions given.

65. (a) $\begin{cases} x+y \le 200{,}000 & (1) \\ 3x-y \le 0 & (2) \\ x \ge 20{,}000 & (3) \\ y \ge 60{,}000 & (4) \end{cases}$

(b)

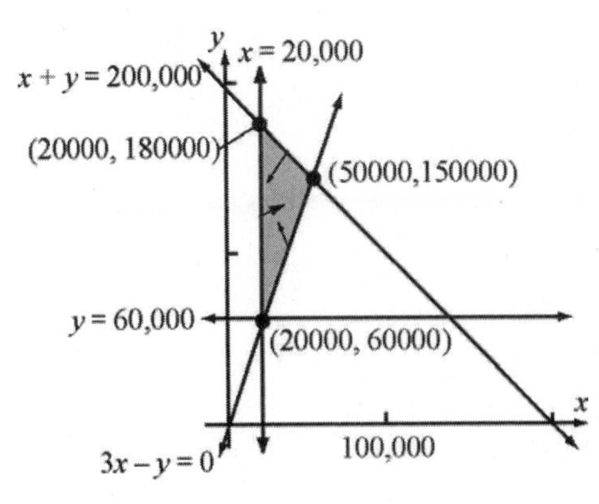

Corner Points: Two of the corner points lie along the line $x = 20{,}000$. So they have values (20000, 60000) and (20000, 180000). The third corner point is found by solving the system of equations formed by (1) and (2).

$$\begin{aligned} x + y &= 200{,}000 \\ 3x - y &= 0 \\ \hline \text{Add.} \quad 4x \quad &= 200{,}000 \\ x &= 50{,}000 \end{aligned}$$

Back-substitute: $y = 3 \cdot 50{,}000 = 150{,}000$

The third corner point is (50000, 150000).

(c) The projected annual rate of return is found by evaluating the expression $0.1333x + 0.3889y$ at each corner point.

$(20000, 60000)$: $(0.1333) \cdot (20{,}000) + (0.3889) \cdot (60{,}000) = \$26{,}000$

$(20000, 180000)$: $(0.1333) \cdot (20{,}000) + (0.3889) \cdot (180{,}000) = \$72{,}668$

$(50000, 150000)$: $(0.1333) \cdot (50{,}000) + (0.3889) \cdot (150{,}000) = \$65{,}000$

3.2 A Geometric Approach to Linear Programming Problems

1. Objective function

3. False

5. Evaluate the objective function at each corner point; then choose the maximum and minimum values.

$z = 2x + 3y$

Point: (2, 2)	Point: (8, 1)	Point: (2, 7)	Point: (7, 8)
$z = 2(2) + 3(2)$	$z = 2(8) + 3(1)$	$z = 2(2) + 3(7)$	$z = 2(7) + 3(8)$
$z = 10$	$z = 19$	$z = 25$	$z = 38$

Maximum: $z = 38$; Minimum: $z = 10$

7. Evaluate the objective function at each corner point; then choose the maximum and minimum values.

$z = x + y$

Point: (2, 2)	Point: (8, 1)	Point: (2, 7)	Point: (7, 8)
$z = 2 + 2$	$z = 8 + 1$	$z = 2 + 7$	$z = 7 + 8$
$z = 4$	$z = 9$	$z = 9$	$z = 15$

Maximum: $z = 15$; Minimum: $z = 4$

9. Evaluate the objective function at each corner point; then choose the maximum and minimum values.

$z = x + 6y$

Point: (2, 2)	Point: (8, 1)	Point: (2, 7)	Point: (7, 8)
$z = 2 + 6(2)$	$z = 8 + 6(1)$	$z = 2 + 6(7)$	$z = 7 + 6(8)$
$z = 14$	$z = 14$	$z = 44$	$z = 55$

Maximum: $z = 55$; Minimum: $z = 14$, all feasible points on line containing the points (2, 2) and (8, 1) will give the minimum value of z.

11. Evaluate the objective function at each corner point; then choose the maximum and minimum values.

$z = 3x + 4y$

Point: (2, 2)	Point: (8, 1)	Point: (2, 7)	Point: (7, 8)
$z = 3(2) + 4(2)$	$z = 3(8) + 4(1)$	$z = 3(2) + 4(7)$	$z = 3(7) + 4(8)$
$z = 14$	$z = 28$	$z = 34$	$z = 53$

Maximum: $z = 53$; Minimum: $z = 14$

13. Evaluate the objective function at each corner point; then choose the maximum and minimum values.

$z = 10x + y$

Point: (2, 2)	Point: (8, 1)	Point: (2, 7)	Point: (7, 8)
$z = 10(2) + 2$	$z = 10(8) + 1$	$z = 10(2) + 7$	$z = 10(7) + 8$
$z = 22$	$z = 81$	$z = 27$	$z = 78$

Maximum: $z = 81$; Minimum: $z = 22$

15. The constraints form the unbounded region that is shaded in the graph.

The corner points are: (0, 4), (3, 0), and (13, 0).

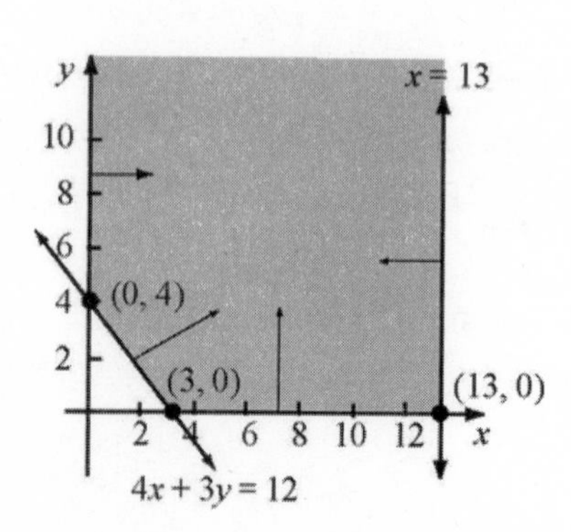

17. The constraints form the region that is shaded in the graph.

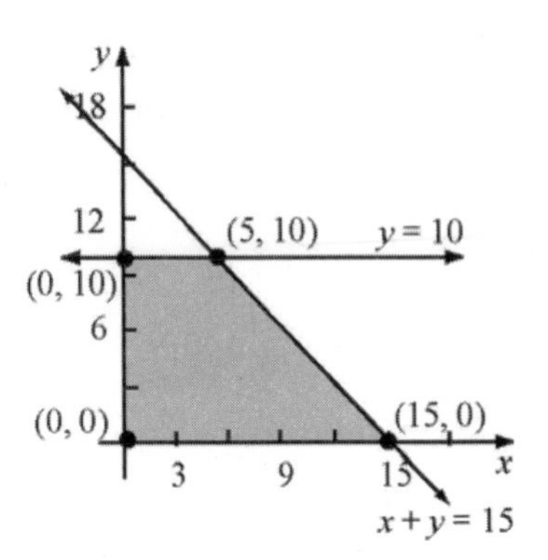

Three of the corner points, (0, 0), (15, 0), and (0, 10) are easy to identify. The fourth corner, (5, 10) is found by solving

$$\begin{cases} y = 10 \\ x + y = 15 \end{cases}$$

Substituting $y = 10$ into the second equation, we find $x = 5$.

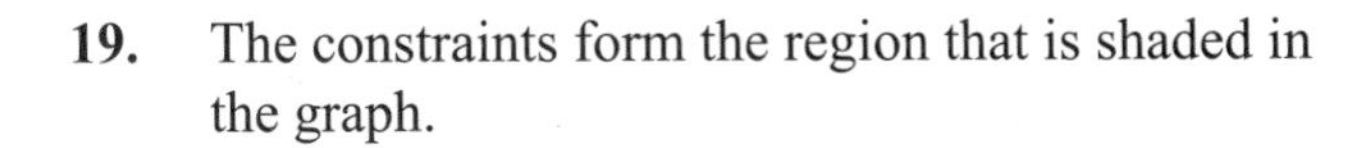

19. The constraints form the region that is shaded in the graph.

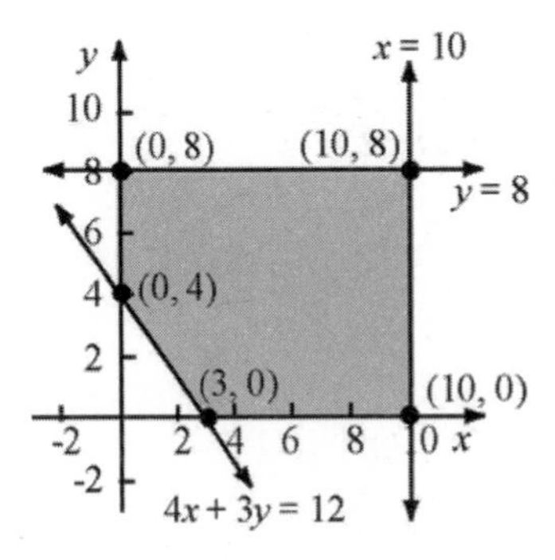

Four of the corner points (3, 0), (10, 0), (0, 4), and (0, 8) are easy to identify. The fifth corner, (10, 8) is found at the intersection of the lines $x = 10$ and $y = 8$.

21. To maximize $z = 5x + 7y$, graph the system of linear inequalities, shade the set of feasible points, and locate the corner points. Then evaluate the objective function at each corner point.

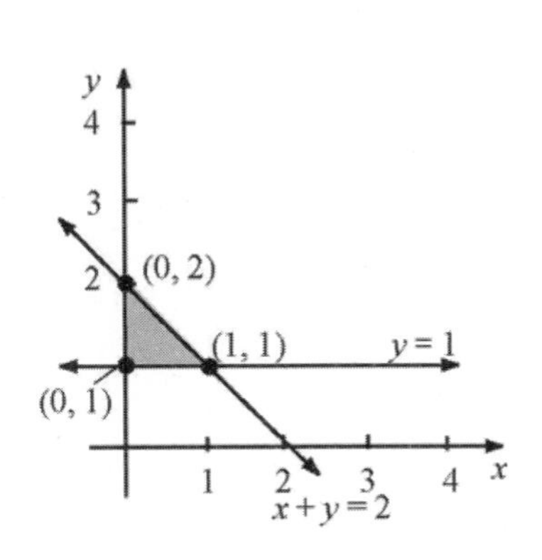

The corner points are (0, 1), (0, 2) and (1, 1). The third point was obtained by solving

$$\begin{cases} x + y = 2 \\ y = 1 \end{cases}$$

Substituting $y = 1$ into the first equation we find $x + 1 = 2$ or $x = 1$.

Corner Point (x, y)	Value of the Objective Function $z = 5x + 7y$
(0, 1)	$z = 5(0) + 7(1) = 7$
(0, 2)	$z = 5(0) + 7(2) = 14$
(1, 1)	$z = 5(1) + 7(1) = 12$

The maximum value of z is 14, and it occurs at the point (0, 2).

23. To maximize $z = 5x + 7y$, graph the system of linear inequalities, shade the set of feasible points, and locate the corner points. Then evaluate the objective function at each corner point.

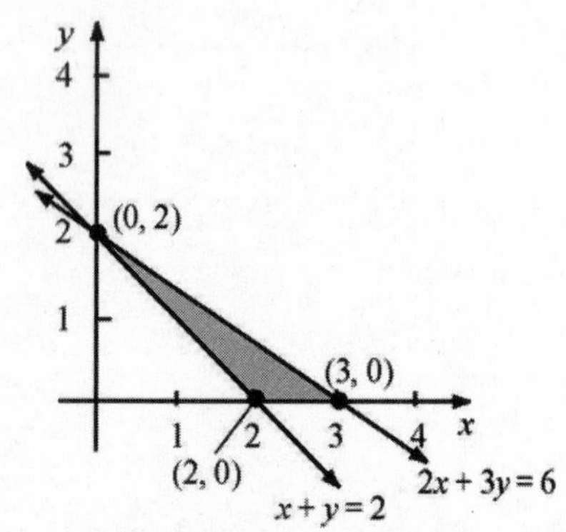

The corner points are (2, 0), (3, 0), and (0, 2).

Corner Point (x, y)	Value of the Objective Function $z = 5x + 7y$
(2, 0)	$z = 5(2) + 7(0) = 10$
(3, 0)	$z = 5(3) + 7(0) = 15$
(0, 2)	$z = 5(0) + 7(2) = 14$

The maximum value of z is 15, and it occurs at the point (3, 0).

25. To maximize $z = 5x + 7y$, graph the system of linear inequalities, shade the set of feasible points, and locate the corner points. Then evaluate the objective function at each corner point.

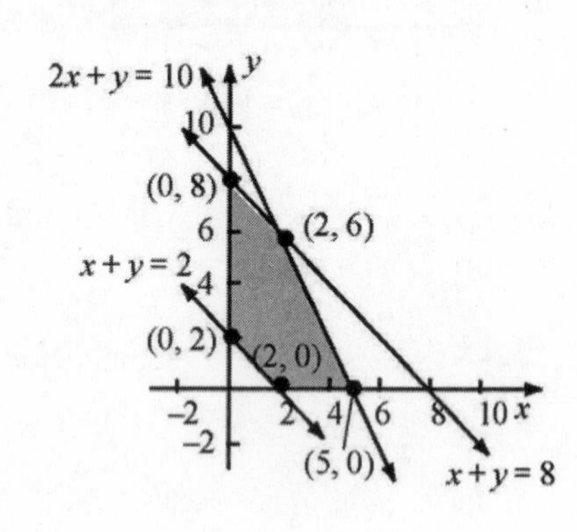

The corner points are (2, 0), (5, 0), (0, 2), (0, 8), and (2,6). The fifth point was obtained by solving

$$\begin{cases} x + y = 8 \\ 2x + y = 10 \end{cases}$$

Subtracting the first equation from the second equation, we find $x = 2$. Back-substituting x into equation 1 we find $y = 6$.

Corner Point (x, y)	Value of the Objective Function $z = 5x + 7y$
(2, 0)	$z = 5(2) + 7(0) = 10$
(5, 0)	$z = 5(5) + 7(0) = 25$
(0, 2)	$z = 5(0) + 7(2) = 14$
(0, 8)	$z = 5(0) + 7(8) = 56$
(2, 6)	$z = 5(2) + 7(6) = 52$

The maximum value of z is 56, and it occurs at the point (0,8).

27. To maximize $z = 5x + 7y$, graph the system of linear inequalities, shade the set of feasible points, and locate the corner points. Then evaluate the objective function at each corner point.

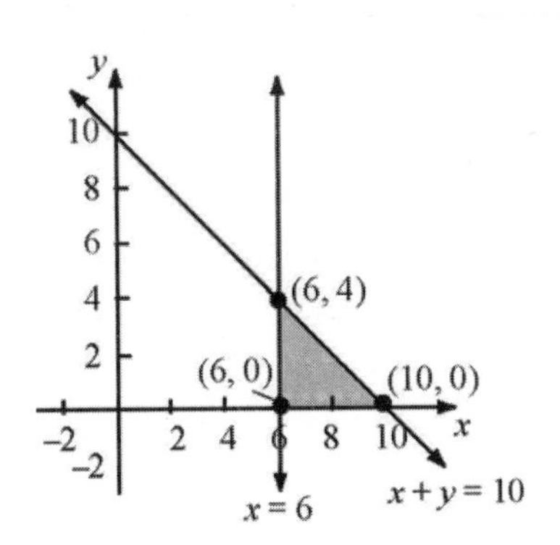

The corner points are (6, 0), (10, 0), and (6, 4). The third point was obtained by solving

$$\begin{cases} x = 6 \\ x + y = 10 \end{cases}$$

Substituting 6 for x in equation 2, we find $y = 4$.

Corner Point (x, y)	Value of the Objective Function $z = 5x + 7y$
(6, 0)	$z = 5(6) + 7(0) = 30$
(10, 0)	$z = 5(10) + 7(0) = 50$
(6, 4)	$z = 5(6) + 7(4) = 58$

The maximum value of z is 58, and it occurs at the point (6, 4).

29. To minimize $z = 2x + 3y$, graph the system of linear inequalities, shade the set of feasible points, and locate the corner points. Then evaluate the objective function at each corner point.

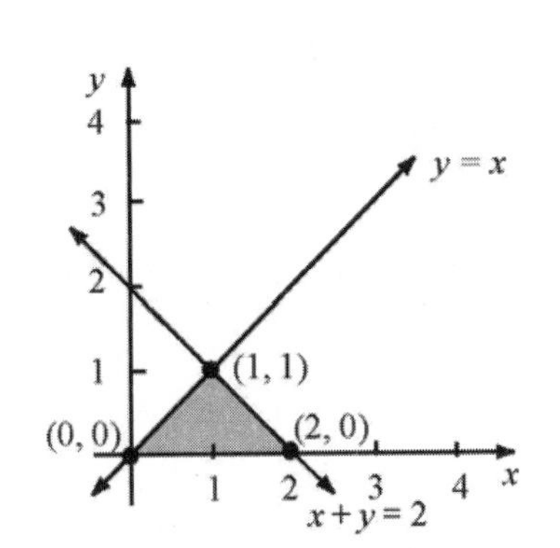

The corner points are (0, 0), (2, 0), and (1, 1). The third point was obtained by solving

$$\begin{cases} x + y = 2 \\ y = x \end{cases}$$

Substituting x for y in equation 1, we find $2x = 2$, or $x = 1$. Back-substituting in equation 2, gives $y = 1$.

Corner Point (x, y)	Value of the Objective Function $z = 2x + 3y$
(0, 0)	$z = 2(0) + 3(0) = 0$
(2, 0)	$z = 2(2) + 3(0) = 4$
(1, 1)	$z = 2(1) + 3(1) = 5$

The minimum value of z is 0, and it occurs at the point (0, 0).

31. To minimize $z = 2x + 3y$, graph the system of linear inequalities, shade the set of feasible points, and locate the corner points. Then evaluate the objective function at each corner point.

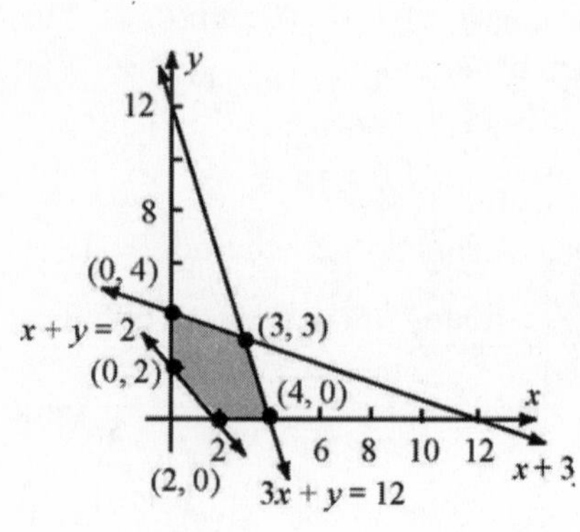

The corner points are (2, 0), (4, 0), (0, 2), (0, 4), and (3, 3). The fifth point was obtained by solving

$$\begin{cases} 3x + y = 12 \\ x + 3y = 12 \end{cases}$$

Subtracting the first equation from three times the second equation gives, $8y = 24$ or $y = 3$. Back-substituting 3 for y in the first equation gives $3x + 3 = 12$ or $3x = 9$ or $x = 3$.

Corner Point (x, y)	Value of the Objective Function $z = 2x + 3y$
(2, 0)	$z = 2(2) + 3(0) = 4$
(4, 0)	$z = 2(4) + 3(0) = 8$
(0, 2)	$z = 2(0) + 3(2) = 6$
(0, 4)	$z = 2(0) + 3(4) = 12$
(3, 3)	$z = 2(3) + 3(3) = 15$

The minimum value of z is 4, and it occurs at the point (2, 0).

33. To minimize $z = 2x + 3y$, graph the system of linear inequalities, shade the set of feasible points, and locate the corner points. Then evaluate the objective function at each corner point.

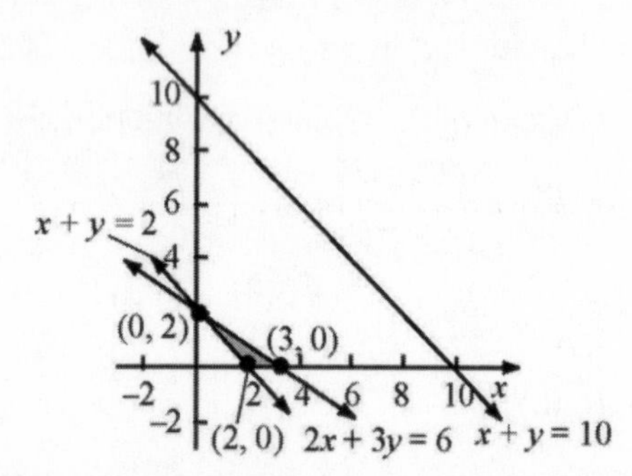

The corner points are (2, 0), (3, 0), and (0, 2).

Corner Point (x, y)	Value of the Objective Function $z = 2x + 3y$
(2, 0)	$z = 2(2) + 3(0) = 4$
(3, 0)	$z = 2(3) + 3(0) = 6$
(0, 2)	$z = 2(0) + 3(2) = 6$

The minimum value of z is 4, and it occurs at the point (2, 0).

35. To minimize $z = 2x + 3y$, graph the system of linear inequalities, shade the set of feasible points, and locate the corner points. Then evaluate the objective function at each corner point.

The corner points are

$$(0, 5), (2, 4), \left(0, \frac{1}{2}\right), \text{ and } \left(\frac{1}{5}, \frac{2}{5}\right).$$

The point (2,4) is found by solving

$$\begin{cases} x + 2y = 10 \\ y = 2x \end{cases}$$

Substituting $2x$ for y in the first equation gives $5x = 10$ or $x = 2$. Back-substituting 2 for x in the second equation gives $y = 4$.

The point $\left(\frac{1}{5}, \frac{2}{5}\right)$ is found by solving

$$\begin{cases} x + 2y = 1 \\ y = 2x \end{cases}$$

Substituting $2x$ for y in the first equation gives $5x = 1$ or $x = \frac{1}{5}$. Back-substituting $\frac{1}{5}$ for x in the second equation gives $y = \frac{2}{5}$.

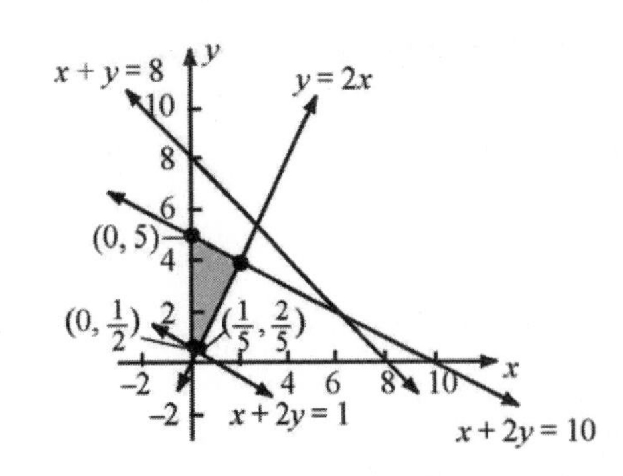

Corner Point (x, y)	Value of the Objective Function $z = 2x + 3y$
(0, 5)	$z = 2(0) + 3(5) = 15$
(2, 4)	$z = 2(2) + 3(4) = 16$
$\left(0, \frac{1}{2}\right)$	$z = 2(0) + 3\left(\frac{1}{2}\right) = \frac{3}{2}$
$\left(\frac{1}{5}, \frac{2}{5}\right)$	$z = 2\left(\frac{1}{5}\right) + 3\left(\frac{2}{5}\right) = \frac{8}{5}$

The minimum value of z is $\frac{3}{2}$, and it occurs at the point $\left(0, \frac{1}{2}\right)$.

37.

Corner Point (x, y)	Value of Objective Function $z = x + y$
(10, 0)	$z = 10 + 0 = 10$
(0, 10)	$z = 0 + 10 = 10$
$\left(\frac{10}{3}, \frac{10}{3}\right)$	$z = \frac{10}{3} + \frac{10}{3} = \frac{20}{3}$

The maximum value of z is 10 at any point on the line $x + y = 10$ between (10, 0) and (0, 10).

The minimum value of z is $\frac{20}{3}$ at $\left(\frac{10}{3}, \frac{10}{3}\right)$.

39.

Corner Point (x, y)	Value of Objective Function $z = 5x + 2y$
(10, 0)	$z = 5(10) + 2(0) = 50$
(0, 10)	$z = 5(0) + 2(10) = 20$
$\left(\frac{10}{3}, \frac{10}{3}\right)$	$z = 5\left(\frac{10}{3}\right) + 2\left(\frac{10}{3}\right) = \frac{70}{3}$

The maximum value of z is 50 at (10, 0). The minimum value of z is 20 at (0, 10).

41.

Corner Point (x, y)	Value of Objective Function $z = 3x + 4y$
$(10, 0)$	$z = 3(10) + 4(0) = 30$
$(0, 10)$	$z = 3(0) + 4(10) = 40$
$\left(\frac{10}{3}, \frac{10}{3}\right)$	$z = 3\left(\frac{10}{3}\right) + 4\left(\frac{10}{3}\right) = \frac{70}{3}$

The maximum value of z is 40 at (0, 10). The minimum value of z is $\frac{70}{3}$ at $\left(\frac{10}{3}, \frac{10}{3}\right)$.

43.

Corner Point (x, y)	Value of Objective Function $z = 10x + y$
$(10, 0)$	$z = 10(10) + 0 = 100$
$(0, 10)$	$z = 10(0) + 10 = 10$
$\left(\frac{10}{3}, \frac{10}{3}\right)$	$z = 10\left(\frac{10}{3}\right) + \frac{10}{3} = \frac{110}{3}$

The maximum value of z is 100 at (10, 0). The minimum value of z is 10 at (0, 10).

45. To find the maximum and minimum values of z, we graph the constraints, shade the set of feasible points, find the corner points, and evaluate the objective function, z, at each corner point.

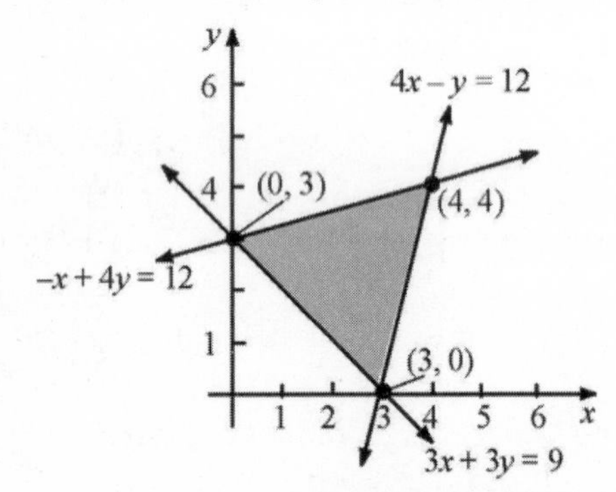

The corner points are (3, 0), (0, 3), and (4, 4). We find (4, 4) by solving $\begin{cases} -x + 4y = 12 & (1) \\ 4x - \ \ y = 12 & (2) \end{cases}$.

Multiplying (1) by 4 and adding it to (2) gives $15y = 60$ or $y = 4$. Back-substituting into (2) gives $4x = 16$ or $x = 4$.

Corner Point	$z = 18x + 30y$
$(3, 0)$	$z = 18(3) + 30(0) = 54$
$(0, 3)$	$z = 18(0) + 30(3) = 90$
$(4, 4)$	$z = 18(4) + 30(4) = 192$

The maximum value of $z = 192$ at the point (4, 4).
The minimum value of $z = 54$ at the point (3, 0).

47. To find the maximum and minimum values of z, we will graph the constraints, shade the set of feasible points, find the corner points, and evaluate the objective function, z, at each corner point.

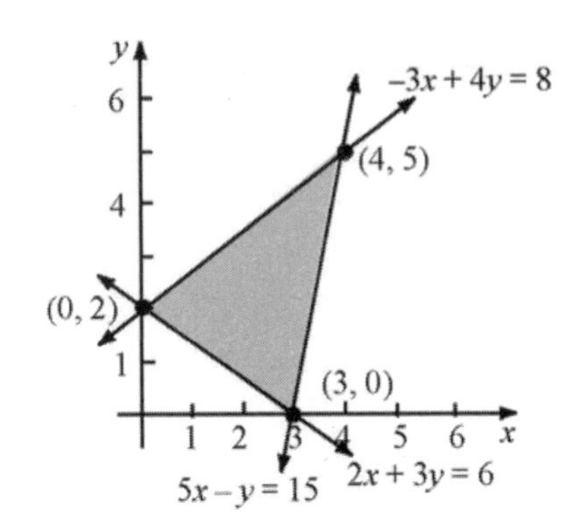

The corner points are (3, 0), (0, 2), and (4, 5). We find (4, 5) by solving

$$\begin{cases} -3x+4y=8 & (2) \\ 5x-y=15 & (3) \end{cases}.$$

Multiplying (3) by 4 and adding it to (2) gives $17x=68$ or $x=4$. Back-substituting into (3) gives $y=5$.

Corner Point	$z=7x+6y$
(3, 0)	$z=7(3)+6(0)=21$
(0, 2)	$z=7(0)+6(2)=12$
(4, 5)	$z=7(4)+6(5)=58$

The maximum value of $z=58$ at the point (4, 5).
The minimum value of $z=12$ at the point (0, 2).

49. To find the maximum value of z, we will graph the constraints, shade the set of feasible points, find the corner points, and evaluate the objective function, z, at each corner point.

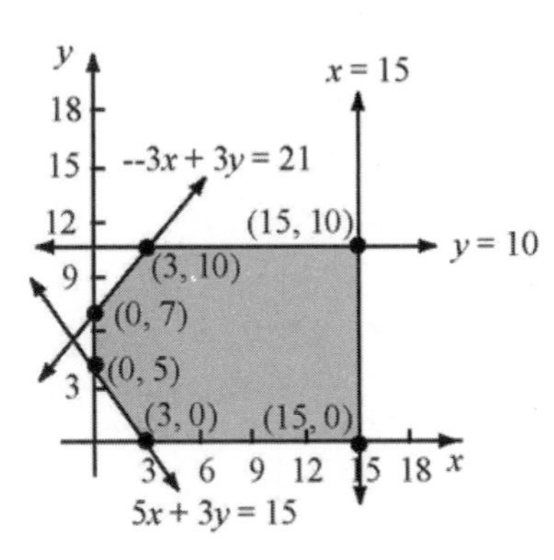

The corner points are (3, 0), (15, 0) (0, 5), (0, 7), (15, 10), and (3, 10).

We find (3, 10) by evaluating $-3x+3y=21$ when $y=10$.

Corner Point	$z=-20x+30y$
(3, 0)	$z=-20(3)+30(0)=-60$
(15, 0)	$z=-20(15)+30(0)=-300$
(0, 5)	$z=-20(0)+30(5)=150$
(0, 7)	$z=-20(0)+30(7)=210$
(15, 10)	$z=-20(15)+30(10)=0$
(3, 10)	$z=-20(3)+30(10)=240$

The maximum value of $z=240$ at (3, 10).

51. To find the maximum value of z, we will graph the constraints, shade the set of feasible points, find the corner points, and evaluate the objective function, z, at each corner point.

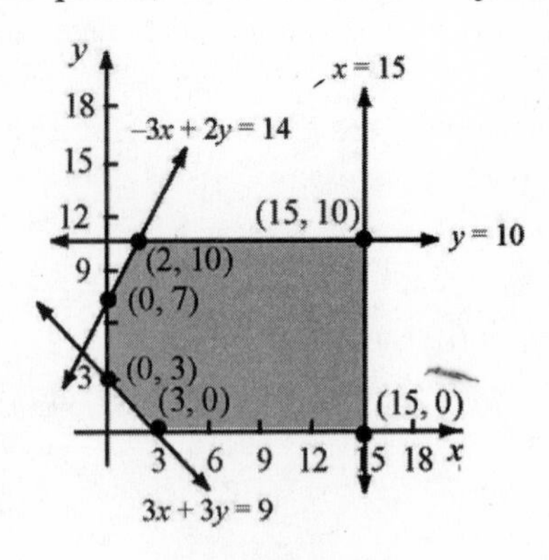

The corner points are (3, 0), (15, 0) (0, 3), (0, 7), (15, 10), and (2, 10).

We find (2, 10) by evaluating $-3x + 2y = 14$ when $y = 10$.

Corner Point	$z = -12x + 24y$
(3, 0)	$z = -12(3) + 24(0) = -36$
(15, 0)	$z = -12(15) + 24(0) = -180$
(0, 3)	$z = -12(0) + 24(3) = 72$
(0, 7)	$z = -12(0) + 24(7) = 168$
(15, 10)	$z = -12(15) + 24(10) = 60$
(2, 10)	$z = -12(2) + 24(10) = 216$

The maximum value: $z = 216$ at the point (2, 10).

53. To answer parts (a), (b), and (c) we set up a linear programming problem and analyze its solution. We begin by naming the variables.

Let x denote the number of acres of soybeans planted, and y denote the number of acres of wheat planted. We want to

maximize $P = 180x + 100y$

subject to the constraints

$$\begin{cases} x + y \le 70 \\ 60x + 30y \le 1800 \\ 3x + 4y \le 120 \\ x \ge 0 \\ y \ge 0 \end{cases} \text{ which simplifies to } \begin{cases} x + y \le 70 \\ 2x + y \le 60 \\ 3x + 4y \le 120 \\ x \ge 0 \\ y \ge 0 \end{cases}$$

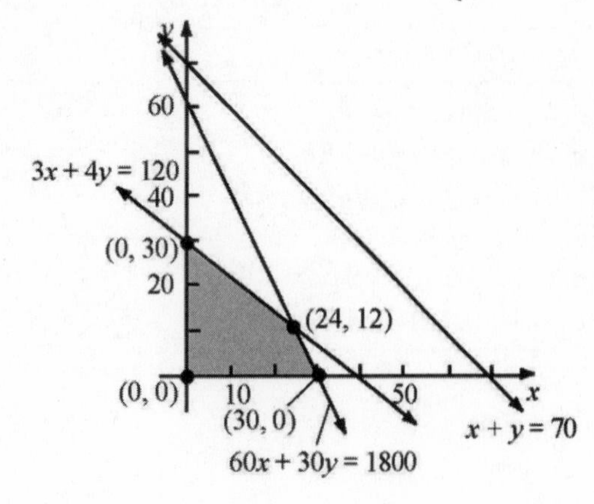

(a) Evaluate the objective function at each corner.

Corner Point	$P = 180x + 100y$
(0, 0)	$180 \cdot 0 + 100 \cdot 0 = 0$
(30, 0)	$180 \cdot 30 + 100 \cdot 0 = 5400$
(24, 12)	$180 \cdot 24 + 100 \cdot 12 = 5520$
(0, 30)	$180 \cdot 0 + 100 \cdot 30 = 3000$

Plant 24 acres of soybeans and 12 acres of wheat.

(b) The maximum profit is \$5520.

(c) To find the maximum if the preparation constraint is raised to \$2400, we rewrite the system of inequalities and re-evaluate the objective function at the corner points.

$$\begin{cases} x + y \le 70 \\ 60x + 30y \le 2400 \\ 3x + 4y \le 120 \\ x \ge 0 \\ y \ge 0 \end{cases} \text{ or } \begin{cases} x + y \le 70 \\ 2x + y \le 80 \\ 3x + 4y \le 120 \\ x \ge 0 \\ y \ge 0 \end{cases}.$$

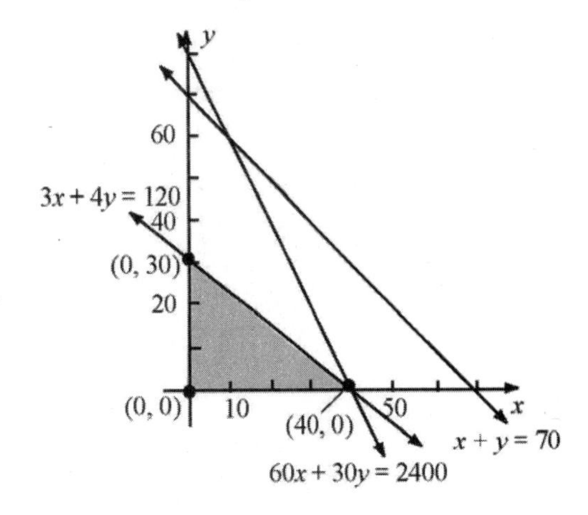

The graph shows that there is only 1 new corner point (40, 0), and that point (24, 12) is not longer a corner. So we evaluate the objective function at (40, 0) and compare its value to that at the other point.

$$P = 180 \cdot 40 + 100 \cdot 0 = 7200$$

The maximum profit is \$7200.00 if the preparation constraint is raised to \$2400.00.

55. Let x = amount invested in type AAA bonds, and

y = amount invested in type BB bonds.

We want to maximize the return on the investment. That is,

maximize $P = 0.10x + 0.15y$

subject to the constraints

$$\begin{cases} x + y \le 20{,}000 \\ x \ge y \\ x \ge 5000 \\ y \le 8000 \\ x \ge 0 \\ y \ge 0 \end{cases}$$

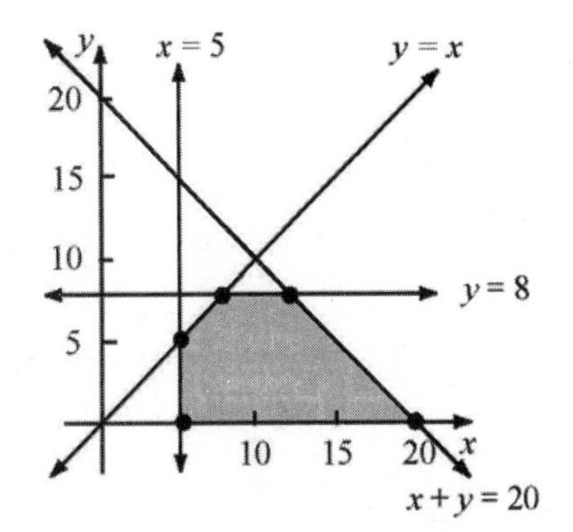

The constraints are graphed in thousands of dollars, and the feasible points are shaded.

We now evaluate the objective function at each corner point to find the amount that should be invested in each type of bond to maximize her return.

Corner Point	$P = 0.10x + 0.15y$
(5000, 0)	$P = 0.10(5000) + 0.15(0) = 500$
(20000, 0)	$P = 0.10(20000) + 0.15(0) = 2000$
(5000, 5000)	$P = 0.10(5000) + 0.15(5000) = 1250$
(8000, 8000)	$P = 0.10(8000) + 0.15(8000) = 2000$
(12000, 8000)	$P = 0.10(12000) + 0.15(8000) = 2400$

She should invest \$12,000 in the AAA bonds and \$8000 in the BB bonds for a maximum return is \$2400.

57. Let x = number (in thousands) of high potency vitamins produced, and

y = number (in thousands) of calcium-enriched vitamins produced.

We want to maximize the profit, P, which is given by: $P = [0.10x + 0.05y] \cdot 1000$

The limited amount of vitamin C and calcium available put the following constraints on the problem:

$$\begin{cases} 500x + 100y \le 300{,}000 & (1) \\ 40x + 400y \le 220{,}000 & (2) \\ x \ge 0 & (3) \\ y \ge 0 & (4) \end{cases} \quad \text{which can be simplified to} \quad \begin{cases} 5x + y \le 3000 \\ x + 10y \le 5500 \\ x \ge 0 \\ y \ge 0 \end{cases}$$

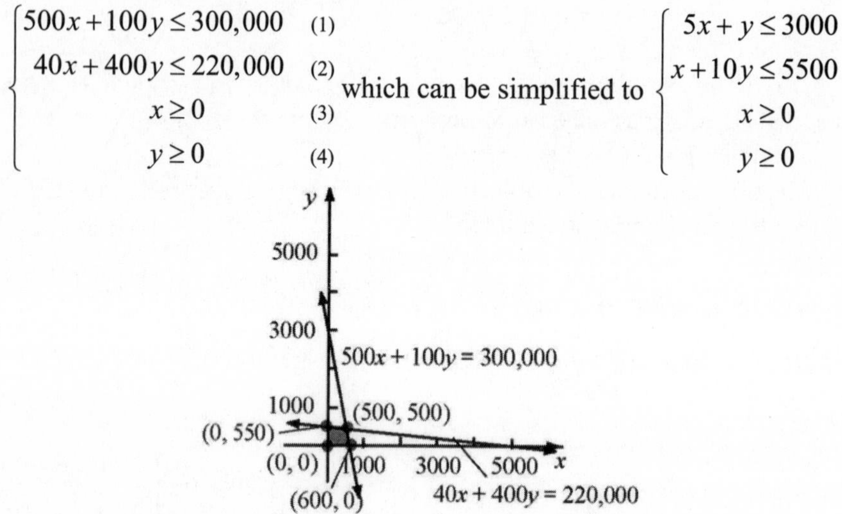

The constraints are graphed and the set of feasible points is shaded. The corner point (500, 500) is found by solving equations (1) and (2).

The objective function P is evaluated at the corner points as shown in the table. Since x and y are in thousands, the profit corresponding to each corner point is:

Corner Point	$P = (0.10x + 0.05y)1000$
(0, 0)	$P = (0.10(0) + 0.05(0))1000 = 0$
(600, 0)	$P = (0.10(600) + 0.05(0))1000 = 60{,}000$
(0, 550)	$P = (0.10(0) + 0.05(550))1000 = 27{,}500$
(500, 500)	$P = (0.10(500) + 0.05(500))1000 = 75{,}000$

The maximum profit of \$75,000 is made when 500,000 high-potency vitamins and 500,000 calcium enriched vitamins are produced.

59. Let x denote the number of rectangular tables rented, and y denote the number of round tables rented.

Kathleen wants to minimize her cost,

$$C = 28x + 52y$$

subject to the following constraints

$$\begin{cases} 6x + 10y = 250 \\ x + \quad y \le 35 \\ x \le 15 \\ x \ge 0 \\ y \ge 0 \end{cases}$$

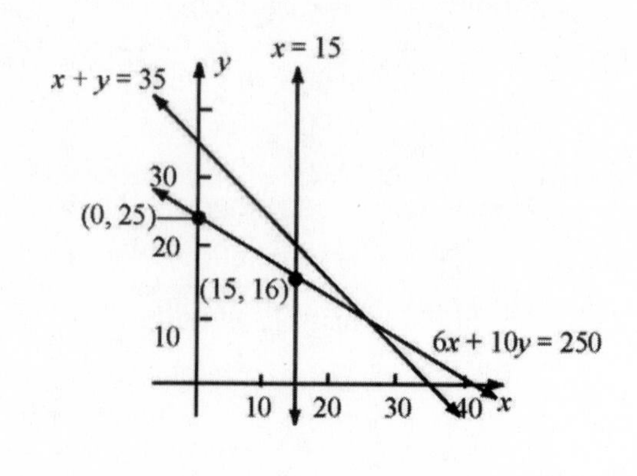

The objective function is evaluated at the corner points and the minimum value is found.

Corner Point	$C = 28x + 52y$
(0, 25)	$28 \cdot 0 + 52 \cdot 25 = 1300$
(15, 16)	$28 \cdot 15 + 52 \cdot 16 = 1252$

Kathleen should rent 15 rectangular tables and 16 round tables for a minimum cost of \$1252.00.

61. Let x denote the number of turkey sandwiches, and
y denote the number of veggie sandwiches that Eric consumes.
Eric wants to minimize his fat intake while assuring that he has sufficient protein, fiber, and carbohydrates. He wants to solve the linear programming problem,

Minimize

$$F = 4.5x + 3y$$

subject to the constraints

$$\begin{cases} 18x + 9y \geq 81 \\ 4x + 4y \geq 28 \\ 46x + 44y \geq 300 \\ x \geq 0 \\ y \geq 0 \end{cases}$$

The constraints are graphed, the feasible region shaded, and the corner points identified. Then the objective function is evaluated at each corner point and the minimum value is chosen.

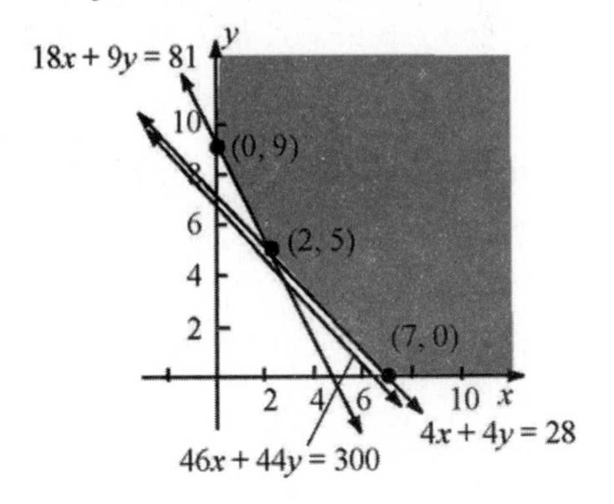

The corner point (2, 5) was found by simplifying the first two equations and solving the system by elimination.

$$\begin{aligned} 2x + y &= 9 \\ x + y &= 7 \\ \text{Subtract:}\quad x \quad &= 2 \end{aligned}$$

Back-substitute into the second equation to get $y = 5$.

Corner Point	$F = 4.5x + 3y$
(7, 0)	$4.5 \cdot 7 + 3 \cdot 0 = 31.5$
(0, 9)	$4.5 \cdot 0 + 3 \cdot 9 = 27$
(2, 5)	$4.5 \cdot 2 + 3 \cdot 5 = 24$

Eric should eat 2 turkey and 5 veggie sandwiches to meet his dietary requirements and to minimize his fat. He will eat 24 grams of fat.

63. Let x = units of the first product produced, and
y = units of the second product produced.
The objective is to maximize the profit,

$$P = 40x + 60y$$

The hours that each machine is available form the constraints on the problem.

$$\begin{cases} 2x + y \leq 70 & \text{hrs. available machine 1} \quad (1) \\ x + y \leq 40 & \text{hrs. available machine 2} \quad (2) \\ x + 3y \leq 90 & \text{hrs. available machine 3} \quad (3) \\ x \geq 0 & \quad (4) \\ y \geq 0 & \quad (5) \end{cases}$$

The constraints are graphed, the feasible region is shaded, and the corner points identified.

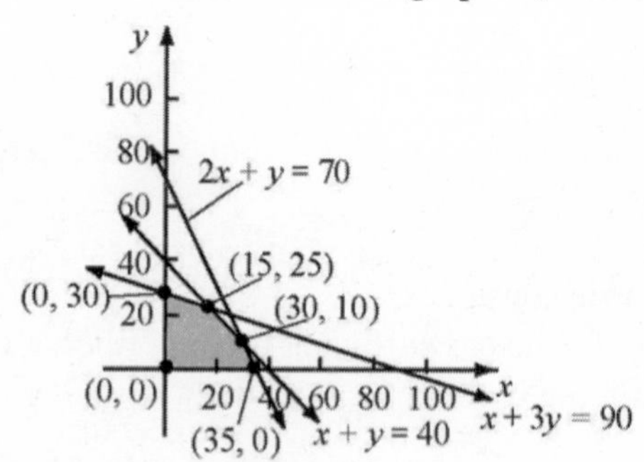

Corner point, (15, 25), is found by writing $x = 40 - y$ and substituting into Equation (3)

$$40 - y + 3y = 90$$
$$2y = 50 \text{ or } y = 25$$

Back-substitute into $x + y = 40$ to get $x = 15$.

Corner point, (30, 10) is found by writing $y = 40 - x$ and substituting into Equation (1)

$$2x + 40 - x = 70$$
$$x = 30$$

Back-substitute into $x + y = 40$ to get $y = 10$.

The objective function is now evaluated at each corner point and the maximum value of P is chosen.

Corner Point	$P = 40x + 60y$
(0, 0)	$P = 40(0) + 60(0) = 0$
(35, 0)	$P = 40(35) + 60(0) = 1400$
(0, 30)	$P = 40(0) + 60(30) = 1800$
(30, 10)	$P = 40(30) + 60(10) = 1800$
(15, 25)	$P = 40(15) + 60(25) = 2100$

The factory should manufacture 15 units of the first product and 25 units of the second product, maximizing the profit at $2100.00.

65. Let x denote the units of Supplement A , and y denote the units of Supplement B in the diet.

Minimize

$$C = 1.5x + y$$

subject to the constraints

$$\begin{cases} 5x + 2y \geq 60 \\ 3x + 2y \geq 45 \\ 4x - y \geq 30 \\ x \geq 0 \\ y \geq 0 \end{cases}$$

The constraints are graphed, the feasible region is shaded, and the corner points identified.

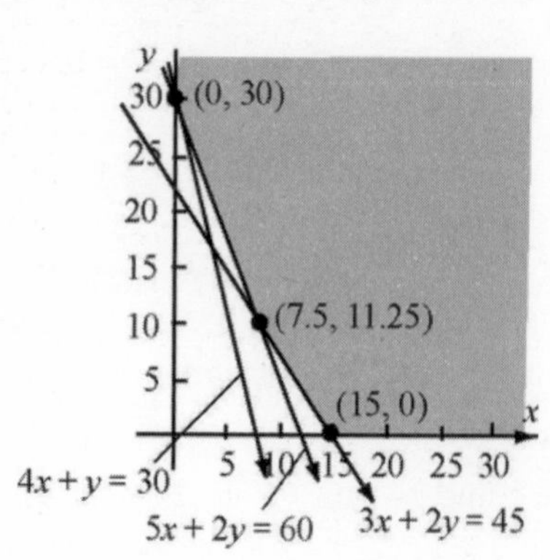

Corner point (7.5, 11.25) was found by substitution

Equation 1: $2y = 60 - 5x$

Equation 2: $2y = 45 - 3x$

Setting these equal, we get

$$60 - 5x = 45 - 3x$$
$$15 = 2x$$
$$x = 7.5$$

Back-substituting, $y = \dfrac{(60 - 5 \cdot 7.5)}{2} = 11.25$

The objective function C is evaluated at the corner points as shown in the table.

Corner Point	$C = 1.5x + y$
$(15, 0)$	$1.5 \cdot 15 + 0 = 22.5$
$(0, 30)$	$1.5 \cdot 0 + 30 = 30$
$(7.5, 11.25)$	$1.5 \cdot 7.5 + 11.25 = 22.5$

One should take any combination of supplements A and B that satisfy the equation $3x + 2y = 45$ between the points (15, 0) and (7.5, 11.25). These combinations minimize the cost at $22.50.

67. Let x denote the amount allocated to the 30 year loans, and y denote the amount allocated to the 15 year loans.

Maximize

$$P = \frac{0.07125}{12}x + \frac{0.06875}{12}y$$

subject to the constraints

$$\begin{cases} x + y \le 72 \\ x \le 2y \\ 0 \le x \le 15 \\ 0 \le y \le 15 \end{cases}$$

The constraints are graphed, the feasible region is shaded and the corner points are identified. Then the objective function is evaluated at the corner points and the maximum value is chosen.

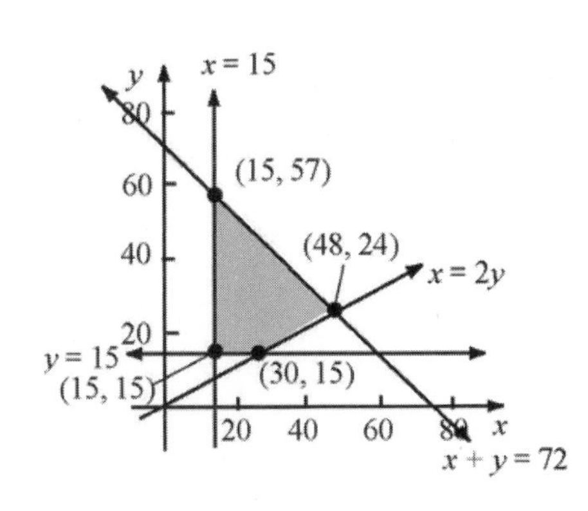

The point (48, 24) is found by substitution. $x = 2y$ is substituted into the first equation

$$2y + y = 72$$
$$3y = 72 \quad \text{or } y = 24$$

Back-substituting, we find $x = 2 \cdot 24 = 48$.

Corner Point	$P = \frac{0.07125}{12}x + \frac{0.06875}{12}y$
$(30, 15)$	$\frac{0.07125}{12} \cdot 30 + \frac{0.06875}{12} \cdot 15 = 0.264065$
$(48, 24)$	$\frac{0.07125}{12} \cdot 48 + \frac{0.06875}{12} \cdot 24 = 0.4225$
$(15, 57)$	$\frac{0.07125}{12} \cdot 15 + \frac{0.06875}{12} \cdot 57 = 0.415625$
$(15, 15)$	$\frac{0.07125}{12} \cdot 15 + \frac{0.06875}{12} \cdot 15 = 0.175$

Remembering the amounts were given in millions, we conclude that Fremont Bank should allocate $48 million to the 30 year loans and $24 million to 15 year loans. The bank will receive $422,500.00 in interest.

69. Let x denote the number of racing skates and y denote the number of figure skates to be manufactured.

Maximize

$$P = 10x + 12y$$

subject to the constraints

$$\begin{cases} 6x + 4y \le 120 \\ x + 2y \le 40 \\ x \ge 0 \\ y \ge 0 \end{cases}$$

The constraints are graphed, the feasible region is shaded and the corner points are identified. Then the objective function is evaluated at the corner points and the maximum value is chosen.

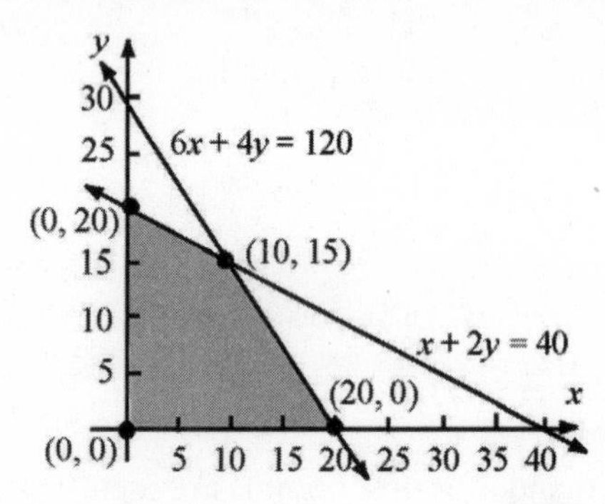

The corner point (10, 15) is found by substitution. The first equation is simplified to $3x + 2y = 60$. Then $2y = 60 - 3x$ is substituted into

$$x + 2y = 40$$
$$x + 60 - 3x = 40$$
$$2x = 20 \quad \text{or} \quad x = 10$$

Back-substitute x, so $y = \frac{40 - x}{2} = \frac{40 - 10}{2} = 15$.

Corner Point	$P = 10x + 12y$
(0, 0)	$10 \cdot 0 + 12 \cdot 0 = 0$
(20, 0)	$10 \cdot 20 + 12 \cdot 0 = 200$
(0, 20)	$10 \cdot 0 + 12 \cdot 20 = 240$
(10, 15)	$10 \cdot 10 + 12 \cdot 15 = 280$

The factory should manufacture 10 pairs of racing skates and 15 pairs of figure skates to obtain a maximum profit of \$280.00.

71. Let x = number of items of A produced, and y = number of items of B produced.

The chemical company wants to maximize profit, P which is given by: $P = 1.5x + 1y$

Maximize

$$P = 1.5x + y$$

subject to the constraints

$$\begin{cases} 2x + 4y \le 3000 & \text{carbon monoxide limits} \\ 6x + 3y \le 5400 & \text{sulfur dioxide limits} \\ x \ge 0 \\ y \ge 0 \end{cases}$$

The constraints are graphed and the set of feasible points is shaded.

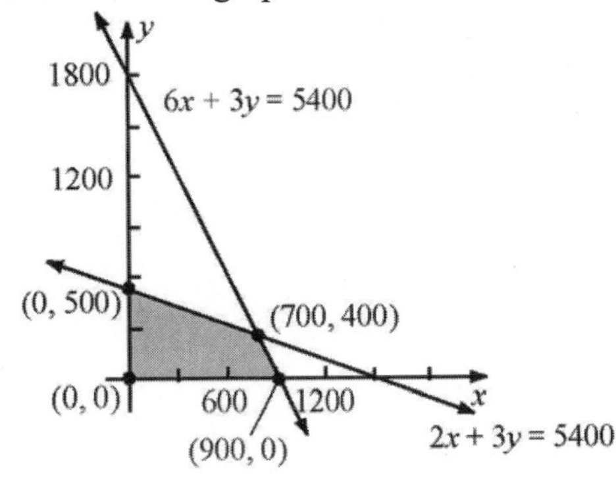

The corner point (700, 400) was found by solving the simplified system of equations

$$\begin{cases} x + 2y = 1500 \\ 2x + y = 1800 \end{cases}$$

by substitution. We wrote the first equation as $x = 1500 - 2y$ and substituted into the second equation.

$$2(1500 - 2y) + y = 1800$$
$$3000 - 4y + y = 1800$$
$$y = 400$$

Back-substituting, we get $x = 1500 - 2(400) = 700$.

The objective function P is evaluated at the corner points as shown in the table:

Corner Point	$P = 1.5x + 1y$
(0, 0)	$P = 1.5(0) + 1(0) = 0$
(900, 0)	$P = 1.5(900) + 1(0) = 1350$
(0, 750)	$P = 1.5(0) + 1(750) = 750$
(700, 400)	$P = 1.5(700) + 1(400) = 1450$

The maximum profit is \$1450, and it is obtained when 700 units of type A compound and 400 units of type B compound are produced.

73. Let x = number of tons of Type 1 steel produced, and y = number of tons of Type 2 steel produced.

The company wants to maximize its profit, which is given by $P = 240x + 80y$.

Maximize

$$P = 240x + 80y$$

subject to the constraints

$$\begin{cases} 2x + 5y \le 40 & \text{melting} & (1) \\ 4x + y \le 20 & \text{cutting} & (2) \\ 10x + 5y \le 60 & \text{rolling} & (3) \\ x \ge 0 & & (4) \\ y \ge 0 & & (5) \end{cases}$$

The constraints are graphed and the set of feasible points is shaded.

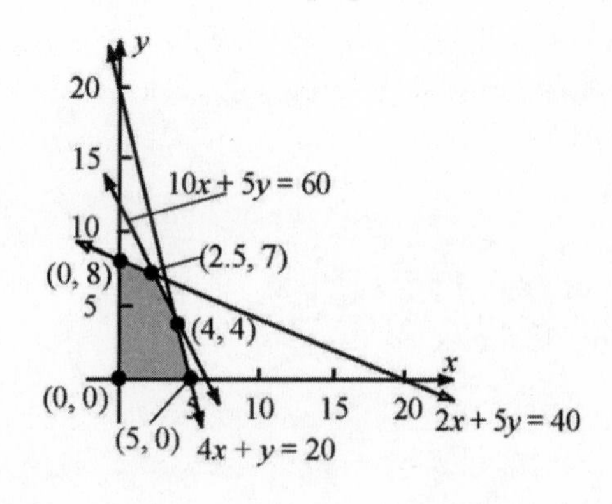

Corner point (4, 4) was found by solving Equations (2) and (3) using substitution.

$$y = 20 - 4x \qquad (2)$$
$$10x + 5(20 - 4x) = 60 \qquad (3)$$
$$10x + 100 - 20x = 60$$
$$x = 4$$

Back-substituting x into Equation (2), we get $y = 4$.

Corner point (2.5, 7) was found by solving Equations (1) and (3) using elimination.

$$10x + 5y = 60 \qquad (3)$$
$$\underline{2x + 5y = 40} \qquad (1)$$
$$\text{Subtract:} \quad 8x \quad = 20 \quad \text{or} \quad x = 2.5$$

Back-substituting x into (3), we get

$$25 + 5y = 60$$
$$5y = 35 \quad \text{or} \quad y = 7$$

The objective function P is evaluated at the corner points as shown in the table:

Corner Point	$P = 240x + 80y$
(0, 0)	$P = 240(0) + 80(0) = 0$
(5, 0)	$P = 240(5) + 80(0) = 1200$
(0, 8)	$P = 240(0) + 80(8) = 640$
$\left(\frac{5}{2}, 7\right)$	$P = 240\left(\frac{5}{2}\right) + 80(7) = 1160$
(4, 4)	$P = 240(4) + 80(4) = 1280$

The maximum profit is \$1280, and it is obtained when 4 tons of Type 1 steel and 4 tons of Type 2 steel are produced.

75. Let x = number of ounces of Supplement I added, and y = number of ounces of Supplement II added. Danny wants to minimize his cost, C, while maintaining a healthy diet for his chickens.

Minimize

$$C = 0.03x + 0.04y$$

subject to the constraints

$$\begin{cases} 5x + 25y \geq 50 & \text{vitamin 1} & (1) \\ 25x + 10y \geq 100 & \text{vitamin 2} & (2) \\ 10x + 10y \geq 60 & \text{vitamin 3} & (3) \\ 35x + 20y \geq 180 & \text{vitamin 4} & (4) \\ x \geq 0 & & (5) \\ y \geq 0 & & (6) \end{cases}$$

The constraints are graphed and the set of feasible points, which is unbounded, is shaded.

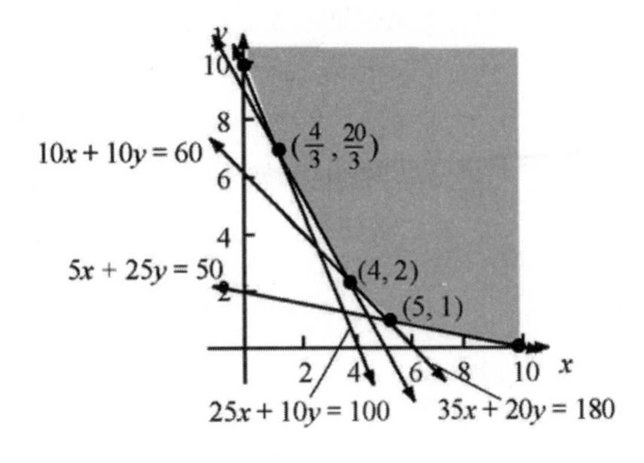

Corner point (5, 1) is found by solving the system of equations formed by Equations (1) and (3). We used substitution.

$$x = 6 - y \qquad (3)$$
$$(6 - y) + 5y = 10 \qquad (1)$$
$$6 + 4y = 10 \quad \text{or} \quad y = 1$$

Back-substituting y into equation (3) yields

$$x = 6 - 1 = 5$$

Corner point (4, 2) is found by solving the system of equations formed by Equations (3) and (4). We used substitution.

$$y = 6 - x \qquad (3)$$
$$7x + 4(6 - x) = 36 \qquad (4)$$
$$7x + 24 - 4x = 36$$
$$3x = 12 \quad \text{or} \quad x = 4$$

Back-substituting x into Equation (3) yields

$$y = 6 - 4 = 2$$

Corner point $\left(\frac{4}{3}, \frac{20}{3}\right)$ is the solution to the system formed by Equations (2) and (4). We used elimination.

$$\begin{array}{lrl} \text{Multiply (2) by 2:} & 10x + 4y = 40 & (2) \\ & \underline{7x + 4y = 36} & (4) \\ \text{Subtract:} & 3x \quad = 4 & \text{or } x = \frac{4}{3} \end{array}$$

Back-substitute x in Equation (2): $10\left(\frac{4}{3}\right) + 4y = 40$

$$y = \frac{40 - \frac{40}{3}}{4} = 10 - \frac{10}{3} = \frac{20}{3}$$

The objective function C is now evaluated at the corner points as shown in the table:

Corner Point	$C = 0.03x + 0.04y$
(10, 0)	$C = 0.03(10) + 0.04(0) = 0.3$
(0, 10)	$C = 0.03(0) + 0.04(10) = 0.4$
(5, 1)	$C = 0.03(5) + 0.04(1) = 0.19$
$\left(\frac{4}{3}, \frac{20}{3}\right)$	$C = 0.03\left(\frac{4}{3}\right) + 0.04\left(\frac{20}{3}\right) = 0.31$
(4, 2)	$C = 0.03(4) + 0.04(2) = 0.20$

The minimum cost is \$0.19 when Danny adds 5 ounces of Supplement I and 1 ounce of Supplement II to every 100 ounces of feed.

77. Let x denote the number of sales reps from NYC attending the meeting, and let y denote the number of sales reps from Chicago attending the meeting. The company wants to minimize travel costs.

Minimize

$$C = 343x + 329y$$

subject to the constraints

$$\begin{cases} x + y \geq 40 \\ x \leq 28 \\ y \leq 22 \\ x \geq 16 \\ y \geq 12 \end{cases}$$

The constraints are graphed, the set of feasible points shaded, and the corner points identified. Then the objective function is evaluated at each corner point and the minimum value chosen.

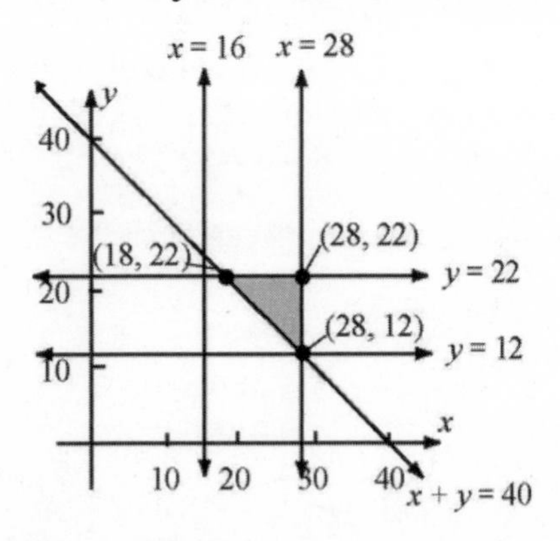

Corner Points	$C = 343x + 329y$
(28, 12)	$343 \cdot 28 + 329 \cdot 12 = 13{,}552$
(28, 22)	$343 \cdot 28 + 329 \cdot 22 = 16{,}842$
(18, 22)	$343 \cdot 18 + 329 \cdot 22 = 13{,}412$

Eighteen sales representatives from New York City and 22 from Chicago should be sent to the meeting to minimize the total airfare. The minimum airfare is \$13,412.00.

79. If the new price of the high-grade carpet is denoted by P, then the income from selling x rolls of high-grade carpet becomes $I_{\text{high-grade}} = Px - 420x$, and the objective function to be maximized becomes:

$$I = Px - 420x + 100y$$

The constraints, graph of the set of feasible points, and the corner points remain the same as in Problem 78. So the new objective function I is evaluated at the corner points as shown in the table:

Corner Point	$I = Px - 420x + 100y$
(0, 0)	$I = P(0) - 420(0) + 100(0) = 0$
(20, 0)	$I = P(20) - 420(20) + 100(0) = 20P - 8400$
(0, 25)	$I = P(0) - 420(0) + 100(25) = 2500$
(15, 10)	$I = P(15) - 420(15) + 100(10) = 15P - 5300$

Since the problem requires that some of each grade carpet be manufactured, the corner point associated with the maximum income must be (15, 10). This means that both of the following inequalities must be true.

$$\begin{aligned} 15P - 5300 &\geq 2500 \\ 15P &\geq 7800 \\ P &\geq 520 \end{aligned} \quad \text{and} \quad \begin{aligned} 15P - 5300 &\geq 20P - 8400 \\ -5P &\geq -3100 \\ P &\leq 620 \end{aligned}$$

So if high-grade carpet is priced between \$520 and \$620 per roll, some of each type of carpeting will be manufactured to maximize income.

Chapter 3 Review

1.

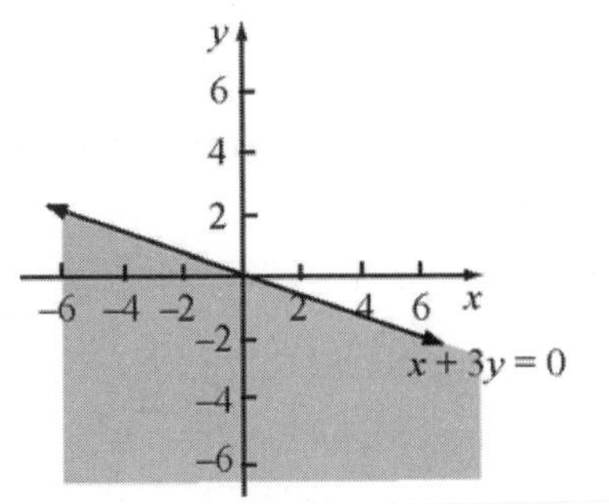

The inequality is nonstrict, so the line is solid. The point (–2, 0) satisfies the inequality, so we shade the side of the graph containing (–2, 0).

3.

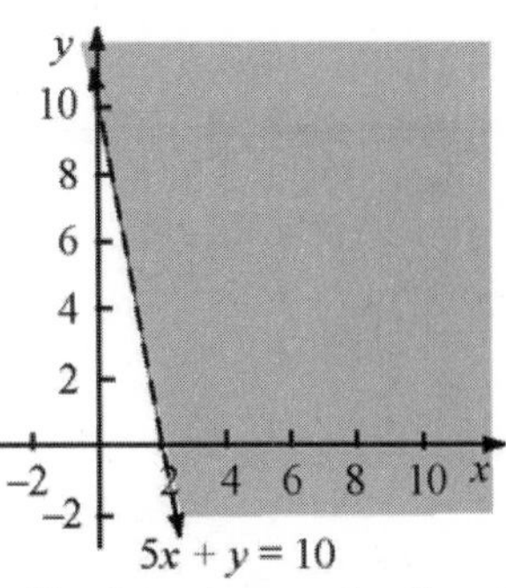

The inequality is strict, so the line is dashed. The point (4, 0) satisfies the inequality, so we shade the side of the graph containing (4, 0).

5. To determine which points are part of the graph of the system check each point to see if it satisfies both of the inequalities:

Point	Inequality 1 $x+2y\le 8$	Inequality 2 $2x-y\ge 4$	Conclusion
$P_1=(4,-3)$	Is $4+2\cdot(-3)\le 8$? $-2<8$	Is $2\cdot 4-(-3)\ge 4$? $11>4$	Both inequalities are satisfied; (4, –3) is part of the graph.
$P_2=(2,-6)$	Is $2+2\cdot(-6)\le 8$? $-10<8$	Is $2\cdot 2-(-6)\ge 4$? $10>4$	Both inequalities are satisfied; (2, –6) is part of the graph.
$P_3=(8,-3)$	Is $8+2\cdot(-3)\le 8$? $2<8$	Is $2\cdot 8-(-3)\ge 4$? $19>4$	Both inequalities are satisfied; (8, –3) is part of the graph.

7. The region that represents the graph of the system is the set of points common to the solutions of each individual inequality. Use the test point (0, 0).

$$\begin{array}{rr} 6x-4y\le 12 & 3x+2y\le 18 \\ 6\cdot 0-4\cdot 0=0 & 3\cdot 0+2\cdot 0=0 \\ 0<12 & 0<18 \end{array}$$

The test point (0, 0) satisfies both inequalities, so the region (*a*) represents the graph of the system.

9.

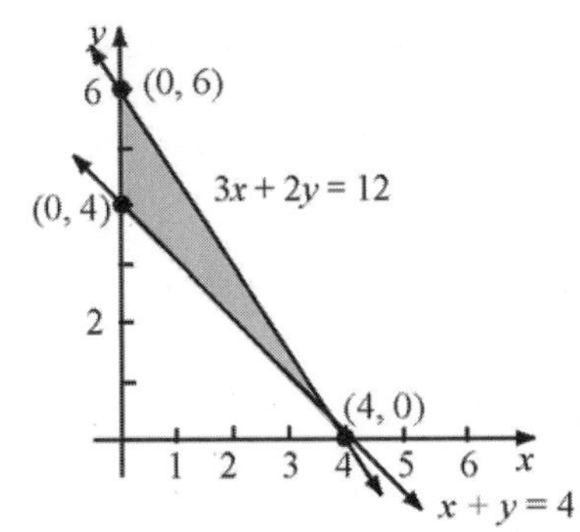

Graph is bounded.
Corner points are (4, 0), (0, 4) and (0, 6).

11.

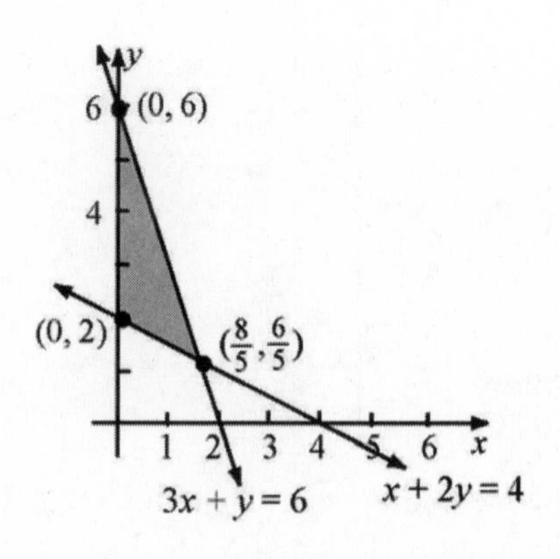

Graph is bounded.

Corner points are (0, 2), (0, 6), and $\left(\frac{8}{5}, \frac{6}{5}\right)$.

To find the third corner point we solved the equations by elimination.

$$\begin{cases} x+2y=4 & (1) \\ 3x+\ \ y=6 & (2) \end{cases}$$

Subtracting (1) from twice (2) gives $5x = 8$ or $x = \frac{8}{5}$.

Back-substituting in (1) we get $y = \frac{6}{5}$.

13.

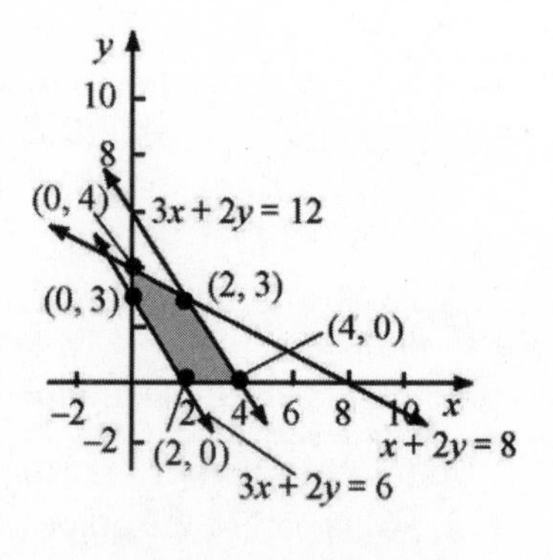

Graph is bounded.

Corner points are (2, 0), (4, 0), (0, 3), (0, 4), and (2, 3). To find (2, 3) solve we solved the system of equations

$$\begin{cases} x+2y=\ 8 & (1) \\ 3x+2y=12 & (2) \end{cases}$$

Subtracting (1) from (2) gives $2x = 4$ or $x = 2$. Back-substituting in (1) gives $2y = 6$ or $y = 3$.

In Problems 15-22, *Use the graph below and its corner points to find the maximum and minimum values of z.*

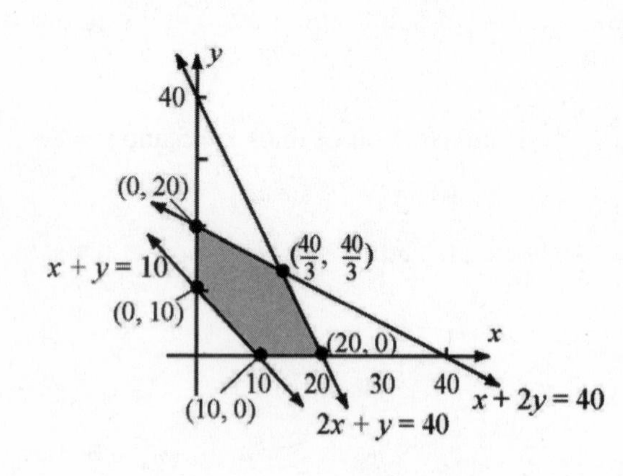

The corner points are (10, 0), (20, 0), (0, 10), (0, 20), and $\left(\frac{40}{3}, \frac{40}{3}\right)$. We found $\left(\frac{40}{3}, \frac{40}{3}\right)$ by solving the system of equations

$$\begin{cases} x+2y=40 & (1) \\ 2x+y=40 & (2) \end{cases}$$

Using elimination, we subtract (2) from two times (1) and find $3y = 40$ or $y = \frac{40}{3}$. Back-substituting y in (2) gives $2x + \frac{40}{3} = 40$ or $x = \frac{40}{3}$.

To find the maximum and minimum values of z, we construct a table and evaluate the objective function at the corner points.

15.

Corner Point	$z = x + y$
(10, 0)	$z = 10 + 0 = 10$
(20, 0)	$z = 20 + 0 = 20$
(0, 10)	$z = 0 + 10 = 10$
(0, 20)	$z = 0 + 20 = 20$
$\left(\frac{40}{3}, \frac{40}{3}\right)$	$z = \frac{40}{3} + \frac{40}{3} = \frac{80}{3}$

The maximum value of z is $\frac{80}{3} \approx 26.667$,

and it occurs at the point $\left(\frac{40}{3}, \frac{40}{3}\right)$.

17.

Corner Point	$z = 5x + 2y$
(10, 0)	$z = 5(10) + 2(0) = 50$
(20, 0)	$z = 5(20) + 2(0) = 100$
(0, 10)	$z = 5(0) + 2(10) = 20$
(0, 20)	$z = 5(0) + 2(20) = 40$
$\left(\frac{40}{3}, \frac{40}{3}\right)$	$z = 5\left(\frac{40}{3}\right) + 2\left(\frac{40}{3}\right) = \frac{280}{3}$

The minimum value of z is 20, and it occurs at the point (0, 10).

19.

Corner Point	$z = 2x + y$
(10, 0)	$z = 2(10) + 0 = 20$
(20, 0)	$z = 2(20) + 0 = 40$
(0, 10)	$z = 2(0) + 10 = 10$
(0, 20)	$z = 2(0) + 20 = 20$
$\left(\frac{40}{3}, \frac{40}{3}\right)$	$z = 2\left(\frac{40}{3}\right) + \frac{40}{3} = 40$

The maximum value of z is 40, and it occurs at the points (20, 0) and $\left(\frac{40}{3}, \frac{40}{3}\right)$ and at all points on the line segment connecting them.

21.

Corner Point	$z = 2x + 5y$
(10, 0)	$z = 2(10) + 5(0) = 20$
(20, 0)	$z = 2(20) + 5(0) = 40$
(0, 10)	$z = 2(0) + 5(10) = 50$
(0, 20)	$z = 2(0) + 5(20) = 100$
$\left(\frac{40}{3}, \frac{40}{3}\right)$	$z = 2\left(\frac{40}{3}\right) + 5\left(\frac{40}{3}\right) = \frac{280}{3}$

The minimum value of z is 20, and it occurs at the point (10, 0).

23. To find the maximum and minimum values of z, we will graph the constraints, shade the set of feasible points, find the corner points, and evaluate the objective function, z, at each corner point.

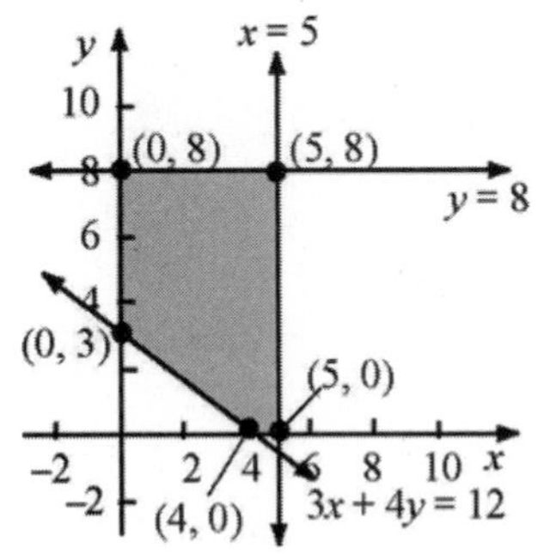

The corner points are (4, 0), (5, 0), (0, 3), (0, 8), and (5, 8).

Point (5, 8) is the intersection of lines $x = 5$ and $y = 8$.

Corner Point	$z = 15x + 20y$
(4, 0)	$z = 15(4) + 20(0) = 60$
(5, 0)	$z = 15(5) + 20(0) = 75$
(0, 3)	$z = 15(0) + 20(3) = 60$
(0, 8)	$z = 15(0) + 20(8) = 160$
(5, 8)	$z = 15(5) + 20(8) = 235$

The maximum value of $z = 235$; it occurs at the point (5, 8).
The minimum value of $z = 60$; it occurs at the points (4, 0), (0, 3), and at all points on the line segment connecting them.

25. To find the maximum and minimum values of z, we will graph the constraints, shade the set of feasible points, find the corner points, and evaluate the objective function, z, at each corner point.

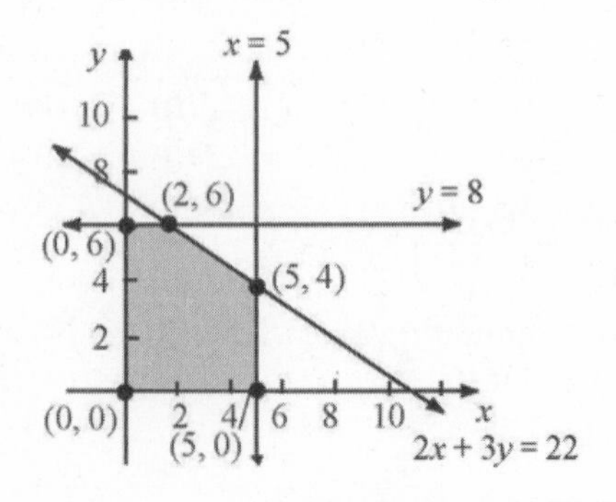

The corner points are (0, 0), (5, 0), (0, 6), (5, 4), and (2, 6).

To find (5, 4) solve $\begin{cases} 2x+3y=22 & (1) \\ x=5 & (2) \end{cases}$. Substitute 5 for x in (1) to get $3y = 12$ or $y = 4$.

To find (2, 6) solve $\begin{cases} 2x+3y=22 & (1) \\ y=6 & (2) \end{cases}$.

Substitute 6 for y in (1) to get $2x = 4$ or $x = 2$.

Corner Point	$z = 15x + 20y$
(0, 0)	$z = 15(0) + 20(0) = 0$
(5, 0)	$z = 15(5) + 20(0) = 75$
(0, 6)	$z = 15(0) + 20(6) = 120$
(5, 4)	$z = 15(5) + 20(4) = 155$
(2, 6)	$z = 15(2) + 20(6) = 150$

The maximum value of $z = 155$; it occurs at the point (5, 4).
The minimum value of $z = 0$; it occurs at the point (0, 0).

27. To solve a linear programming problem graph the constraints, identify the corner points, evaluate the objective function at each corner point, and choose the largest -value of z.

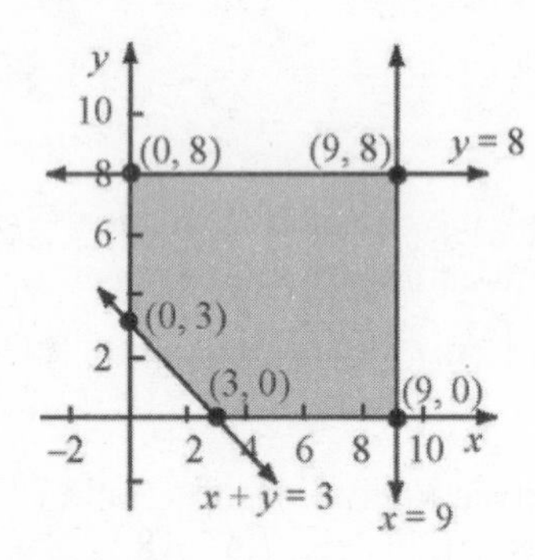

The corner points are (3, 0), (9, 0), (0, 3), (0, 8), and (9, 8).

Corner Point	$z = 2x + 3y$
(3, 0)	$z = 2(3) + 3(0) = 6$
(9, 0)	$z = 2(9) + 3(0) = 18$
(0, 3)	$z = 2(0) + 3(3) = 9$
(0, 8)	$z = 2(0) + 3(8) = 24$
(9, 8)	$z = 2(9) + 3(8) = 42$

The maximum value of $z = 42$; it occurs at the point (9, 8).

29. To solve a linear programming problem graph the constraints, identify the corner points, evaluate the objective function at each corner point, and choose the largest value of z.

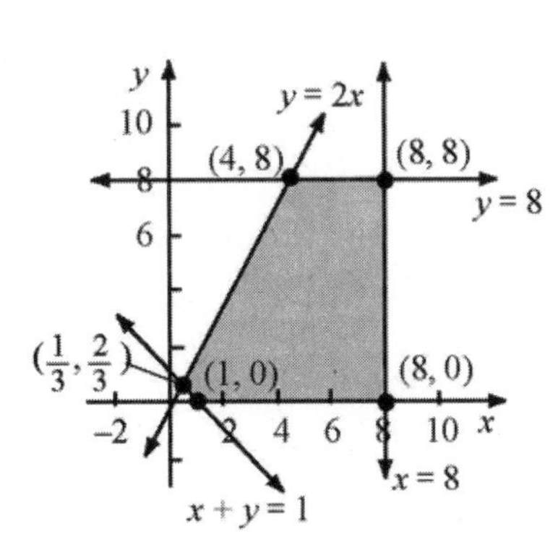

The corner points are (1, 0), (8, 0), (4, 8), $\left(\frac{1}{3}, \frac{2}{3}\right)$, and (8, 8).

$\left(\frac{1}{3}, \frac{2}{3}\right)$ is the solution of $\begin{cases} x + y = 1 & (1) \\ y = 2x & (2) \end{cases}$

By substituting $2x$ for y in (1), we get $x = \frac{1}{3}$, then back–substituting into (2) gives $y = \frac{2}{3}$.

Point (4, 8) is the intersection of the lines $y = 8$ and $y = 2x$.

Corner Point	$z = x + 2y$
(1, 0)	$z = 1 + 2(0) = 1$
(8, 0)	$z = 8 + 2(0) = 8$
$\left(\frac{1}{3}, \frac{2}{3}\right)$	$z = \frac{1}{3} + 2\left(\frac{2}{3}\right) = \frac{5}{3}$
(4, 8)	$z = 4 + 2(8) = 20$
(8, 8)	$z = 8 + 2(8) = 24$

The maximum value of $z = 24$; it occurs at the point (8, 8).

31. To solve a linear programming problem graph the constraints, identify the corner points, evaluate the objective function at each corner point, and choose the smallest value of z.

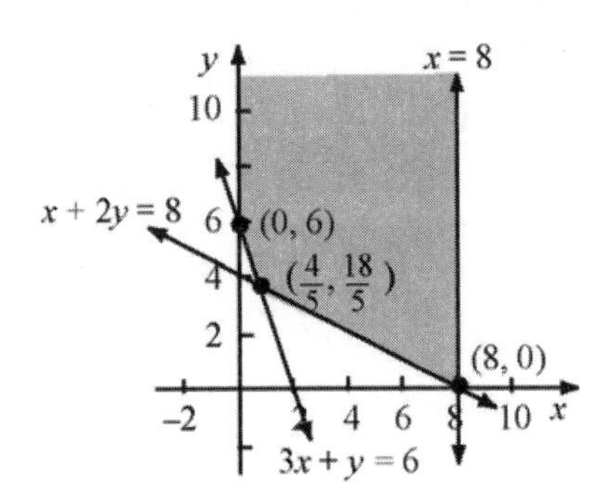

The corner points are (8, 0), (0, 6), and $\left(\frac{4}{5}, \frac{18}{5}\right)$.

Solve $\begin{cases} x + 2y = 8 & (1) \\ 3x + y = 6 & (2) \end{cases}$ to find $\left(\frac{4}{5}, \frac{18}{5}\right)$.

Subtracting (1) from 2 times (2) gives $5x = 4$, or $x = \frac{4}{5}$. Back-substituting into (1), we find $y = \frac{18}{5}$.

Corner Point	$z = 3x + 2y$
(8, 0)	$z = 3(8) + 2(0) = 24$
$\left(\frac{4}{5}, \frac{18}{5}\right)$	$z = 3\left(\frac{4}{5}\right) + 2\left(\frac{18}{5}\right) = \frac{48}{5} = 9.6$
(0, 6)	$z = 3(0) + 2(6) = 12$

The minimum value of $z = \frac{48}{5}$; it occurs at the point $\left(\frac{4}{5}, \frac{18}{5}\right)$.

33. (a) $\begin{cases} \frac{1}{4}x+\frac{1}{2}y \le 75 & (1) \\ \frac{3}{4}x+\frac{1}{2}y \le 120 & (2) \\ x \ge 0 & (3) \\ y \ge 0 & (4) \end{cases}$ or simplified $\begin{cases} x+2y \le 300 & (1) \\ 3x+2y \le 480 & (2) \\ x \ge 0 & (3) \\ y \ge 0 & (4) \end{cases}$

(b)

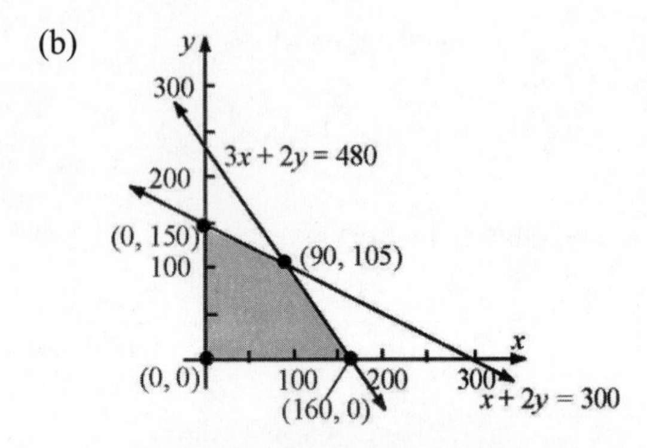

(c) $P = 0.3x + 0.4y$

Evaluate the objective function at each corner point, and choose the maximum.

Corner Point	$P = 0.3x + 0.4y$
$(0, 0)$	$0.3 \cdot 0 + 0.4 \cdot 0 = 0$
$(160, 0)$	$0.3 \cdot 160 + 0.4 \cdot 0 = 48$
$(0, 150)$	$0.3 \cdot 0 + 0.4 \cdot 150 = 60$
$(90, 105)$	$0.3 \cdot 90 + 0.4 \cdot 105 = 69$

Bill should prepare 90 packages of economy blend and 105 packages of superior blend to maximize his profit. The maximum profit will be \$69.00.

35. Let x = number of downhill skis made, and
y = number of cross-country skis made.

The objective is to
Maximize

$$P = 70x + 50y$$

subject to the constraints

$$\begin{cases} 2x + y \le 40 & (1) \\ x + y \le 32 & (2) \\ x \ge 0 & (3) \\ y \ge 0 & (4) \end{cases}$$

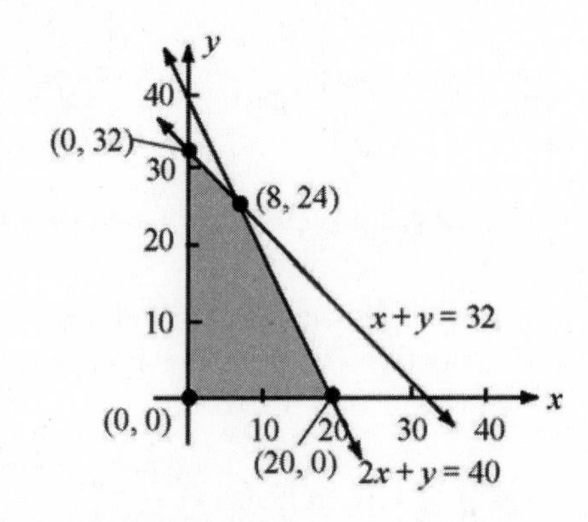

The constraints are graphed, the set of feasible points is shaded, and the corner points are identified. The corner point (8, 24) is found by solving the system of equations $\begin{cases} 2x + y = 40 & (1) \\ x + y = 32 & (2) \end{cases}$. Subtracting Equation (2) from (1), gives $x = 8$. Back-substituting, get $y = 24$.

The objective function is evaluated at the corner points and the maximum value of P is chosen.

Corner Point	$P = 70x + 50y$
(0, 0)	$P = 70(0) + 50(0) = 0$
(20, 0)	$P = 70(20) + 50(0) = 1400$
(0, 32)	$P = 70(0) + 50(32) = 1600$
(8, 24)	$P = 70(8) + 50(24) = 1760$

Produce 8 pairs of downhill skis and 24 of cross-country skis for a maximum profit of \$1760.

37. Let x = servings of Banana Oatmeal & Peach, and y = servings of Mixed Fruit Juice consumed. The objective is to minimize the cost, C, of the baby food while maintaining nutritional requirements.

Minimize

$$C = 0.79x + 0.65y$$

subject to the constraints

$$\begin{cases} 90x + 60y \geq 130 & (1) \\ 19x + 15y \geq 30 & (2) \\ 0.45x + y \geq 0.60 & (3) \\ x \geq 0 & (4) \\ y \geq 0 & (5) \end{cases}$$

The constraints are graphed, the set of feasible points, which is unbounded, is shaded, and the corner points are identified.

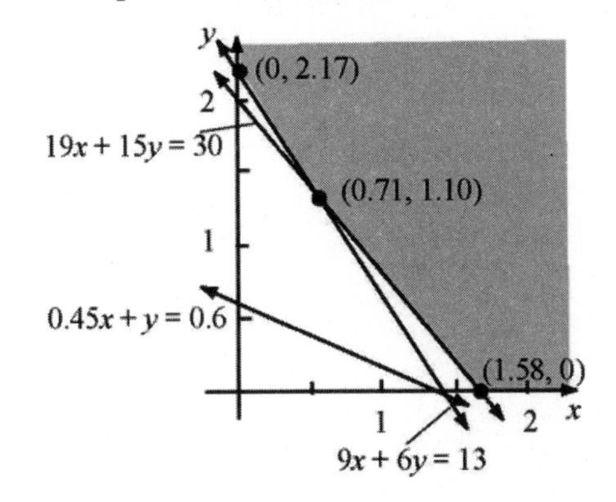

The point (0.71, 1.10) is found by solving the system $\begin{cases} 90x + 60y = 130 & (1) \\ 19x + 15y = 30 & (2) \end{cases}$.

When 4 times (2) is subtracted from (1), we get $14x = 10$ or $x = \frac{5}{7} \approx 0.71$. Back-substituting x into (2) gives $y = \frac{23}{21} \approx 1.10$.

The objective function is evaluated at the corner points and the minimum value of C is chosen.

Corner Point	$C = 0.79x + 0.65y$
(1.58, 0)	$C = 0.79(1.58) + 0.65(0) = 1.248$
(0.71, 1.10)	$C = 0.79(0.71) + 0.65(1.10) = 1.276$
(0, 2.17)	$C = 0.79(0) + 0.65(2.17) = 1.411$

Give the child 1.58 servings of Gerber Banana Oatmeal and Peach and no Gerber Mixed Fruit Juice for a minimum cost is \$1.25.

39. Let x = number (in thousands) of high-potency vitamins to be produced, and

y = number (in thousands) of calcium-enriched vitamins to be produced.

The amount of available ingredients is given in kilograms. 1kg = 1,000,000 mg.

Maximize

$$P = 0.10(1000)x + 0.05(1000)y$$

subject to the constraints

$$\begin{cases} 500{,}000x + 100{,}000y \leq 300{,}000{,}000 \\ 40{,}000x + 400{,}000y \leq 122{,}000{,}000 \\ 100{,}000x + 40{,}000y \leq 65{,}000{,}000 \\ x \geq 0 \\ y \geq 0 \end{cases} \quad \text{or} \quad \begin{cases} 5x + y \leq 3000 \\ 4x + 40y \leq 12{,}200 \\ 10x + 4y \leq 6500 \\ x \geq 0 \\ y \geq 0 \end{cases}$$

The constraints are graphed, the feasible region is shaded, and the corner points are identified.

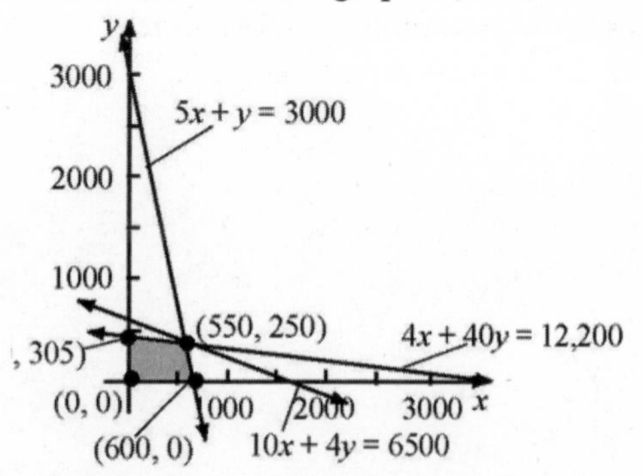

The corner point (550, 250) is found by solving the system of equations formed by any two of the first three constraints.

The objective function is evaluated at the corner points and the maximum value of P is chosen.

Corner Point	$P = 100x + 50y$
(0, 0)	$P = 100(0) + 50(0) = 0$
(600, 0)	$P = 100(600) + 50(0) = 60{,}000$
(0, 305)	$P = 100(0) + 50(305) = 15{,}250$
(550, 250)	$P = 100(550) + 50(250) = 67{,}500$

The maximum profit is \$67,500. It is attained when 550,000 high-potency vitamins and 250,000 calcium-enriched vitamins are produced.

41. Let x denote the amount of money invested in the ING account, and let y denote the amount invested in the HSBC account. The couple want to maximize their annual yield.

Maximize

$$Z = 0.0435x + 0.0505y$$

subject to the constraints

$$\begin{cases} x + y \le 50{,}000 & (1) \\ x \ge 10{,}000 & (2) \\ x \le 15{,}000 & (3) \\ y \ge 10{,}000 & (4) \\ y \le 2x & (5) \end{cases}$$

The constraints are graphed, the set of feasible points is shaded, and the corner points are identified. Note: The graph is drawn in thousands of dollars.

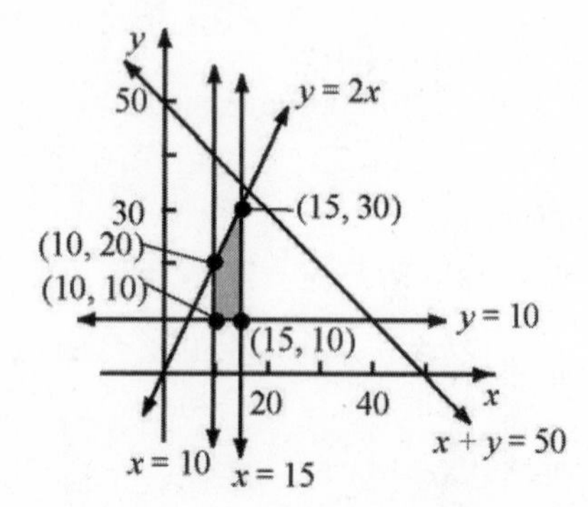

The objective function is now evaluated at each of the corner points, and the maximum value of P is chosen.

Corner Point	$Z = 0.0435x + 0.0505y$
(10, 10)	$0.0435 \cdot 10 + 0.0505 \cdot 10 = 0.94$
(15, 10)	$0.0435 \cdot 15 + 0.0505 \cdot 10 = 1.1575$
(10, 20)	$0.0435 \cdot 10 + 0.0505 \cdot 20 = 1.445$
(15, 30)	$0.0435 \cdot 15 + 0.0505 \cdot 30 = 2.1675$

Remember, these values must be multiplied by 1000.

The couple should deposit \$15,000.00 in the ING account and \$30,000.00 in the HSBC account. Their interest income will be \$2167.50.

43. Let x denote the number of shares of Chevron to be purchased, and y denote the number of shares of Bank of America to be purchased. The investor wants to maximize the quarterly yield.

Maximize

$$Y = 0.52x + 0.50y$$

subject to the constraints

$$\begin{cases} 66x + 48y \le 500{,}000 & (1) \\ x \ge 2{,}000 & (2) \\ y \ge 2{,}000 & (3) \\ 66x \le 300{,}000 & (4) \\ 48y \le 300{,}000 & (5) \end{cases}$$

The constraints are graphed, the set of feasible points is shaded, and the corner points are identified.

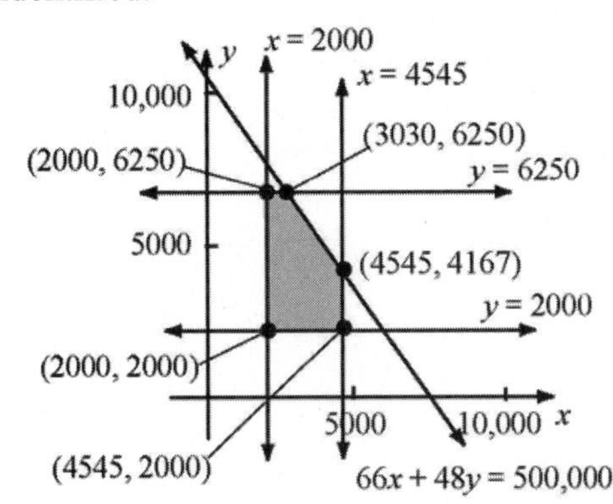

The objective function is evaluated at each the corner points, and the maximum value of R is chosen.

Corner Point	$Y = 0.52x + 0.50y$
(2000, 2000)	$0.52 \cdot 2000 + 0.50 \cdot 2000 = 2040$
(4545, 2000)	$0.52 \cdot 4545 + 0.50 \cdot 2000 = 3363.4$
(4545, 4167)	$0.52 \cdot 4545 + 0.50 \cdot 4167 = 4446.9$
(2000, 6250)	$0.52 \cdot 2000 + 0.50 \cdot 6250 = 4165$
(3030, 6250)	$0.52 \cdot 3030 + 0.50 \cdot 6250 = 4700.6$

To maximize the projected dividend the fund should purchase 3030 shares of Chevron and 6250 shares of Bank of America. The projected dividend is $4700.60.

Chapter 3 Project

1. Objective function: $c = 31.4x + 114.74y$

3.

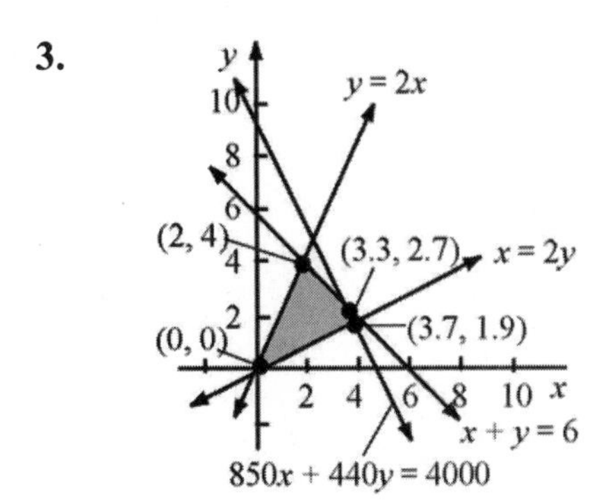

The corner point (3.3, 2.7) is the solution to the system

$$\begin{cases} x + \quad y = \quad 6 & (1) \\ 850x + 440y = 4000 & (2) \end{cases}.$$

Substituting $6 - x$ for y in (2), gives $410x = 1360$ or $x = 3.3$. Back-substitution into (1) gives $y = 2.7$.

The point (3.7, 1.9) is the solution to the system

$$\begin{cases} 850x + 440y = 4000 \\ x = 2y \end{cases}$$

5. The total mix contains 6 cups, so there are $521.76 \div 6 = 86.96$ grams of carbohydrates per cup.

Since each cup of raisins contain 440 calories and each cup of peanuts have 850 calories, the calorie count per cup is $(4 \cdot 440 + 2 \cdot 850) \div 6 = 576.67$ calories per cup of mix.

7. To determine the recipe needed to minimize the fat in the mix, construct a new objective function and evaluate it at the corner points.

Objective function: $f = 72.5x + 0.67y$

Corner Point	$f = 72.5x + 0.67y$
(0, 0)	$f = 72.5(0) + 0.67(0) = 0$
(3.7, 1.9)	$f = 72.5(3.7) + 0.67(1.9) = 269.52$
(3.3, 2.7)	$f = 72.5(3.3) + 0.67(2.7) = 241.06$
(2, 4)	$f = 72.5(2) + 0.67(4) = 147.68$

Since you intend to eat at least 3 cups of mix, we cannot use the corner point (0, 0). So you would minimize the fat content by making a mix that consists of 2 cups of peanuts and 4 cups of raisins.

9. This low–fat mix has 87.82 grams of protein. See problem 6.

Mathematical Questions from Professional Exams

1. (b) **3.** (c)

5. Profit is the difference between revenue and cost, so profit for product $Q = 20 - 12 = 8$, and profit for product $P = 17 - 13 = 4$. The objective function is **(d).**

7. **(c)**

9. **(b)**

11. **(b)**

13. **(c)**

Chapter 4

Linear Programming: Simplex Method

4.1 The Simplex Tableau; Pivoting

3. True

5. False. Slack variables are always nonnegative.

Problems 7-13: A maximum problem is in standard form if both of the following conditions are met.

Condition 1 All the variables are nonnegative.
Condition 2 Every other constraint is written as a linear expression that is less than or equal to a positive constant.

7. The problem is in standard form since both conditions are met.

9. The problem is not in standard form since variables x_2 and x_3 are not given as nonnegative.

11. The problem is not in standard form. The constraint $2x_1 + x_2 + 4x_3 \geq 6$ is not a linear expression that is less than or equal to a positive constant.

13. The problem is not in standard form. The constraint $2x_1 + x_2 \geq -6$ is not a linear expression that is less than or equal to a positive constant.

15. The problem is in standard form since both conditions are met.

17. The maximum problem cannot be modified so to be in standard form.

19. The maximum problem cannot be modified so to be in standard form.

21. The problem can be modified so it is in standard form. The first two constraints are multiplied by −1. The problem becomes

Maximize

$$P = 2x_1 + x_2 + 3x_3$$

subject to the constraints

$$\begin{aligned} x_1 - x_2 - x_3 &\le 6 \\ -2x_1 + 3x_2 &\le 12 \\ x_3 &\le 2 \end{aligned}$$

$x_1 \ge 0 \quad x_2 \ge 0 \quad x_3 \ge 0$

23. We write the objective function as

$$P - 2x_1 - x_2 - 3x_3 = 0$$

subject to the constraints

$$\begin{aligned} 5x_1 + 2x_2 + x_3 + s_1 &= 20 \\ 6x_1 - x_2 + 4x_3 + s_2 &= 24 \\ x_1 + x_2 + 4x_3 + s_3 &= 16 \end{aligned}$$

$$x_1 \ge 0 \quad x_2 \ge 0 \quad x_3 \ge 0 \quad s_1 \ge 0 \quad s_2 \ge 0 \quad s_3 \ge 0$$

The initial tableau is

BV	P	x_1	x_2	x_3	s_1	s_2	s_3	RHS
s_1	0	5	2	1	1	0	0	20
s_2	0	6	−1	4	0	1	0	24
s_3	0	1	1	4	0	0	1	16
P	1	−2	−1	−3	0	0	0	0

25. We write the objective function as

$$P - 3x_1 - 5x_2 = 0$$

subject to the constraints

$$\begin{aligned} 2.2x_1 - 1.8x_2 + s_1 &= 5 \\ 0.8x_1 + 1.2x_2 + s_2 &= 2.5 \\ x_1 + x_2 + s_3 &= 0.1 \end{aligned}$$

$$x_1 \ge 0 \quad x_2 \ge 0 \quad s_1 \ge 0 \quad s_2 \ge 0 \quad s_3 \ge 0$$

The initial tableau is

BV	P	x_1	x_2	s_1	s_2	s_3	RHS
s_1	0	2.2	−1.8	1	0	0	5
s_2	0	0.8	1.2	0	1	0	2.5
s_3	0	1	1	0	0	1	0.1
P	1	−3	−5	0	0	0	0

27. We write the objective function as

$$P - 2x_1 - 3x_2 - x_3 = 0$$

subject to the constraints

$$\begin{aligned} x_1 + x_2 + x_3 + s_1 \quad &= 50 \\ 3x_1 + 2x_2 + x_3 \quad + s_2 &= 10 \end{aligned}$$

$$x_1 \ge 0 \quad x_2 \ge 0 \quad x_3 \ge 0 \quad s_1 \ge 0 \quad s_2 \ge 0$$

The initial tableau is

BV	P	x_1	x_2	x_3	s_1	s_2	RHS
s_1	0	1	1	1	1	0	50
s_2	0	3	2	1	0	1	10
P	1	−2	−3	−1	0	0	0

29. We write the objective function as

$$P - 3x_1 - 4x_2 - 2x_3 = 0$$

subject to the constraints

$$\begin{aligned} 3x_1 + x_2 + 4x_3 + s_1 \quad &= 5 \\ x_1 - x_2 \quad + s_2 \quad &= 5 \\ 2x_1 - x_2 + x_3 \quad + s_3 &= 6 \end{aligned}$$

$$x_1 \ge 0 \quad x_2 \ge 0 \quad x_3 \ge 0 \quad s_1 \ge 0 \quad s_2 \ge 0 \quad s_3 \ge 0$$

The initial tableau is

BV	P	x_1	x_2	x_3	s_1	s_2	s_3	RHS
s_1	0	3	1	4	1	0	0	5
s_2	0	1	−1	0	0	1	0	5
s_3	0	2	−1	1	0	0	1	6
P	1	−3	−4	−2	0	0	0	0

31. We modify the problem by writing each linear constraint as an inequality less than or equal to a positive constant. The modified maximum problem in standard form is

Maximize

$$P = x_1 + 2x_2 + 5x_3$$

subject to the constraints:

$$\begin{aligned} x_1 - 2x_2 - 3x_3 &\le 10 \\ -3x_1 - x_2 + x_3 &\le 12 \\ x_1 \ge 0,\ x_2 \ge 0,\ x_3 &\ge 0 \end{aligned}$$

We then introduce slack variables and set up the initial simplex tableau.

$$\begin{aligned} P - x_1 - 2x_2 - 5x_3 \quad &= 0 \\ x_1 - 2x_2 - 3x_3 + s_1 \quad &= 10 \\ -3x_1 - x_2 + x_3 \quad + s_2 &= 12 \end{aligned}$$

$$x_1 \ge 0,\ x_2 \ge 0,\ x_3 \ge 0,\ s_1 \ge 0,\ s_2 \ge 0$$

The initial tableau is

BV	P	x_1	x_2	x_3	s_1	s_2	RHS
s_1	0	1	–2	–3	1	0	10
s_2	0	–3	–1	1	0	1	12
P	1	–1	–2	–5	0	0	0

33. We modify the problem by writing each linear constraint as an inequality less than or equal to a positive constant. The modified maximum problem in standard form is

Maximize

$$P = 2x_1 + 3x_2 + x_3 + 6x_4$$

subject to the constraints:

$$\begin{aligned} -x_1 + x_2 + 2x_3 + x_4 &\le 10 \\ -x_1 + x_2 - x_3 + x_4 &\le 8 \\ x_1 + x_2 + x_3 + x_4 &\le 9 \\ x_1 \ge 0 \quad x_2 \ge 0 \quad x_3 &\ge 0 \quad x_4 \ge 0 \end{aligned}$$

We then introduce slack variables and set up the initial simplex tableau.

$$\begin{aligned} P - 2x_1 - 3x_2 - x_3 - 6x_4 &= 0 \\ -x_1 + x_2 + 2x_3 + x_4 + s_1 &= 10 \\ -x_1 + x_2 - x_3 + x_4 + s_2 &= 8 \\ x_1 + x_2 + x_3 + x_4 + s_3 &= 9 \\ x_1 \ge 0 \quad x_2 \ge 0 \quad x_3 \ge 0 \quad x_4 \ge 0 \quad s_1 \ge 0 \quad s_2 &\ge 0 \quad s_3 \ge 0 \end{aligned}$$

The initial tableau is

BV	P	x_1	x_2	x_3	x_4	s_1	s_2	s_3	RHS
s_1	0	–1	1	2	1	1	0	0	10
s_2	0	–1	1	–1	1	0	1	0	8
s_3	0	1	1	1	1	0	0	1	9
P	1	–2	–3	–1	–6	0	0	0	0

35. Pivoting: Step 1: Divide each entry in the *pivot row* by the *pivot element*.

BV	P	x_1	x_2	s_1	s_2	RHS
s_1	0	1	2	1	0	300
s_2	0	[3]	2	0	1	480
P	1	–1	–2	0	0	0

→

BV	P	x_1	x_2	s_1	s_2	RHS
s_1	0	1	2	1	0	300
x_1	0	1	$\frac{2}{3}$	0	$\frac{1}{3}$	160
P	1	–1	–2	0	0	0

Step 2: Obtain 0s elsewhere in the *pivot column* by performing row operations using the revised *pivot row*.

$\xrightarrow[R_3 = R_2 + r_3]{R_1 = -R_2 + r_1}$

BV	P	x_1	x_2	s_1	s_2	RHS
s_1	0	0	$\frac{4}{3}$	1	$-\frac{1}{3}$	140
x_1	0	1	$\frac{2}{3}$	0	$\frac{1}{3}$	160
P	1	0	$-\frac{4}{3}$	0	$\frac{1}{3}$	160

The system of equations corresponding to the new tableau is

$$\begin{cases} s_1 = -\frac{4}{3}x_2 + \frac{1}{3}s_2 + 140 \\ x_1 = -\frac{2}{3}x_2 - \frac{1}{3}s_2 + 160 \\ P = \frac{4}{3}x_2 - \frac{1}{3}s_2 + 160 \end{cases}$$

The current values are $P = 160$, $x_1 = 160$, and $s_1 = 140$.

37. Pivoting: Step 1: Divide each entry in the *pivot row* by the *pivot element*.

$$\begin{array}{c} \text{BV} \\ s_1 \\ s_2 \\ s_3 \\ P \end{array} \begin{array}{c} \begin{array}{cccccccc} P & x_1 & x_2 & x_3 & s_1 & s_2 & s_3 & \text{RHS} \end{array} \\ \left[\begin{array}{ccccccc|c} 0 & 1 & 2 & 4 & 1 & 0 & 0 & 24 \\ 0 & 2 & -1 & 1 & 0 & 1 & 0 & 32 \\ 0 & \boxed{3} & 2 & 4 & 0 & 0 & 1 & 18 \\ \hline 1 & -1 & -2 & -3 & 0 & 0 & 0 & 0 \end{array}\right] \end{array} \quad \begin{array}{c} \text{BV} \\ s_1 \\ s_2 \\ \to x_1 \\ P \end{array} \begin{array}{c} \begin{array}{cccccccc} P & x_1 & x_2 & x_3 & s_1 & s_2 & s_3 & \text{RHS} \end{array} \\ \left[\begin{array}{ccccccc|c} 0 & 1 & 2 & 4 & 1 & 0 & 0 & 24 \\ 0 & 2 & -1 & 1 & 0 & 1 & 0 & 32 \\ 0 & 1 & \frac{2}{3} & \frac{4}{3} & 0 & 0 & \frac{1}{3} & 6 \\ \hline 1 & -1 & -2 & -3 & 0 & 0 & 0 & 0 \end{array}\right] \end{array}$$

Step 2: Obtain 0s elsewhere in the *pivot column* by performing row operations using the revised *pivot row*.

$$\xrightarrow[\substack{R_1 = -R_3 + r_1 \\ R_2 = -2R_3 + r_2 \\ R_4 = R_3 + r_4}]{} \begin{array}{c} \text{BV} \\ s_1 \\ s_2 \\ x_1 \\ P \end{array} \begin{array}{c} \begin{array}{cccccccc} P & x_1 & x_2 & x_3 & s_1 & s_2 & s_3 & \text{RHS} \end{array} \\ \left[\begin{array}{ccccccc|c} 0 & 0 & \frac{4}{3} & \frac{8}{3} & 1 & 0 & -\frac{1}{3} & 18 \\ 0 & 0 & -\frac{7}{3} & -\frac{5}{3} & 0 & 1 & -\frac{2}{3} & 20 \\ 0 & 1 & \frac{2}{3} & \frac{4}{3} & 0 & 0 & \frac{1}{3} & 6 \\ \hline 1 & 0 & -\frac{4}{3} & -\frac{5}{3} & 0 & 0 & \frac{1}{3} & 6 \end{array}\right] \end{array}$$

The system of equations corresponding to the new tableau is

$$\begin{cases} s_1 = -\frac{4}{3}x_2 - \frac{8}{3}x_3 + \frac{1}{3}s_3 + 18 \\ s_2 = \frac{7}{3}x_2 + \frac{5}{3}x_3 + \frac{2}{3}s_3 + 20 \\ x_1 = -\frac{2}{3}x_2 - \frac{4}{3}x_3 - \frac{1}{3}s_3 + 6 \\ P = \frac{4}{3}x_2 + \frac{5}{3}x_3 - \frac{1}{3}s_3 + 6 \end{cases}$$

The current values are $P = 6$, $x_1 = 6$, $s_1 = 18$, and $s_2 = 20$.

39. Pivoting: Step 1: Divide each entry in the *pivot row* by the *pivot element*.

$$\begin{array}{c} \text{BV} \\ s_1 \\ s_2 \\ s_3 \\ s_4 \\ P \end{array}\begin{array}{c} \begin{array}{cccccccccc} P & x_1 & x_2 & x_3 & x_4 & s_1 & s_2 & s_3 & s_4 & \text{RHS} \end{array} \\ \left[\begin{array}{ccccccccc|c} 0 & -3 & 0 & 1 & 0 & 1 & 0 & 0 & 0 & 20 \\ 0 & \boxed{2} & 0 & 0 & 1 & 0 & 1 & 0 & 0 & 24 \\ 0 & 0 & -3 & 1 & 0 & 0 & 0 & 1 & 0 & 28 \\ 0 & 0 & -3 & 0 & 1 & 0 & 0 & 0 & 1 & 24 \\ \hline 1 & -1 & -2 & -3 & -4 & 0 & 0 & 0 & 0 & 0 \end{array}\right] \end{array}$$

$$\begin{array}{c} \text{BV} \\ s_1 \\ \to x_1 \\ s_3 \\ s_4 \\ P \end{array}\begin{array}{c} \begin{array}{cccccccccc} P & x_1 & x_2 & x_3 & x_4 & s_1 & s_2 & s_3 & s_4 & \text{RHS} \end{array} \\ \left[\begin{array}{ccccccccc|c} 0 & -3 & 0 & 1 & 0 & 1 & 0 & 0 & 0 & 20 \\ 0 & 1 & 0 & 0 & \frac{1}{2} & 0 & \frac{1}{2} & 0 & 0 & 12 \\ 0 & 0 & -3 & 1 & 0 & 0 & 0 & 1 & 0 & 28 \\ 0 & 0 & -3 & 0 & 1 & 0 & 0 & 0 & 1 & 24 \\ \hline 1 & -1 & -2 & -3 & -4 & 0 & 0 & 0 & 0 & 0 \end{array}\right] \end{array}$$

Step 2: Obtain 0s elsewhere in the *pivot column* by performing row operations using the revised *pivot row*.

$$\xrightarrow[R_5=R_2+r_5]{R_1=3R_2+r_1} \begin{array}{c} \text{BV} \\ s_1 \\ x_1 \\ s_3 \\ s_4 \\ P \end{array}\begin{array}{c} \begin{array}{cccccccccc} P & x_1 & x_2 & x_3 & x_4 & s_1 & s_2 & s_3 & s_4 & \text{RHS} \end{array} \\ \left[\begin{array}{ccccccccc|c} 0 & 0 & 0 & 1 & \frac{3}{2} & 1 & \frac{3}{2} & 0 & 0 & 56 \\ 0 & 1 & 0 & 0 & \frac{1}{2} & 0 & \frac{1}{2} & 0 & 0 & 12 \\ 0 & 0 & -3 & 1 & 0 & 0 & 0 & 1 & 0 & 28 \\ 0 & 0 & -3 & 0 & 1 & 0 & 0 & 0 & 1 & 24 \\ \hline 1 & 0 & -2 & -3 & -\frac{7}{2} & 0 & \frac{1}{2} & 0 & 0 & 12 \end{array}\right] \end{array}$$

The system of equations corresponding to the new tableau is

$$\begin{cases} s_1 = -x_3 - \dfrac{3}{2}x_4 - \dfrac{3}{2}s_2 + 56 \\ x_1 = -\dfrac{1}{2}x_4 - \dfrac{1}{2}s_2 + 12 \\ s_3 = 3x_2 - x_3 + 28 \\ s_4 = 3x_2 - x_4 + 24 \\ P = 2x_2 + 2x_3 + \dfrac{7}{2}x_4 - \dfrac{1}{2}s_2 + 12 \end{cases}$$

The current values are $P = 12$, $x_1 = 12$, $s_1 = 56$, $s_3 = 28$, and $s_4 = 24$.

41. (a) P is the profit, x_1 is the number of sleeping dolls, x_2 is the number of talking dolls, and x_3 is the number of walking dolls.

(b) To find the profit, we subtract the cost of each doll from its selling price.
Sleeping baby dolls: $12 – $6 = $6; Talking baby dolls: $13.50 – $7.50 = $6; Walking baby dolls: $17 – $9 = $8. We use these values in the objective function.

Maximize

$$P = 6x_1 + 6x_2 + 8x_3$$

subject to the constraints

$$\begin{aligned} x_1 + x_2 + x_3 &\le 60 \\ 6x_1 + 7.5x_2 + 9x_3 &\le 405 \\ x_1 \ge 0 \quad x_2 \ge 0 \quad x_3 &\ge 0 \end{aligned}$$

(c) System with slack variables added:

$$\begin{aligned} x_1 + x_2 + x_3 + s_1 \qquad &= 60 \\ 6x_1 + 7.5x_2 + 9x_3 \qquad + s_2 &= 405 \\ P - 6x_1 - 6x_2 - 8x_3 \qquad &= 0 \\ x_1 \ge 0 \quad x_2 \ge 0 \quad x_3 \ge 0 \quad s_1 \ge 0 \quad s_2 &\ge 0 \end{aligned}$$

(d) The initial tableau

BV	P	x_1	x_2	x_3	s_1	s_2	RHS
s_1	0	1	1	1	1	0	60
s_2	0	6	7.5	9	0	1	405
P	1	−6	−6	−8	0	0	0

43. (a) P is the daily profit, x_1 is the number of shirts produced, x_2 is the number of jackets produced, and x_3 is the number of pairs of pants produced each day.

(b) Maximize

$$P = 19x_1 + 34x_2 + 15x_3$$

subject to the constraints

$$\begin{aligned} 10x_1 + 20x_2 + 20x_3 &\le 500 \\ 20x_1 + 40x_2 + 30x_3 &\le 900 \\ x_1 + x_2 + x_3 &\le 30 \\ x_1 \ge 0 \quad x_2 \ge 0 \quad x_3 &\ge 0 \end{aligned}$$

(c) System with slack variables added:

$$\begin{aligned} 10x_1 + 20x_2 + 20x_3 + s_1 \qquad &= 500 \\ 20x_1 + 40x_2 + 30x_3 \qquad + s_2 \qquad &= 900 \\ x_1 + x_2 + x_3 \qquad + s_3 &= 30 \\ P - 19x_1 - 34x_2 - 15x_3 \qquad &= 0 \\ x_1 \ge 0 \quad x_2 \ge 0 \quad x_3 \ge 0 \quad s_1 \ge 0 \quad s_2 \ge 0 \quad s_3 &\ge 0 \end{aligned}$$

(d) The initial tableau:

$$\begin{array}{c} \text{BV} \\ s_1 \\ s_2 \\ s_3 \\ P \end{array} \begin{array}{c} \begin{matrix} P & x_1 & x_2 & x_3 & s_1 & s_2 & s_3 & \text{RHS} \end{matrix} \\ \left[\begin{array}{ccccccc|c} 0 & 10 & 20 & 20 & 1 & 0 & 0 & 500 \\ 0 & 20 & 40 & 30 & 0 & 1 & 0 & 900 \\ 0 & 1 & 1 & 1 & 0 & 0 & 1 & 30 \\ \hline 1 & -19 & -34 & -15 & 0 & 0 & 0 & 0 \end{array}\right] \end{array}$$

45. (a) P is the return on her investments;
x_1 is the amount invested in the money market,
x_2 is the amount invested in the mutual fund, and
x_3 is the amount invested in the CD.

(b) Maximize

$$P = 0.0525x_1 + 0.075x_2 + 0.04x_3$$

Subject to the constraints

$$\begin{aligned} x_1 + x_2 + x_3 &\le 25{,}000 \\ x_2 + x_3 &\le 10{,}000 \\ -x_1 \qquad + x_3 &\le 1{,}500 \\ x_1 \ge 0 \quad x_2 \ge 0 \quad x_3 &\ge 0 \end{aligned}$$

(c) System with slack variables added:

$$\begin{aligned} x_1 + x_2 + x_3 + s_1 \qquad\qquad &= 25{,}000 \\ x_2 + x_3 \qquad + s_2 \qquad &= 10{,}000 \\ -x_1 \qquad + x_3 \qquad\qquad + s_3 &= 1{,}500 \\ P - 0.0525x_1 - 0.075x_2 - 0.04x_3 \qquad &= 0 \\ x_1 \ge 0 \quad x_2 \ge 0 \quad x_3 \ge 0 \quad s_1 \ge 0 \quad s_2 \ge 0 \quad s_3 &\ge 0 \end{aligned}$$

(d) Initial tableau:

$$\begin{array}{c} \text{BV} \\ s_1 \\ s_2 \\ s_3 \\ P \end{array} \begin{array}{c} \begin{matrix} P & x_1 & x_2 & x_3 & s_1 & s_2 & s_3 & \text{RHS} \end{matrix} \\ \left[\begin{array}{ccccccc|c} 0 & 1 & 1 & 1 & 1 & 0 & 0 & 25{,}000 \\ 0 & 0 & 1 & 1 & 0 & 1 & 0 & 10{,}000 \\ 0 & -1 & 0 & 1 & 0 & 0 & 1 & 1{,}500 \\ \hline 1 & -0.0525 & -0.075 & -0.04 & 0 & 0 & 0 & 0 \end{array}\right] \end{array}$$

4.2 The Simplex Method: Solving Maximum Problems in Standard Form

1. False. The pivot element is never in the objective row. The pivot element is chosen from the positive entries above the objective row in the pivot column.

3. False. The maximum problem is unbounded and has no solution only if all the entries in the pivot column are zero or negative.

5. (b) It requires further pivoting. Since there is still a negative entry in the objective row, the tableau needs further pivoting. We find the pivot element by selecting the first column containing a negative entry in the objective row. Here it is column x_1. We select the pivot row by computing the quotients formed by dividing the entry in the right hand side by the corresponding positive entry of the pivot column. The pivot row has the smallest nonnegative quotient.

$$20 \div 1 = 20 \text{ and } 30 \div \frac{1}{2} = 60$$

The pivot element is 1 in row s_1, column x_1.

7. (a) This is the final tableau since there are no negative entries in the objective row. The maximum value is $P = \frac{256}{7}$. It occurs when $x_1 = \frac{32}{7}$ and $x_2 = 0$.

9. (c) There is no solution to this problem. Although there is a negative entry in the objective row, all entries in the pivot column are negative, so the problem is unbounded and has no solution.

11. (b) This tableau requires further pivoting since there are still negative entries in the objective row. We find the pivot element by selecting the first column containing a negative entry in the objective row. Here it is column s_1. We select the pivot row by computing the quotients formed by dividing the entry in the right hand side by the corresponding positive entry of the pivot column. The pivot row has the smallest nonnegative quotient.

$$6 \div 1 = 6 \text{ and } 1 \div 1 = 1.$$

The new pivot element is 1 in row x_1, column s_1.

13. To solve the problem using the simplex method, we first must introduce slack variables and construct the initial tableau.

Maximize

$$P - 5x_1 - 7x_2 = 0$$

subject to the constraints

$$\begin{aligned} 2x_1 + 3x_2 + s_1 \quad\quad &= 12 \\ 3x_1 + x_2 \quad\quad + s_2 &= 12 \\ x_1 \geq 0 \;\; x_2 \geq 0 \;\; s_1 \geq 0 \;\; s_2 &\geq 0 \end{aligned}$$

The initial simplex tableau with the pivot element marked is below. The first negative entry in the objective row identifies the pivot column. The smallest nonnegative quotient formed by the RHS and positive entries in the pivot column identifies the pivot row. Here the pivot element is in row s_2, column x_1.

$$\begin{array}{c} \text{BV} \\ s_1 \\ s_2 \\ P \end{array} \begin{array}{c} \begin{array}{cccccc} P & x_1 & x_2 & s_1 & s_2 & \text{RHS} \end{array} \\ \left[\begin{array}{ccccc|c} 0 & 2 & 3 & 1 & 0 & 12 \\ 0 & \boxed{3} & 1 & 0 & 1 & 12 \\ \hline 1 & -5 & -7 & 0 & 0 & 0 \end{array}\right] \end{array}$$

We pivot using row operations

$$\xrightarrow[R_2=\frac{1}{3}r_2]{} \begin{array}{c} \text{BV} \\ s_1 \\ x_1 \\ P \end{array} \begin{array}{c} \begin{array}{cccccc} P & x_1 & x_2 & s_1 & s_2 & \text{RHS} \end{array} \\ \left[\begin{array}{ccccc|c} 0 & 2 & 3 & 1 & 0 & 12 \\ 0 & 1 & \frac{1}{3} & 0 & \frac{1}{3} & 4 \\ \hline 1 & -5 & -7 & 0 & 0 & 0 \end{array}\right] \end{array} \xrightarrow[\substack{R_1=-2R_2+r_1 \\ R_3=5R_2+r_3}]{} \begin{array}{c} \text{BV} \\ s_1 \\ x_1 \\ P \end{array} \begin{array}{c} \begin{array}{cccccc} P & x_1 & x_2 & s_1 & s_2 & \text{RHS} \end{array} \\ \left[\begin{array}{ccccc|c} 0 & 0 & \boxed{\frac{7}{3}} & 1 & -\frac{2}{3} & 4 \\ 0 & 1 & \frac{1}{3} & 0 & \frac{1}{3} & 4 \\ \hline 1 & 0 & -\frac{16}{3} & 0 & \frac{5}{3} & 20 \end{array}\right] \end{array}$$

Since there is still a negative entry in the objective row, the problem needs further pivoting. We choose the pivot element in row s_1, column x_2, and pivot using row operations.

$$\xrightarrow[R_1=\frac{3}{7}r_1]{} \begin{array}{c} \text{BV} \\ x_2 \\ x_1 \\ P \end{array} \begin{array}{c} \begin{array}{cccccc} P & x_1 & x_2 & s_1 & s_2 & \text{RHS} \end{array} \\ \left[\begin{array}{ccccc|c} 0 & 0 & 1 & \frac{3}{7} & -\frac{2}{7} & \frac{12}{7} \\ 0 & 1 & \frac{1}{3} & 0 & \frac{1}{3} & 4 \\ \hline 1 & 0 & -\frac{16}{3} & 0 & \frac{5}{3} & 20 \end{array}\right] \end{array} \xrightarrow[\substack{R_2=-\frac{1}{3}R_1+r_2 \\ R_3=\frac{16}{3}R_1+r_3}]{} \begin{array}{c} \text{BV} \\ x_1 \\ x_1 \\ P \end{array} \begin{array}{c} \begin{array}{cccccc} P & x_1 & x_2 & s_1 & s_2 & \text{RHS} \end{array} \\ \left[\begin{array}{ccccc|c} 0 & 0 & 1 & \frac{3}{7} & -\frac{2}{7} & \frac{12}{7} \\ 0 & 1 & 0 & -\frac{1}{7} & \frac{3}{7} & \frac{24}{7} \\ \hline 1 & 0 & 0 & \frac{16}{7} & \frac{1}{7} & \frac{204}{7} \end{array}\right] \end{array}$$

This is the final tableau since all entries in the objective row are positive. The solution is $P=\frac{204}{7}$, obtained when $x_1=\frac{24}{7}$ and $x_2=\frac{12}{7}$.

15. To solve the problem using the simplex method, we first must introduce slack variables and construct the initial tableau.

Maximize

$$P-5x_1-7x_2=0$$

subject to the constraints

$$\begin{aligned} x_1+2x_2+s_1 \quad &=2 \\ 2x_1+\ x_2 \quad +s_2 &=2 \\ x_1\ge 0\ x_2\ge 0\ s_1\ge 0\ s_2 &\ge 0 \end{aligned}$$

The initial Simplex tableau with the pivot element marked is below. The first negative entry in the objective row identifies the pivot column. The smallest nonnegative quotient formed by the RHS and positive entries in the pivot column identifies the pivot row. Here the pivot element is in row s_2, column x_1. We then use row operations to make the pivot element 1 and all the other entries in the pivot column 0.

$$\begin{array}{c} \text{BV} \\ s_1 \\ s_2 \\ P \end{array} \begin{array}{c} \begin{array}{cccccc} P & x_1 & x_2 & s_1 & s_2 & \text{RHS} \end{array} \\ \left[\begin{array}{ccccc|c} 0 & 1 & 2 & 1 & 0 & 2 \\ 0 & \boxed{2} & 1 & 0 & 1 & 2 \\ \hline 1 & -5 & -7 & 0 & 0 & 0 \end{array}\right] \end{array} \xrightarrow[\substack{R_2=\frac{1}{2}r_2 \\ R_1=-R_2+r_1 \\ R_3=5R_2+r_3}]{} \begin{array}{c} \text{BV} \\ s_1 \\ x_1 \\ P \end{array} \begin{array}{c} \begin{array}{cccccc} P & x_1 & x_2 & s_1 & s_2 & \text{RHS} \end{array} \\ \left[\begin{array}{ccccc|c} 0 & 0 & \boxed{\frac{3}{2}} & 1 & -\frac{1}{2} & 1 \\ 0 & 1 & \frac{1}{2} & 0 & \frac{1}{2} & 1 \\ \hline 0 & 0 & -\frac{9}{2} & 0 & \frac{5}{2} & 5 \end{array}\right] \end{array}$$

Since there is still a negative entry in the objective row, we need to pivot again. We choose the pivot entry as before. The pivot column will be x_2, the pivot row will be s_1 because

$$\frac{1}{\frac{3}{2}}=\frac{2}{3}<\frac{1}{\frac{1}{2}}=2$$

$\xrightarrow[R_3=\frac{9}{2}R_1+r_3]{R_1=\frac{2}{3}r_1,\ R_2=-\frac{1}{2}R_1+r_2}$

BV	P	x_1	x_2	s_1	s_2	RHS
x_2	0	0	1	$\frac{2}{3}$	$-\frac{1}{3}$	$\frac{2}{3}$
x_1	0	1	0	$-\frac{1}{3}$	$\frac{2}{3}$	$\frac{2}{3}$
P	0	0	0	3	1	8

This is the final tableau since all entries in the objective row are nonnegative. The solution is $P = 8$, obtained when $x_1 = \frac{2}{3}$ and $x_2 = \frac{2}{3}$.

17. To solve the problem using the simplex method, we first must introduce slack variables and construct the initial tableau.

Maximize

$$P - 3x_1 - x_2 = 0$$

subject to the constraints

$$\begin{aligned} x_1 + x_2 + s_1 &= 2 \\ 2x_1 + 3x_2 + s_2 &= 12 \\ 3x_1 + x_2 + s_3 &= 12 \\ x_1 \geq 0 \quad x_2 \geq 0 \quad s_1 \geq 0 \quad s_2 &\geq 0 \quad s_3 \geq 0 \end{aligned}$$

The initial Simplex tableau with the pivot element marked is below. The first negative entry in the objective row identifies the pivot column. The smallest nonnegative quotient formed by the RHS and positive entries in the pivot column identifies the pivot row. Here the pivot element is in row s_1, column x_1. We then use row operations to make all entries in the pivot column other than the pivot element 0.

BV	P	x_1	x_2	s_1	s_2	s_3	RHS
s_1	0	[1]	1	1	0	0	2
s_2	0	2	3	0	1	0	12
s_3	0	3	1	0	0	1	12
P	1	−3	−1	0	0	0	0

$\xrightarrow{R_2=-2R_1+r_2,\ R_3=-3R_1+r_3,\ R_4=3R_1+r_4}$

BV	P	x_1	x_2	s_1	s_2	s_3	RHS
x_1	0	1	1	1	0	0	2
s_2	0	0	1	−2	1	0	8
s_3	0	0	−2	−3	0	1	6
P	1	0	2	3	0	0	6

This is the final tableau since all entries in the objective row are nonnegative. The solution is $P = 6$, obtained when $x_1 = 2$ and $x_2 = 0$.

19. To solve the problem using the simplex method, we first must introduce slack variables and construct the initial tableau.

Maximize

$$P - 2x_1 - x_2 - x_3 = 0$$

subject to the constraints

$$\begin{aligned} -2x_1 + x_2 - 2x_3 + s_1 &= 4 \\ x_1 - 2x_2 + x_3 + s_2 &= 2 \\ x_1 \geq 0 \quad x_2 \geq 0 \quad x_3 \geq 0 \quad s_1 &\geq 0 \quad s_2 \geq 0 \end{aligned}$$

The initial Simplex tableau with the pivot element marked is below. The first negative entry in the objective row identifies the pivot column. The smallest nonnegative quotient formed by the RHS and positive entries in the pivot column identifies the pivot row. Here the pivot element is in row s_2, column x_1. We then use row operations to make all entries in the pivot column other than the pivot element 0.

BV	P	x_1	x_2	x_3	s_1	s_2	RHS
s_1	0	−2	1	−2	1	0	4
s_2	0	[1]	−2	1	0	1	2
P	1	−2	−1	−1	0	0	0

$\xrightarrow{R_1 = 2R_2 + r_1,\ R_3 = 2R_2 + r_3}$

BV	P	x_1	x_2	x_3	s_1	s_2	RHS
s_1	0	0	−3	0	1	2	8
x_1	0	1	−2	1	0	1	2
P	1	0	−5	1	0	2	4

The new pivot column is x_2, since −5 is the only negative entry in the objective row. But all the entries in the pivot column are negative. This problem is unbounded and has no solution.

21. To solve the problem using the simplex method, we first must introduce slack variables and construct the initial tableau.

Maximize

$$P - 2x_1 - x_2 - 3x_3 = 0$$

subject to the constraints

$$\begin{aligned} x_1 + 2x_2 + x_3 + s_1 \quad &= 25 \\ 3x_1 + 2x_2 + 3x_3 \quad + s_2 &= 30 \\ x_1 \ge 0 \quad x_2 \ge 0 \quad x_3 \ge 0 \quad s_1 &\ge 0 \quad s_2 \ge 0 \end{aligned}$$

The initial Simplex tableau with the pivot element marked is below. The first negative entry in the objective row identifies the pivot column. The smallest nonnegative quotient formed by the RHS and positive entries in the pivot column identifies the pivot row. Here the pivot element is in row s_2, column x_1. We then use row operations to make all entries in the pivot column other than the pivot element 0.

BV	P	x_1	x_2	x_3	s_1	s_2	RHS
s_1	0	1	2	1	1	0	25
s_2	0	[3]	2	3	0	1	30
P	1	−2	−1	−3	0	0	0

$\xrightarrow{R_2 = \frac{1}{3}r_2,\ R_1 = -R_2 + r_1,\ R_3 = 2R_2 + r_3}$

BV	P	x_1	x_2	x_3	s_1	s_2	RHS
s_1	0	0	$\frac{4}{3}$	0	1	$-\frac{1}{3}$	15
x_1	0	1	$\frac{2}{3}$	[1]	0	$\frac{1}{3}$	10
P	1	0	$\frac{1}{3}$	−1	0	$\frac{2}{3}$	20

Since there is still a negative entry in the objective row, we need to pivot again. We choose the pivot entry as before. The pivot column will be x_3, the pivot row will be x_1.

$\xrightarrow{R_3 = R_2 + r_3}$

BV	P	x_1	x_2	x_3	s_1	s_2	RHS
s_1	0	0	$\frac{4}{3}$	0	1	$-\frac{1}{3}$	15
x_3	0	1	$\frac{2}{3}$	1	0	$\frac{1}{3}$	10
P	1	1	1	0	0	1	30

Since all the entries in the objective row are nonnegative, this is the final tableau. The solution is $P = 30$, obtained when $x_1 = 0$, $x_2 = 0$, and $x_3 = 10$.

23. To solve the problem using the simplex method, we first must introduce slack variables and construct the initial tableau.

Maximize

$$P - 2x_1 - 4x_2 - x_3 - x_4 = 0$$

subject to the constraints

$$\begin{aligned} 2x_1 + x_2 + 2x_3 + 3x_4 + s_1 \qquad &= 12 \\ 2x_2 + x_3 + 2x_4 \quad + s_2 \quad &= 20 \\ 2x_1 + x_2 + 4x_3 \qquad + s_3 &= 16 \\ x_1 \ge 0 \quad x_2 \ge 0 \quad x_3 \ge 0 \quad x_4 \ge 0 \quad s_1 \ge 0 \quad s_2 &\ge 0 \quad s_3 \ge 0 \end{aligned}$$

The initial tableau is below. The pivot element is 2 (row s_1, column x_1), making x_1 a basic variable.

BV	P	x_1	x_2	x_3	x_4	s_1	s_2	s_3	RHS
s_1	0	**[2]**	1	2	3	1	0	0	12
s_2	0	0	2	1	2	0	1	0	20
s_3	0	2	1	4	0	0	0	1	16
P	1	−2	−4	−1	−1	0	0	0	0

$\xrightarrow{R_1=\frac{1}{2}r_1,\ R_3=-2R_1+r_3,\ R_4=2R_1+r_4}$

BV	P	x_1	x_2	x_3	x_4	s_1	s_2	s_3	RHS
x_1	0	1	$\frac{1}{2}$	1	$\frac{3}{2}$	$\frac{1}{2}$	0	0	6
s_2	0	0	**[2]**	1	2	0	1	0	20
s_3	0	0	0	2	−3	−1	0	1	4
P	1	0	−3	1	2	1	0	0	12

The new tableau needs further pivoting. The new pivot column is x_2, the new pivot row is x_2 making 2 the pivot element.

$\xrightarrow{R_2=\frac{1}{2}r_2,\ R_1=-\frac{1}{2}R_2+r_1,\ R_4=3R_2+r_4}$

BV	P	x_1	x_2	x_3	x_4	s_1	s_2	s_3	RHS
x_1	0	1	0	$\frac{3}{4}$	1	$\frac{1}{2}$	$-\frac{1}{4}$	0	1
x_2	0	0	1	$\frac{1}{2}$	1	0	$\frac{1}{2}$	0	10
s_3	0	0	0	2	−3	−1	0	1	4
P	1	0	0	$\frac{5}{2}$	5	1	$\frac{3}{2}$	0	42

Since there are no negative entries in the objective row, this is the final tableau. The solution is $P = 42$, obtained when $x_1 = 1$, $x_2 = 10$, $x_3 = 0$, and $x_4 = 0$.

25. To solve the problem using the simplex method, we first must introduce slack variables and construct the initial tableau.

Maximize

$$P - 2x_1 - x_2 - x_3 = 0$$

subject to the constraints

$$\begin{aligned} x_1 + 2x_2 + 4x_3 + s_1 \qquad\qquad &= 20 \\ 2x_1 + 4x_2 + 4x_3 \qquad + s_2 \qquad &= 60 \\ 3x_1 + 4x_2 + \ x_3 \qquad\qquad + s_3 &= 90 \\ x_1 \geq 0 \quad x_2 \geq 0 \quad x_3 \geq 0 \quad s_1 \geq 0 \quad s_2 \geq 0 \quad s_3 &\geq 0 \end{aligned}$$

The initial tableau is below. The pivot element is 1; the entering variable is x_1.

BV	P	x_1	x_2	x_3	s_1	s_2	s_3	RHS
s_1	0	**[1]**	2	4	1	0	0	20
s_2	0	2	4	4	0	1	0	60
s_3	0	3	4	1	0	0	1	90
P	1	−2	−1	−1	0	0	0	0

$\xrightarrow{R_2=-2r_1+r_2,\ R_3=-3r_1+r_3,\ R_4=2r_1+r_4}$

BV	P	x_1	x_2	x_3	s_1	s_2	s_3	RHS
x_1	0	1	2	4	1	0	0	20
s_2	0	0	0	−4	−2	1	0	20
s_3	0	0	−2	−11	−3	0	1	30
P	1	0	3	7	2	0	0	40

Since all entries in the new tableau are nonnegative, it is the final tableau. The solution is $P = 40$, obtained when $x_1 = 20$, $x_2 = 0$, and $x_3 = 0$.

27. To solve the problem using the simplex method, we first must introduce slack variables and construct the initial tableau.

Maximize

$$P - x_1 - 2x_2 - 4x_3 + x_4 = 0$$

subject to the constraints

$$\begin{aligned} 5x_1 \quad\quad + 4x_3 + 6x_4 + s_1 \quad\quad &= 20 \\ 4x_1 + 2x_2 + 2x_3 + 8x_4 \quad\quad + s_2 &= 40 \\ x_1 \ge 0 \;\; x_2 \ge 0 \;\; x_3 \ge 0 \;\; x_4 \ge 0 \;\; s_1 &\ge 0 \;\; s_2 \ge 0 \end{aligned}$$

The initial tableau is below. The pivot element is 5. The new basic variable will be x_1.

BV	P	x_1	x_2	x_3	x_4	s_1	s_2	RHS
s_1	0	[5]	0	4	6	1	0	20
s_2	0	4	2	2	8	0	1	40
P	1	−1	−2	−4	1	0	0	0

$\xrightarrow{R_1=\frac{1}{5}r_1,\; R_2=-4R_1+r_2,\; R_3=R_1+r_3}$

BV	P	x_1	x_2	x_3	x_4	s_1	s_2	RHS
x_1	0	1	0	$\frac{4}{5}$	$\frac{6}{5}$	$\frac{1}{5}$	0	4
s_2	0	0	[2]	$-\frac{6}{5}$	$\frac{16}{5}$	$-\frac{4}{5}$	1	24
P	1	0	−2	$-\frac{16}{5}$	$\frac{11}{5}$	$\frac{1}{5}$	0	4

Since there is still a negative entry in the objective row, this tableau needs further pivoting. The new pivot will be 2, x_2 will be the entering basic variable.

$\xrightarrow{R_2=\frac{1}{2}r_2,\; R_3=2R_2+r_3}$

BV	P	x_1	x_2	x_3	x_4	s_1	s_2	RHS
x_1	0	1	0	[$\frac{4}{5}$]	$\frac{6}{5}$	$\frac{1}{5}$	0	4
x_2	0	0	1	$-\frac{3}{5}$	$\frac{8}{5}$	$-\frac{2}{5}$	$\frac{1}{2}$	12
P	1	0	0	$-\frac{22}{5}$	$\frac{27}{5}$	$-\frac{3}{5}$	1	28

$\xrightarrow{R_1=\frac{5}{4}r_1,\; R_2=\frac{3}{5}R_1+r_2,\; R_3=\frac{22}{5}R_1+r_3}$

BV	P	x_1	x_2	x_3	x_4	s_1	s_2	RHS
x_3	0	$\frac{5}{4}$	0	1	$\frac{3}{2}$	$\frac{1}{4}$	0	5
x_2	0	$\frac{3}{4}$	1	0	$\frac{5}{2}$	$-\frac{1}{4}$	$\frac{1}{2}$	15
P	1	$\frac{11}{2}$	0	0	12	$\frac{1}{2}$	1	50

This is the final tableau. The solution $P = 50$, is obtained when $x_1 = 0$, $x_2 = 15$, $x_3 = 5$, and $x_4 = 0$.

29. (a) We let x_1, x_2, and x_3 represent the number of type I jeans, type II jeans and type III jeans produced. We want to

Maximize

$$P = 4x_1 + 4.50x_2 + 6x_3$$

subject to (manufacturing) constraints

$$\begin{aligned} 8x_1 + 12x_2 + 18x_3 + &\le 5200 \quad \text{Cutting constraint} \\ 12x_1 + 18x_2 + 24x_3 + &\le 6000 \quad \text{Sewing constraint} \\ 4x_1 + 8x_2 + 12x_3 + &\le 2200 \quad \text{Finishing constraint} \\ x_1 \ge 0 \;\; x_2 \ge 0 \;\; x_3 &\ge 0 \end{aligned}$$

(b) Slack variables are added and the initial tableau is constructed and shown below.

BV	P	x_1	x_2	x_3	s_1	s_2	s_3	RHS
s_1	0	8	12	18	1	0	0	5200
s_2	0	**[12]**	18	24	0	1	0	6000
s_3	0	4	8	12	0	0	1	2200
P	1	−3	−4.5	−6	0	0	0	0

Using the standard pivoting procedure, the pivot element is chosen from column x_1. The quotients of the RHS and the entries in column x_1 are computed and compared.

$$\frac{5200}{8}=650,\ \frac{6000}{12}=500,\text{ and } \frac{2200}{4}=550$$

The smallest quotient, 500, indicates that the pivot element is 12. So x_1 will enter the basis and s_2 will exit the basis. The new tableau will be

$R_2=\frac{1}{12}r_2$, $R_1=-8R_2+r_1$, $R_3=-4R_2+r_3$, $R_4=3R_2+r_4$ →

BV	P	x_1	x_2	x_3	s_1	s_2	s_3	RHS
s_1	0	0	0	2	1	$-\frac{2}{3}$	0	1200
x_1	0	1	$\frac{3}{2}$	2	0	$\frac{1}{12}$	0	500
s_3	0	0	2	4	0	$-\frac{1}{3}$	1	200
P	1	0	1.5	2	0	$\frac{1}{3}$	0	2000

This is the final tableau. The solution is $P = 2000$, obtained when $x_1 = 500$, $x_2 = 0$, and $x_3 = 0$.

(d) The manufacturer should produce 500 pairs of type I jeans and no pairs of type II or type III jeans for a maximum profit of \$2000.

31. (a) Let P be the weekly profit, x_1 be the number of sets of steak knives sold per week, x_2 be the number of carving knives sold weekly, and x_3 be the number of butcher knives sold per week.

Maximize

$$P = x_1 + x_2 + 2x_3$$

subject to the constraints

$$\begin{aligned} 3x_1 + 5x_2 + 4x_3 &\le 500 \\ 10x_1 + 15x_2 + 12x_3 &\le 1800 \\ 0.50x_1 + \phantom{15x_2 +{}} x_3 &\le 75 \\ x_1 \ge 0 \quad x_2 \ge 0 \quad x_3 &\ge 0 \end{aligned}$$

(b) Slack variables are added and the initial tableau is constructed and shown below. The pivot element is 0.5.

BV	P	x_1	x_2	x_3	s_1	s_2	s_3	RHS
s_1	0	3	5	4	1	0	0	500
s_2	0	10	15	12	0	1	0	1800
s_3	0	**[0.5]**	0	1	0	0	1	75
P	1	−1	−1	−2	0	0	0	0

$$\begin{array}{c} \\ R_3=2r_3 \\ R_1=-3R_3+r_1 \\ R_2=-10R_3+r_2 \\ R_4=R_3+r_2 \end{array} \longrightarrow \begin{array}{c} \text{BV} \\ s_1 \\ s_2 \\ x_1 \\ P \end{array} \begin{array}{c} \begin{array}{cccccccc} P & x_1 & x_2 & x_3 & s_1 & s_2 & s_3 & \text{RHS} \end{array} \\ \left[\begin{array}{ccccccc|c} 0 & 0 & \boxed{5} & -2 & 1 & 0 & -6 & 50 \\ 0 & 0 & 15 & -8 & 0 & 1 & -20 & 300 \\ 0 & 1 & 0 & 2 & 0 & 0 & 2 & 150 \\ \hline 1 & 0 & -1 & 0 & 0 & 0 & 2 & 150 \end{array}\right] \end{array}$$

Since there is still a negative entry in the objective row, it denotes the new pivot column. The pivot row is determined by choosing the smallest quotient when the RHS is divided by the elements in the pivot column. The new pivot element is 5.

$$\begin{array}{c} \\ R_1=\frac{1}{5}r_1 \\ R_2=-15R_1+r_2 \\ R_4=R_1+r_4 \end{array} \longrightarrow \begin{array}{c} \text{BV} \\ x_2 \\ s_2 \\ x_1 \\ P \end{array} \begin{array}{c} \begin{array}{cccccccc} P & x_1 & x_2 & x_3 & s_1 & s_2 & s_3 & \text{RHS} \end{array} \\ \left[\begin{array}{ccccccc|c} 0 & 0 & 1 & -\frac{2}{5} & \frac{1}{5} & 0 & -\frac{6}{5} & 10 \\ 0 & 0 & 0 & -2 & -3 & 1 & -2 & 150 \\ 0 & 1 & 0 & 2 & 0 & 0 & -2 & 150 \\ \hline 1 & 0 & 0 & -\frac{2}{5} & \frac{1}{5} & 0 & \frac{4}{5} & 160 \end{array}\right] \end{array}$$

$$\begin{array}{c} \\ R_3=\frac{1}{2}r_3 \\ R_1=\frac{2}{5}R_3+r_1 \\ R_2=2R_3+r_2 \\ R_4=\frac{2}{5}R_3+r_4 \end{array} \longrightarrow \begin{array}{c} \text{BV} \\ x_2 \\ s_2 \\ x_3 \\ P \end{array} \begin{array}{c} \begin{array}{cccccccc} P & x_1 & x_2 & x_3 & s_1 & s_2 & s_3 & \text{RHS} \end{array} \\ \left[\begin{array}{ccccccc|c} 0 & \frac{1}{5} & 1 & 0 & \frac{1}{5} & 0 & -\frac{4}{5} & 40 \\ 0 & 1 & 0 & 0 & -3 & 1 & 0 & 300 \\ 0 & \frac{1}{2} & 0 & 1 & 0 & 0 & 1 & 75 \\ \hline 1 & \frac{1}{5} & 0 & 0 & \frac{1}{5} & 0 & \frac{6}{5} & 190 \end{array}\right] \end{array}$$

Since all the entries in the objective row are nonnegative, this is the final tableau. The solution is $P = 190$, obtained when $x_1 = 0$, $x_2 = 40$, and $x_3 = 75$.

(c) The sales person makes a maximum profit of \$190 a week when constrained as in the problem by selling no steak knives, 40 carving knives, and 75 butcher knives.

33. (a) We let P denote the company's revenue and we let x_1, x_2, and x_3 represent the number of gallons of regular, premium, and super premium gasoline, respectively to be refined. The company wants to refine amounts that will maximize its revenue subject to its available resources.

Maximize

$$P = 2.20x_1 + 2.30x_2 + 2.40x_3$$

subject to the constraints

$$\begin{aligned} 0.6x_1 + 0.7x_2 + 0.8x_3 &\le 140{,}000 \\ 0.4x_1 + 0.3x_2 + 0.2x_3 &\le 120{,}000 \\ x_1 + x_2 + x_3 &\le 225{,}000 \\ x_1 \ge 0 \quad x_2 \ge 0 \quad x_3 &\ge 0 \end{aligned}$$

(b) Slack variables are added and the initial tableau is constructed and shown below. The pivot element is in column x_1 and row s_3.

BV	P	x_1	x_2	x_3	s_1	s_2	s_3	RHS
s_1	0	0.6	0.7	0.8	1	0	0	140,000
s_2	0	0.4	0.3	0.2	0	1	0	120,000
s_3	0	$\boxed{1}$	1	1	0	0	1	225,000
P	1	−2.2	−2.3	−2.4	0	0	0	0

$\xrightarrow{\substack{R_1=-0.6R_3+r_1 \\ R_2=-0.4R_3+r_2 \\ R_4=2.2R_3+r_4}}$

BV	P	x_1	x_2	x_3	s_1	s_2	s_3	RHS
s_1	0	0	$\boxed{0.1}$	0.2	1	0	−0.6	5000
s_2	0	0	−0.1	−0.2	0	1	−0.4	30,000
x_1	0	1	1	1	0	0	1.0	225,000
P	1	0	−0.1	−0.2	0	0	2.2	495,000

$\xrightarrow{\substack{R_1=10r_1 \\ R_2=\frac{1}{10}R_1+r_2 \\ R_3=-R_1+r_3 \\ R_4=0.1R_1+r_4}}$

BV	P	x_1	x_2	x_3	s_1	s_2	s_3	RHS
x_2	0	0	1	2	10	0	−6	50,000
s_2	0	0	0	0	1	1	−1	35,000
x_1	0	1	0	−1	−10	0	7	175,000
P	1	0	0	0	1	0	1.6	500,000

Since all the entries in the objective row are nonnegative, this is the final tableau. The solution is $P = 500{,}000$, obtained when $x_1 = 175{,}000$, $x_2 = 50{,}000$, and $x_3 = 0$.

(c) To achieve a maximize revenue of $500,000, the company should refine 175,000 gallons of regular gasoline, 50,000 gallons of premium gasoline, and no super premium gasoline.

35. (a) Let P be the total yield, x_1 be the amount invested in stocks, x_2 be the amount invested in corporate bonds, and x_3 be the amount invested in municipal bonds.

Maximize

$$P = 0.1x_1 + 0.08x_2 + 0.06x_3$$

subject to the constraints

$$\begin{aligned} x_1 + x_2 + x_3 &\le 90{,}000 \\ x_1 \qquad\qquad &\le 45{,}000 \\ x_2 - x_3 &\le 18{,}000 \\ x_1 \ge 0 \quad x_2 \ge 0 \quad x_3 &\ge 0 \end{aligned}$$

(b) Slack variables are added and the initial tableau is constructed and shown below. The pivot element is 1.

BV	P	x_1	x_2	x_3	s_1	s_2	s_3	RHS
s_1	0	1	1	1	1	0	0	90,000
s_2	0	$\boxed{1}$	0	0	0	1	0	45,000
s_3	0	0	1	−1	0	0	1	18,000
P	1	−0.1	−0.08	−0.06	0	0	0	0

$R_1 = -R_2 + r_1$, $R_4 = 0.1R_2 + r_4$

BV	P	x_1	x_2	x_3	s_1	s_2	s_3	RHS
s_1	0	0	1	1	1	−1	0	45,000
x_1	0	[1]	0	0	0	1	0	45,000
s_3	0	0	1	−1	0	0	1	18,000
P	1	0	−0.08	−0.06	0	0.1	0	4500

This still is not the final tableau, we choose 1 in row s_3, column x_2 and pivot again.

$R_1 = -r_3 + r_1$, $R_4 = .08r_3 + r_4$

BV	P	x_1	x_2	x_3	s_1	s_2	s_3	RHS
s_1	0	0	0	2	1	−1	−1	27,000
x_1	0	1	0	0	0	1	0	45,000
x_2	0	0	1	−1	0	0	1	18,000
P	1	0	0	−0.14	0	0.1	0.08	5940

Pivoting again, we introduce variable x_3 as a basic variable.

$R_1 = \frac{1}{2}r_1$, $R_3 = r_1 + r_3$, $R_4 = .14r_1 + r_4$

BV	P	x_1	x_2	x_3	s_1	s_2	s_3	RHS
x_3	0	0	0	1	0.5	−0.5	−0.5	13,500
x_1	0	1	0	0	0	1	0	45,000
x_2	0	0	1	0	0.5	−0.5	0.5	31,500
P	1	0	0	0	0.07	0.03	0.01	7830

Since all the entries in the objective row are nonnegative, this is the final tableau. The maximum $P = 7830$, obtained when $x_1 = 45{,}000$, $x_2 = 31{,}500$, and $x_3 = 13{,}500$.

(c) The financial planner can maximize her client's investment income, while maintaining her investment strategy by investing \$45,000 in stocks, \$31,500 in corporate bonds, and \$13,500 in municipal bonds. The maximum return will be \$7830.

37. (a) Let P denote the profit, x_1 denote the number of acres of soy planted, x_2 denote the number of acres of corn planted, and x_3 denote the number of acres of wheat planted.

Maximize

$$P = 70x_1 + 90x_2 + 50x_3$$

subject to the constraints

$$\begin{aligned} x_1 + x_2 + x_3 &\le 200 \\ 40x_1 + 50x_2 + 30x_3 &\le 18{,}000 \\ 20x_1 + 30x_2 + 15x_3 &\le 4{,}200 \\ x_1 \ge 0 \quad x_2 \ge 0 \quad x_3 &\ge 0 \end{aligned}$$

(b) Slack variables are added and the initial tableau is constructed and shown below.

BV	P	x_1	x_2	x_3	s_1	s_2	s_3	RHS
s_1	0	[1]	1	1	1	0	0	200
s_2	0	40	50	30	0	1	0	18,000
s_3	0	20	30	15	0	0	1	4,200
P	1	−70	−90	−50	0	0	0	0

$R_2 = -40R_1 + r_2$, $R_3 = -20R_1 + r_3$, $R_4 = 70R_1 + r_4$

BV	P	x_1	x_2	x_3	s_1	s_2	s_3	RHS
x_1	0	1	1	1	1	0	0	200
→ s_2	0	0	10	−10	−40	1	0	10,000
s_3	0	0	[10]	−5	−20	0	1	200
P	1	0	−20	20	70	0	0	14,000

$R_3 = \frac{1}{10}r_3$, $R_1 = -R_3 + r_1$, $R_2 = -10R_3 + r_2$, $R_4 = 20R_3 + r_4$

BV	P	x_1	x_2	x_3	s_1	s_2	s_3	RHS
x_1	0	1	0	$\frac{3}{2}$	3	0	$-\frac{1}{10}$	180
→ s_2	0	0	0	−5	−20	1	−1	9,800
x_2	0	0	1	$-\frac{1}{2}$	−20	0	$\frac{1}{10}$	20
P	1	0	0	10	30	0	2	14,400

Since all the entries in the objective row are nonnegative, this is the final tableau. The solution is $P = 14{,}400$, obtained when $x_1 = 180$, $x_2 = 20$, and $x_3 = 0$.

(c) The farmer realizes a maximum profit of \$14,400 when planting 180 acres of soy, 20 acres of corn, and no wheat.

39. (a) Let P denote the revenue, x_1 denote the number of Can I nuts packaged, x_2 denote the number of Can II nuts packaged, and x_3 denote the number of Can III nuts packaged.

Maximize

$$P = 28x_1 + 24x_2 + 21x_3$$

subject to the constraints

$$\begin{aligned} 3x_1 + 4x_2 + 5x_3 &\le 500 \\ x_1 + \tfrac{1}{2}x_2 &\le 100 \\ x_1 + \tfrac{1}{2}x_2 &\le 50 \\ x_1 \ge 0 \quad x_2 \ge 0 \quad x_3 &\ge 0 \end{aligned}$$

Slack variables are added and the initial tableau is constructed and shown below.

BV	P	x_1	x_2	x_3	s_1	s_2	s_3	RHS
s_1	0	3	4	5	1	0	0	500
s_2	0	1	$\frac{1}{2}$	0	0	1	0	100
s_3	0	[1]	$\frac{1}{2}$	0	0	0	1	50
P	1	−28	−24	−21	0	0	0	0

$R_1 = -3R_3 + r_1$, $R_2 = -R_3 + r_2$, $R_4 = 28R_3 + r_4$

BV	P	x_1	x_2	x_3	s_1	s_2	s_3	RHS
s_1	0	0	$\frac{5}{2}$	5	1	0	−3	350
→ s_2	0	0	0	0	0	1	−1	50
x_1	0	1	[$\frac{1}{2}$]	0	0	0	1	50
P	1	0	−10	−21	0	0	28	1400

$$\begin{array}{c} \\ \\ R_3 = 2r_3 \\ R_1 = -\frac{5}{2}R_3 + r_1 \\ R_4 = 10R_3 + r_4 \end{array} \longrightarrow \begin{array}{c} \text{BV} \\ s_1 \\ s_2 \\ x_2 \\ P \end{array} \begin{array}{c} \begin{array}{cccccccc|c} P & x_1 & x_2 & x_3 & s_1 & s_2 & s_3 & & \text{RHS} \end{array} \\ \left[\begin{array}{ccccccc|c} 0 & -5 & 0 & \boxed{5} & 1 & 0 & -8 & 100 \\ 0 & 0 & 0 & 0 & 0 & 1 & -1 & 50 \\ 0 & 2 & 1 & 0 & 0 & 0 & 2 & 100 \\ \hline 1 & 20 & 0 & -21 & 0 & 0 & 48 & 2400 \end{array}\right] \end{array}$$

$$\begin{array}{c} \\ \\ R_1 = \frac{1}{5}r_1 \\ R_4 = 21R_1 + r_4 \end{array} \longrightarrow \begin{array}{c} \text{BV} \\ x_3 \\ s_2 \\ x_2 \\ P \end{array} \begin{array}{c} \begin{array}{cccccccc|c} P & x_1 & x_2 & x_3 & s_1 & s_2 & s_3 & & \text{RHS} \end{array} \\ \left[\begin{array}{ccccccc|c} 0 & -1 & 0 & 1 & \frac{1}{5} & 0 & -\frac{8}{5} & 20 \\ 0 & 0 & 0 & 0 & 0 & 1 & -1 & 50 \\ 0 & \boxed{2} & 1 & 0 & 0 & 0 & 2 & 100 \\ \hline 1 & -1 & 0 & 0 & \frac{21}{5} & 0 & \frac{72}{5} & 2820 \end{array}\right] \end{array}$$

$$\begin{array}{c} \\ \\ R_3 = \frac{1}{2}r_3 \\ R_1 = R_3 + r_1 \\ R_4 = R_3 + r_4 \end{array} \longrightarrow \begin{array}{c} \text{BV} \\ x_3 \\ s_2 \\ x_1 \\ P \end{array} \begin{array}{c} \begin{array}{cccccccc|c} P & x_1 & x_2 & x_3 & s_1 & s_2 & s_3 & & \text{RHS} \end{array} \\ \left[\begin{array}{ccccccc|c} 0 & 0 & \frac{1}{2} & 1 & \frac{1}{5} & 0 & -\frac{8}{5} & 70 \\ 0 & 0 & 0 & 0 & 0 & 1 & -1 & 50 \\ 0 & 1 & \frac{1}{2} & 0 & 0 & 0 & 1 & 50 \\ \hline 1 & 0 & \frac{1}{2} & 0 & \frac{21}{5} & 0 & \frac{77}{5} & 2870 \end{array}\right] \end{array}$$

Since all the entries in the objective row are nonnegative, this is the final tableau. The maximum $P = 2870$, obtained when $x_1 = 50$, $x_2 = 0$, and $x_3 = 70$.

(c) Revenue is maximized at $2870.00 when 50 packages of Can I nuts, 70 packages of Can III nuts, and no packages of Can II nuts are produced.

41. (a) Let P denote the profit, x_1 denote the number television cabinets made, x_2 denote the number of stereo cabinets made, and x_3 denote the number of radio cabinets made.

Maximize

$$P = 10x_1 + 25x_2 + 3x_3$$

subject to the constraints

$$\begin{aligned} 3x_1 + 10x_2 + x_3 &\le 30{,}000 \\ 5x_1 + 8x_2 + x_3 &\le 40{,}000 \\ 0.1x_1 + 0.6x_2 + 0.1x_3 &\le 120 \\ x_1 \ge 0 \quad x_2 \ge 0 \quad x_3 &\ge 0 \end{aligned}$$

(b) Slack variables are added and the initial tableau is constructed and shown below. Before setting up the initial tableau, we multiplied the crating constraint by 10 to remove decimals.

BV	P	x_1	x_2	x_3	s_1	s_2	s_3	RHS
s_1	0	3	10	1	1	0	0	30,000
s_2	0	5	8	1	0	1	0	40,000
s_3	0	[1]	6	1	0	0	1	1,200
P	1	−10	−25	−3	0	0	0	0

$R_1 = -3R_3 + r_1$
$R_2 = -5R_3 + r_2$
$R_4 = 10R_4 + r_4$

BV	P	x_1	x_2	x_3	s_1	s_2	s_3	RHS
s_1	0	0	−8	−2	1	0	−3	26,400
s_2	0	0	−22	−4	0	1	−5	34,000
x_1	0	1	6	1	0	0	1	1,200
P	1	0	35	7	0	0	10	12,000

Since all the entries in the objective row are nonnegative, this is the final tableau. The solution is $P = 12{,}000$, obtained when $x_1 = 1200$, $x_2 = 0$, and $x_3 = 0$.

(c) The maximum profit is $12,000 obtained when 1200 television cabinets and no stereo cabinets or radio cabinets are made.

43. (a) Let P denote the profit, x_1 denote the number televisions shipped from Chicago, x_2 denote the number shipped from New York, and x_3 denote the number shipped from Denver.

Maximize

$$P = 70x_1 + 80x_2 + 40x_3$$

subject to the constraints

$$\begin{aligned} x_1 + x_2 + x_3 &\le 400 \\ 20x_1 + 20x_2 + 40x_3 &\le 10{,}000 \\ 6x_1 + 8x_2 + 4x_3 &\le 3{,}000 \\ x_1 \ge 0 \quad x_2 \ge 0 \quad x_3 &\ge 0 \end{aligned}$$

Slack variables are added and the initial tableau is constructed and shown below.

BV	P	x_1	x_2	x_3	s_1	s_2	s_3	RHS
s_1	0	[1]	1	1	1	0	0	400
s_2	0	20	20	40	0	1	0	10,000
s_3	0	6	8	4	0	0	1	3,000
P	1	−70	−80	−40	0	0	0	0

$R_2 = -20R_1 + r_2$
$R_3 = -6R_1 + r_3$
$R_4 = 70R_1 + r_4$

BV	P	x_1	x_2	x_3	s_1	s_2	s_3	RHS
x_1	0	1	1	1	1	0	0	400
s_2	0	0	0	20	−20	1	0	2,000
s_3	0	0	[2]	−2	−6	0	1	600
P	1	0	−10	30	70	0	0	28,000

$$\xrightarrow[\substack{R_3=\frac{1}{2}r_3 \\ R_1=-R_3+r_1 \\ R_4=10R_3+r_4}]{} \begin{array}{c} \text{BV} \\ x_1 \\ s_2 \\ x_2 \\ P \end{array} \begin{array}{c} \begin{array}{cccccccc|c} P & x_1 & x_2 & x_3 & s_1 & s_2 & s_3 & & \text{RHS} \end{array} \\ \left[\begin{array}{ccccccc|c} 0 & 1 & 0 & 2 & 4 & 0 & -\frac{1}{2} & 100 \\ 0 & 0 & 0 & 20 & -20 & 1 & 0 & 2{,}000 \\ 0 & 0 & 1 & -1 & -3 & 0 & \frac{1}{2} & 300 \\ \hline 1 & 0 & 0 & 20 & 40 & 0 & 5 & 31{,}000 \end{array}\right] \end{array}$$

Since all the entries in the objective row are nonnegative, this is the final tableau. The maximum $P = 31{,}000$, obtained when $x_1 = 100$, $x_2 = 300$, and $x_3 = 0$.

(c) The manufacturer will obtain a maximum profit of \$31,000, if it ships 100 televisions from Chicago, 300 televisions from New York and none from Denver.

45. P is the profit, x_1 is the number of sleeping dolls, x_2 is the number of talking dolls, and x_3 is the number of walking dolls.

Maximize

$$P = 6x_1 + 6x_2 + 8x_3$$

subject to the constraints

$$\begin{aligned} x_1 + x_2 + x_3 &\le 60 \\ 6x_1 + 7.5x_2 + 9x_3 &\le 405 \\ x_1 \ge 0 \quad x_2 \ge 0 \quad x_3 &\ge 0 \end{aligned}$$

We begin to the Simplex method using the initial tableau from Problem 41, Section 4.1. The first pivot element is in row s_1, column x_1.

$$\begin{array}{c} \text{BV} \\ s_1 \\ s_2 \\ P \end{array} \begin{array}{c} \begin{array}{ccccccc} P & x_1 & x_2 & x_3 & s_1 & s_2 & \text{RHS} \end{array} \\ \left[\begin{array}{cccccc|c} 0 & \boxed{1} & 1 & 1 & 1 & 0 & 60 \\ 0 & 6 & 7.5 & 9 & 0 & 1 & 405 \\ \hline 1 & -6 & -6 & -8 & 0 & 0 & 0 \end{array}\right] \end{array}$$

$$\xrightarrow[\substack{R_2=-6R_1+r_2 \\ R_3=6R_1+r_3}]{} \begin{array}{c} \text{BV} \\ x_1 \\ s_2 \\ P \end{array} \begin{array}{c} \begin{array}{ccccccc} P & x_1 & x_2 & x_3 & s_1 & s_2 & \text{RHS} \end{array} \\ \left[\begin{array}{cccccc|c} 0 & 1 & 1 & 1 & 1 & 0 & 60 \\ 0 & 0 & 1.5 & \boxed{3} & -6 & 1 & 45 \\ \hline 1 & 0 & 0 & -2 & 6 & 0 & 360 \end{array}\right] \end{array}$$

$$\xrightarrow[\substack{R_2=\frac{1}{3}r_2 \\ R_1=-R_2+r_1 \\ R_3=2R_2+r_3}]{} \begin{array}{c} \text{BV} \\ x_1 \\ x_2 \\ P \end{array} \begin{array}{c} \begin{array}{ccccccc} P & x_1 & x_2 & x_3 & s_1 & s_2 & \text{RHS} \end{array} \\ \left[\begin{array}{cccccc|c} 0 & 1 & \frac{1}{2} & 0 & 3 & -\frac{1}{3} & 45 \\ 0 & 0 & \frac{1}{2} & 1 & -2 & \frac{1}{3} & 15 \\ \hline 1 & 0 & 1 & 0 & 2 & \frac{2}{3} & 390 \end{array}\right] \end{array}$$

(a) The store should order 45 sleeping dolls, 15 walking dolls and no talking dolls to maximize profit.

(d) The maximum profit is \$390.00.

47. We begin the Simplex method using the initial tableau from Problem 43, Section 4.1. The first pivot element is marked in the initial tableau

BV	P	x_1	x_2	x_3	s_1	s_2	s_3	RHS
s_1	0	10	20	20	1	0	0	500
s_2	0	20	40	30	0	1	0	900
s_3	0	[1]	1	1	0	0	1	30
P	1	−19	−34	−15	0	0	0	0

$\xrightarrow[\substack{R_1=-20R_3+r_1\\ R_2=-40R_3+r_2\\ R_4=19R_3+r_4}]{}$

BV	P	x_1	x_2	x_3	s_1	s_2	s_3	RHS
s_1	0	0	10	10	1	0	−10	200
s_2	0	0	[20]	10	0	1	−20	300
x_1	0	1	1	1	0	0	1	30
P	1	0	−15	4	0	0	19	570

$\xrightarrow[\substack{R_2=\frac{1}{20}r_2\\ R_1=-10R_2+r_1\\ R_3=-R_2+r_3\\ R_4=15R_2+r_4}]{}$

BV	P	x_1	x_2	x_3	s_1	s_2	s_3	RHS
s_1	0	0	0	5	1	−0.5	0	50
x_2	0	0	1	0.5	0	0.05	−1	15
x_1	0	1	0	0.5	0	−0.05	2	15
P	1	0	0	11.5	0	0.75	4	795

There are no longer any negative entries in the objective row indicating that we have found the maximum P.

(a) Steven should produce 15 shirts, 15 jackets, and no pants to maximize his profit.

(b) The maximum profit is $795.00.

49. We begin the Simplex method using the initial tableau from Problem 45, Section 4.1. The first pivot element is marked on the initial tableau.

BV	P	x_1	x_2	x_3	s_1	s_2	s_3	RHS
s_1	0	[1]	1	1	1	0	0	25,000
s_2	0	0	1	1	0	1	0	10,000
s_3	0	−1	0	1	0	0	1	1,500
P	1	−0.0525	−0.075	−0.04	0	0	0	0

$\xrightarrow[\substack{R_3=R_1+r_3\\ R_4=0.0525R_1+r_4}]{}$

BV	P	x_1	x_2	x_3	s_1	s_2	s_3	RHS
x_1	0	1	1	1	1	0	0	25,000
s_2	0	0	[1]	1	0	1	0	10,000
s_3	0	0	1	2	1	0	1	26,500
P	1	0	−0.023	0.012	0.052	0	0	1312.50

$\xrightarrow[\substack{R_1=-R_2+r_1\\ R_3=-R_2+r_3\\ R_4=0.023R_1+r_4}]{}$

BV	P	x_1	x_2	x_3	s_1	s_2	s_3	RHS
x_1	0	1	0	0	1	−1	0	15,000
x_2	0	0	1	1	0	1	0	10,000
s_3	0	0	0	1	1	−1	1	16,500
P	1	0	0	0.035	0.052	0.023	0	1537.50

There are no negative entries in the objective row. So we have found the maximum value of P.

(a) Kami should invest \$15,000 in the money market, \$10,000 in the mutual fund and nothing in the CD.

(b) The maximum return on Kami's investments is \$1537.50.

4.3 Solving Minimum Problems Using the Duality Principle

3. True

5. True

7. Duality Principle

For problems 9-13 use the following conditions when determining if the minimum problems are in standard form.

9. The minimum problem is in standard form.

11. The minimum problem is not in standard form. Condition 3 is not met; the objective function has a negative coefficient.

13. The minimum problem is not in standard form. Condition 2 is not met; one of the constraints is an expression less than or equal to a constant.

In Problems 15-19 use the following 4 steps to write the Dual Problem.

STEP 1 Write the minimum problem in standard form.
STEP 2 Construct a matrix that represents the constraints and the objective function, placing the objective function in the bottom row.
STEP 3 Interchange the rows and columns to form the matrix of the dual problem.
STEP 4 Translate this matrix into a maximum problem in standard form.

15. STEP 1 The problem is in standard form.

STEP 2 The matrix with the objective function in the bottom row:

$$\begin{array}{c} \begin{matrix} x_1 & x_2 & \end{matrix} \\ \left[\begin{array}{cc|c} 1 & 1 & 2 \\ 2 & 3 & 6 \\ 2 & 3 & 0 \end{array}\right] \end{array}$$

STEP 3 The matrix with the rows and columns interchanged:

$$\left[\begin{array}{cc|c} 1 & 2 & 2 \\ 1 & 3 & 3 \\ 2 & 6 & 0 \end{array}\right]$$

STEP 4 The corresponding maximum problem in standard form:

Maximize

$$P = 2y_1 + 6y_2$$

subject to the constraints

$$\begin{aligned} y_1 + 2y_2 &\le 2 \\ y_1 + 3y_2 &\le 3 \\ y_1 \ge 0 \quad y_2 &\ge 0 \end{aligned}$$

This maximum problem is the dual of the minimum problem.

17. **STEP 1** The problem is in standard form.

STEP 2 The matrix with the objective function in the bottom row:

$$\begin{array}{c} \begin{array}{ccc} x_1 & x_2 & x_3 \end{array} \\ \left[\begin{array}{ccc|c} 1 & 1 & 1 & 5 \\ 2 & 1 & 0 & 4 \\ 3 & 1 & 1 & 0 \end{array}\right] \end{array}$$

STEP 3 The matrix with the rows and columns interchanged:

$$\left[\begin{array}{cc|c} 1 & 2 & 3 \\ 1 & 1 & 1 \\ 1 & 0 & 1 \\ 5 & 4 & 0 \end{array}\right]$$

STEP 4 The corresponding maximum problem in standard form:

Maximize

$$P = 5y_1 + 4y_2$$

subject to the constraints

$$\begin{aligned} y_1 + 2y_2 &\le 3 \\ y_1 + y_2 &\le 1 \\ y_1 &\le 1 \\ y_1 \ge 0 \quad y_2 &\ge 0 \end{aligned}$$

This maximum problem is the dual of the minimum problem.

19. **STEP 1** The problem is in standard form.

STEP 2 The matrix with the objective function in the bottom row:

$$\begin{array}{c} \begin{array}{cccc} x_1 & x_2 & x_3 & x_4 \end{array} \\ \left[\begin{array}{cccc|c} 1 & 1 & 1 & 2 & 60 \\ 3 & 2 & 1 & 2 & 90 \\ 3 & 4 & 1 & 2 & 0 \end{array}\right] \end{array}$$

STEP 3 The matrix with the rows and columns interchanged:

$$\left[\begin{array}{cc|c} 1 & 3 & 3 \\ 1 & 2 & 4 \\ 1 & 1 & 1 \\ 2 & 2 & 2 \\ 60 & 90 & 0 \end{array}\right]$$

STEP 4 The corresponding maximum problem in standard form:

Maximize

$$P = 60y_1 + 90y_2$$

subject to the constraints

$$\begin{aligned} y_1 + 3y_2 &\le 3 \\ y_1 + 2y_2 &\le 4 \\ y_1 + y_2 &\le 1 \\ 2y_1 + 2y_2 &\le 2 \\ y_1 \ge 0 \quad y_2 &\ge 0 \end{aligned}$$

This maximum problem is the dual of the minimum problem.

21. **STEP 1** Write the dual problem.

$$\text{The matrix: } \begin{array}{c} \begin{matrix} x_1 & x_2 & \end{matrix} \\ \left[\begin{array}{cc|c} 1 & 1 & 2 \\ 2 & 6 & 6 \\ 6 & 3 & 0 \end{array}\right] \end{array}; \text{ its transpose: } \left[\begin{array}{cc|c} 1 & 2 & 6 \\ 1 & 6 & 3 \\ 2 & 6 & 0 \end{array}\right];$$

Maximize

$$P = 2y_1 + 6y_2$$

subject to the constraints

$$\begin{aligned} y_1 + 2y_2 &\le 6 \\ y_1 + 6y_2 &\le 3 \\ y_1 \ge 0 \quad y_2 &\ge 0 \end{aligned}$$

STEP 2 Set up the initial tableau and use the simplex method to solve the dual problem.

$$\begin{array}{c} \text{BV} \\ s_1 \\ s_2 \\ P \end{array} \begin{array}{c} \begin{matrix} P & y_1 & y_2 & s_1 & s_2 & \text{RHS} \end{matrix} \\ \left[\begin{array}{ccccc|c} 0 & 1 & 2 & 1 & 0 & 6 \\ 0 & \boxed{1} & 6 & 0 & 1 & 3 \\ \hline 1 & -2 & -6 & 0 & 0 & 0 \end{array}\right] \end{array} \xrightarrow[R_3 = 2R_2 + r_3]{R_1 = -R_2 + r_1} \begin{array}{c} \text{BV} \\ s_1 \\ y_1 \\ P \end{array} \begin{array}{c} \begin{matrix} P & y_1 & y_2 & s_1 & s_2 & \text{RHS} \end{matrix} \\ \left[\begin{array}{ccccc|c} 0 & 0 & -4 & 1 & -1 & 3 \\ 0 & 1 & 6 & 0 & 1 & 3 \\ \hline 1 & 0 & 6 & 0 & 2 & 6 \end{array}\right] \end{array}$$

STEP 3 This is the final tableau. From it we read that the minimum cost $C = 6$ is obtained when $x_1 = 0$ and $x_2 = 2$.

23. **STEP 1** Write the dual problem.

$$\text{The matrix: } \begin{array}{c} \begin{matrix} x_1 & x_2 & \end{matrix} \\ \left[\begin{array}{cc|c} 1 & 1 & 4 \\ 3 & 4 & 12 \\ 6 & 3 & 0 \end{array}\right] \end{array}; \text{ its transpose: } \left[\begin{array}{cc|c} 1 & 3 & 6 \\ 1 & 4 & 3 \\ 4 & 12 & 0 \end{array}\right];$$

Maximize

$$P = 4y_1 + 12y_2$$

subject to the constraints

$$\begin{aligned} y_1 + 3y_2 &\le 6 \\ y_1 + 4y_2 &\le 3 \\ y_1 \ge 0 \quad y_2 &\ge 0 \end{aligned}$$

STEP 2 Set up the initial tableau and use the simplex method to solve the dual problem.

$$\begin{array}{c} \text{BV} \\ s_1 \\ s_2 \\ P \end{array}\begin{array}{c} \begin{array}{cccccc} P & y_1 & y_2 & s_1 & s_2 & \text{RHS} \end{array} \\ \left[\begin{array}{ccccc|c} 0 & 1 & 3 & 1 & 0 & 6 \\ 0 & \boxed{1} & 4 & 0 & 1 & 3 \\ \hline 1 & -4 & -12 & 0 & 0 & 0 \end{array}\right] \end{array} \xrightarrow[R_3=4R_2+r_3]{R_1=-R_2+r_1} \begin{array}{c} \text{BV} \\ s_1 \\ y_1 \\ P \end{array}\begin{array}{c} \begin{array}{cccccc} P & y_1 & y_2 & s_1 & s_2 & \text{RHS} \end{array} \\ \left[\begin{array}{ccccc|c} 0 & 0 & -1 & 1 & -1 & 3 \\ 0 & 1 & 4 & 0 & 1 & 3 \\ \hline 1 & 0 & 4 & 0 & 4 & 12 \end{array}\right] \end{array}$$

STEP 3 This is the final tableau. From it we read that the minimum cost $C = 12$ is obtained when $x_1 = 0$ and $x_2 = 4$.

25. **STEP 1** Write the dual problem.

$$\text{The matrix: } \begin{array}{c} \begin{array}{cccc} x_1 & x_2 & x_3 & \end{array} \\ \left[\begin{array}{ccc|c} 1 & -3 & 4 & 12 \\ 3 & 1 & 2 & 10 \\ 1 & -1 & -1 & -8 \\ 1 & 2 & 1 & 0 \end{array}\right] \end{array}; \text{ its transpose: } \left[\begin{array}{ccc|c} 1 & 3 & 1 & 1 \\ -3 & 1 & -1 & 2 \\ 4 & 2 & -1 & 1 \\ 12 & 10 & -8 & 0 \end{array}\right]$$

Maximize

$$P = 12y_1 + 10y_2 - 8y_3$$

subject to the constraints

$$\begin{aligned} y_1 + 3y_2 + y_3 &\le 1 \\ -3y_1 + y_2 - y_3 &\le 2 \\ 4y_1 + 2y_2 - y_3 &\le 1 \\ y_1 \ge 0 \quad y_2 \ge 0 \quad & y_3 \ge 0 \end{aligned}$$

STEP 2 Set up the initial tableau and use the simplex method to solve the dual problem.

$$\begin{array}{c} \text{BV} \\ s_1 \\ s_2 \\ s_3 \\ P \end{array}\begin{array}{c} \begin{array}{cccccccc} P & y_1 & y_2 & y_3 & s_1 & s_2 & s_3 & \text{RHS} \end{array} \\ \left[\begin{array}{ccccccc|c} 0 & 1 & 3 & 1 & 1 & 0 & 0 & 1 \\ 0 & -3 & 1 & -1 & 0 & 1 & 0 & 2 \\ 0 & \boxed{4} & 2 & -1 & 0 & 0 & 1 & 1 \\ \hline 1 & -12 & -10 & 8 & 0 & 0 & 0 & 0 \end{array}\right] \end{array}$$

$$\xrightarrow[\substack{R_1=-R_3+r_1 \\ R_2=3R_3+r_2 \\ R_4=12R_3+r_4}]{R_3=\frac{1}{4}r_3} \begin{array}{c} \text{BV} \\ s_1 \\ s_2 \\ y_1 \\ P \end{array}\begin{array}{c} \begin{array}{cccccccc} P & y_1 & y_2 & y_3 & s_1 & s_2 & s_3 & \text{RHS} \end{array} \\ \left[\begin{array}{ccccccc|c} 0 & 0 & \boxed{\frac{5}{2}} & \frac{5}{4} & 1 & 0 & -\frac{1}{4} & \frac{3}{4} \\ 0 & 0 & \frac{5}{2} & -\frac{7}{4} & 0 & 1 & \frac{3}{4} & \frac{11}{4} \\ 0 & 1 & \frac{1}{2} & -\frac{1}{4} & 0 & 0 & \frac{1}{4} & \frac{1}{4} \\ \hline 1 & 0 & -4 & 5 & 0 & 0 & 3 & 3 \end{array}\right] \end{array}$$

$$\xrightarrow[\substack{R_2=-\frac{5}{2}R_1+r_2 \\ R_3=-\frac{1}{2}R_1+r_3 \\ R_4=4R_1+r_4}]{R_1=\frac{2}{5}r_1} \begin{array}{c} \text{BV} \\ y_2 \\ s_2 \\ y_1 \\ P \end{array}\begin{array}{c} \begin{array}{cccccccc} P & y_1 & y_2 & y_3 & s_1 & s_2 & s_3 & \text{RHS} \end{array} \\ \left[\begin{array}{ccccccc|c} 0 & 0 & 1 & \frac{1}{2} & \frac{2}{5} & 0 & -\frac{1}{10} & \frac{3}{10} \\ 0 & 0 & 0 & -3 & -1 & 1 & 1 & 2 \\ 0 & 1 & 0 & -\frac{1}{2} & -\frac{1}{5} & 0 & \frac{3}{10} & \frac{1}{10} \\ \hline 1 & 0 & 0 & 7 & \frac{8}{5} & 0 & \frac{13}{5} & \frac{21}{5} \end{array}\right] \end{array}$$

STEP 3 This is the final tableau. From it we read that the minimum cost $C = \frac{21}{5} = 4.20$ is obtained when $x_1 = \frac{8}{5}$, $x_2 = 0$, and $x_3 = \frac{13}{5}$.

27. **STEP 1** Write the dual problem.

The matrix:
$$\begin{array}{cccc|c} x_1 & x_2 & x_3 & x_4 & \\ 1 & 0 & 1 & 0 & 1 \\ 0 & 1 & 0 & 1 & 1 \\ -1 & -1 & -1 & -1 & -3 \\ 1 & 4 & 2 & 4 & 0 \end{array}$$
; its transpose:
$$\left[\begin{array}{ccc|c} 1 & 0 & -1 & 1 \\ 0 & 1 & -1 & 4 \\ 1 & 0 & -1 & 2 \\ 0 & 1 & -1 & 4 \\ 1 & 1 & -3 & 0 \end{array}\right]$$

Maximize
$$P = y_1 + y_2 - 3y_3$$
subject to the constraints
$$\begin{aligned} y_1 \quad - y_3 &\le 1 \\ y_2 - y_3 &\le 4 \\ y_1 \quad - y_3 &\le 2 \\ y_2 - y_3 &\le 4 \\ y_1 \ge 0 \quad y_2 \ge 0 \quad & y_3 \ge 0 \end{aligned}$$

STEP 2 Set up the initial tableau and use the simplex method to solve the dual problem.

BV	P	y_1	y_2	y_3	s_1	s_2	s_3	s_4	RHS
s_1	0	$\boxed{1}$	0	−1	1	0	0	0	1
s_2	0	0	1	−1	0	1	0	0	4
s_3	0	1	0	−1	0	0	1	0	2
s_4	0	0	1	−1	0	0	0	1	4
P	1	−1	−1	3	0	0	0	0	0

$\xrightarrow[R_5 = R_1 + r_5]{R_3 = -R_1 + r_3}$

BV	P	y_1	y_2	y_3	s_1	s_2	s_3	s_4	RHS
y_1	0	1	0	−1	1	0	0	0	1
s_2	0	0	$\boxed{1}$	−1	0	1	0	0	4
s_3	0	0	0	0	−1	0	1	0	1
s_4	0	0	1	−1	0	0	0	1	4
P	1	0	−1	2	1	0	0	0	1

$\xrightarrow[R_5 = R_2 + r_5]{R_4 = -R_2 + r_4}$

BV	P	y_1	y_2	y_3	s_1	s_2	s_3	s_4	RHS
y_1	0	1	0	−1	1	0	0	0	1
y_2	0	0	1	−1	0	1	0	0	4
s_3	0	0	0	0	−1	0	1	0	1
s_4	0	0	0	0	0	−1	0	1	0
P	1	0	0	1	1	1	0	0	5

STEP 3 This is the final tableau. From it we read that the minimum cost $C = 5$ is obtained when $x_1 = 1$, $x_2 = 1$, $x_3 = 0$ and $x_4 = 0$.

29. (a) Let C be the cost of the supplements; x_1 be the number of pill P needed and x_2 the number of pill Q needed. He wishes to minimize his cost while meeting the nutritional requirements.

Minimize

$$C = 3x_1 + 4x_2$$

subject to the constraints

$$5x_1 + 10x_2 \geq 50$$
$$2x_1 + x_2 \geq 8$$
$$x_1 \geq 0 \quad x_2 \geq 0$$

The matrix representing the system and its transpose are

$$\text{matrix: } \begin{array}{c} \begin{matrix} x_1 & x_2 & \end{matrix} \\ \left[\begin{array}{cc|c} 5 & 10 & 50 \\ 2 & 1 & 8 \\ 3 & 4 & 0 \end{array}\right] \end{array} \qquad \text{transpose: } \left[\begin{array}{cc|c} 5 & 2 & 3 \\ 10 & 1 & 4 \\ 50 & 8 & 0 \end{array}\right]$$

(b) The dual of the problem is

Maximize

$$P = 50y_1 + 8y_2$$

subject to the constraints

$$5y_1 + 2y_2 \leq 3$$
$$10y_1 + y_2 \leq 4$$
$$y_1 \geq 0 \quad y_2 \geq 0$$

We set up the initial simplex tableau and solve the maximum problem.

$$\begin{array}{c} \text{BV} \\ s_1 \\ s_2 \\ P \end{array} \begin{array}{c} \begin{matrix} P & y_1 & y_2 & s_1 & s_2 & \text{RHS} \end{matrix} \\ \left[\begin{array}{cccccc|c} 0 & 5 & 2 & 1 & 0 & 3 \\ 0 & \boxed{10} & 1 & 0 & 1 & 4 \\ \hline 1 & -50 & -8 & 0 & 0 & 0 \end{array}\right] \end{array} \xrightarrow[\substack{R_2 = \frac{1}{10}r_2 \\ R_1 = -5R_2 + r_1 \\ R_3 = 50R_2 + r_3}]{} \begin{array}{c} \text{BV} \\ s_1 \\ y_1 \\ P \end{array} \begin{array}{c} \begin{matrix} P & y_1 & y_2 & s_1 & s_2 & \text{RHS} \end{matrix} \\ \left[\begin{array}{ccccc|c} 0 & 0 & \boxed{\frac{3}{2}} & 1 & -\frac{1}{2} & 1 \\ 0 & 1 & \frac{1}{10} & 0 & \frac{1}{10} & \frac{2}{5} \\ \hline 1 & 0 & -3 & 0 & 5 & 20 \end{array}\right] \end{array}$$

$$\xrightarrow[\substack{R_1 = \frac{2}{3}r_1 \\ R_2 = -\frac{1}{10}R_1 + r_2 \\ R_3 = 3R_1 + r_3}]{} \begin{array}{c} \text{BV} \\ y_2 \\ y_1 \\ P \end{array} \begin{array}{c} \begin{matrix} P & y_1 & y_2 & s_1 & s_2 & \text{RHS} \end{matrix} \\ \left[\begin{array}{ccccc|c} 0 & 0 & 1 & \frac{2}{3} & -\frac{1}{3} & \frac{2}{3} \\ 0 & 1 & 0 & -\frac{1}{15} & \frac{2}{15} & \frac{1}{3} \\ \hline 1 & 0 & 0 & 2 & 4 & 22 \end{array}\right] \end{array}$$

Minimum $C = 22$; when $x_1 = 2$ and $x_2 = 4$

(c) Mr. Jones minimizes his cost at \$0.22 when he adds 2 vitamin P pills and 4 vitamin Q pills to his diet.

31. Let C represent the weekly payroll, and let x_1, x_2, and x_3 denote the number of employees scheduled to work on Friday, Saturday, and Sunday respectively.

Minimize

$$C = 100x_1 + 150x_2 + 150x_3$$

subject to the constraints

$$x_1 \geq 6$$
$$x_2 \geq 15$$
$$x_3 \geq 8$$

Minimize cost by assigning 7 employees to work on Friday, 15 to work on Saturday, and 8 to work on Sunday. The minimum cost is \$4150.00. No one is scheduled to work a Friday/Sunday schedule.

33. (a) Let C represent the cost of the order, and let x_1, x_2, and x_3 represent the number of lunch #1, lunch #2, and lunch #3 respectively that Mrs. Mintz purchases. She and her friends want to order the foods they need at minimum cost. They want to

Minimize

$$C = 6.20x_1 + 7.40x_2 + 9.10x_3$$

subject to the constraints

$$\begin{aligned} x_1 \qquad\qquad &\geq 4 \\ x_1 + x_2 + x_3 &\geq 9 \\ x_1 \qquad + x_3 &\geq 6 \\ x_2 + x_3 &\geq 5 \\ x_1 \geq 0 \quad x_2 \geq 0 \quad x_3 &\geq 0 \end{aligned}$$

The matrix (on the left) representing the system and its transpose (on the right) are:

$$\begin{array}{c} \begin{array}{ccc} x_1 & x_2 & x_3 \end{array} \\ \left[\begin{array}{ccc|c} 1 & 0 & 0 & 4 \\ 1 & 1 & 1 & 9 \\ 1 & 0 & 1 & 6 \\ 0 & 1 & 1 & 5 \\ 6.2 & 7.4 & 9.1 & 0 \end{array}\right] \end{array} \qquad \begin{array}{c} \begin{array}{cccc} y_1 & y_2 & y_3 & y_4 \end{array} \\ \left[\begin{array}{cccc|c} 1 & 1 & 1 & 0 & 6.2 \\ 0 & 1 & 0 & 1 & 7.4 \\ 0 & 1 & 1 & 1 & 9.1 \\ 4 & 9 & 6 & 5 & 0 \end{array}\right] \end{array}$$

The dual of the problem is

Maximize

$$P = 4y_1 + 9y_2 + 6y_3 + 5y_4$$

subject to the constraints

$$\begin{aligned} y_1 + y_2 + y_3 \qquad &\leq 6.2 \\ y_2 + \qquad y_4 &\leq 7.4 \\ y_2 + y_3 + y_4 &\leq 9.1 \\ y_1 \geq 0 \quad y_2 \geq 0 \quad y_3 \geq 0 \quad y_4 &\geq 0 \end{aligned}$$

We set up the initial simplex tableau and solve the maximum problem.

$$\begin{array}{c} \text{BV} \\ s_1 \\ s_2 \\ s_3 \\ P \end{array} \begin{array}{c} \begin{array}{ccccccccc} P & y_1 & y_2 & y_3 & y_4 & s_1 & s_2 & s_3 & \text{RHS} \end{array} \\ \left[\begin{array}{cccccccc|c} 0 & \boxed{1} & 1 & 1 & 0 & 1 & 0 & 0 & 6.2 \\ 0 & 0 & 1 & 0 & 1 & 0 & 1 & 0 & 7.4 \\ 0 & 0 & 1 & 1 & 1 & 0 & 0 & 1 & 9.1 \\ \hline 1 & -4 & -9 & -6 & -5 & 0 & 0 & 0 & 0 \end{array}\right] \end{array}$$

$$\xrightarrow{R_4 = 4R_1 + r_4} \begin{array}{c} \text{BV} \\ y_1 \\ s_2 \\ s_3 \\ P \end{array} \begin{array}{c} \begin{array}{ccccccccc} P & y_1 & y_2 & y_3 & y_4 & s_1 & s_2 & s_3 & \text{RHS} \end{array} \\ \left[\begin{array}{cccccccc|c} 0 & 1 & \boxed{1} & 1 & 0 & 1 & 0 & 0 & 6.2 \\ 0 & 0 & 1 & 0 & 1 & 0 & 1 & 0 & 7.4 \\ 0 & 0 & 1 & 1 & 1 & 0 & 0 & 1 & 9.1 \\ 1 & 0 & -5 & -2 & -5 & 4 & 0 & 0 & 24.80 \end{array}\right] \end{array}$$

$$\xrightarrow[\substack{R_2 = -R_1 + r_2 \\ R_3 = -R_1 + r_3 \\ R_4 = 5R_1 + r_4}]{} \begin{array}{c} \text{BV} \\ y_2 \\ s_2 \\ s_3 \\ P \end{array} \begin{array}{c} \begin{array}{ccccccccc} P & y_1 & y_2 & y_3 & y_4 & s_1 & s_2 & s_3 & \text{RHS} \end{array} \\ \left[\begin{array}{cccccccc|c} 0 & 1 & 1 & 1 & 0 & 1 & 0 & 0 & 6.2 \\ 0 & -1 & 0 & -1 & \boxed{1} & -1 & 1 & 0 & 1.2 \\ 0 & -1 & 0 & 0 & 1 & -1 & 0 & 1 & 2.9 \\ 1 & 5 & 0 & 3 & -5 & 9 & 0 & 0 & 55.80 \end{array}\right] \end{array}$$

$\xrightarrow[R_4 = 5R_2 + r_4]{R_3 = -R_2 + r_3}$

BV	P	y_1	y_2	y_3	y_4	s_1	s_2	s_3	RHS
y_2	0	1	1	1	0	1	0	0	6.2
y_4	0	−1	0	−1	1	−1	1	0	1.2
s_3	0	0	0	[1]	0	0	−1	1	1.7
P	1	0	0	−2	0	4	5	0	61.80

$\xrightarrow[R_1 = -R_3 + r_1,\; R_2 = R_3 + r_2,\; R_4 = 5R_2 + r_4]{}$

BV	P	y_1	y_2	y_3	y_4	s_1	s_2	s_3	RHS
y_2	0	1	1	0	0	1	1	−1	4.5
y_4	0	−1	0	0	1	−1	0	1	2.9
y_3	0	0	0	1	0	0	−1	1	1.7
P	1	0	0	0	0	4	3	2	65.20

Minimum $C = 65.20$ when $x_1 = 4$, $x_2 = 3$, and $x_3 = 2$.

(c) Mrs. Mintz spends the least amount, $65.20, when she orders 4 of Lunch #1, 3 of Lunch #2, and 2 of Lunch #3.

35. Let C represent the amount of sodium in the mix, and let x_1, x_2, and x_3 represent the amount (in 100 gram units) of apricots, peaches, and pears respectively in the mix.

Minimize

$$C = 26x_1 + 16x_2 + 7x_3$$

subject to the constraints

$$\begin{aligned} x_1 + x_2 + x_3 &\geq 20 \\ 67x_1 + 48x_2 + 35x_3 &\geq 400 \\ 12x_1 + 18x_2 + 7x_3 &\geq 250 \\ x_1 + x_3 &\geq 5 \\ x_1 \geq 0 \quad x_2 \geq 0 \quad x_3 &\geq 0 \end{aligned}$$

We will solve the minimum problem using the Duality Principle.

STEP 1 The matrix representing the system written with the objective row on the bottom and its transpose are below.

$$\left[\begin{array}{ccc|c} 1 & 1 & 1 & 20 \\ 67 & 48 & 35 & 400 \\ 12 & 18 & 7 & 250 \\ 1 & 0 & 1 & 5 \\ 26 & 16 & 7 & 0 \end{array}\right]; \text{ The transpose matrix: } \left[\begin{array}{cccc|c} 1 & 67 & 12 & 1 & 26 \\ 1 & 48 & 18 & 0 & 16 \\ 1 & 35 & 7 & 1 & 7 \\ 20 & 400 & 250 & 5 & 0 \end{array}\right]$$

STEP 2 The initial tableau of the dual problem with the first pivot marked is shown below.

BV	P	y_1	y_2	y_3	y_4	s_1	s_2	s_3	RHS
s_1	0	1	67	12	1	1	0	0	26
s_2	0	1	48	18	0	0	1	0	16
s_3	0	[1]	35	7	1	0	0	1	7
P	1	−20	−400	−250	−5	0	0	0	0

$$\xrightarrow[\substack{R_1=-R_3+r_1\\ R_2=-R_3+r_2\\ R_4=20R_3+r_4}]{}$$

BV	P	y_1	y_2	y_3	y_4	s_1	s_2	s_3	RHS
s_1	0	0	32	5	0	1	0	−1	19
s_2	0	0	13	[11]	-1	0	1	−1	9
y_1	0	1	35	7	1	0	0	1	7
P	1	0	300	−110	15	0	0	20	140

$$\xrightarrow[\substack{R_2=\frac{1}{11}r_2\\ R_1=-5R_2+r_1\\ R_3=-7R_2+r_3\\ R_4=110R_2+r_4}]{}$$

BV	P	y_1	y_2	y_3	y_4	s_1	s_2	s_3	RHS
s_1	0	0	26.091	0	0.455	1	−0.455	−0.545	14.909
y_3	0	0	1.182	1	−0.091	0	0.091	−0.091	0.818
y_1	0	1	26.727	0	1.636	0	−0.636	1.636	1.273
P	1	0	430	0	5	0	10	10	230

STEP 3 Katy should mix 1 kilogram of dried peaches and 1 kilogram of dried pears to meet her requirements and to minimize the sodium content. The minimum sodium is 230 milligrams.

4.4 The Simplex Method for Problems Not in Standard Form

1. False. All variables, including slack variables, must be nonnegative.

3. True

5. Rewrite the constraints:

$$\begin{aligned} x_1 + x_2 &\le 12 \\ -5x_1 - 2x_2 &\le -36 \\ -7x_1 - 4x_2 &\le -14 \\ x_1 \ge 0,\ x_2 &\ge 0 \end{aligned}$$

Introduce slack variables:

$$\begin{aligned} x_1 + x_2 + s_1 &= 12 \\ -5x_1 - 2x_2 + s_2 &= -36 \\ -7x_1 - 4x_2 + s_3 &= -14 \\ x_1 \ge 0,\ x_2 \ge 0,\ s_1 \ge 0,\ s_2 \ge 0,\ s_3 &\ge 0 \end{aligned}$$

We now set up the initial tableau and use the alternate pivoting method as long as there are negative entries in the RHS. When all the entries in the RHS are positive, we use the standard way of choosing a pivot.

BV	P	x_1	x_2	s_1	s_2	s_3	RHS
s_1	0	1	1	1	0	0	12
s_2	0	[−5]	−2	0	1	0	−36
s_3	0	−7	−4	0	0	1	−14
P	1	−3	−4	0	0	0	0

$$\xrightarrow{\text{Alternative Pivoting Strategy}}$$

BV	P	x_1	x_2	s_1	s_2	s_3	RHS
s_1	0	0	[$\frac{3}{5}$]	1	$\frac{1}{5}$	0	$\frac{24}{5}$
x_1	0	1	$\frac{2}{5}$	0	$-\frac{1}{5}$	0	$\frac{36}{5}$
s_3	0	0	$-\frac{6}{5}$	0	$-\frac{7}{5}$	1	$\frac{182}{5}$
P	1	0	$-\frac{14}{5}$	0	$-\frac{3}{5}$	0	$\frac{108}{5}$

$$\xrightarrow{\text{Standard Pivoting Strategy}}$$

BV	P	x_1	x_2	s_1	s_2	s_3	RHS
x_2	0	0	1	$\frac{5}{3}$	$\frac{1}{3}$	0	8
x_1	0	1	0	$-\frac{2}{3}$	$-\frac{1}{3}$	0	4
s_3	0	0	0	2	−1	1	46
P	1	0	0	$\frac{14}{3}$	$\frac{1}{3}$	0	44

The maximum $P = 44$, obtained when $x_1 = 4$ and $x_2 = 8$.

7. Rewrite the constraints:

$$\begin{aligned} x_1 + 3x_2 + x_3 &\le 9 \\ -2x_1 - 3x_2 + x_3 &\le -2 \\ -3x_1 + 2x_2 - x_3 &\le -5 \\ x_1 \ge 0,\ x_2 \ge 0,\ x_3 &\ge 0 \end{aligned}$$

Introduce slack variables:

$$\begin{aligned} x_1 + 3x_2 + x_3 + s_1 \qquad\qquad &= 9 \\ -2x_1 - 3x_2 + x_3 \qquad + s_2 \qquad &= -2 \\ -3x_1 + 2x_2 - x_3 \qquad\qquad + s_3 &= -5 \\ x_1 \ge 0,\ x_2 \ge 0,\ x_3 \ge 0,\ s_1 \ge 0,\ s_2 \ge 0,\ s_3 &\ge 0 \end{aligned}$$

We now set up the initial tableau and use the alternate pivoting method as long as there are negative entries in the RHS. When all the entries in the RHS are positive, we use the standard way of choosing a pivot.

BV	P	x_1	x_2	x_3	s_1	s_2	s_3	RHS
s_1	0	1	3	1	1	0	0	9
s_2	0	−2	−3	1	0	1	0	−2
s_3	0	$\boxed{-3}$	2	−1	0	0	1	−5
P	1	−3	−2	1	0	0	0	0

Alternate Pivoting: $R_3 = -\frac{1}{3}r_3$, $R_1 = -R_3 + r_1$, $R_2 = 2R_3 + r_2$, $R_4 = 3R_3 + r_4$

BV	P	x_1	x_2	x_3	s_1	s_2	s_3	RHS
s_1	0	0	$\boxed{\frac{11}{3}}$	$\frac{2}{3}$	1	0	$\frac{1}{3}$	$\frac{22}{3}$
s_2	0	0	$-\frac{13}{3}$	$\frac{5}{3}$	0	1	$-\frac{2}{3}$	$\frac{4}{3}$
x_1	0	1	$-\frac{2}{3}$	$\frac{1}{3}$	0	0	$-\frac{1}{3}$	$\frac{5}{3}$
P	1	0	−4	2	0	0	−1	5

Standard Pivoting Strategy: $R_1 = \frac{3}{11}r_1$, $R_2 = \frac{2}{3}R_1 + r_2$, $R_3 = \frac{13}{3}R_1 + r_3$, $R_4 = 4R_1 + r_4$

BV	P	x_1	x_2	x_3	s_1	s_2	s_3	RHS
x_2	0	0	1	$\frac{2}{11}$	$\frac{3}{11}$	0	$\boxed{\frac{1}{11}}$	2
s_2	0	0	0	$\frac{27}{11}$	$\frac{13}{11}$	1	$-\frac{3}{11}$	10
x_1	0	1	0	$\frac{5}{11}$	$\frac{2}{11}$	0	$-\frac{3}{11}$	3
P	1	0	0	$\frac{30}{11}$	$\frac{12}{11}$	0	$-\frac{7}{11}$	13

$R_1 = 11r_1$, $R_2 = \frac{3}{11}R_1 + r_2$, $R_3 = \frac{3}{11}R_1 + r_3$, $R_4 = \frac{7}{11}R_1 + r_4$

BV	P	x_1	x_2	x_3	s_1	s_2	s_3	RHS
s_3	0	0	11	2	3	0	1	22
s_2	0	0	3	3	2	1	0	16
x_1	0	1	3	1	1	0	0	9
P	1	0	7	4	3	0	0	27

Maximum $P = 27$, obtained when $x_1 = 9$, $x_2 = 0$ and $x_3 = 0$.

9. One constraint is an equation, so it gets no slack variable. The equation is the first row to be pivoted.

$$\begin{array}{c} \text{BV} \\ s_1 \\ \\ P \end{array}\begin{array}{c} \begin{array}{ccccc} P & x_1 & x_2 & s_1 & \text{RHS} \end{array} \\ \left[\begin{array}{cccc|c} 0 & 2 & 1 & 1 & 4 \\ 0 & \boxed{1} & 1 & 0 & 3 \\ \hline 1 & -3 & -2 & 0 & 0 \end{array}\right] \end{array} \xrightarrow[R_3=3R_2+r_3]{R_1=-2R_2+r_1} \begin{array}{c} \text{BV} \\ s_1 \\ x_1 \\ P \end{array}\begin{array}{c} \begin{array}{ccccc} P & x_1 & x_2 & s_1 & \text{RHS} \end{array} \\ \left[\begin{array}{cccc|c} 0 & 0 & -1 & 1 & -2 \\ 0 & 1 & 1 & 0 & 3 \\ \hline 1 & 0 & 1 & 0 & 9 \end{array}\right] \end{array}$$

We now have a negative value in the RHS, so we use the alternative pivoting strategy.

$$\xrightarrow[\substack{R_1=-r_1 \\ R_2=-R_1+r_2 \\ R_3=-R_1+r_3}]{\text{Alternative Pivoting Strategy}} \begin{array}{c} \text{BV} \\ x_2 \\ x_1 \\ P \end{array}\begin{array}{c} \begin{array}{ccccc} P & x_1 & x_2 & s_1 & \text{RHS} \end{array} \\ \left[\begin{array}{cccc|c} 0 & 0 & 1 & -1 & 2 \\ 0 & 1 & 0 & 1 & 1 \\ \hline 1 & 0 & 0 & 1 & 7 \end{array}\right] \end{array}$$

This is the final tableau. Maximum $P = 7$, obtained when $x_1 = 1$ and $x_2 = 2$.

11. We will solve the minimum problem by changing it to a maximum problem. Since it is a maximum problem all the constraints must be written as less than or equal inequalities. We will maximize $P = -z = -6x_1 - 8x_2 - x_3$

Rewrite the constraints:

$$\begin{aligned} -3x_1 - 5x_2 - 3x_3 &\le -20 \\ -x_1 - 3x_2 - 2x_3 &\le -9 \\ -6x_1 - 2x_2 - 5x_3 &\le -30 \\ x_1 + x_2 + x_3 &\le 10 \\ x_1 \ge 0,\ x_2 \ge 0,\ x_3 &\ge 0 \end{aligned}$$

Introduce slack variables:

$$\begin{aligned} -3x_1 - 5x_2 - 3x_3 + s_1 \qquad\qquad &= -20 \\ -x_1 - 3x_2 - 2x_3 \quad + s_2 \qquad &= -9 \\ -6x_1 - 2x_2 - 5x_3 \qquad + s_3 \quad &= -30 \\ x_1 + x_2 + x_3 \qquad\qquad + s_4 &= 10 \\ x_1 \ge 0,\ x_2 \ge 0,\ x_3 \ge 0,\ s_1 \ge 0,\ s_2 \ge 0,\ s_3 \ge 0,\ s_4 &\ge 0 \end{aligned}$$

We now set up the initial tableau and use the Alternate Pivoting Strategy as long as there are negative entries in the RHS. When all the entries in the RHS are positive, we use the standard way of choosing a pivot.

$$\begin{array}{c} \text{BV} \\ s_1 \\ s_2 \\ s_3 \\ s_4 \\ P \end{array}\begin{array}{c} \begin{array}{cccccccccc} P & x_1 & x_2 & x_3 & s_1 & s_2 & s_3 & s_4 & \text{RHS} \end{array} \\ \left[\begin{array}{cccccccc|c} 0 & -3 & -5 & -3 & 1 & 0 & 0 & 0 & -20 \\ 0 & -1 & -3 & -2 & 0 & 1 & 0 & 0 & -9 \\ 0 & \boxed{-6} & -2 & -5 & 0 & 0 & 1 & 0 & -30 \\ 0 & 1 & 1 & 1 & 0 & 0 & 0 & 1 & 10 \\ \hline 1 & 6 & 8 & 1 & 0 & 0 & 0 & 0 & 0 \end{array}\right] \end{array}$$

$$\xrightarrow[\substack{R_3=-\frac{1}{6}r_3 \\ R_1=3R_3+r_1 \\ R_2=R_3+r_2 \\ R_4=-R_3+r_4 \\ R_5=-6R_3+r_5}]{\text{Alternative Pivoting Strategy}} \begin{array}{c} \text{BV} \\ s_1 \\ s_2 \\ x_1 \\ s_4 \\ P \end{array}\begin{array}{c} \begin{array}{cccccccccc} P & x_1 & x_2 & x_3 & s_1 & s_2 & s_3 & s_4 & \text{RHS} \end{array} \\ \left[\begin{array}{cccccccc|c} 0 & 0 & \boxed{-4} & -\frac{1}{2} & 1 & 0 & -\frac{1}{2} & 0 & -5 \\ 0 & 0 & -\frac{8}{3} & -\frac{7}{6} & 0 & 1 & -\frac{1}{6} & 0 & -4 \\ 0 & 1 & \frac{1}{3} & \frac{5}{6} & 0 & 0 & -\frac{1}{6} & 0 & 5 \\ 0 & 0 & \frac{2}{3} & \frac{1}{6} & 0 & 0 & \frac{1}{6} & 1 & 5 \\ \hline 1 & 0 & 6 & -4 & 0 & 0 & 1 & 0 & -30 \end{array}\right] \end{array}$$

Alternative Pivoting Strategy: $R_1 = -\frac{1}{4}r_1$, $R_2 = \frac{8}{3}R_1 + r_2$, $R_3 = -\frac{1}{3}R_1 + r_3$, $R_4 = -\frac{2}{3}R_1 + r_4$, $R_5 = -6R_1 + r_5$

BV	P	x_1	x_2	x_3	s_1	s_2	s_3	s_4	RHS
x_2	0	0	1	$\frac{1}{8}$	$-\frac{1}{4}$	0	$\frac{1}{8}$	0	$\frac{5}{4}$
s_2	0	0	0	$\boxed{-\frac{5}{6}}$	$-\frac{2}{3}$	1	$\frac{1}{6}$	0	$-\frac{2}{3}$
x_1	0	1	0	$\frac{19}{24}$	$\frac{1}{12}$	0	$-\frac{5}{24}$	0	$\frac{55}{12}$
s_4	0	0	0	$\frac{1}{12}$	$\frac{1}{6}$	0	$\frac{1}{12}$	1	$\frac{25}{6}$
P	1	0	0	$-\frac{19}{4}$	$\frac{3}{2}$	0	$\frac{1}{4}$	0	$-\frac{75}{2}$

Alternative Pivoting Strategy: $R_2 = -\frac{6}{5}r_2$, $R_1 = -\frac{1}{8}R_2 + r_1$, $R_3 = -\frac{19}{24}R_2 + r_3$, $R_4 = -\frac{1}{12}R_2 + r_4$, $R_5 = \frac{19}{4}R_2 + r_5$

BV	P	x_1	x_2	x_3	s_1	s_2	s_3	s_4	RHS
x_2	0	0	1	0	$-\frac{7}{20}$	$\frac{3}{20}$	$\frac{3}{20}$	0	$\frac{23}{20}$
x_3	0	0	0	1	$\frac{4}{5}$	$-\frac{6}{5}$	$-\frac{1}{5}$	0	$\frac{4}{5}$
x_1	0	1	0	0	$-\frac{11}{20}$	$\boxed{\frac{19}{20}}$	$-\frac{1}{20}$	0	$\frac{79}{20}$
s_4	0	0	0	0	$\frac{1}{10}$	$\frac{1}{10}$	$\frac{1}{10}$	1	$\frac{41}{10}$
P	1	0	0	0	$\frac{53}{10}$	$-\frac{57}{10}$	$-\frac{7}{10}$	0	$-\frac{337}{10}$

Standard Pivoting Strategy: $R_3 = \frac{20}{19}r_3$, $R_1 = -\frac{3}{20}R_3 + r_1$, $R_2 = \frac{6}{5}R_3 + r_2$, $R_4 = -\frac{1}{10}R_3 + r_4$, $R_5 = \frac{57}{10}R_3 + r_5$

BV	P	x_1	x_2	x_3	s_1	s_2	s_3	s_4	RHS
x_2	0	$\frac{3}{19}$	1	0	$-\frac{5}{19}$	0	$\boxed{\frac{3}{19}}$	0	$\frac{10}{19}$
x_3	0	$\frac{24}{19}$	0	1	$\frac{2}{19}$	0	$-\frac{5}{19}$	0	$\frac{110}{19}$
s_2	0	$\frac{20}{19}$	0	0	$-\frac{11}{19}$	1	$-\frac{1}{19}$	0	$\frac{79}{19}$
s_4	0	$\frac{2}{19}$	0	0	$\frac{3}{19}$	0	$\frac{2}{19}$	1	$\frac{70}{19}$
P	1	6	0	0	2	0	−1	0	−10

Standard Pivoting Strategy: $R_1 = \frac{19}{3}r_1$, $R_2 = \frac{5}{19}R_1 + r_2$, $R_3 = \frac{1}{19}R_1 + r_3$, $R_4 = -\frac{2}{19}R_1 + r_4$, $R_5 = R_1 + r_5$

BV	P	x_1	x_2	x_3	s_1	s_2	s_3	s_4	RHS
s_3	0	−1	$\frac{19}{3}$	0	$-\frac{5}{3}$	0	1	0	$\frac{10}{3}$
x_3	0	1	$\frac{5}{3}$	1	$-\frac{1}{3}$	0	0	0	$\frac{20}{3}$
s_2	0	1	$\frac{1}{3}$	0	$-\frac{2}{3}$	1	0	0	$\frac{13}{3}$
s_4	0	0	$-\frac{2}{3}$	0	$\frac{1}{3}$	0	0	1	$\frac{10}{3}$
P	1	5	$\frac{19}{3}$	0	$\frac{1}{3}$	0	0	0	$-\frac{20}{3}$

The maximum value of $P = -\frac{20}{3}$, so the minimum value of $z = \frac{20}{3}$, and is obtained when $x_1 = 0,\ x_2 = 0$, and $x_3 = \frac{20}{3}$.

13. Let P denote the total number of shares Judith purchases, and let x_1, x_2, and x_3 denote the number of shares of Ameren Corp., United Utilities, and Progressive Energy respectively that she buys. This is a maximum problem that is not in standard form.

Maximize

$$P = x_1 + x_2 + x_3$$

subject to the constraints

$$\begin{aligned} 53.44x_1 + 29.85x_2 + 47.28x_3 &= 10{,}000 \\ 2.54x_1 + 1.15x_2 + 2.44x_3 &\geq 450 \\ x_1 - 2x_2 \quad &\geq 0 \\ x_1 \geq 0 \quad x_2 \geq 0 \quad x_3 &\geq 0 \end{aligned}$$

We rewrite the constraints, making the greater than constraints less than and adding nonnegative slack variables to the two inequality constraints.

$$\begin{aligned} 53.44x_1 + 29.85x_2 + 47.28x_3 \quad &= 10{,}000 \\ -2.54x_1 - 1.15x_2 - 2.44x_3 + s_1 \quad &= -450 \\ -x_1 + 2x_2 \quad + s_2 &= 0 \\ x_1 \geq 0 \quad x_2 \geq 0 \quad x_3 \geq 0 \quad s_1 \geq 0 \quad & s_2 \geq 0 \end{aligned}$$

We next set up the initial tableau. The first pivot will come from the equation. We will make x_1 the entering basic variable.

BV	P	x_1	x_2	x_3	s_1	s_2	RHS
	0	[53.44]	29.85	47.28	0	0	10,000
s_1	0	−2.54	−1.15	−2.44	1	0	−450
s_2	0	−1	2		0	0	1	0
P	1	−1	−1	−1	0	0	0

Pivoting Strategy for Equations →
$R_1 = \frac{1}{53.44} r_1$
$R_2 = 2.54R_1 + r_2$
$R_3 = R_1 + r_3$
$R_4 = R_1 + r_4$

BV	P	x_1	x_2	x_3	s_1	s_2	RHS
x_1	0	1	0.559	0.885	0	0	187.126
s_1	0	0	0.269	−0.193	1	0	25.299
s_2	0	0	[2.559]	0.885	0	1	187.126
P	1	0	−0.441	−0.115	0	0	187.126

Standard Pivoting Strategy →

BV	P	x_1	x_2	x_3	s_1	s_2	RHS
x_1	0	1	0	0.692	0	−0.218	146.274
s_1	0	0	0	−0.286	1	−0.105	5.643
x_2	0	0	1	0.346	0	0.391	73.137
P	1	0	0	0.037	0	0.173	219.411

This is the final tableau because all entries in the RHS and in the objective row are positive.

Judith should purchase 146.27 shares of Ameren Corp., 73.14 shares of United Utilities, and no Progressive Energy. The maximum number of shares is 219.41.

15. (a) We define the variables x_1, x_2, x_3, and x_4 so that

$x_1 =$ the number of units shipped from M1 to A1,

$x_2 =$ the number of units shipped from M1 to A2,

$x_3 =$ the number of units shipped from M2 to A1, and

$x_4 =$ the number of units shipped from M2 to A2.

The objective is to minimize shipping costs, C.

Minimize

$$C = 400x_1 + 100x_2 + 200x_3 + 300x_4$$

subject to the constraints

$$\begin{aligned} x_1 + x_2 \qquad\qquad &\le 600 \\ x_3 + x_4 &\le 400 \\ x_1 \qquad + x_3 \qquad &\ge 500 \\ x_2 \qquad + x_4 &\ge 300 \\ x_1 \ge 0,\ x_2 \ge 0,\ x_3 \ge 0,\ x_4 &\ge 0 \end{aligned}$$

(b) We change the problem to a maximization problem and write the constraints as less than or equal to inequalities.

Maximize

$$P = -C = -400x_1 - 100x_2 - 200x_3 - 300x_4$$

subject to the constraints

$$\begin{aligned} x_1 + x_2 \qquad\qquad &\le 600 \\ x_3 + x_4 &\le 400 \\ -x_1 \qquad - x_3 \qquad &\le -500 \\ -x_2 \qquad - x_4 &\le -300 \\ x_1 \ge 0,\ x_2 \ge 0,\ x_3 \ge 0,\ x_4 &\ge 0 \end{aligned}$$

We introduce nonnegative slack variables and set up the initial tableau.

BV	P	x_1	x_2	x_3	x_4	s_1	s_2	s_3	s_4	RHS
s_1	0	1	1	0	0	1	0	0	0	600
s_2	0	0	0	1	1	0	1	0	0	400
s_3	0	[−1]	0	−1	0	0	0	1	0	−500
s_4	0	0	−1	0	−1	0	0	0	1	−300
P	1	400	100	200	300	0	0	0	0	0

Since there are negative entries in the RHS, we use the Alternative Pivoting Strategy.

Alternative Pivoting Strategy → x_1

BV	P	x_1	x_2	x_3	x_4	s_1	s_2	s_3	s_4	RHS
s_1	0	0	1	−1	0	1	0	1	0	100
s_2	0	0	0	1	1	0	1	0	0	400
x_1	0	1	0	1	0	0	0	−1	0	500
s_4	0	0	[−1]	0	−1	0	0	0	1	−300
P	1	0	100	−200	300	0	0	400	0	−200,000

Alternative Pivoting Strategy → x_1

BV	P	x_1	x_2	x_3	x_4	s_1	s_2	s_3	s_4	RHS
s_1	0	0	0	[−1]	−1	1	0	1	1	−200
s_2	0	0	0	1	1	0	1	0	0	400
x_1	0	1	0	1	0	0	0	−1	0	500
x_2	0	0	1	0	1	0	0	0	−1	300
P	1	0	0	−200	200	0	0	400	100	−230,000

Alternative Pivoting Strategy → x_1

BV	P	x_1	x_2	x_3	x_4	s_1	s_2	s_3	s_4	RHS
x_3	0	0	0	1	1	−1	0	−1	−1	200
s_2	0	0	0	0	0	[1]	1	1	1	200
x_1	0	1	0	0	−1	1	0	0	1	300
x_2	0	0	1	0	1	0	0	0	−1	300
P	1	0	0	0	400	−200	0	200	−100	−190,000

Standard Pivoting Strategy → x_1

BV	P	x_1	x_2	x_3	x_4	s_1	s_2	s_3	s_4	RHS
x_3	0	0	0	1	1	0	1	0	0	400
s_1	0	0	0	0	0	1	1	1	1	200
x_1	0	1	0	0	−1	0	−1	−1	0	100
x_2	0	0	1	0	1	0	0	0	−1	300
P	1	0	0	0	400	0	200	400	100	−150,000

The maximum $P = -150{,}000$, so the minimum cost = 150,000 obtained when $x_1 = 100$, $x_2 = 300$, $x_3 = 400$ and $x_4 = 0$.

(c) Private motors should ship 100 engines from M1 to A1, 300 engines from M1 to A2, 400 engines from M2 to A1, and no engines from M2 to A2 for a minimum shipping charge of \$150,000.

17. (a) Let C denote the total shipping cost, and x_1 the number of motorcycles shipped from W_1 to D_1; x_2 the number shipped from W_1 to D_2; x_3 the number shipped from W_2 to D_1; and x_4 the number of motorcycles shipped from W_2 to D_2.

The manufacturer wants to fill the orders at the lowest possible cost.

Minimize

$$C = 15x_1 + 13x_2 + 14x_3 + 16x_4$$

subject to the constraints

$$\begin{aligned} x_1 + x_3 &= 20 \\ x_2 + x_4 &= 30 \\ x_1 + x_2 &\le 40 \\ x_3 + x_4 &\le 15 \end{aligned}$$

$$x_1 \ge 0 \quad x_2 \ge 0 \quad x_3 \ge 0 \quad x_4 \ge 0$$

(b) This is a minimum problem that is not in standard form. We will solve the maximum problem formed by writing $P = -C = -15x_1 - 13x_2 - 14x_3 - 16x_4$. Add nonnegative slack variables and set up the initial tableau. Since there are equality constraints, we first pivot on the nonbasic variables.

BV	P	x_1	x_2	x_3	x_4	s_1	s_2	RHS
	0	[1]	0	1	0	0	0	20
	0	0	1	0	1	0	0	30
s_1	0	1	1	0	0	1	0	40
s_2	0	0	0	1	1	0	1	15
P	1	15	13	14	16	0	0	0

Nonbasic Variable Pivoting Strategy →

BV	P	x_1	x_2	x_3	x_4	s_1	s_2	RHS
x_1	0	1	0	1	0	0	0	20
	0	0	[1]	0	1	0	0	30
s_1	0	0	1	−1	0	1	0	20
s_2	0	0	0	1	1	0	1	15
P	1	0	13	−1	16	0	0	−300

Nonbasic Variable Pivoting Strategy →

BV	P	x_1	x_2	x_3	x_4	s_1	s_2	RHS
x_1	0	1	0	1	0	0	0	20
x_2	0	0	1	0	1	0	0	30
s_1	0	0	0	[−1]	−1	1	0	−10
s_2	0	0	0	1	1	0	1	15
P	1	0	0	−1	3	0	0	−690

Since the initial tableau has negative entries in the RHS, we use the Alternative Pivoting Strategy.

Alternative Pivoting Strategy →

BV	P	x_1	x_2	x_3	x_4	s_1	s_2	RHS
x_1	0	1	0	0	−1	1	0	10
x_2	0	0	1	0	1	0	0	30
x_3	0	0	0	1	1	−1	0	10
s_2	0	0	0	0	0	[1]	1	5
P	1	0	0	0	4	−1	0	−680

We now use the Standard Pivoting Strategy since there are no negative entries in the RHS.

Standard Pivoting Strategy →

BV	P	x_1	x_2	x_3	x_4	s_1	s_2	RHS
x_1	0	1	0	0	−1	0	−1	5
x_2	0	0	1	0	1	0	0	30
x_3	0	0	0	1	1	0	1	15
s_1	0	0	0	0	0	[1]	1	5
P	1	0	0	0	4	0	1	−675

Minimum $C = 675$ when $x_1 = 5, x_2 = 30, x_3 = 15, x_4 = 0$.

(c) The motorcycle manufacturer minimizes shipping costs at $675 by shipping 5 motorcycles from warehouse W_1 to dealer D_1, shipping 30 motorcycles from W_1 to D_2, and 15 motorcycles from W_2 to D_1.

19. (a) Let x_1 represent the amount spent on newspaper advertising, and x_2 represent the amount spent on radio advertising. C is the total cost of advertising.

Minimize

$$C = x_1 + x_2$$

subject to the constraints

$$\begin{aligned} 50x_1 + 70x_2 &\geq 100{,}000 \\ 40x_1 + 20x_2 &\geq 120{,}000 \\ x_1 \geq 0,\ x_2 &\geq 0 \end{aligned}$$

(b) We change the problem to a maximization problem and write the constraints as less than or equal to inequalities.

Maximize

$$P = -C = -x_1 - x_2$$

subject to the constraints

$$\begin{aligned} -50x_1 - 70x_2 &\leq -100{,}000 \\ -40x_1 - 20x_2 &\leq -120{,}000 \\ x_1 \geq 0,\ x_2 \geq 0,\ x_3 &\geq 0 \end{aligned}$$

We introduce nonnegative slack variables and set up the initial tableau. Since there are negative entries in the RHS, we use the Alternative Pivoting Strategy.

$$\begin{array}{c} \\ s_1 \\ s_2 \\ P \end{array}\begin{array}{c} \begin{array}{cccccc|c} P & x_1 & x_2 & s_1 & s_2 & & \text{RHS} \end{array} \\ \left[\begin{array}{cccccc|c} 0 & -50 & -70 & 1 & 0 & & -100{,}000 \\ 0 & \boxed{-40} & -20 & 0 & 1 & & -120{,}000 \\ \hline 1 & 1 & 1 & 0 & 0 & & 0 \end{array}\right] \end{array} \xrightarrow{\text{Alternative Pivoting Strategy}} \begin{array}{c} \\ s_1 \\ x_1 \\ P \end{array}\begin{array}{c} \begin{array}{cccccc|c} P & x_1 & x_2 & s_1 & s_2 & & \text{RHS} \end{array} \\ \left[\begin{array}{cccccc|c} 0 & 0 & \boxed{-45} & 1 & -\frac{6}{5} & & 50{,}000 \\ 0 & 1 & \frac{1}{2} & 0 & -\frac{1}{40} & & 3{,}000 \\ \hline 1 & 0 & \frac{1}{2} & 0 & \frac{1}{40} & & -3{,}000 \end{array}\right] \end{array}$$

Maximum $P = -3000$, so minimum $C = 3000$, which is obtained when $x_1 = 3000$ and $x_2 = 0$.

(c) The appliance store will spend the least on advertising, \$3000, and reach the intended audience, if it spends all \$3000 on newspaper advertising and nothing on radio advertising.

21. (a) Let P represent the total sales representatives that are sent to the trade shows, and let x_1 be the number sent from New York to Dallas, x_2 be the number sent from New York to Chicago, x_3 be the number sent from San Francisco to Dallas, and x_4 be the number sent from San Francisco to Chicago.

Maximize

$$P = x_1 + x_2 + x_3 + x_4$$

subject to the constraints

$$\begin{aligned} x_1 + x_2 \qquad\qquad &\leq 12 \\ x_3 + x_4 &\leq 18 \\ x_1 + x_3 \qquad &\geq 5 \\ x_2 + x_4 &\geq 5 \\ 280x_1 + 180x_2 + 340x_3 + 200x_4 &\leq 4830 \\ x_1 \geq 0 \quad x_2 \geq 0 \quad x_3 \geq 0 \quad x_4 &\geq 0 \end{aligned}$$

(b) This is a maximum problem that is not in standard form. Write the constraints as less than or equal inequalities, add nonnegative slack variables and set up the initial tableau.

$$\begin{aligned}
x_1 + x_2 &\le 12 \\
x_3 + x_4 &\le 18 \\
-x_1 - x_3 &\le -5 \\
-x_2 - x_4 &\le -5 \\
280x_1 + 180x_2 + 340x_3 + 200x_4 &\le 4830 \\
x_1 \ge 0 \quad x_2 \ge 0 \quad x_3 \ge 0 &\quad x_4 \ge 0
\end{aligned}$$

BV	P	x_1	x_2	x_3	x_4	s_1	s_2	s_3	s_4	s_5	RHS
s_1	0	1	1	0	0	1	0	0	0	0	12
s_2	0	0	0	1	1	0	1	0	0	0	18
s_3	0	$\boxed{-1}$	0	−1	0	0	0	1	0	0	−5
s_4	0	0	−1	0	−1	0	0	0	1	0	−5
s_5	0	280	180	340	200	0	0	0	0	1	4830
P	1	−1	−1	−1	−1	0	0	0	0	0	0

Since there are negative entries in the RHS we use the Alternative Pivoting Strategy.

BV	P	x_1	x_2	x_3	x_4	s_1	s_2	s_3	s_4	s_5	RHS
s_1	0	0	1	−1	0	1	0	1	0	0	7
s_2	0	0	0	1	1	0	1	0	0	0	18
x_1	0	1	0	1	0	0	0	−1	0	0	5
s_4	0	0	$\boxed{-1}$	0	−1	0	0	0	1	0	−5
s_5	0	0	180	60	200	0	0	280	0	1	3430
P	1	0	−1	0	−1	0	0	−1	0	0	5

BV	P	x_1	x_2	x_3	x_4	s_1	s_2	s_3	s_4	s_5	RHS
s_1	0	0	0	−1	−1	1	0	$\boxed{1}$	1	0	2
s_2	0	0	0	1	1	0	1	0	0	0	18
x_1	0	1	0	1	0	0	0	−1	0	0	5
x_2	0	0	1	0	1	0	0	0	−1	0	5
s_5	0	0	0	60	20	0	0	280	180	1	2530
P	1	0	0	0	0	0	0	−1	−1	0	10

Now that all the entries in the RHS are nonnegative, we use the Standard Pivoting Strategy.

BV	P	x_1	x_2	x_3	x_4	s_1	s_2	s_3	s_4	s_5	RHS
s_3	0	0	0	−1	−1	1	0	1	1	0	2
s_2	0	0	0	1	1	0	1	0	0	0	18
x_1	0	1	0	0	−1	1	0	0	1	0	7
x_2	0	0	1	0	1	0	0	0	−1	0	5
s_5	0	0	0	$\boxed{340}$	300	−280	0	0	−100	1	1970
P	1	0	0	−1	−1	1	0	−1	0	0	12

BV	P	x_1	x_2	x_3	x_4	s_1	s_2	s_3	s_4	s_5	RHS
s_3	0	0	0	0	−0.118	0.176	0	1	0.706	0.003	7.794
s_2	0	0	0	0	0.118	0.824	1	0	0.294	−0.003	12.206
x_1	0	1	0	0	−1	1	0	0	1	0	7
x_2	0	0	1	0	[1]	0	0	0	−1	0	5
x_3	0	0	0	1	0.882	−0.824	0	0	−0.294	0.003	5.794
P	1	0	0	0	−0.118	0.176	0	0	−0.294	0.003	17.794

BV	P	x_1	x_2	x_3	x_4	s_1	s_2	s_3	s_4	s_5	RHS
s_3	0	0	0.118	0	0	0.176	0	1	0.588	0.003	8.382
s_2	0	0	−0.118	0	0	0.824	1	0	0.412	−0.003	11.618
x_1	0	1	1	0	0	1	0	0	0	0	12
x_4	0	0	1	0	1	0	0	0	−1	0	5
x_3	0	0	−0.882	1	0	−0.824	0	0	0.588	0.003	1.382
P	1	0	0.118	0	0	0.176	0	0	−0.412	0.003	18.382

BV	P	x_1	x_2	x_3	x_4	s_1	s_2	s_3	s_4	s_5	RHS
s_3	0	0	1	−1	0	1	0	1	0	0	7
s_2	0	0	0.5	−0.7	0	1.4	1	0	0	−0.005	10.65
x_1	0	1	1	0	0	1	0	0	0	0	12
x_4	0	0	−0.5	1.7	1	−1.4	0	0	0	0.005	7.35
s_4	0	0	−1.5	1.7	0	−1.4	0	0	1	0.005	2.35
P	1	0	−0.5	0.7	0	−0.4	0	0	0	0.005	19.35

BV	P	x_1	x_2	x_3	x_4	s_1	s_2	s_3	s_4	s_5	RHS
x_2	0	0	1	−1	0	1	0	1	0	0	7
s_2	0	0	0	−0.2	0	0.9	1	−0.5	0	−0.005	7.15
x_1	0	1	0	1	0	0	0	−1	0	0	5
x_4	0	0	0	1.2	1	−0.9	0	0.5	0	0.005	10.85
s_4	0	0	0	0.2	0	0.1	0	1.5	1	0.005	12.85
P	1	0	0	0.2	0	0.1	0	0.5	0	0.005	22.85

Maximum $P = 22.85$ when $x_1 = 5$, $x_2 = 7$, $x_3 = 0$, and $x_4 = 10.85$.

(c) The company can maximize the number of people sent to the trade shows by sending 5 representatives from New York to Dallas at a cost of \$1400; 7 from NY to Chicago at a cost of \$1269; no one from San Francisco to Dallas, and 11 from San Francisco to Chicago at a cost of \$2200. The total cost is \$4860, only \$30 over budget.

23. Let z denote the average price/earning ratio of the stocks purchased, and let x_1, x_2, x_3, and x_4 denote the number of shares of Duke Energy, H.J. Heinz, General Electric, and Ferrellgas Partners respectively purchased. The aim is to

Minimize

$$z = 10.3x_1 + 29.6x_2 + 18.1x_3 + 80.6x_4$$

subject to the constraints

$$\begin{aligned} 19x_1 + 46.5x_2 + 36x_3 + 21.5x_4 &= 50{,}000 \\ 46.5x_2 &\geq 5{,}000 \\ 19x_1 &\geq 10{,}000 \\ 36x_3 &\geq 10{,}000 \\ 21.5x_4 &\geq 10{,}000 \\ 19x_1 + 36x_3 &\leq 25{,}000 \\ x_1 \geq 0 \quad x_2 \geq 0 \quad x_3 \geq 0 \quad x_4 &\geq 0 \end{aligned}$$

This is a minimum linear programming problem with mixed constraints. So we will maximize $P = -z = -10.3x_1 - 29.6x_2 - 18.1x_3 - 80.6x_4$ subject to the above constraints, and solve the mixed constraint maximum problem.

Step 1 Write the $\geq$ inequalities as $\leq$ inequalities.

$$\begin{aligned} 19x_1 + 46.5x_2 + 36x_3 + 21.5x_4 &= 50{,}000 \\ -46.5x_2 &\leq -5{,}000 \\ -19x_1 &\leq -10{,}000 \\ -36x_3 &\leq -10{,}000 \\ -21.5x_4 &\leq -10{,}000 \\ 19x_1 + 36x_3 &\leq 25{,}000 \end{aligned}$$

Step 2 Introduce nonnegative slack variables to inequalities 2 through 6.

$$\begin{aligned} 19x_1 + 46.5x_2 + 36x_3 + 21.5x_4 &= 50{,}000 \\ -46.5x_2 + s_1 &= -5{,}000 \\ -19x_1 + s_2 &= -10{,}000 \\ -36x_3 + s_3 &= -10{,}000 \\ -21.5x_4 + s_4 &= -10{,}000 \\ 19x_1 + 36x_3 + s_5 &= 25{,}000 \end{aligned}$$

Step 3 Set up the tableau.

BV	P	x_1	x_2	x_3	x_4	s_1	s_2	s_3	s_4	s_5	RHS
	0	19	46.5	36	21.5	0	0	0	0	0	50,000
s_1	0	0	–46.5	0	0	1	0	0	0	0	–5,000
s_2	0	–19	0	0	0	0	1	0	0	0	–10,000
s_3	0	0	0	–36	0	0	0	1	0	0	–10,000
s_4	0	0	0	0	–21.5	0	0	0	1	0	–10,000
s_5	0	19	0	36	0	0	0	0	0	1	25,000
P	1	10.3	29.6	18.1	80.6	0	0	0	0	0	0

The first row represents a constraint that is an equation. We pivot on that row, choosing x_1 as the pivot column.

BV	P	x_1	x_2	x_3	x_4	s_1	s_2	s_3	s_4	s_5	RHS
x_1	0	1	2.447	1.895	1.132	0	0	0	0	0	2631.579
s_1	0	0	−46.5	0	0	1	0	0	0	0	−5,000
s_2	0	0	46.5	36	21.5	0	1	0	0	0	40,000
s_3	0	0	0	−36	0	0	0	1	0	0	−10,000
s_4	0	0	0	0	−21.5	0	0	0	1	0	−10,000
s_5	0	0	−46.5	0	−21.5	0	0	0	0	1	−25,000
P	1	0	4.392	−1.416	68.945	0	0	0	0	0	−27,000

Now we have the initial simplex tableau.

Step 4 Since there are negative entries in the RHS column, we follow the Alternative Pivoting Strategy. The pivot row will be s_5; the pivot column, x_2.

BV	P	x_1	x_2	x_3	x_4	s_1	s_2	s_3	s_4	s_5	RHS
x_1	0	1	0	1.895	0	0	0	0	0	0.053	1315.789
s_1	0	0	0	0	21.5	1	0	0	0	−1	20,000
s_2	0	0	0	36	0	0	1	0	0	1	15,000
s_3	0	0	0	−36	0	0	0	1	0	0	−10,000
s_4	0	0	0	0	−21.5	0	0	0	1	0	−10,000
x_2	0	0	1	0	0.462	0	0	0	0	−0.022	537.634
P	1	0	0	−1.416	66.914	0	0	0	0	0.094	−29,467

There are still negative entries in the RHS column, so we continue using the Alternative Pivoting Strategy. In this next pivot, the pivot row will be s_3 and the pivot column is x_3.

BV	P	x_1	x_2	x_3	x_4	s_1	s_2	s_3	s_4	s_5	RHS
x_1	0	1	0	0	0	0	0	0.053	0	0.053	789.474
s_1	0	0	0	0	21.5	1	0	0	0	−1	20,000
s_2	0	0	0	0	0	0	1	1	0	1	5,000
x_3	0	0	0	1	0	0	0	−0.028	0	0	277.778
s_4	0	0	0	0	−21.5	0	0	0	1	0	−10,000
x_2	0	0	1	0	0.462	0	0	0	0	−0.022	537.634
P	1	0	0	0	66.914	0	0	−0.039	0	0.094	−29,073

Pivoting again using the Alternative Pivoting Strategy, we use row s_4 and column x_4.

BV	P	x_1	x_2	x_3	x_4	s_1	s_2	s_3	s_4	s_5	RHS
x_1	0	1	0	0	0	0	0	0.053	0	0.053	789.474
s_1	0	0	0	0	0	1	0	0	1	−1	10,000
s_2	0	0	0	0	0	0	1	1	0	1	5,000
x_3	0	0	0	1	0	0	0	−0.028	0	0	277.778
x_4	0	0	0	0	1	0	0	0	−0.047	0	465.116
x_2	0	0	1	0	0	0	0	0	0.022	−0.022	322.581
P	1	0	0	0	0	0	0	−0.039	3.112	0.094	−60,196

Step 5 This tableau has only positive entries in the RHS. So it is a maximum problem in standard form. We pivot using the Standard Pivoting Strategy. We choose column s_3 as the pivot column and s_2 as the pivot row.

BV	P	x_1	x_2	x_3	x_4	s_1	s_2	s_3	s_4	s_5	RHS
x_1	0	1	0	0	0	0	−0.053	0	0	0	526.316
s_1	0	0	0	0	0	1	0	0	1	−1	10,000
s_3	0	0	0	0	0	0	1	1	0	1	5,000
x_3	0	0	0	1	0	0	0.028	0	0	0.028	416.667
x_4	0	0	0	0	1	0	0	0	−0.047	0	465.116
x_2	0	0	1	0	0	0	0	0	0.022	−0.022	322.581
P	1	0	0	0	0	0	0.039	0	3.112	0.13	−59,999

This is a final tableau. The maximum value of P is −59,999, so the minimum value of z is 59,999. It occurs when 526.3 shares of Duke Energy, 322.6 shares of H.J. Heinz, 416.7 shares of General Electric, and 465.1 shares of Ferrellgas Partners are purchased.

The minimum average price/earning ratio is

$$\frac{59{,}999}{526.3+322.6+416.7+465.1}=\frac{59{,}999}{1730.7}\approx 34.67$$

The annual yield for this optimal investment strategy is

$$0.067(19\cdot 526.3)+0.03(46.50\cdot 322.6)+0.031(36\cdot 416.7)+0.093(21.50\cdot 465.1)=\$2515.0$$

Chapter 4 Review

In Problems 1-7 use the following definition..

A maximum problem is in standard form if the following conditions are met.

Condition 1 All the variables are nonnegative.

Condition 2 Every other constraint is written as a linear expression that is less than or equal to a positive constant.

1. The problem is in standard form. Both conditions are met.

3. The problem is in standard form. Both conditions are met.

5. This problem is not in standard form. Condition 2 is not met. One of the constraints is written as a greater than or equal to inequality.

7. This problem is not in standard form. Condition 1 is not met. x_2 can be negative.

9. We add slack variables and rewrite the problem as

Maximize

$$P - 2x_1 - x_2 - 3x_3 = 0$$

subject to the constraints

$$\begin{aligned} 2x_1 + 5x_2 + x_3 + s_1 &= 100 \\ x_1 + 3x_2 + x_3 + s_2 &= 80 \\ 2x_1 + 3x_2 + 3x_3 + s_3 &= 120 \end{aligned}$$

$$x_1 \geq 0 \quad x_2 \geq 0 \quad x_3 \geq 0 \quad s_1 \geq 0 \quad s_2 \geq 0 \quad s_3 \geq 0$$

We can then write the initial tableau as

BV	P	x_1	x_2	x_3	s_1	s_2	s_3	RHS
s_1	0	2	5	1	1	0	0	100
s_2	0	1	3	1	0	1	0	80
s_3	0	2	3	3	0	0	1	120
P	1	–2	–1	–3	0	0	0	0

11. We add slack variables and rewrite the problem as

Maximize

$$P - 6x_1 - 3x_2 = 0$$

subject to the constraints

$$\begin{aligned} x_1 + 5x_2 + s_1 &= 200 \\ 5x_1 + 3x_2 + s_2 &= 450 \\ x_1 + x_2 + s_3 &= 120 \end{aligned}$$

$$x_1 \geq 0 \quad x_2 \geq 0 \quad s_1 \geq 0 \quad s_2 \geq 0 \quad s_3 \geq 0$$

We can then write the initial tableau as

BV	P	x_1	x_2	s_1	s_2	s_3	RHS
s_1	0	1	5	1	0	0	200
s_2	0	5	3	0	1	0	450
s_3	0	1	1	0	0	1	120
P	1	–6	–3	0	0	0	0

13. We add slack variables and rewrite the problem as

Maximize

$$P - x_1 - 2x_2 - x_3 - 4x_4 = 0$$

subject to the constraints

$$\begin{aligned} x_1 + 3x_2 + x_3 + 2x_4 + s_1 &= 20 \\ 4x_1 + x_2 + x_3 + 6x_4 + s_2 &= 80 \end{aligned}$$

$$x_1 \geq 0 \quad x_2 \geq 0 \quad x_3 \geq 0 \quad x_4 \geq 0 \quad s_1 \geq 0 \quad s_2 \geq 0$$

We can then write the initial tableau as

BV	P	x_1	x_2	x_3	x_4	s_1	s_2	RHS
s_1	0	1	3	1	2	1	0	20
s_2	0	4	1	1	6	0	1	80
P	1	−1	−2	−1	−4	0	0	0

In Problems 15(a)-22(a) we use the following steps to choose the pivot element.

STEP 1 Find the first negative entry in the objective row. It identifies the pivot column.

STEP 2 For each positive entry above the objective row in the pivot column, we form the quotient of the corresponding RHS entry divided by the positive entry.

STEP 3 Use the smallest quotient from step 2 to identify the pivot row and the pivot element.

15. (a) We choose the pivot element

Step 1 The pivot column is column x_2.

Step 2 The quotients:

$$40 \div 5 = 8 \qquad 10 \div 2 = 5$$

Step 3 The smallest quotient is 5 so the pivot element is 2.

The original tableau with the pivot element marked is shown on the left. The tableau after pivoting is on the right.

BV	P	x_1	x_2	s_1	s_2	RHS
x_1	0	1	5	1	0	40
s_2	0	0	[2]	2	1	10
P	1	0	−1	0	3	120

BV	P	x_1	x_2	s_1	s_2	RHS
x_1	0	1	0	−4	$-\frac{5}{2}$	15
x_2	0	0	1	1	$\frac{1}{2}$	5
P	1	0	0	1	$\frac{7}{2}$	125

(b) The resulting system of equations is

$$x_1 = 15 + 4s_1 + \frac{5}{2}s_2$$

$$x_2 = 5 - s_1 - \frac{1}{2}s_2$$

$$P = 125 - s_1 - \frac{7}{2}s_2$$

(c) The new tableau is the final tableau. The solution is maximum $P = 125$ obtained when $x_1 = 15$ and $x_2 = 5$.

17. (a) We choose the pivot element

Step 1 The pivot column is column x_1.

Step 2 In this column there is only one positive entry, so it becomes the pivot.
The original tableau with the pivot element marked is shown below.

BV	P	x_1	x_2	x_3	s_1	s_2	RHS
s_1	0	[1]	1	−1	1	0	10
s_2	0	0	1	1	0	1	4
P	1	−2	−1	−3	0	0	0

The tableau after pivoting:

BV	P	x_1	x_2	x_3	s_1	s_2	RHS
x_1	0	1	1	−1	1	0	10
s_2	0	0	1	1	0	1	4
P	1	0	1	−5	2	0	20

(b) The resulting system of equations is

$$\begin{aligned} x_1 &= 10 - x_2 + x_3 - s_1 \\ s_2 &= 4 - x_2 - x_3 \\ P &= 20 - x_1 + 5x_3 - 2s_1 \end{aligned}$$

(c) This tableau needs further pivoting, since there is still a negative entry in the objective row. We find the new pivot element by using the steps.
Step 1: x_3 forms the pivot column.
Step 2: In this tableau there is only one positive entry so row s_2 is the new pivot row. The new pivot element is 1.

19. (a) We choose the pivot element
Step 1 The pivot column is column x_1.
Step 2 The quotients:

$$1 \div 0.5 = 2 \qquad 3 \div 1 = 3$$

Step 3 The smallest quotient is 2 so the pivot row is s_1.
The original tableau with the pivot element marked is

BV	P	x_1	x_2	s_1	s_2	RHS
s_1	0	[0.5]	0.5	1	0	1
s_2	0	1	1.5	0	1	3
P	1	−2.5	−2	0	0	0

The tableau after pivoting is

BV	P	x_1	x_2	s_1	s_2	RHS
x_1	0	1	1	2	0	2
s_2	0	0	0.5	−2	1	1
P	1	0	0.5	5	0	5

(b) The resulting system of equations is

$$\begin{aligned} x_1 &= 2 - x_2 - 2s_1 \\ s_2 &= 1 - 0.5x_2 + 2s_1 \\ P &= 5 - 0.5x_2 - 5s_2 \end{aligned}$$

(c) This is the final tableau. The solution to the maximum problem is maximum $P = 5$ when $x_1 = 2$ and $x_2 = 0$.

21. (a) We choose the pivot element

Step 1 The pivot column is column x_1.

Step 2 In this column there is only one positive entry, so it becomes the pivot.

The original tableau with the pivot element marked is

BV	P	x_1	x_2	x_3	s_1	s_2	s_3	RHS
s_1	0	−1	0	1	1	−1	0	7
x_2	0	−1	1	5	0	1	0	5
s_3	0	$\boxed{1}$	0	3	0	−5	1	3
P	1	−3	0	4	0	0	0	5

The tableau after pivoting is

BV	P	x_1	x_2	x_3	s_1	s_2	s_3	RHS
s_1	0	0	0	4	1	−6	1	10
x_2	0	0	1	8	0	−4	1	8
x_1	0	1	0	3	0	−5	1	3
P	1	0	0	13	0	−15	3	14

(b) The resulting system of equations is

$$
\begin{aligned}
s_1 &= 10 - 4x_3 + 6s_2 - s_3 \\
x_2 &= 8 - 8x_3 + 4s_2 - s_3 \\
x_1 &= 3 - 3x_3 + 5s_2 - s_3 \\
P &= 14 - 13x_3 + 15s_2 - 3s_3
\end{aligned}
$$

(c) This tableau indicates no solution exists for the problem. The pivot column would be column s_3, but every entry in the pivot column is negative.

23.

BV	P	x_1	x_2	x_3	s_1	s_2	s_3	RHS
s_1	0	5	5	10	1	0	0	1000
s_2	0	10	8	5	0	1	0	2000
s_3	0	$\boxed{10}$	5	0	0	0	1	500
P	1	−100	−200	−50	0	0	0	0

$\rightarrow$

BV	P	x_1	x_2	x_3	s_1	s_2	s_3	RHS
s_1	0	0	$\boxed{2.5}$	10	1	0	−0.5	750
s_2	0	0	3	5	0	1	−1	1500
x_1	0	1	0.5	0	0	0	0.1	50
P	1	0	−150	−50	0	0	10	5000

$\rightarrow$

BV	P	x_1	x_2	x_3	s_1	s_2	s_3	RHS
x_2	0	0	1	4	0.4	0	−0.2	300
s_2	0	0	0	−7	−1.2	1	−0.4	600
x_1	0	1	0	$\boxed{-2}$	−0.2	0	0.2	−100
P	1	0	0	550	60	0	−20	50,000

Since there is a negative entry in RHS, the pivot is chosen using the Alternative Pivoting Strategy.

$$\begin{array}{c} \text{BV} \\ x_2 \\ s_2 \\ x_3 \\ P \end{array} \begin{array}{c} \begin{array}{ccccccc|c} P & x_1 & x_2 & x_3 & s_1 & s_2 & s_3 & \text{RHS} \end{array} \\ \left[\begin{array}{ccccccc|c} 0 & 2 & 1 & 0 & 0 & 0 & 0.2 & 100 \\ 0 & -3.5 & 0 & 0 & -0.5 & 1 & -1.1 & 950 \\ 0 & -0.5 & 0 & 1 & 0.1 & 0 & -0.1 & 50 \\ \hline 1 & 275 & 0 & 0 & 5 & 0 & 35 & 22{,}500 \end{array}\right] \end{array}$$

The maximum value for P is 22,500, obtained when $x_1 = 0$, $x_2 = 100$ and $x_3 = 50$.

25.

$$\begin{array}{c} \text{BV} \\ s_1 \\ s_2 \\ P \end{array} \begin{array}{c} \begin{array}{cccccc|c} P & x_1 & x_2 & x_3 & s_1 & s_2 & \text{RHS} \end{array} \\ \left[\begin{array}{cccccc|c} 0 & \boxed{2} & 2 & 1 & 1 & 0 & 8 \\ 0 & 1 & -4 & 3 & 0 & 1 & 12 \\ \hline 1 & -40 & -60 & -50 & 0 & 0 & 0 \end{array}\right] \end{array} \rightarrow \begin{array}{c} \text{BV} \\ x_1 \\ s_2 \\ P \end{array} \begin{array}{c} \begin{array}{cccccc|c} P & x_1 & x_2 & x_3 & s_1 & s_2 & \text{RHS} \end{array} \\ \left[\begin{array}{cccccc|c} 0 & 1 & \boxed{1} & \frac{1}{2} & \frac{1}{2} & 0 & 4 \\ 0 & 0 & -5 & \frac{5}{2} & -\frac{1}{2} & 1 & 8 \\ \hline 1 & 0 & -20 & -30 & 20 & 0 & 160 \end{array}\right] \end{array}$$

$$\begin{array}{c} \text{BV} \\ x_2 \\ s_2 \\ P \end{array} \begin{array}{c} \begin{array}{cccccc|c} P & x_1 & x_2 & x_3 & s_1 & s_2 & \text{RHS} \end{array} \\ \left[\begin{array}{cccccc|c} 0 & 1 & 1 & \frac{1}{2} & \frac{1}{2} & 0 & 4 \\ 0 & 5 & 0 & \boxed{5} & 2 & 1 & 28 \\ \hline 1 & 20 & 0 & -20 & 30 & 0 & 240 \end{array}\right] \end{array} \rightarrow \begin{array}{c} \text{BV} \\ x_2 \\ x_3 \\ P \end{array} \begin{array}{c} \begin{array}{cccccc|c} P & x_1 & x_2 & x_3 & s_1 & s_2 & \text{RHS} \end{array} \\ \left[\begin{array}{cccccc|c} 0 & \frac{1}{2} & 1 & 0 & \frac{3}{10} & -\frac{1}{10} & \frac{6}{5} \\ 0 & 1 & 0 & 1 & \frac{2}{5} & \frac{1}{5} & \frac{28}{5} \\ \hline 1 & 40 & 0 & 0 & 38 & 4 & 352 \end{array}\right] \end{array}$$

The maximum value for P is 352, obtained when $x_1 = 0$, $x_2 = \frac{6}{5}$ and $x_3 = \frac{28}{5}$.

In Problems 27-32 use the following conditions to determine if the minimum problem is in standard form.

There are three conditions to be met for a minimum problem to be in standard form:

CONDITION 1: All the variables must be nonnegative.

CONDITION 2: All other constraints are written as linear expressions that are greater than or equal to a constant.

CONDITION 3: The objective function is a linear expression with nonnegative coefficients.

27. The minimum problem is in standard form.

29. The minimum problem is not in standard form. Condition 2 is not met; the constraints are not written as greater than or equal to inequalities.

31. The minimum problem is not in standard form. Condition 2 is not met; one of the constraints is not written as greater than or equal to inequality.

33. To write the dual of the minimum problem we first write the matrix that represents the constraints and the objective function and then we write its transpose.

$$\text{The matrix: } \begin{array}{c} \begin{array}{cc} x_1 & x_2 \end{array} \\ \left[\begin{array}{cc|c} 2 & 2 & 8 \\ 1 & -1 & 2 \\ 2 & 1 & 0 \end{array}\right] \end{array} \qquad \text{The transpose of the matrix: } \left[\begin{array}{cc|c} 2 & 1 & 2 \\ 2 & -1 & 1 \\ 8 & 2 & 0 \end{array}\right]$$

From the transpose we create the maximum problem.

Maximize

$$P = 8y_1 + 2y_2$$

subject to the conditions

$$\begin{aligned} 2y_1 + y_2 &\le 2 \\ 2y_1 - y_2 &\le 1 \\ y_1 \ge 0 \quad y_2 &\ge 0 \end{aligned}$$

This maximum problem is the dual of the minimum problem.

35. To write the dual of the minimum problem we first write the matrix that represents the constraints and the objective function and then we write its transpose.

$$\text{The matrix: } \begin{array}{c} x_1 \quad x_2 \quad x_3 \qquad \\ \left[\begin{array}{ccc|c} 1 & 1 & 1 & 100 \\ 2 & 1 & 0 & 50 \\ 5 & 4 & 2 & 0 \end{array}\right] \end{array} \quad \text{The transpose of the matrix: } \left[\begin{array}{cc|c} 1 & 2 & 5 \\ 1 & 1 & 4 \\ 1 & 0 & 2 \\ 100 & 50 & 0 \end{array}\right]$$

From the transpose we create the maximum problem.

Maximize

$$P = 100y_1 + 50y_2$$

subject to the constraints

$$\begin{aligned} y_1 + 2y_2 &\le 5 \\ y_1 + \ \ y_2 &\le 4 \\ y_1 \qquad &\le 2 \\ y_1 \ge 0 \quad y_2 &\ge 0 \end{aligned}$$

This maximum problem is the dual of the minimum problem.

37. Using the dual found in Problem 33, we get

$$\begin{array}{c} \\ s_1 \\ s_2 \\ P \end{array} \begin{array}{c} \begin{array}{cccccc} P & y_1 & y_2 & s_1 & s_2 & \text{RHS} \end{array} \\ \left[\begin{array}{ccccc|c} 0 & 2 & 1 & 1 & 0 & 2 \\ 0 & \boxed{2} & -1 & 0 & 1 & 1 \\ \hline 1 & -8 & -2 & 0 & 0 & 0 \end{array}\right] \end{array} \rightarrow \begin{array}{c} \\ s_1 \\ y_1 \\ P \end{array} \begin{array}{c} \begin{array}{cccccc} P & y_1 & y_2 & s_1 & s_2 & \text{RHS} \end{array} \\ \left[\begin{array}{ccccc|c} 0 & 0 & \boxed{2} & 1 & -1 & 1 \\ 0 & 1 & -\frac{1}{2} & 0 & \frac{1}{2} & \frac{1}{2} \\ \hline 1 & 0 & -6 & 0 & 4 & 4 \end{array}\right] \end{array}$$

$$\rightarrow \begin{array}{c} \\ y_2 \\ y_1 \\ P \end{array} \begin{array}{c} \begin{array}{cccccc} P & y_1 & y_2 & s_1 & s_2 & \text{RHS} \end{array} \\ \left[\begin{array}{ccccc|c} 0 & 0 & 1 & \frac{1}{2} & -\frac{1}{2} & \frac{1}{2} \\ 0 & 1 & 0 & \frac{1}{4} & \frac{1}{4} & \frac{3}{4} \\ \hline 1 & 0 & 0 & 3 & 1 & 7 \end{array}\right] \end{array}$$

The minimum value for C is 7, when $x_1 = 3$ and $x_2 = 1$.

39. Using the dual found in Problem 35, we get

$$\begin{array}{c} \\ s_1 \\ s_2 \\ s_3 \\ P \end{array} \begin{array}{c} \begin{array}{cccccccc} P & y_1 & y_2 & s_1 & s_2 & s_3 & \text{RHS} \end{array} \\ \left[\begin{array}{cccccc|c} 0 & 1 & 2 & 1 & 0 & 0 & 5 \\ 0 & 1 & 1 & 0 & 1 & 0 & 4 \\ 0 & \boxed{1} & 0 & 0 & 0 & 1 & 2 \\ \hline 1 & -100 & -50 & 0 & 0 & 0 & 0 \end{array}\right] \end{array} \rightarrow \begin{array}{c} \\ s_1 \\ s_2 \\ y_1 \\ P \end{array} \begin{array}{c} \begin{array}{cccccccc} P & y_1 & y_2 & s_1 & s_2 & s_3 & \text{RHS} \end{array} \\ \left[\begin{array}{cccccc|c} 0 & 0 & \boxed{2} & 1 & 0 & -1 & 3 \\ 0 & 0 & 1 & 0 & 1 & -1 & 2 \\ 0 & 1 & 0 & 0 & 0 & 1 & 2 \\ \hline 1 & 0 & -50 & 0 & 0 & 100 & 200 \end{array}\right] \end{array}$$

BV	P	y_1	y_2	s_1	s_2	s_3	RHS
y_2	0	0	1	0.5	0	–0.5	1.5
→ s_2	0	0	0	–0.5	1	–0.5	0.5
y_1	0	1	0	0	0	1	2
P	1	0	0	25	0	75	275

The minimum value for C is 275 when $x_1 = 25$, $x_2 = 0$ and $x_3 = 75$.

41. **Step 1** Write the constraints as less than or equal to inequalities.

$$-x_1 - x_2 \le -2$$
$$2x_1 + 3x_2 \le 12$$
$$3x_1 + 2x_2 \le 12$$

Step 2 Introduce nonnegative slack variables

$$\begin{aligned} -x_1 - x_2 + s_1 \quad &= -2 \\ 2x_1 + 3x_2 \quad + s_2 \quad &= 12 \\ 3x_1 + 2x_2 \quad + s_3 &= 12 \end{aligned}$$

Step 3 Set up the initial tableau

BV	P	x_1	x_2	s_1	s_2	s_3	RHS
s_1	0	–1	–1	1	0	0	–2
s_2	0	2	3	0	1	0	12
s_3	0	3	2	0	0	1	12
P	1	–3	–5	0	0	0	0

Step 4 The RHS has a negative entry, so we use the Alternative Pivoting Strategy. The first negative entry is –1 in column x_1. It is the pivot element.
Step 5 Pivot

BV	P	x_1	x_2	s_1	s_2	s_3	RHS
s_1	0	[–1]	–1	1	0	0	–2
s_2	0	2	3	0	1	0	12
s_3	0	3	2	0	0	1	12
P	1	–3	–5	0	0	0	0

→

BV	P	x_1	x_2	s_1	s_2	s_3	RHS
x_1	0	1	[1]	–1	0	0	2
s_2	0	0	1	2	1	0	8
s_3	0	0	–1	3	0	1	6
P	1	0	–2	–3	0	0	6

The new tableau has only nonnegative entries in the RHS, it represents a maximum problem in standard form. Since the objective row has negative entries, we use the standard pivoting strategy. The pivot column is x_2. We form the quotients

$$2 \div 1 = 2 \qquad 8 \div 1 = 8$$

The smaller of these is 2, so the pivot row is row x_1.

$$\begin{array}{c} \text{BV} \\ x_1 \\ \rightarrow s_2 \\ s_3 \\ P \end{array}\begin{array}{c} \begin{array}{ccccccc} P & x_1 & x_2 & s_1 & s_2 & s_3 & \text{RHS} \end{array} \\ \left[\begin{array}{cccccc|c} 0 & 1 & 1 & -1 & 0 & 0 & 2 \\ 0 & -1 & 0 & 3 & 1 & 0 & 6 \\ 0 & 1 & 0 & 2 & 0 & 1 & 8 \\ \hline 1 & 2 & 0 & -5 & 0 & 0 & 10 \end{array}\right] \end{array} \rightarrow \begin{array}{c} \text{BV} \\ x_2 \\ s_1 \\ s_3 \\ P \end{array}\begin{array}{c} \begin{array}{ccccccc} P & x_1 & x_2 & s_1 & s_2 & s_3 & \text{RHS} \end{array} \\ \left[\begin{array}{cccccc|c} 0 & \frac{2}{3} & 1 & 0 & -\frac{1}{3} & 0 & 4 \\ 0 & -\frac{1}{3} & 0 & 1 & \frac{1}{3} & 0 & 2 \\ 0 & \frac{1}{6} & 0 & 0 & -\frac{2}{3} & 1 & 4 \\ \hline 1 & \frac{1}{3} & 0 & 0 & \frac{5}{3} & 0 & 20 \end{array}\right] \end{array}$$

The maximum value of P is 20, and it is achieved when $x_1 = 0$ and $x_2 = 4$.

43. This is a minimum problem not in standard form. We will let $P = -C$ and then maximize P.

Maximize

$$P = -C = -2x_1 - 3x_2$$

Step 1 Write the constraints as less than or equal to inequalities.

$$\begin{aligned} -x_1 - x_2 &\le -3 \\ x_1 + x_2 &\le 9 \end{aligned}$$

Step 2 Introduce nonnegative slack variables.

$$\begin{aligned} -x_1 - x_2 + s_1 \qquad &= -3 \\ x_1 + x_2 \qquad + s_2 &= 9 \end{aligned}$$

Step 3 Set up the initial tableau.

$$\begin{array}{c} \text{BV} \\ s_1 \\ s_2 \\ P \end{array}\begin{array}{c} \begin{array}{cccccc} P & x_1 & x_2 & s_1 & s_2 & \text{RHS} \end{array} \\ \left[\begin{array}{ccccc|c} 0 & -1 & -1 & 1 & 0 & -3 \\ 0 & 1 & 1 & 0 & 1 & 9 \\ \hline 1 & 2 & 3 & 0 & 0 & 0 \end{array}\right] \end{array}$$

Step 4 The RHS has a negative entry, so we use the Alternative Pivoting Strategy. The first negative entry is -1 in column x_1. It is the pivot element.

Step 5 Pivot

$$\begin{array}{c} \text{BV} \\ s_1 \\ s_2 \\ P \end{array}\begin{array}{c} \begin{array}{cccccc} P & x_1 & x_2 & s_1 & s_2 & \text{RHS} \end{array} \\ \left[\begin{array}{ccccc|c} 0 & \boxed{-1} & -1 & 1 & 0 & -3 \\ 0 & 1 & 1 & 0 & 1 & 9 \\ \hline 1 & 2 & 3 & 0 & 0 & 0 \end{array}\right] \end{array} \rightarrow \begin{array}{c} \text{BV} \\ x_1 \\ s_2 \\ P \end{array}\begin{array}{c} \begin{array}{cccccc} P & x_1 & x_2 & s_1 & s_2 & \text{RHS} \end{array} \\ \left[\begin{array}{ccccc|c} 0 & 1 & 1 & -1 & 0 & 3 \\ 0 & 0 & 0 & 1 & 1 & 6 \\ \hline 1 & 0 & 1 & 2 & 0 & -6 \end{array}\right] \end{array}$$

This is the final tableau. The maximum value of P is -6, so the minimum value of C is 6 when $x_1 = 3$ and $x_2 = 0$.

45. Introduce a nonnegative slack variable to the inequality constraint.

$$\begin{aligned} 4x_1 + 3x_2 + 5x_3 + s_1 &= 140 \\ x_1 + x_2 + x_3 \qquad &= 30 \end{aligned}$$

The objective function is $P = -300x_1 - 200x_2 - 450x_3$

Set up the initial tableau

$$\begin{array}{c} \text{BV} \\ s_1 \\ s_2 \\ P \end{array} \begin{array}{c} \begin{array}{ccccc|c} P & x_1 & x_2 & x_3 & s_1 & \text{RHS} \end{array} \\ \left[\begin{array}{ccccc|c} 0 & 4 & 3 & 5 & 1 & 140 \\ 0 & \boxed{1} & 1 & 1 & 0 & 30 \\ \hline 1 & -300 & -200 & -450 & 0 & 0 \end{array}\right] \end{array}$$

Pivot on a nonbasic variable in the row with the equation.

$$\begin{array}{c} \text{BV} \\ s_1 \\ x_1 \\ P \end{array} \begin{array}{c} \begin{array}{ccccc|c} P & x_1 & x_2 & x_3 & s_1 & \text{RHS} \end{array} \\ \left[\begin{array}{ccccc|c} 0 & 0 & -1 & \boxed{1} & 1 & 20 \\ 0 & 1 & 1 & 1 & 0 & 30 \\ \hline 1 & 0 & 100 & -150 & 0 & 9000 \end{array}\right] \end{array}$$

Now we pivot using the Standard Pivoting Strategy.

$$\begin{array}{c} \text{BV} \\ x_3 \\ x_1 \\ P \end{array} \begin{array}{c} \begin{array}{ccccc|c} P & x_1 & x_2 & x_3 & s_1 & \text{RHS} \end{array} \\ \left[\begin{array}{ccccc|c} 0 & 0 & -1 & 1 & 1 & 20 \\ 0 & 1 & \boxed{2} & 0 & -1 & 10 \\ \hline 1 & 0 & -50 & 0 & 150 & 12{,}000 \end{array}\right] \end{array} \rightarrow \begin{array}{c} \text{BV} \\ x_3 \\ x_2 \\ P \end{array} \begin{array}{c} \begin{array}{ccccc|c} P & x_1 & x_2 & x_3 & s_1 & \text{RHS} \end{array} \\ \left[\begin{array}{ccccc|c} 0 & \frac{1}{2} & 0 & 1 & \frac{1}{2} & 25 \\ 0 & \frac{1}{2} & 1 & 0 & -\frac{1}{2} & 5 \\ \hline 1 & 25 & 0 & 0 & 125 & 12{,}250 \end{array}\right] \end{array}$$

This is the final tableau. The maximum value of P is 12,250, obtained when $x_1 = 0$, $x_2 = 5$ and $x_3 = 25$.

47. Let x_1, x_2, and x_3 represent the number of vats of lite, regular, and dark beer manufactured, respectively. The objective is to maximize profit, P.

Maximize

$$P = 10x_1 + 20x_2 + 30x_3$$

subject to the constraints

$$\begin{cases} 6x_1 + 4x_2 + 2x_3 \le 800 \quad \text{or} \quad 3x_1 + 2x_2 + x_3 \le 400 \\ x_1 + 3x_2 + 2x_3 \le 600 \\ x_1 + x_2 + 4x_3 \le 300 \\ x_1 \ge 0, x_2 \ge 0, x_3 \ge 0 \end{cases}$$

The constraints are imposed by the availability of barley, sugar, and hops.

This is a maximum problem in standard form. The initial tableau is

$$\begin{array}{c} \text{BV} \\ s_1 \\ s_2 \\ s_3 \\ P \end{array} \begin{array}{c} \begin{array}{ccccccc|c} P & x_1 & x_2 & x_3 & s_1 & s_2 & s_3 & \text{RHS} \end{array} \\ \left[\begin{array}{ccccccc|c} 0 & \boxed{3} & 2 & 1 & 1 & 0 & 0 & 400 \\ 0 & 1 & 3 & 2 & 0 & 1 & 0 & 600 \\ 0 & 1 & 1 & 4 & 0 & 0 & 1 & 300 \\ \hline 1 & -10 & -20 & -30 & 0 & 0 & 0 & 0 \end{array}\right] \end{array} \rightarrow$$

$$\begin{array}{c} \text{BV} \\ x_1 \\ s_2 \\ x_3 \\ P \end{array} \begin{array}{c} \begin{array}{cccccccc} P & x_1 & x_2 & x_3 & s_1 & s_2 & s_3 & \text{RHS} \end{array} \\ \left[\begin{array}{ccccccc|c} 0 & 1 & \frac{2}{3} & \frac{1}{3} & \frac{1}{6} & 0 & 0 & \frac{400}{3} \\ 0 & 0 & \boxed{\frac{7}{3}} & \frac{5}{3} & -\frac{1}{6} & 1 & 0 & \frac{1400}{3} \\ 0 & 0 & \frac{1}{3} & \frac{11}{3} & -\frac{1}{6} & 0 & 1 & \frac{500}{3} \\ \hline 1 & 0 & -\frac{40}{3} & -\frac{80}{3} & \frac{1}{6} & 0 & 0 & \frac{4000}{3} \end{array}\right] \end{array} \rightarrow \begin{array}{c} \text{BV} \\ x_1 \\ x_2 \\ s_3 \\ P \end{array} \begin{array}{c} \begin{array}{cccccccc} P & x_1 & x_2 & x_3 & s_1 & s_2 & s_3 & \text{RHS} \end{array} \\ \left[\begin{array}{ccccccc|c} 0 & 1 & 0 & -\frac{1}{7} & \frac{3}{14} & -\frac{2}{7} & 0 & 0 \\ 0 & 0 & 1 & \frac{5}{7} & \frac{1}{14} & \frac{3}{7} & 0 & 200 \\ 0 & 0 & 0 & \boxed{\frac{24}{7}} & -\frac{1}{7} & -\frac{1}{7} & 1 & 100 \\ \hline 1 & 0 & 0 & -\frac{120}{7} & \frac{5}{7} & \frac{40}{7} & 0 & 4000 \end{array}\right] \end{array}$$

$$\begin{array}{c} \text{BV} \\ x_1 \\ x_2 \\ x_3 \\ P \end{array} \begin{array}{c} \begin{array}{cccccccc} P & x_1 & x_2 & x_3 & s_1 & s_2 & s_3 & \text{RHS} \end{array} \\ \left[\begin{array}{ccccccc|c} 0 & 1 & 0 & 0 & \frac{5}{24} & -\frac{7}{24} & \frac{1}{24} & \frac{25}{6} \\ 0 & 0 & 1 & 0 & -\frac{1}{24} & \frac{11}{24} & \frac{5}{24} & \frac{1075}{6} \\ 0 & 0 & 0 & 1 & -\frac{1}{24} & -\frac{1}{24} & \frac{7}{24} & \frac{175}{6} \\ \hline 1 & 0 & 0 & 0 & 0 & 5 & 5 & 4500 \end{array}\right] \end{array}$$

This is the final tableau.

The brewer should brew $4\frac{1}{6}$ beer, $179\frac{1}{6}$ vats of regular beer, and $29\frac{1}{6}$ vats of dark beer to attain a maximum profit of \$4500.

49. Let x_1 and x_2 represent the number of pounds of ground beef and ground pork, respectively, to be combined in a one pound package of meat loaf. The objective is to

Minimize

$$C = 0.7x_1 + 0.5x_2$$

subject to the constraints

$$\begin{aligned} x_1 + \quad x_2 &= 1 \\ 0.75x_1 + 0.6x_2 &\geq 0.7 \\ x_1 \geq 0, \quad x_2 &\geq 0 \end{aligned}$$

This is a minimum problem with mixed constraints, we will transform it into a maximization problem:

Step 1 Write the greater than or equal to constraint as less than or equal to inequality.

$$\begin{aligned} x_1 + \quad x_2 &= 1 \\ -0.75x_1 - 0.6x_2 &\leq -0.7 \end{aligned}$$

Step 2 Introduce a nonnegative slack variable.

$$\begin{aligned} x_1 + \quad x_2 \quad &= 1 \\ -0.75x_1 - 0.6x_2 + s_1 &= -0.7 \end{aligned}$$

The objective function is

$$\text{Maximize } P = -C = -0.7x_1 - 0.5x_2$$

Step 3 Set up the initial tableau:

$$\begin{array}{c} \text{BV} \\ \\ s_1 \\ P \end{array} \begin{array}{c} \begin{array}{cccccc} P & x_1 & x_2 & s_1 & \text{RHS} \end{array} \\ \left[\begin{array}{cccc|c} 0 & 1 & 1 & 0 & 1 \\ 0 & -0.75 & -0.6 & 1 & -0.7 \\ \hline 1 & 0.7 & 0.5 & 0 & 0 \end{array}\right] \end{array}$$

Step 4 There is an equation in the first row, so the first pivot must be in that row. We will use the x_1 column.

$$\begin{array}{c} \text{BV} \\ x_1 \\ s_1 \\ P \end{array} \begin{array}{c} \begin{array}{cccccc} P & x_1 & x_2 & s_1 & \text{RHS} \end{array} \\ \left[\begin{array}{cccc|c} 0 & 1 & 1 & 0 & 1 \\ 0 & 0 & 0.15 & 1 & 0.05 \\ \hline 1 & 0 & -0.2 & 0 & -0.7 \end{array}\right] \end{array}$$

The new tableau represents a maximum problem in standard form. Since the objective row has a negative entry, we use the standard pivoting strategy. The pivot column is x_2. We form the quotients and find the pivot row. It is row s_1.

Step 5 Pivot

$$\begin{array}{c} \text{BV} \\ x_1 \\ x_2 \\ P \end{array} \begin{array}{c} \begin{array}{cccccc} P & x_1 & x_2 & s_1 & \text{RHS} \end{array} \\ \left[\begin{array}{cccc|c} 0 & 1 & 0 & -6.667 & 0.667 \\ 0 & 0 & 1 & 6.667 & 0.333 \\ \hline 1 & 0 & 0 & 1.333 & -0.633 \end{array}\right] \end{array}$$

This is the final tableau. The maximum $P = -0.633$, so $C = 0.633$ and it is attained $x_1 = 0.667$ and $x_2 = 0.333$. The butcher should combine $\frac{2}{3}$ of a pound of ground beef with $\frac{1}{3}$ pound ground pork for a meatloaf mixture of minimum cost of 63¢ per pound.

Chapter 4 Project

1. The objective is to maximize the carbohydrates in the trail mix. The objective function is

$$\text{Maximize } C = 31.4x_1 + 114.74x_2 + 148.12x_3 + 33.68x_4$$

3. This is a maximum problem with mixed constraints, so we will rewrite all the constraints but the nonnegativity inequalities to be less than or equal to inequalities.

$$\begin{array}{rrrrl} x_1 + & x_2 + & x_3 + & x_4 \le & 10 \\ -0.9x_1 & +0.1x_2 + & 0.1\,x_3 + & 0.1\,x_4 \le & 0 \\ +0.1x_1 & -0.9x_2 + & 0.1\,x_3 + & 0.1\,x_4 \le & 0 \\ +0.1x_1 & +0.1x_2 - & 0.9\,x_3 + & 0.1\,x_4 \le & 0 \\ +0.1x_1 & +0.1x_2 + & 0.1\,x_3 - & 0.9\,x_4 \le & 0 \\ 854x_1 & +435x_2 + & 1023.96x_3 & +162.02x_4 \le & 7000 \\ x_1 \ge 0 & x_2 \ge 0 & x_3 \ge 0 & x_4 \ge 0 & \end{array}$$

Since the problem is so large we will use Excel to find the solution.
As illustrated in the text

1. Enter the variables and their initial values of 0 (since they are nonbasic in the initial tableau).
2. Enter the objective function.
3. Enter the constraints, and the initial RHS
4. Go to the solver, set the target cell equal to the max(imum).
   then enter the constraints including the nonnegativity constraints.
5. Check the options to be sure a "assume linear model" is checked.
6. Solve and highlight answer.

We find that the trail mix will maximize carbohydrates while meeting the constraints on the mix if 1 cup of peanuts, 3.7 cups of raisins, 4.3 cups of M&M's, and 1 cup of pretzels are used in the mix. Then the maximum carbohydrates will be 1124.9 grams.

5. To find the mix that will maximize the protein, we change the objective function. We

maximize $P = 34.57x_1 + 4.67x_2 + 9.01x_3 + 3.87x_4$

Again we use Excel, following the steps in the text, and we find that the trail mix which maximizes protein while meeting the constraints will contain 6.5 cups of peanuts, 0.9 cups of raisins, 0.9 cups of M&M's, and 0.9 cups of pretzels. The maximum protein in the mix will be 239.06 grams.

This trail mix contains 9.2 cups of mixture. So there are $\frac{231.06}{9.2} = 25.9$ grams of protein per cup.

There are $\frac{(854.10)(6.5)+(435)(0.9)+(1023.96)(0.9)+(162.02)(0.9)}{9.2} = 762.01$ calories per cup of mix.

7. The mix that minimizes the fat contains

$$\frac{(72.5)(1)+(0.67)(7)+(43.95)(1)+(1.49)(1)}{10} = 12.26 \text{ grams of fat per cup of mix.}$$

$$\frac{(34.57)(1)+(4.67)(7)+(9.01)(1)+(3.87)(1)}{10} = 8.0 \text{ grams of protein per cup of mix.}$$

$$\frac{(31.4)(1)+(114.74)(7)+(148.12)(1)+(33.68)(1)}{10} = 101.64 \text{ grams of carbohydrates}$$

per cup of mix.

Mathematical Questions from Professional Exams

1. (c)

3. (c)

5. (b)

7. (c)

9. (d)

11. (d)

Chapter 5

Finance

5.1 Interest

1. Prt

3. False. For a simple interest loan with interest rate r, the amount A due at the end of t years on a principal P borrowed is $A = P(1 + rt)$.

5. $0.60 = \frac{60}{100} = 60\%$

7. $1.1 = \frac{110}{100} = 110\%$

9. $0.06 = \frac{6}{100} = 6\%$

11. $0.0025 = \frac{25}{10000} = \frac{0.25}{100} = 0.25\%$

13. $25\% = \frac{25}{100} = 0.25$

15. $100\% = \frac{100}{100} = 1.00$

17. $6.5\% = \frac{6.5}{100} = 0.065$

19. $73.4\% = \frac{73.4}{100} = 0.734$

21. 15% of 1000 = $(0.15) \cdot (1000) = 150$

23. 18% of 100 = $(0.18) \cdot (100) = 18$

25. 210% of 50 = $(2.10) \cdot (50) = 105$

27. $x\%$ of $80 = 4$

$$\frac{x}{100}\cdot 80 = 4$$
$$80x = 400$$
$$x = 5$$

4 is 5% of 80

29. $x\%$ of $5 = 8$

$$\frac{x}{100}\cdot 5 = 8$$
$$5x = 800$$
$$x = 160$$

8 is 160% of 5

31. $20 = 8\%$ of x

$$20 = 0.08x$$
$$x = \frac{20}{0.08} = 250$$

20 is 8% of 250

33. $50 = 15\%$ of x

$$50 = 0.15x$$
$$x = \frac{50}{0.15} = 333.333$$

50 is 15% of 333.33

35.
$$I = Prt$$
$$= (\$1000)(0.04)\left(\frac{3}{12}\right)$$
$$= \$10$$

37.
$$I = Prt$$
$$= (\$500)(0.12)\left(\frac{9}{12}\right)$$
$$= \$45$$

39.
$$I = Prt$$
$$= (\$1000)(0.10)\left(\frac{18}{12}\right)$$
$$= \$150$$

41.
$$A = P + Prt$$
$$1050 = 1000 + 1000r\left(\frac{6}{12}\right)$$
$$50 = 500r$$
$$r = \frac{50}{500} = 0.1$$

The per annum rate of interest is 10%.

43.
$$A = P + Prt$$
$$400 = 300 + 300r\left(\frac{12}{12}\right)$$
$$100 = 300r$$
$$r = \frac{100}{300} = 0.3333$$

The per annum rate of interest is 33.3%.

45.
$$A = P + Prt$$
$$1000 = 900 + 900r\left(\frac{10}{12}\right)$$
$$100 = 750r$$
$$r = \frac{100}{750} = 0.1333$$

The per annum rate of interest is 13.3%.

47.
$$R = L - Lrt$$
$$= 1200 - 1200(0.10)\left(\frac{6}{12}\right)$$
$$= 1140$$

The proceeds of the loan is \$1140.

49.
$$R = L - Lrt$$
$$= 2000 - 2000(0.08)(2)$$
$$= 1680$$

The proceeds of the loan is \$1680.

51.

$$R = L - Lrt = L(1 - rt)$$

$$1200 = L\left[1-(0.10)\left(\frac{1}{2}\right)\right]$$

$$L = \frac{1200}{.95} = 1263.16$$

You must repay $1263.16 for the discounted loan.

The equivalent simple interest on the loan

$$A = P + Prt$$

$$1263.16 = 1200 + 1200r\left(\frac{1}{2}\right)$$

$$63.16 = 600r$$

$$r = 0.1052667$$

The simple rate of interest for the loan is 10.53%.

53.

$$R = L - Lrt = L(1 - rt)$$

$$2000 = L\left[1-(0.08)(2)\right]$$

$$L = \frac{2000}{0.84} = 2380.95$$

You must repay $2380.95 for the discounted loan.

The equivalent simple interest on the loan

$$A = P + Prt$$

$$2380.95 = 2000 + 2000r(2)$$

$$380.95 = 4000r$$

$$r = 0.0952375$$

The simple rate of interest for the loan is 9.52%.

55. We know the amount $A = \$500$, $t = 9$ months $= \frac{9}{12}$ of a year, and $r = 3\%$ simple interest. We want to know P, the principal.

$$A = P + Prt = P(1 + rt)$$

$$500 = P\left[1 + 0.03\left(\frac{9}{12}\right)\right]$$

$$500 = 1.0225P$$

$$P = 489.00$$

Madalyn should invest $489.00 if she wants to buy the stereo.

57. The sheets that list for $130 are selling for $84.50. The difference between the list price and the sale price is the discount. Let x denote the discount written as a decimal.

$$130 - 130x = 84.50$$

$$130(1 - x) = 84.50$$

$$(1 - x) = \frac{84.50}{130}$$

$$x = 1 - \frac{84.50}{130} = \frac{7}{20} = 0.35$$

The list price of the sheets was reduced 35%.

59. Let L denote the amount that Sarah Jane borrows. We know $R = \$5000$, $r = 0.0899$, and $t = \frac{18}{12} = 1.5$.

$$R = L(1 - rt)$$

$$5000 = L(1 - 0.899 \cdot 1.5)$$

$$5000 = L(0.86515)$$

$$L = \frac{5000}{0.86515} = 5779.3446$$

Sarah Jane should get a loan of $5779.34.

61. We are told that this is a discounted loan with $R = 250$, $L = 296$, and $t = 2$ weeks $= \frac{2}{52} = \frac{1}{26}$ of a year. We seek r.

$$R = L(1 - rt)$$
$$250 = 296\left(1 - \frac{1}{26}r\right)$$
$$\frac{250}{296} = 1 - \frac{1}{26}r$$
$$\frac{1}{26}r = 1 - \frac{250}{296}$$
$$r = 26\left(1 - \frac{250}{296}\right) = 4.0405$$

The per annum interest rate of this loan is 404%.

63. We need to compare both loans and choose the one with the least interest due.

Discounted Loan: $R = 1000$, $r = 0.09$, $t = \frac{1}{2}$.

$$R = L - Lrt = L(1 - rt)$$
$$1000 = L\left[(1 - (0.09)\left(\frac{1}{2}\right)\right]$$
$$1000 = 0.955L$$
$$L = 1047.1204$$

The interest on the loan is
$\$1047.12 - \$1000 = \$\,47.12$.

Simple Interest Loan: $P = 1000$, $r = 0.10$, $t = \frac{1}{2}$.

$$I = Prt$$
$$I = 1000(0.10)\left(\frac{1}{2}\right)$$
$$= 50$$

The interest on the loan is $50.00.

You should choose the discounted loan at 9% per annum. You will save $2.88.

65. We need to compare both loans and choose the one with the least interest due.

Discounted Loan: $R = 4000$, $r = 0.06$, $t = 1$.

$$R = L - Lrt = L(1 - rt)$$
$$4000 = L\left[(1 - (0.06)(1)\right]$$
$$4000 = 0.94L$$
$$L = 4255.319$$

The interest on the loan is
$\$4255.32 - \$4000 = \$\,255.32$.

Simple Interest Loan: $P = 4000$, $r = 0.063$, $t = 1$.

$$I = Prt$$
$$I = 4000(0.063)(1)$$
$$= 252.0$$

The interest on the loan is $252.00.

You should choose the simple interest loan at 6.3% per annum. You will save $ 3.32.

67. (a) $P = \$8000$, $r = 0.074$, $t = 3$. The interest charged is
$$I = Prt = (\$8000)(0.074)(3) = \$1776$$

(b) The total amount of the loan is $A = P + I = \$8000 + \$1776 = \$9776.00$.

(c) Assuming that the loan is repaid in 36 equal monthly payments, each payment will be
$$\frac{\$9776}{36} = \$271.56$$

69. This is an example of a discounted loan. The bank's $1 million is L; the amount the bank pays (its bid) is R. In this problem $r = 5\% = 0.05$ and $t = 3$ months $= \frac{3}{12}$ year.

$$R = L(1 - rt)$$

$$R = 1{,}000{,}000\left[1 - 0.05\left(\frac{3}{12}\right)\right] = 1{,}000{,}000\ (0.9875) = 987{,}500$$

The bank should bid $987,500 for the treasury bill.

71. Since the price of the T-bills were $996.18 per $1000, and the investor purchased $10,000 worth of bills, the investor paid ($996.18)(10) = $9961.80 for the treasuries.

The interest earned on the investment was $10,000 – $9961.80 = $38.20.

The per annum simple interest rate of the investment is

$$12\left(\frac{38.20}{9961.80}\right) = 0.04602 = 4.602\%$$

73. Since the price of the treasuries was $996.08 per $1000, and the investor purchased $5000 worth of bills, the investor paid ($996.08)(5) = $4980.40 for the treasury bills.
The interest earned on the maturity date was $5000.00 – $4980.40 = $19.60.
The per annum simple interest rate on this investment is

$$12\left(\frac{19.60}{4980.40}\right) = 0.04723 = 4.723\%$$

75. No.
Let x denote the original price of the stock. If the price drops by 10%, the new price of the stock is $0.90x$. When the price later improves by 10%, the final price is

$$1.1(0.90x) = 0.99x \neq x$$

5.2 Compound Interest

5. True

7. Compounded continuously

In Problems 9 – 17 use the formula $A = P\left(1 + \frac{r}{n}\right)^{nt}$ *if the interest is compounded n times per annum and* $A = Pe^{rt}$ *if the interest is compounded continuously.*

9. $P = \$100;\ r = 0.04;\ n = 4;\ t = 2;$

$$A = 100\left(1 + \frac{0.04}{4}\right)^{4 \cdot 2} = 100(1.01)^8 = 108.2857$$

The amount accumulated in the investment is $A = \$108.29$.

11. $P = \$500;\ r = 0.08;\ n = 4;\ t = 2.5;$

$$A = 500\left(1+\frac{0.08}{4}\right)^{4\cdot 2.5} = 500(1.02)^{10} = 609.4972$$

The amount accumulated in the investment is $A = \$609.50$.

13. $P = \$600;\ r = 0.05;\ n = 365,\ t = 3;$

$$A = 600\left(1+\frac{0.05}{365}\right)^{1095} = 697.0934$$

The amount accumulated in the investment is $A = \$697.09$.

15. $P = \$10;\ r = 11\%$; compounded continuously; $t = 2$.

$$A = 10e^{(0.11)(2)} = 12.4608$$

The amount accumulated in the investment is $A = \$12.46$.

17. $P = \$100;\ r = 10\%$; compounded continuously; $t = 2.25$.

$$A = 100e^{(0.10)(2.25)} = 125.2323$$

The amount accumulated in the investment is $A = \$125.23$.

In Problems 19 – 27, use the formula $P = A\cdot\left(1+\frac{r}{n}\right)^{-nt}$ *if the interest is compounded n times per annum and the formula* $P = Ae^{-rt}$ *if the interest is compounded continuously.*

19. $A = \$100,\ r = 0.06;\ n = 12;\ t = 2;$

$$P = 100\left(1+\frac{0.06}{12}\right)^{-12\cdot 2} = 88.7186$$

The present value of the investment is \$88.72.

21. $A = \$1000;\ r = 0.06;\ n = 365;\ t = 2.5;$

$$P = 1000\left(1+\frac{0.06}{365}\right)^{-365\cdot 2.5} = 860.7186$$

The present value of the investment is \$860.72.

23. $A = \$600;\ r = 0.04;\ n = 4;\ t = 2;$

$$P = 600\left(1+\frac{0.04}{4}\right)^{-4\cdot 2} = 600(1.01)^{-8} = 554.0899$$

The present value of the investment is \$554.09.

25. $A = \$80;\ r = 0.09$ compounded continuously; $t = 3.25;$

$$P = 80e^{-0.09\cdot 3.25} = 59.7116$$

The present value of the investment is \$59.71.

27. $A = \$400$; $r = 0.10$ compounded continuously; $t = 1$;

$$P = 400e^{-0.10 \cdot 1} = 361.9349$$

The present value of the investment is \$361.93.

29. To decide which rate yields the larger amount in 1 year, we use $A = P\left(1+\frac{r}{n}\right)^{nt}$ when $t = 1$ for each rate and compare the results. We use $P = \$10,000$ as suggested in the text.

$r = 6\%$; $n = 4$:

$$A = \$10,000\left(1+\frac{0.06}{4}\right)^4 = \$10,613.64$$

$r = 6\frac{1}{4}\%$; $n = 1$:

$$A = \$10,000(1 + 0.0625)^1 = \$10,625.00$$

$6\frac{1}{4}\%$ compounded annually has a greater yield.

31. To decide which rate yields the larger amount in 1 year, we use $A = P\left(1+\frac{r}{n}\right)^{nt}$ when $t = 1$ for each rate and compare the results. We use $P = \$10,000$ as suggested in the text.

$r = 9\%$; $n = 12$:

$$A = \$10,000\left(1+\frac{0.09}{12}\right)^{12} = \$10,938.07$$

$r = 8.8\%$; $n = 365$:

$$A = \$10,000\left(1+\frac{0.088}{365}\right)^{365} = \$10,919.77$$

9% compounded monthly has a greater yield.

33. The effective rate of interest for $r = 0.05$; $n = 4$ is

$$\begin{aligned} Eff &= \left(1+\frac{0.05}{4}\right)^4 - 1 \\ &= 1.0125^4 - 1 \\ &= 0.05095 \text{ or } 5.095\% \end{aligned}$$

35. The effective rate of interest for $r = 0.05$ compounded continuously is

$$\begin{aligned} Eff &= e^{0.05} - 1 \\ &= 1.05127 - 1 \\ &= 0.05127 \text{ or } 5.127\% \end{aligned}$$

37. To double an investment in 3 years, let $P = 1$, $A = 2P = 2$, $n = 1$, and $t = 3$. Find r.

$$\begin{aligned} A &= P(1+r)^t \\ 2 &= (1+r)^3 \\ 1+r &= \sqrt[3]{2} \\ r &= \sqrt[3]{2} - 1 \approx 1.25992 - 1 = 0.25992 \end{aligned}$$

The annual rate of interest needed to double the principal in 3 years is 25.99%.

39. To triple an investment in 5 years, let $P = 1$, $A = 3P = 3$, $n = 1$, and $t = 5$. Find r.

$$A = P(1+r)^t$$
$$3 = (1+r)^5$$
$$1+r = \sqrt[5]{3}$$
$$r = \sqrt[5]{3} - 1 \approx 1.24573 = 0.24573$$

The annual rate of interest needed to triple the principal in 5 years is 24.573%.

41. (a) To double an investment when $r = 0.08$ and $n = 12$, let $P = 1$ and $A = 2P = 2$. Find t.

$$A = P\left(1+\frac{r}{n}\right)^{nt}$$
$$2 = \left(1+\frac{0.08}{12}\right)^{12t}$$

We change the exponential equation to a logarithm and then use the change of base formula.

$$12t = \log_{1.00667} 2$$
$$t = \frac{\log 2}{12 \cdot \log 1.00667} = 8.688$$

It will take approximately 8.69 years to double money invested at 8% compounded monthly.

(b) To double an investment for $r = 0.08$ compounded continuously, let $P = 1$ and $A = 2P = 2$. Find t.

$$A = Pe^{rt}$$
$$2 = e^{0.08t}$$

We change the exponential equation to a logarithm and simplify.

$$0.08t = \ln 2$$
$$t = \frac{\ln 2}{0.08} = 8.664$$

It will take approximately 8.66 years to double money invested at 8% compounded continuously.

43.

$$Eff = \left(1+\frac{r}{4}\right)^4 - 1 = 0.07$$
$$\left(1+\frac{r}{4}\right)^4 = 1.07$$
$$r = 4 \cdot \left(\sqrt[4]{1.07} - 1\right)$$
$$r \approx 0.06823$$

6.823% compounded quarterly has an effective rate of 7%.

45. The principal $P = \$1000$, and the interest rate $r = 0.04$. The investment earns interest for $t = 3$ years.

(a) For annual compounding $n = 1$.

$$\begin{aligned} A &= P\left(1+\frac{r}{n}\right)^{nt} \\ &= (\$1000)(1.04)^3 \\ &= \$1124.86 \end{aligned}$$

The interest earned is $A - P = \$1124.86 - \$1000.00 = \$124.86$.

(b) For monthly compounding $n = 12$.

$$\begin{aligned} A &= P\left(1+\frac{r}{n}\right)^{nt} \\ &= (\$1000)\left(1+\frac{0.04}{12}\right)^{36} \\ &= \$1127.27 \end{aligned}$$

The interest earned is $A - P = \$1127.27 - \$1000.00 = \$127.27$.

47. The principal $P = \$1000$, $r = 0.06$, and $n = 4$.

(a) The amount A after 2 years is

$$\begin{aligned} A &= (\$1000)\left(1+\frac{0.06}{4}\right)^{8} \\ &= \$1126.49 \end{aligned}$$

(b) The amount A after 3 years is

$$\begin{aligned} A &= (\$1000)\left(1+\frac{0.06}{4}\right)^{12} \\ &= \$1195.62 \end{aligned}$$

(c) The amount A after 4 years is

$$\begin{aligned} A &= (\$1000)\left(1+\frac{0.06}{4}\right)^{16} \\ &= \$1268.99 \end{aligned}$$

49. When $r = 0.03$ and $n = 2$, the deposit necessary to have \$5000

(a) in 4 years is

$$P = (\$5000)\left(1+\frac{0.03}{2}\right)^{-8} = \$4438.56$$

(b) in 8 years is

$$P = (\$5000)\left(1+\frac{0.03}{2}\right)^{-16} = \$3940.16$$

51. We find the interest due on each loan and compare the amounts. We have $P = \$1000$, $t = 2$ years. With a simple interest rate $r = 12\% = 0.12$, the interest due after two years is

$$I = Prt = (\$1000)(0.12)(2) = \$240$$

With an $r = 10\% = 0.10$ compounded monthly, the interest due after 2 years is

$$I = A - P = (\$1000)\left(1 + \frac{0.10}{12}\right)^{24} - \$1000 = \$220.39$$

Mr. Nielsen will pay less interest with a loan charging 10% interest compounded monthly.

53. $P = \$100$, $r = 0.08$, $A = \$150$.
Find t when the interest is compounded monthly ($n = 12$).

$$A = P\left(1 + \frac{r}{n}\right)^{nt}$$

$$150 = 100\left(1 + \frac{0.08}{12}\right)^{12t}$$

$$1.5 = \left(1 + \frac{0.08}{12}\right)^{12t}$$

$$12t = \log_{1.00666667} 1.5$$

$$t = \frac{\log 1.5}{12 \cdot \log 1.00666667}$$

$$t = 5.085 \text{ years}$$

Find t when the interest is compounded continuously.

$$A = Pe^{rt}$$

$$150 = 100e^{0.08t}$$

$$1.5 = e^{0.08t}$$

$$0.08t = \ln 1.5$$

$$t = \frac{\ln 1.5}{0.08} = 5.068 \text{ years}$$

55. $P = \$10{,}000$, $A = \$25{,}000$, $r = 0.06$ compounded continuously.

$$A = Pe^{rt}$$
$$25{,}000 = 10{,}000e^{0.06t}$$
$$2.5 = e^{0.06t}$$
$$0.06t = \ln 2.5$$
$$t = \frac{\ln 2.5}{0.06} = 15.27$$

It takes 15.27 years for \$10,000 to grow to \$25,000 at 6% compounded continuously.

57. In this problem we are looking for the amount of a \$90,000 investment after 5 years if interest is 3% compounded annually. So, $P = 90{,}000$; $r = 0.03$; $n = 1$; $t = 5$.

$$A = P\left(1 + \frac{r}{n}\right)^{nt}$$
$$A = 90{,}000(1.03)^5 = 104{,}334.667$$

The house will appreciate to \$104,335 after 5 years.

59. We are looking for the present value of \$15,000, if $r = 0.05$ compounded continuously and $t = 3$.

$$P = Ae^{-rt}$$
$$P = 15{,}000 \cdot e^{-0.05 \cdot 3} = 12{,}910.619$$

Jerome should ask his parents for \$12,910.62.

61. Assuming that the stock continues to grow at 15%, we are looking for A when $P = 100 \cdot \$15 = \$1500$, $r = 0.15$, $n = 1$, and $t = 5$.

$$A = P\left(1 + \frac{r}{n}\right)^{nt}$$
$$A = 1500 \cdot 1.15^5 = 3017.0357$$

The 100 shares of the stock should be worth \$3017.04 in 5 years.

63. Find A if $P = \$1000$, $r = 0.056$ compounded continuously, and $t = 1$ year.

$$A = Pe^{rt}$$
$$A = 1000e^{0.056} = 1057.5977$$

Jim's investment will accumulate to \$1057.60. He will not have enough to purchase the computer.

If $r = 0.059$ and $n = 12$, then A will become

$$A = P\left(1 + \frac{r}{n}\right)^{nt}$$
$$A = 1000\left(1 + \frac{0.059}{12}\right)^{12} = 1060.6219$$

The new interest rate is a better deal. Jim's investment will accumulate to \$1060.62, and he will be able to purchase his computer.

65. Will: $P = \$2000$, $r = 0.090$, $n = 2$, and $t = 20$.

$$A = P\left(1+\frac{r}{n}\right)^{nt}$$

$$A = 2000\left(1+\frac{0.09}{2}\right)^{2\cdot 20}$$

$$A = 2000\cdot 1.045^{40} = 11{,}632.72908$$

After 20 years, Will will have \$11,632.73 in his IRA.

Henry: $P = \$2000$, $r = 0.085$ compounded continuously, and $t = 20$.

$$A = Pe^{rt}$$

$$A = 2000e^{0.085\cdot 20} = 10{,}947.89478$$

After 20 years, Henry will have \$10,947.89 in his IRA.

After 20 years, Will will have \$684.84 more than Henry.

67. We need the present value of \$40,000 after 4 years. We are told $A = \$40{,}000$, $r = 0.08$, $n = 4$, and $t = 4$.

$$P = A\left(1+\frac{r}{n}\right)^{-nt}$$

$$P = \$40{,}000\left(1+\frac{0.08}{4}\right)^{-16}$$

$$P = \$29{,}137.83$$

Tami and Todd should deposit \$29,137.83 now if they want the \$40,000 down payment for the house.

69. We want the accumulated value of the \$6000 investment after 25 years. $P = \$6000$, $r = 0.08$, $n = 2$, and $t = 25$.

$$A = P\left(1+\frac{r}{n}\right)^{nt}$$

$$A = 6000\left(1+\frac{0.08}{2}\right)^{2\cdot 25}$$

$$A = 6000\cdot 1.04^{50} = 42{,}640.10008$$

The child will have \$42,640.10 when she is 25 years old.

71. $P = \$10{,}000$, $A = \$25{,}000$, $r = 0.06$, $n = 365$. Find t.

$$\$25{,}000 = \$10{,}000\left(1+\frac{0.06}{365}\right)^{365t}$$

$$2.5 = \left(1+\frac{0.06}{365}\right)^{365t}$$

$$2.5 = (1.000164384)^{365t}$$

$$365t = \log_{1.000164384}(2.5)$$

$$t = \left(\frac{1}{365}\right)\cdot\frac{\log 2.5}{\log(1.000164384)}$$

$$t = 15.2727$$

It will take about 15.27 years for \$10,000 to grow to \$25,000 at 6% compounded daily.

73. $P = \$4000$, $r = 0.07$, $n = 1$, $t = 30$ years.

$$A = \$4000(1.07)^{30} = \$30{,}449.02$$

75. We want the present value P of $A = \$25{,}922$ when $r = 0.08$ compounded continuously and $t = 18$ years.

$$P = Ae^{-rt}$$

$$P = 25{,}922\, e^{-0.08\cdot 18} = 6141.64$$

Tom and Anita should invest \$6141.46 now to have the money needed in 18 years.

77. (a) We are looking for r the annual growth rate of the debt, given $P = \$5.61$ trillion, $A = \$8.68$ trillion, $t = 8$.

$$A = P(1+r)^t$$

$$8.68 = 5.61(1+r)^8$$

$$(1+r)^8 = \frac{8.68}{5.61}$$

$$r = \sqrt[8]{\frac{8.68}{5.61}} - 1 \approx 0.05607$$

The national debt grew at an annualized rate of 5.607%.

(b) Assuming the growth rate continues, then in 2015 $t = 16$, and the national debt will be

$$A = 5.61(1.05607)^{16} = 13.429$$

The national debt will be 13.43 trillion on January 1, 2015.

79. (a) $P = \$29{,}026$, $r = 0.055$ $t = 10$. At this rate of increase, the cost of a private 4-year college in 2015 will be

$$A = \$29{,}026(1.055)^{10} = \$49{,}580.60$$

(b) The present value of $A = \$49{,}580.60$ when $r = 0.04$ compounded continuously and $t = 10$ years is

$$P = \$49{,}580.60\ e^{0.04 \cdot 10} = \$33{,}234.87$$

\$33,234.87 should be invested in the college savings plan.

81. The budget deficit is $P = \$319$ billion, $r = 0.032$, $n = 2$, $t = 20$ years. The amount A is the amount to be paid back in 2025.

$$A = \$319\left(1 + \frac{0.032}{2}\right)^{2 \cdot 20}$$

$$A = \$319(1.016)^{40} = \$601.92$$

If the government funded the deficit with EE bonds, it will need to pay back \$601.92 billion in 2025.

83. The purchasing power of $P = \$1000$ after $n = 2$ years when the inflation rate $r = 3\% = 0.03$ is

$$A = \$1000(1 - 0.03)^2 = 940.9$$

With a 3% inflation rate, in 2 years \$1000 will buy \$940.90 worth of goods and services.

85. We are given $A = \$950$, $P = \$1000$, and $n = 2$. We solve for r.

$$A = P \cdot (1 - r)^n$$

$$950 = 1000 \cdot (1 - r)^2$$

$$1 - r = \sqrt{0.950}$$

$$r = 1 - \sqrt{0.950} = 0.0253$$

The inflation rate is 2.5%.

87. We are given $r = 0.02$; we seek n. Let $P = 2$ and $A = \frac{1}{2}P = 1$.

$$A = P \cdot (1 - r)^n$$

$$1 = 2 \cdot 0.98^n$$

$$0.5 = 0.98^n$$

$$n = \log_{0.98} 0.5$$

$$n = \frac{\log 0.5}{\log 0.98} = 34.3$$

At a 2% annual inflation rate, money's value is halved in 34.3 years.

89. $A = \$10,000$, $t = 20$, we need to find P, the present value of the bond.

(a) When the return is 10% compounded monthly, $r = 0.10$ and $n = 12$.

$$P = A\left(1+\frac{r}{n}\right)^{-nt}$$

$$P = 10,000\left(1+\frac{0.10}{12}\right)^{-12 \cdot 20}$$

$$P = 1364.6151$$

You should pay \$1364.62 for the bond if the interest is 10% compounded monthly.

(b) When the return is 10% compounded continuously, $r = 0.10$.

$$P = Ae^{-rt}$$

$$P = 10,000e^{-0.10 \cdot 20} = 1353.3528$$

You should pay \$1353.35 for the bond if the interest is 10% compounded continuously.

91. $A = \$10,000$, $t = 10$, $r = 0.08$, and $n = 1$, we need to find P, the present value of the bond.

$$P = A\left(1+\frac{r}{n}\right)^{-nt}$$

$$P = 10,000(1.08)^{-10} = 4631.934881$$

The \$10,000 bond should be sold for \$4631.93.

93. (a) $m = 2$, $r = 0.12$, $n = 1$;

$$t = \frac{\ln m}{n\ln\left(1+\frac{r}{n}\right)}$$

$$t = \frac{\ln 2}{1 \cdot \ln\left(1+\frac{0.12}{1}\right)}$$

$$t = \frac{\ln 2}{\ln(1.12)} = 6.116$$

It takes about 6.12 years to double an investment that earns 12% per annum.

(b) $m = 3$, $r = 0.06$, $n = 4$;

$$t = \frac{\ln m}{n\ln\left(1+\frac{r}{n}\right)}$$

$$t = \frac{\ln 3}{4\ln\left(1+\frac{0.06}{4}\right)}$$

$$t = \frac{\ln 3}{4 \cdot \ln 1.015} = 18.447$$

It takes about 18.45 years to triple an investment that earns 6% compounded quarterly.

(c) Given a principal P and interest rate r compounded n times per annum, the time t it takes the investment to multiply the investment m times is given by

$$A = mP = P\left(1+\frac{r}{n}\right)^{nt}$$

$$m = \left(1+\frac{r}{n}\right)^{nt} \quad \text{Divide both sides by } P.$$

$$nt = \log_{\left(1+\frac{r}{n}\right)} m \quad \text{Write the exponential equation as a logarithm.}$$

$$nt = \frac{\ln m}{\ln\left(1+\frac{r}{n}\right)}$$ Use the change of base formula.

$$t = \frac{\ln m}{n\ln\left(1+\frac{r}{n}\right)}$$ Divide both sides by n.

95. (a) There are 10 years between 1995 and 2005. To find the inflation rate we solve for r.

$$\text{CPI} = \text{CPI}_0\left(1+\frac{r}{100}\right)^n$$

$$195.3 = 152.4\left(1+\frac{r}{100}\right)^{10}$$

$$1+\frac{r}{100} = \sqrt[10]{\frac{195.3}{152.4}}$$

$$r = 100\left(\sqrt[10]{\frac{195.3}{152.4}}-1\right) = 2.51$$

The average rate of inflation from 1995 to 2005 was 2.51%.

(b) Using $r = 2.5\%$ and CPI = 300, we solve for t.

$$\text{CPI} = \text{CPI}_0\left(1+\frac{r}{100}\right)^n$$

$$300 = 152.4\left(1+\frac{2.5}{100}\right)^t$$

$$\frac{300}{152.4} = 1.025^t$$

$$t = \log_{1.025}\left(\frac{300}{152.4}\right)$$

$$t = \frac{\log\left(\frac{300}{152.4}\right)}{\log 1.025} = 27.428$$

The CPI will reach 300 in the year $1995 + 27 = 2022$

97. With $r = 3.1\%$, the CPI will double

$$\text{CPI} = \text{CPI}_0\left(1+\frac{r}{100}\right)^n$$

$$2 = 1 \cdot \left(1+\frac{3.1}{100}\right)^n$$

$$2 = 1.031^n$$

$$n = \log_{1.031} 2$$

$$n = \frac{\log 2}{\log 1.031} = 22.7$$

It will take 22.7 years for the CPI to double if the inflation rate is 3.1%.

5.3 Annuities; Sinking Funds

3. Annuity

5. The deposit is $P = \$100$. The number of deposits is $n = 10$ and the interest per payment period is $i = 0.10$. We use the formula

$$A = P\frac{(1+i)^n - 1}{i} = \$100 \cdot \frac{(1+0.10)^{10} - 1}{0.10} = (\$100)(15.93742) = \$1593.74$$

There is $1593.74 in the account after 10 years.

7. The deposit is $P = \$400$. The number of deposits is $n = 12$ and the interest per payment period is $i = \frac{0.12}{12} = 0.01$. We use the formula

$$A = P\frac{(1+i)^n - 1}{i} = \$400 \cdot \frac{(1+0.01)^{12} - 1}{0.01} = (\$400)(12.682503) = \$5073.00$$

There is $5073.00 in the account after 12 months (1 year).

9. The deposit is $P = \$200$. The number of deposits is $n = 36$ and the interest per payment period is $i = \frac{0.06}{12} = 0.005$. We use the formula

$$A = P\frac{(1+i)^n - 1}{i} = \$200 \cdot \frac{(1+0.005)^{36} - 1}{0.005} = (\$200)(39.3361) = \$7867.22$$

There is $7867.22 in the account after 36 months (3 years).

11. The deposit is $P = \$100$. The number of deposits is $n = 60$ and the interest per payment period is $i = \frac{0.06}{12} = 0.005$. We use the formula

$$A = P\frac{(1+i)^n - 1}{i} = \$100 \cdot \frac{(1+0.005)^{60} - 1}{0.005} = (\$100)(69.77003) = \$6977.00$$

There is $6977.00 in the account after 60 months (5 years).

13. The deposit is $P = \$9000$. The number of deposits is $n = 10$ and the interest per payment period is $i = 0.05$. We use the formula

$$A = P\frac{(1+i)^n - 1}{i} = \$9000 \cdot \frac{(1+0.05)^{10} - 1}{0.05} = (\$9000)(12.57789) = \$113{,}201.03$$

There is $113,201.03 in the account after 10 years.

15. The amount required is $A = \$10{,}000$ after $n = 12 \cdot 5 = 60$ monthly payments. The interest rate per payment period is $i = \frac{0.5}{12}$. To find the monthly payment use $A = P\frac{(1+i)^n - 1}{i}$ and solve for P.

$$\$10{,}000 = P\,\frac{\left(1+\frac{0.05}{12}\right)^{60} - 1}{\frac{0.05}{12}}$$

$$\$10{,}000 = P(68.0061)$$

$$P = \$147.05$$

Sixty monthly payments of $147.05 are needed to accumulate $10,000.

17. The amount required is $A = \$20{,}000$ after $n = 4 \cdot 2.5 = 10$ quarterly payments. The interest rate per payment period is $i = \frac{0.06}{4} = 0.015$. To find the quarterly payment use $A = P\frac{(1+i)^n - 1}{i}$ and solve for P.

$$\$20{,}000 = P\,\frac{(1+0.015)^{10} - 1}{0.015}$$
$$\$20{,}000 = P(10.7027)$$
$$P = \$1868.68$$

Ten quarterly payments of \$1868.68 are needed to accumulate \$20,000.

19. The amount required is $A = \$25{,}000$ after $n = 6$ monthly payments. The interest rate per payment period is $i = \frac{0.055}{12}$. To find the monthly payment we use $A = P\frac{(1+i)^n - 1}{i}$ and solve for P.

$$\$25{,}000 = P\,\frac{\left(1+\frac{0.055}{12}\right)^{6} - 1}{\frac{0.055}{12}}$$
$$\$25{,}000 = P(6.06917)$$
$$P = \$4119.178$$

Six monthly payments of \$4119.18 are needed to accumulate \$25,000.

21. The amount required is $A = \$5000$ after $n = 12 \cdot 2 = 24$ monthly payments. The interest rate per payment period is $i = \frac{0.04}{12}$. To find the monthly payment use $A = P\frac{(1+i)^n - 1}{i}$ and solve for P.

$$\$5000 = P\,\frac{\left(1+\frac{0.04}{12}\right)^{24} - 1}{\frac{0.04}{12}}$$
$$\$5000 = P(24.94289)$$
$$P = \$200.458$$

Twenty-four monthly payments of \$200.46 are needed to accumulate \$5000.

23. The amount required is $A = \$9000$ after $n = 4$ annual payments. The interest rate per payment period is $i = 0.05$. To find the monthly payment we use $A = P\frac{(1+i)^n - 1}{i}$ and solve for P.

$$\$9000 = P\,\frac{(1+0.05)^{4} - 1}{0.05}$$
$$\$9000 = P(4.31013)$$
$$P = \$2088.106$$

Four annual payments of \$2088.11 are needed to accumulate \$9000.

25. Al's investment is an example of an annuity. The deposit is $P = \$2500$, the number of deposits is $n = 15$ and the interest per payment period is $i = 0.07$. To find the value of the fund after 15 deposits we use the formula

$$A = P\frac{(1+i)^n - 1}{i} = \$2500\frac{(1+0.07)^{15} - 1}{0.07} = \$62{,}822.555$$

The mutual fund is worth \$62,822.56.

27. Todd and Tami have set up an annuity. The deposit is $P = \$300$, the number of deposits is $n = 4 \cdot 6 = 24$, and the interest per payment period is $i = \frac{0.08}{4} = 0.02$. To find the value of the fund after 24 quarterly deposits we use the formula

$$A = P\frac{(1+i)^n - 1}{i} = \$300\frac{(1+0.02)^{24} - 1}{0.02} = \$9126.559$$

Todd and Tami's annuity will be worth \$9126.56.

29. Dan's pension fund can be thought of as a sinking fund. The amount required is $A = \$350{,}000$ after $n = 12 \cdot 20 = 240$ monthly payments. The interest rate per payment period is $i = \frac{0.09}{12} = 0.0075$. To find the monthly payment we use $A = P\frac{(1+i)^n - 1}{i}$ and solve for P.

$$\$350{,}000 = P\,\frac{(1+0.0075)^{240} - 1}{0.0075}$$
$$\$350{,}000 = P(667.88687)$$
$$P = \$524.04$$

Dan needs to save \$524.04 per month to have \$350,000 in 20 years.

31. In this sinking fund the amount required is $A = \$100{,}000$ after $n = 4$ annual payments. The interest rate per payment period is $i = 0.08$. To find the annual payment we use $A = P\frac{(1+i)^n - 1}{i}$ and solve for P.

$$\$100{,}000 = P\,\frac{(1+0.08)^{4} - 1}{0.08}$$
$$\$100{,}000 = P(4.506112)$$
$$P = \$22{,}192.08$$

An annual payment of \$22,192.08 is necessary to accumulate the \$100,000 needed.

The table shows the growth of the sinking fund over time.

Payment Number	Deposit $	Cumulative Deposits	Accumulated Interest $	Total $
1	22,192.08	22,192.08	0	22,192.08
2	22,192.08	44,384.16	1775.37	46,159.53
3	22,192.08	66,576.24	5468.13	72,044.37
4	22,192.08	88,768.32	11,231.68	100,000.00

33. Let x denote the price to be paid for the oil well. Then $0.14x$ represents a 14% annual Return on Investment (ROI). The annual Sinking Fund Contribution (SFC) needed to recover the purchase price x in 30 years can be calculated letting $A = x$, $n = 30$, $i = 0.10$, and $P = \text{SFC}$.

$$A = P\frac{(1+i)^n - 1}{i}$$

$$x = \text{SFC}\frac{(1+0.10)^{30} - 1}{0.10}$$

$$x = \text{SFC}(164.4940227)$$

$$\text{SFC} = 0.0060792483x$$

The required annual ROI plus the annual SFC equals the annual net income of \$30,000.

$$0.14x + 0.006079x = \$30{,}000$$

$$0.146079x = \$30{,}000$$

$$x = \$205{,}368.33$$

The investor should pay \$205,368 for the oil well.

35. In this sinking fund the amount required is $A = \$1{,}000{,}000$ after $n = 4 \cdot 10 = 40$ quarterly payments. The interest rate per payment period is $i = \frac{0.08}{4} = 0.02$. To find the quarterly sinking fund payment we use $A = P\frac{(1+i)^n - 1}{i}$ and solve for P.

$$\$1{,}000{,}000 = P\,\frac{(1+0.02)^{40} - 1}{0.02}$$

$$\$1{,}000{,}000 = P(60.40198)$$

$$P = \$16{,}555.75$$

Forty quarterly payments of \$16,555.75 are necessary to accumulate \$1,000,000.

37. (a) The projected cost of the roof in 20 years is given by $A = P(1 + i)^n$. We have $P = \$100{,}000$, $i = 0.03$, and $n = 20$. The roof will cost $A = \$100{,}000(1.03)^{20} = \$180{,}611.12$.

(b) If the Condo Association makes $n = 2 \cdot 20 = 40$ payments earning interest at $i = \frac{0.06}{2} = 0.03$ per payment period, then each payment should be

$$A = P\frac{(1+i)^n - 1}{i}$$

$$\$180{,}611.12 = P\,\frac{(1+0.03)^{40} - 1}{0.03}$$

$$\$180{,}611.12 = P(75.40126)$$

$$P = \$2395.33$$

39. This is an annuity with monthly payments. $A = \$1,000,000$, $P = \$600$, $i = \dfrac{0.07}{12}$.

$$A = P\frac{(1+i)^n - 1}{i}$$

$$\$1,000,000 = \$600\frac{\left(1+\frac{0.07}{12}\right)^n - 1}{\frac{0.07}{12}}$$

$$\frac{1,000,000}{600}\left(\frac{0.07}{12}\right) + 1 = \left(1+\frac{0.07}{12}\right)^n$$

$$10.72222 = (1.0058333)^n$$

$$\log_{1.0058333}(10.72222) = n = \frac{\log 10.72222}{\log 1.0058333} = 407.868$$

It will take approximately 407.868 months or almost 34 years to accumulate \$1,000,000.

41. (a) Angie is setting up an sinking fund where $A = \$9500$, $n = 2 \cdot 12 = 24$ monthly payments, and $i = \dfrac{0.0675}{12}$. We need to find P.

$$A = P\frac{(1+i)^n - 1}{i}$$

$$\$9500 = P\frac{\left(1+\frac{0.0675}{12}\right)^{24} - 1}{\frac{0.0675}{12}}$$

$$P = \$9500\left[\frac{\frac{0.0675}{12}}{\left(1+\frac{0.0675}{12}\right)^{24} - 1}\right] = \$370.83$$

Angie should save \$370.83 each month to pay for the trip.

(b) There are 24 monthly deposits.

43. In this sinking fund $A = \$1,000,000$, $n = 4 \cdot 5 = 20$ payments, and $i = \dfrac{0.09}{4} = 0.0225$. We need P.

$$A = P\frac{(1+i)^n - 1}{i}$$

$$\$1,000,000 = P\left[\frac{1.0225^{20} - 1}{0.0225}\right]$$

$$P = \$1,000,000\left[\frac{0.0225}{1.0225^{20} - 1}\right] = \$40,142.07$$

The school district should make quarterly deposits of \$40,142.07 into the fund.

45. The value of the plan after 180 deposits is the amount accumulated by the initial investment of \$400 invested for 180 interest periods plus the accumulated amount of an annuity with 179 monthly deposits of \$100.

$$A = 400\left(1+\frac{0.08}{12}\right)^{180} + 100\frac{\left(1+\frac{0.08}{12}\right)^{179} - 1}{\frac{0.08}{12}}$$

$$A = \$1322.77 + \$34{,}275.32 = \$35{,}598.09$$

47. Option 1: $P = \$1{,}326{,}000$ invested for $t = 25$ years at $r = 0.085$ compounded annually.

$$A = \$1{,}326{,}000(1.085)^{25} = \$10{,}192{,}646.88$$

Option 2: An annuity with $P = \$100{,}000$, $i = 0.085$, $n = 25$

$$A = \$100{,}000\frac{(1.085)^{26} - 1}{0.085} = \$8{,}635{,}455.48$$

Dan should select the lump sum payment.

49. (a) $P = \$22{,}550$, $r = 0.0323$, $n = 2011 - 2007 = 4$. We want A

$$A = \$22{,}550(1.0323)^{4} = \$25{,}607.68$$

The projected cost of a Honda Accord in 2011 is \$25,607.68.

(b) The final projected cost, including sales tax is \$25,607.68(1.0925) = \$27,976.39.

(c) The monthly payment into a sinking fund needed to have the projected purchase price when $i = \frac{0.0275}{12}$, $n = 12 \cdot 4 = 48$, is given by

$$\$27{,}976.39 = P\frac{\left(1+\frac{0.0275}{12}\right)^{48} - 1}{\frac{0.0275}{12}}$$

$$P = \$27{,}976.39\left[\frac{\frac{0.0275}{12}}{\left(1+\frac{0.0275}{12}\right)^{48} - 1}\right] = \$552.04$$

51. (a) If $P = \$5492$ and $r = 0.07$, then the projected annual undergraduate college tuition and fees for public 4-year institutions are

Year	n	$A = \$5492(1.07)^n$
2018 – 2019	13	\$13,234.87
2019 – 2020	14	\$14,161.31
2020 – 2021	15	\$15,152.60
2021 – 2022	16	\$16,213.28

(b) The total projected cost of the four years tuition is \$58,762.06.

$$A = P\frac{(1+i)^n - 1}{i}$$

$$\$58,762.06 = P\left(\frac{1.0035^{45} - 1}{0.0035}\right)$$

$$P = \$58,762.06\left(\frac{0.0035}{1.0035^{45} - 1}\right) = \$1207.97$$

The quarterly sinking fund payments are \$1207.97

5.4 Present Value of an Annuity; Amortization

1. $P = \$500;\ n = 36;\ i = \frac{0.10}{12}$

$$V = P\frac{1-(1+i)^{-n}}{i} = \$500\frac{1-\left(1+\frac{0.10}{12}\right)^{-36}}{\frac{0.10}{12}} = \$500(30.991236) = \$15,495.62$$

3. $P = \$100;\ n = 9;\ i = \frac{0.12}{12} = 0.01$

$$V = P\frac{1-(1+i)^{-n}}{i} = \$100\frac{1-(1+.01)^{-9}}{0.01} = \$100(8.56602) = \$856.60$$

5. $P = \$10,000;\ n = 20;\ i = 0.10$

$$V = P\frac{1-(1+i)^{-n}}{i} = \$10,000\frac{1-(1+0.10)^{-20}}{0.10}$$
$$= \$10,000(8.513564) = \$85,135.64$$

7. $V = \$10,000;\ n = 48;\ i = \frac{0.08}{12}$

$$P = V\left[\frac{i}{1-(1+i)^{-n}}\right] = \$10,000\left[\frac{\frac{0.08}{12}}{1-\left(1+\frac{0.08}{12}\right)^{-48}}\right] = \$244.13$$

9. $V = \$500,000;\ n = 360;\ i = \frac{0.10}{12}$

$$P = V\left[\frac{i}{1-(1+i)^{-n}}\right] = \$500,000\left[\frac{\frac{0.10}{12}}{1-\left(1+\frac{0.10}{12}\right)^{-360}}\right] = \$4387.86$$

11. $V = \$1{,}000{,}000;\ n = 360;\ i = \dfrac{0.08}{12}$

$$P = V\left[\frac{i}{1-(1+i)^{-n}}\right] = \$1{,}000{,}000\left[\frac{\frac{0.08}{12}}{1-\left(1+\frac{0.08}{12}\right)^{-360}}\right] = \$7337.65$$

13. For the 2 year loan $V = \$10{,}000$, $n = 12 \cdot 2 = 24$, and $i = \dfrac{0.12}{12} = 0.01$. The monthly payment P is

$$P = V\frac{i}{1-(1+i)^{-n}} = \$10{,}000\,\frac{0.01}{1-(1.01)^{-24}} = \$470.73$$

15. Mr. Doody is interested in finding the present value of an annuity that has an interest rate of $i = \dfrac{0.10}{12}$ and from which he withdraws $P = \$250$ each month for $n = 12 \cdot 20 = 240$ months.

$$V = P\frac{1-(1+i)^{-n}}{i} = \$250\frac{1-\left(1+\frac{0.10}{12}\right)^{-240}}{\frac{0.10}{12}} = \$250(103.62462) = \$25{,}906.15$$

Mr. Doody needs \$25,906.15 now to guarantee his future income.

17. $V = \$8000;\ i = \dfrac{0.074}{12};\ n = 36$ months.

(a) The monthly car payment is

$$P = V\frac{i}{1-(1+i)^{-n}} = \$8000\frac{\frac{0.074}{12}}{1-\left(1+\frac{0.074}{12}\right)^{-36}} = \$248.48$$

(b) The total amount you pay is $n \cdot P = 36 \cdot 248.48 = \$8945.28$.

(c) The interest, I, you pay is the difference between the total amount paid and the amount of the loan V.

$$I = n \cdot P - V = \$8945.28 - \$8000 = \$945.28$$

You pay \$945.28 in interest on the loan.

19. **Option 1:** 20 year mortgage at 8% interest; $i = \dfrac{0.08}{12}$, and $n = 12 \cdot 20 = 240$ monthly payments,

Their monthly payments will be

$$P = V\frac{i}{1-(1+i)^{-n}} = \$160{,}000\,\frac{\frac{0.08}{12}}{1-\left(1+\frac{0.08}{12}\right)^{-240}} = \$1338.30$$

The total interest paid on this loan is (\$1338.30)(240) – \$160,000 = \$161,192.

The equity in the house is the sum of the down payment and the amount paid on the loan. After 10 years the couple still owes 120 payments. The amount still owed on the loan is the present value of these payments.

$$V = P\frac{1-(1+i)^{-n}}{i} = \$1338.30\frac{1-\left(1+\frac{0.08}{12}\right)^{-120}}{\frac{0.08}{12}} = \$110{,}304.67$$

meaning the amount paid on the loan \$160,000 – \$110,304.67 = \$49,695.33.

The couple's equity after 10 years is \$40,000 + \$49,695.33 = \$89,695.33.

Option 2: 25 year mortgage at 9% interest, $i = \frac{0.09}{12} = 0.0075$, and $n = 12 \cdot 25 = 300$ monthly payments. Their monthly payments will be

$$P = V\frac{i}{1-(1+i)^{-n}} = \$160{,}000\ \frac{0.0075}{1-(1.0075)^{-300}} = \$1342.71$$

The total interest paid on this loan is (\$1342.71)(300) – \$160,000 = \$242,813.

The equity in the house is the sum of the down payment and the amount paid on the loan. After 10 years the couple still owes 180 payments. The amount still owed on the loan is the present value of these payments.

$$V = P\frac{1-(1+i)^{-n}}{i} = \$1342.71\frac{1-(1.0075)^{-180}}{0.0075} = \$132{,}382.36$$

meaning the amount paid on the loan \$160,000 – \$132,382.36 = \$27,617.64.

The couple's equity after 10 years is \$40,000 + \$27,617.64 = \$67,617.64.

(a) The 25 year loan at 9% interest has a larger monthly payment.

(b) The total interest paid on the 25 year loan is \$81,621 more than that on the 20 year loan.

(c) After 10 years the couple would have more equity in their purchase if they had the 8% loan.

21. We first find how much John needs to accumulate prior to retirement. This is the present value V of $n = 12 \cdot 30 = 360$ payments of \$300 at $i = \frac{0.09}{12} = 0.0075$ per month.

$$V = P\frac{1-(1+i)^{-n}}{i} = \$300\frac{1-(1.0075)^{-360}}{0.0075} = \$37{,}284.56$$

We can now find the $n = 12 \cdot 20 = 240$ monthly payments necessary to meet this goal.

$$A = P\frac{(1+i)^{n}-1}{i}; \qquad \$37{,}284.56 = P\frac{(1.0075)^{240}-1}{0.0075}$$

$$\$37{,}284.56 = P(637.88687)$$

$$P = \$55.82$$

John should save \$55.82 each month to achieve his retirement goal.

23. (a) Dan's accumulated savings is actually an annuity with $P = \$100$, $n = 12$, and $i = \dfrac{0.06}{52}$ per week. After 12 weeks Dan has

$$A = P\frac{(1+i)^n - 1}{i} = \$100\frac{\left(1+\dfrac{0.06}{52}\right)^{12} - 1}{\dfrac{0.06}{52}} = \$1207.64$$

(b) Dan now amortizes his savings, he needs to find P if $V = \$1207.64$, $n = 34$, and $i = \dfrac{0.06}{52}$.

$$P = V\frac{i}{1-(1+i)^{-n}} = \$1207.64\ \frac{\dfrac{0.06}{52}}{1-\left(1+\dfrac{0.06}{52}\right)^{-34}} = \$36.24$$

Dan can withdraw \$36.24 each week for 34 weeks.

25. (a) Mike and Yola's down payment is 20% of \$200,000.

$$(0.20)(\$200{,}000) = \$40{,}000$$

(b) Their loan amount is the purchase price less the down payment.

$$\$200{,}000 - \$40{,}000 = \$160{,}000$$

(c) The 30 year loan will have $n = 12 \cdot 30 = 360$ monthly payments with $i = \dfrac{0.09}{12} = 0.0075$ per month. Their monthly payments will be

$$P = V\frac{i}{1-(1+i)^{-n}} = \$160{,}000\frac{0.0075}{1-(1.0075)^{-360}} = \$1287.40$$

(d) The total interest paid is the difference between the total payments and the amount borrowed. Mike and Yola will pay (\$1287.40)(360) – \$160,000 = \$303,464 in interest.

(e) If they pay an additional \$100 each month, the payment $P = \$1387.40$. The present value V of the loan remains \$160,000 and the interest rate $i = 0.0075$ per month. We solve for n.

$$V = P\frac{1-(1+i)^{-n}}{i}$$

$$160{,}000 = \$1387.40\ \frac{1-(1.0075)^{-n}}{0.0075}$$

$$1-\frac{160{,}000}{1387.40}\cdot(0.0075) = (1+0.0075)^{-n}$$

$$0.1350728 = 1.0075^{-n}$$

$$-n = \log_{1.0075}0.1350728$$

$$n = -\frac{\log 0.1350728}{\log 1.0075} = 267.93$$

If Mike and Yola increase their monthly payments by \$100, they will pay off the loan in 268 months or 22 years and 4 months.

(f) The total interest paid on the loan with the revised monthly payment of \$1387.40 is

$$(\$1387.40)(268) - \$160{,}000 = \$211{,}823.20$$

27. The down payment on the car is 20% of \$12,000 or (0.20)(\$12,000) = \$2400.
The amount of money borrowed is \$12,000 – \$2400 = \$9600.

The monthly payment on $V = \$9600$ when $n = 12 \cdot 3 = 36$ and $i = \frac{0.15}{12} = 0.0125$ is

$$P = V\frac{i}{1-(1+i)^{-n}} = \$9600\frac{0.0125}{1-(1.0125)^{-36}} = \$332.79$$

29. (a) The down payment on the equipment is 10% of \$20,000 or (0.10)(\$20,000) = \$2000.
The amount of money the restaurant owner borrowed is \$20,000 – \$2000 = \$18,000.

The monthly payment P, on $V = \$18,000$ when $n = 12 \cdot 4 = 48$ months and $i = \frac{0.12}{12} = 0.01$ per month is

$$P = V\frac{i}{1-(1+i)^{-n}} = \$18,000\frac{0.01}{1-(1.01)^{-48}} = \$474.01$$

(b) The restaurant owner will pay a total of (\$474.01)(48) – \$18,000 = \$4752.43 in interest on the loan.

31. The home buyer made a down payment of 20% of \$140,000 or (0.20)(\$140,000) = \$28,000 on the house, and financed the remainder (\$140,000 – \$28,000 = \$112,000). The terms of the mortgage were $n = 12 \cdot 30 = 360$ payments and $i = \frac{0.098}{12}$ per month. The monthly payment is

$$P = V\frac{i}{1-(1+i)^{-n}} = \$112,000\frac{\frac{0.098}{12}}{1-\left(1+\frac{0.098}{12}\right)^{-360}} = \$966.37$$

The total interest paid I = (\$966.37)(360) – \$112,000 = \$235,893.20

If the home buyer chose a 15 year mortgage, then $n = 12 \cdot 15 = 180$ monthly payments.
The monthly payment would be

$$P = V\frac{i}{1-(1+i)^{-n}} = \$112,000\frac{\frac{0.098}{12}}{1-\left(1+\frac{0.098}{12}\right)^{-180}} = \$1189.89$$

and the total interest paid I = (\$1189.89)(180) – \$112,000 = \$102,180.20

Decreasing the term of the loan from 30 years to 15 years increases the monthly payment by \$223.52, but saves \$133,713 in interest.

33. We have $V = \$100{,}000$, $P = \$2000$ per month and $i = \dfrac{0.05}{12}$ per month. To determine how long the IRA payments will last need to find n.

$$V = P\frac{1-(1+i)^{-n}}{i}$$

$$\$100{,}000 = \$2000\frac{1-\left(1+\frac{0.05}{12}\right)^{-n}}{\frac{0.05}{12}}$$

$$1-\left(\frac{0.05}{12}\right)50 = \left(1+\frac{0.05}{12}\right)^{-n}$$

$$0.7916667 = (1.0041667)^{-n}$$

$$-n = \log_{1.0041667}(0.7916667)$$

$$n = -\frac{\log 0.7916667}{\log 1.0041667} = 56.18$$

The payments will last 56.19 months, or about 4 years and 8 months.

35. We need the present value V of an annuity, if $P = \$1422$, $n = 9$ months, and $i = \dfrac{0.048}{12} = 0.004$.

$$V = P\left[\frac{1-(1+i)^{-n}}{i}\right] = \$1422\left[\frac{1-1.004^{-9}}{0.004}\right] = \$12{,}545.75$$

There must be \$12,545.75 in the account on August 1.

37. Each withdrawal is $\dfrac{\$1{,}000{,}000}{20} = \$50{,}000$. We need the present value of the annuity when $n = 20$ and $i = \dfrac{0.078}{4} = 0.0195$.

$$V = P\left[\frac{1-(1+i)^{-n}}{i}\right] = \$50{,}000\left[\frac{1-1.0195^{-20}}{0.0195}\right] = \$821{,}530.33$$

The foundation must invest \$821,530.33 to meet the project's goal.

39. The original loan had 360 payments at an interest rate $i = \dfrac{0.06825}{12} = 0.0056875$ per month. The loan payments were

$$P = V\frac{i}{1-(1+i)^{-n}} = \$312{,}000\frac{0.0056875}{1-(1+0.0056875)^{-360}} = \$2039.20$$

After 60 months there are 300 payments still to be made, and the present value of the loan is

$$V = P\frac{1-(1+i)^{-n}}{i} = \$2039.20\frac{1-(1+0.0056875)^{-300}}{0.0056875} = \$293{,}133.88$$

Amount of the debt paid: \$312,000 – \$293,133.88 = \$18,866.12
The interest paid: (\$2039.20)(60) – \$18,866.12 = \$122,352 – \$18,866.12 = \$103,485.88

The new loan has a present value $V = \$293{,}133.88$, $n = 300$ monthly payments at an interest rate $i = \dfrac{0.06125}{12}$ per month. The new monthly payment is

$$P = V\frac{i}{1-(1+i)^{-n}} = \$293{,}133.88\frac{\frac{0.06125}{12}}{1-\left(1+\frac{0.06125}{12}\right)^{-300}} = \$1911.13$$

With the refinancing the monthly payments are reduced by $2039.20 – $1911.13 = $128.07 per month.

The interest that will be saved by refinancing at a lower interest rate is the difference between the interest that would have been paid without refinancing and the interest that will be paid with refinancing.

Interest without refinancing: ($2039.20)(360) – $312,000 = $422,112

Interest with refinancing: interest paid on the old loan + interest on new loan

= $103,486.18 + [($1911.13)(300) – $293,133.81]

= $383,691.37

The interest saved by refinancing: $422,112 – $383,697.37 = $38,420.63

41. The $V = \$235{,}000$ loan had $n = 12 \cdot 30 = 360$ payment at an interest rate $i = \dfrac{0.06125}{12}$ per month. The minimum monthly payment on the loan was

$$P = V\frac{i}{1-(1+i)^{-n}} = \$235{,}000\frac{\frac{0.06125}{12}}{1-\left(1+\frac{0.06125}{12}\right)^{-360}} = \$1427.88$$

After paying the loan for 4 years (48 payments) there is

$$V = P\frac{1-(1+i)^{-n}}{i} = \$1427.88\frac{1-\left(1+\frac{0.06125}{12}\right)^{-312}}{\frac{0.06125}{12}} = \$222{,}611.64 \text{ remaining on the loan.}$$

Amount of debt paid: $235,000 – $222,611.64 = $12,388.36

Interest paid: ($1427.88)(48) – $12,388.36 = $68,538.24 – $12,388.36 = $56,149.88

To find the number of months that the term of the mortgage will be reduced, we find n with the new payment $P = \$1427.88 + \$150 = \$1577.88$

$$\$222{,}611.64 = \$1577.88\frac{1-\left(1+\frac{0.06125}{12}\right)^{-n}}{\frac{0.06125}{12}}$$

$$\left(1+\frac{0.06125}{12}\right)^{-n} = 1-\left(\frac{222{,}611.64}{1577.88}\right)\left(\frac{0.06125}{12}\right)$$

$$(1.005104167)^{-n} = 0.27989$$

$$-n = \log_{1.005104167}(0.27989)$$

$$n = -\frac{\log 0.27989}{\log 1.005104167} = 250.1$$

The term of the loan will be reduced by 312 – 250.1 = 61.9 months.

The interest that will be saved by increasing the monthly payment is the difference between the interest that would have been paid if the couple continued to make minimum payments and the interest that will be paid with their revised payment.

Interest on the existing loan: ($1427.88)(360) – $235,000 = $279,036.80
Interest with increased payments: interest paid before + interest paid with new payments
= $56,149.88 + [($1577.88)(250.1) – $222,611.64]
= $228,166.03

The interest saved by increasing monthly payments:
$279,036.80 – $228,166.03 = $50,870.77

43. We compare the present value of leasing using the money which is invested at $r = 5.4\ \%$ compounded monthly with the cost of buying the copier outright.

$$V = P\left[\frac{1-(1+i)^{-36}}{i}\right] = \$482.28\left[\frac{1+1.0045^{-36}}{0.0045}\right] = \$15{,}995.59$$

The state should lease the equipment because the V is less than the purchase price of $16,140.00.

45. The present value of a 4 year annuity invested at 5% per annum with annual payments of $50,000 is

$$V = P\,\frac{1-(1+i)^{-n}}{i} = \$50{,}000\,\frac{1-(1+0.05)^{-4}}{0.05} = \$177{,}297.53$$

Since the present value of the annuity exceeds the purchase price of the trucks, purchasing the trucks is preferable.

47. (a) We first find the equivalent annual cost of each machine.
Machine A: Machine A has an expected life of 8 years. At 10% interest the annual cost of Machine A is

$$P = \$10{,}000\left[\frac{0.10}{1-1.10^{-8}}\right] = \$1874.44$$

Machine B: Machine B has an expected life of 6 years. At 10% interest, the annual cost of Machine B is

$$P = \$8000\left[\frac{0.10}{1-1.10^{-6}}\right] = \$1836.86$$

The net annual savings of each machine is

Machine	A	B
Labor Savings	$2000.00	$1800.00
Equivalent Annual Cost	1875.44	1836.86
Net Savings	$ 125.56	($36.86)

Machine A has a greater net savings (Machine B results in a net loss), so Machine A is preferable.

(b) If the time value of money is 14% per annum, then
Machine A: Machine A has an average annual cost of

$$P = \$10{,}000\left[\frac{0.14}{1-1.14^{-8}}\right] = \$2155.70$$

Machine B: Machine B has an average annual cost of

$$P = \$8000\left[\frac{0.14}{1-1.14^{-6}}\right] = \$2057.56$$

The net annual savings of each machine is

Machine	A	B
Labor Savings	$2000.00	$1800.00
Equivalent Annual Cost	2155.17	2057.26
Net Savings	($ 155.17)	($ 257.26)

Neither machine results in a savings. Of the two, Machine A has a smaller increased cost and so is the preferable option.

49. We have a $10,000 bond with a nominal interest rate of 6.25% that matures in 8 years. We want to find the price that will yield a true rate of 6.5%. To find the price we follow the steps outlined on pp. 337–338 in the text.
STEP 1. Calculate the semi-annual interest payments,

$$I = Prt = (\$10{,}000)(0.0625)\left(\frac{1}{2}\right) = \$312.50$$

STEP 2. Calculate the present value of the semi-annual payments; $P = \$312.50$, $n = 2 \cdot 8 = 16$, $i = \frac{0.065}{2} = 0.0325$,

$$PV_1 = P\,\frac{1-(1+i)^{-n}}{i} = \$312.50 \cdot \frac{1-(1+0.0325)^{-16}}{0.0325} = \$3851.36$$

STEP 3. Calculate the present value of the bond at maturity.

$$PV_2 = A(1+i)^{-n} = \$10{,}000(1.0325)^{-16} = \$5994.58$$

STEP 4. Calculate the price of the bond by summing the results of steps 2 and 3.

$$\$3851.36 + \$5994.58 = \$9845.94$$

51. The face value: $1000, 5-year bond; nominal interest rate: $r = 0.0475$, $t = \frac{1}{2}$; true interest rate: 4.855%, $n = 2$, $i = \frac{0.04855}{2} = 0.024275$

STEP 1. Calculate the semi-annual interest payments,

$$I = Prt = (\$1000)(0.0475)\left(\frac{1}{2}\right) = \$23.75$$

STEP 2. Calculate the present value of the semi-annual payments;

$$PV_1 = P\frac{1-(1+i)^{-n}}{i} = \$23.75\left[\frac{1-1.024275^{-10}}{0.024275}\right] = \$208.64$$

STEP 3. Calculate the present value of the bond at maturity.
$$PV_2 = A(1+i)^{-n} = \$1000(1.024275^{-10}) = \$786.75$$

STEP 4. Calculate the price of the bond by summing the results of steps 2 and 3.
$$\$208.64 + \$786.75 = \$995.39$$

53. The face value: \$1000, 5-year bond; nominal interest rate: $r = 0.045$, $t = \frac{1}{2}$; true interest rate: 4.569, $n = 2$, $i = \frac{0.04569}{2} = 0.022845$

STEP 1. Calculate the semi-annual interest payments,
$$I = Prt = (\$1000)(0.045)\left(\frac{1}{2}\right) = \$22.50$$

STEP 2. Calculate the present value of the semi-annual payments;
$$PV_1 = P\frac{1-(1+i)^{-n}}{i} = \$22.50\left[\frac{1-1.022845^{-10}}{0.022845}\right] = \$199.13$$

STEP 3. Calculate the present value of the bond at maturity.
$$PV_2 = A(1+i)^{-n} = \$1000(1.022845^{-10}) = \$797.81$$

STEP 4. Calculate the price of the bond by summing the results of steps 2 and 3.
$$\$199.13 + \$797.81 = \$996.94$$

5.5 Annuities and Amortization Using Recursive Sequences

3. If $B_0 = \$3000$ and $B_n = 1.01B_{n-1} - 100$, then

(a) John's balance after 1 payment is $B_1 = 1.01(\$3000) - \$100 = \$2930$

(b) Using the sequence mode, with n Min = 0
$$u(n) = 1.01u(n-1) - 100$$
$$u(n\text{ Min}) = 3000$$
and TABLE we find the balance is below \$2000 after 14 payments. The balance is \$1953.70.

(c) John will pay off the debt on the 36th payment of \$84.62.
The total of all payments: (35)(\$100) + \$84.62 = \$3584.62

(d) John's interest expense was \$3584.62 – \$3000 = \$584.62

5. If $p_0 = 2000$, and $p_n = 1.03p_{n-1} + 20$, then

(a) After 2 months there are $p_2 = 1.03p_1 + 20$, where $p_1 = 1.03p_0 + 20 = 1.03(2000) + 20$
So, $p_1 = 2080$, and $p_2 = 1.03(2080) + 20 = 2162$ trout in the pond.

(b) Using the sequence mode with nMin = 0
$$u(n) = 1.03u(n-1) + 20$$
$$u(n\text{Min}) = \{2000\}$$
and TABLE we find that the trout population reaches 5000 after 26 months (population p_{26} = 5084.2).

7. (a) If the payments are $P = \$500$ per quarter and the interest rate per payment period is $i = \dfrac{0.08}{4} = 0.02$, then a recursive formula that represents the balance at the end of each quarter is

$$A_0 = \$500; \qquad A_n = (1 + 0.02)A_{n-1} + P = 1.02A_{n-1} + \$500$$

(b) After 80 payments (80 quarters) the fund will have \$101,810.

(c) The value of the account will be $A = \$159{,}738.48$.

9. (a) Since Bill and Laura borrowed \$150,000 at an interest rate $i = \dfrac{0.06}{12} = 0.005$ per month and are repaying the loan with $n = 360$ monthly payments of \$899.33, a recursive formula which represents their balance after each payment is

$$A_0 = \$150{,}000; \qquad A_n = (1 + 0.005)A_{n-1} - P = 1.005A_{n-1} - \$899.33$$

(b) After 1 payment, Bill and Laura's balance is

$$A_1 = (1.005)(150{,}000) - \$899.33 = \$149{,}850.67.$$

(c) To create a table showing Bill and Laura's balance after each payment, set the graphing utility in sequence mode. Then enter the recursive formula

$$n\text{Min} = 0$$
$$u(n) = 1.005u(n-1) - 899.33$$
$$u(n\text{Min}) = \{150{,}000\}$$

to obtain the TABLE reflecting the balance after each monthly payment.

(d) Using the recursive formula from part (c) and the TABLE, we find that after 4 years and 10 months (58 payments) the balance on the loan is \$139,981.

(e) Bill and Laura pay off the balance with the 360th payment. It is a reduced payment of $\$890.65(1.005) = 895.10$

(f) Bill and Laura's interest expense is

$$I = [(359)(899.33) + 895.10] - 150{,}000 = \$173{,}754.57$$

(ga) If Bill and Laura make payments of \$999.33 instead of \$899.33, the recursive formula representing their balance after each payment is

$$A_0 = \$150{,}000; \qquad A_n = (1 + 0.005)A_{n-1} - P = 1.005A_{n-1} - \$999.33$$

(gb) After 1 payment, Bill and Laura's balance is

$$A_1 = (1.005)(150{,}000) - \$999.33 = \$149{,}750.67.$$

(gc) To create a table showing Bill and Laura's balance after each payment, set the graphing utility in sequence mode. Then enter the recursive formula

$$n\text{Min} = 0$$
$$u(n) = 1.005u(n-1) - 999.33$$
$$u(n\text{Min}) = \{150{,}000\}$$

to obtain the TABLE reflecting the balance after each monthly payment.

(gd) Using the recursive formula from part (c) and the TABLE, we find that after 3 years and one month (37 payments) the balance on the loan is \$139,894.

(ge) Bill and Laura pay off the balance with the 279th payment. It is a reduced payment of $\$353.69(1.005) = \$355.37$

(gf) Bill and Laura's interest expense is

$$I = [(278)(999.33) + 355.37] - 150{,}000 = \$128{,}169.11.$$

Chapter 5 Review

1. 3% of $500 = (0.03)\cdot(500) = 15$

3. 140% of $250 = (1.40)\cdot(250) = 350$

5. $x\%$ of $350 = 75$

$$\frac{x}{100}\cdot 350 = 75$$
$$\frac{7}{20}x = 75$$
$$x = \frac{75\cdot 20}{7} = \frac{150}{7} = 21\frac{3}{7}$$

75 is $21\frac{3}{7}\%$ of 350.

7. $12 = 15\%$ of x

$$12 = 0.15x$$
$$x = \frac{12}{0.15} = 80$$

12 is 15% of 80.

9. $11 = 0.5\%$ of x

$$11 = 0.005x$$
$$x = \frac{11}{0.005} = 2200$$

11 is 0.5% of 2200.

11. Dan will have to pay 6% of \$330.00 in sales tax.

$$(0.06)(\$330.00) = \$19.80$$

Dan pays \$19.80 in sales tax.

13. If Dan borrows \$500 for 1 year and 2 months (14 months) at 9% simple interest, the interest charged is

$$I = Prt = (\$500)(0.09)\left(\frac{14}{12}\right) = \$52.50$$

and the amount due is $A = P + I = \$500.00 + \$52.50 = \$552.50$.

15. The proceeds of Warren's loan is \$15,000. If the loan has an interest rate of 12%, Warren must repay L at the end of 2 years where

$$R = L(1 - rt)$$
$$\$15{,}000 = L[1 - (0.12)(2)] = L(0.76)$$
$$L = \$19{,}736.84$$

17. If $P = \$100$ is invested for 2 years and 3 months ($n = 27$ months) at an interest rate $i = \frac{0.10}{12}$ per month, it will accumulate to

$$A = P(1+i)^n = \$100\left(1 + \frac{0.10}{12}\right)^{27} = \$125.12$$

19. Mike is choosing between two loans.
Loan (a): \$3000 for $t = 3$ years at $r = 12\%$ simple interest. This loan will cost Mike

$$A = P(1 + rt) = \$3000[1 + (0.12)(3)] = \$4080.00.$$

Loan (b): \$3000 loan for $n = 12 \cdot 3 = 36$ months at an interest rate $i = \frac{0.10}{12}$ per month. This loan will cost

$$A = P(1 + i)^n = \$3000\left(1 + \frac{0.10}{12}\right)^{36} = \$4044.55$$

Loan (b) costs Mike less.

21. Katy wants the amount $A = \$75$ in $n = 6$ months. If her money can earn an interest rate of $i = \frac{0.10}{12}$ per month then Katy must invest $P = \$71.36$.

$$P = A(1 + i)^{-n}$$

$$P = \$75\left(1 + \frac{0.10}{12}\right)^{-6} = \$71.36$$

23. Money doubles when $A = 2P$. The interest rate i, that allows money to double in $n = 12$ years, is

$$2P = P(1 + i)^n.$$
$$2 = (1 + i)^{12}$$
$$i = \sqrt[12]{2} - 1 = 0.0595$$

Money will double in 12 years when invested at 5.95% per annum.

25. The effective rate of interest is the simple interest rate that yields the same amount as the actual compounded rate. We want the interest rate compounded quarterly that is equivalent to the simple interest rate $r = 0.06$.

$$r_{\text{effective}} = (1 + i)^4 - 1$$
$$0.06 = (1 + i)^4 - 1$$
$$(1 + i)^4 = 1.06$$
$$i = \sqrt[4]{1.06} - 1 = 0.0146738 \text{ per quarter}$$

The annual interest rate compounded quarterly equivalent to 6% simple interest is $4i = 4(0.0146738) = 0.05870 = 5.87\%$.

27. Mr. and Mrs. Corey are setting up a sinking fund to save the down payment for the house, where $A = \$40{,}000$, $n = 12 \cdot 2 = 24$ months, and the interest rate per payment period is $i = \frac{0.03}{12} = 0.0025$

We are looking for P, the Corey's monthly payment.

$$A = P\frac{(1+i)^n - 1}{i}; \qquad \$40{,}000 = P\frac{(1.0025)^{24} - 1}{0.0025}$$
$$\$40{,}000 = P(24.702818)$$
$$P = \$1619.25$$

Mr. and Mrs. Corey must save \$1619.25 per month in order to have the \$40,000 they need for the down payment.

29. Mr. and Mrs. Ostedt's new house cost \$400,000. They made a \$100,000 down payment, and are financing \$300,000 at 10% for 25 years.

(a) The monthly payments are found by evaluating $P = V\frac{i}{1-(1+i)^{-n}}$. In this problem $V =$ \$300,000, $n = 12 \cdot 25 = 300$ months, and the interest rate per month is $i = \frac{0.10}{12}$.

$$P = \$300{,}000\frac{\frac{0.10}{12}}{1-\left(1+\frac{0.10}{12}\right)^{-300}} = \$2726.10$$

The Ostedt's monthly payments are \$2726.10.

(b) The Ostedt's will pay (300)(\$2726.10) – \$300,000 = \$517,830 in interest.

(c) The equity in a house is the sum of the down payment and the amount paid on the loan. After 5 years, the Ostedt's have made 60 payments. The present value of the remaining 240 payments is

$$V = P\frac{1-(1+i)^{-n}}{i} = \$2726.10\frac{1-\left(1+\frac{0.10}{12}\right)^{-240}}{\frac{0.10}{12}} = \$282{,}491.07$$

which indicates that the Ostedt's have paid \$300,000 – \$282,491.07 = \$17,508.93 of the principal. Their equity is then

\$100,000 + \$17,508.93 = \$117,508.93

31. If \$125,000 is to be amortized at a periodic interest rate $i = \frac{0.09}{12} = 0.0075$ per month over $n = 12 \cdot 25 = 300$ months, the monthly payment will be

$$P = V\frac{i}{1-(1+i)^{-n}} = \$125{,}000\frac{0.0075}{1-(1.0075)^{-300}} = \$1049.00$$

Equity is the down payment (if any) plus the amount of the loan already paid. After 10 years, 120 payments have been made and 180 remain to be made on the loan. The present value of the remaining payments is

$$V = P\frac{1-(1+i)^{-n}}{i} = \$1049.00\frac{1-(1.0075)^{-180}}{0.0075} = \$103{,}424.49$$

meaning that \$125,000 – \$103,424.49 = \$21,575.51 of the loan has been paid.

The equity in the property after 10 years is \$21,576 plus any down payment.

33. Let x denote the price Mr. Graf should pay for the gold mine. Then $0.15x$ represents a 15% annual Return on Investment (ROI). The annual Sinking Fund Contribution (SFC) needed to recover the purchase price x in 20 years can be calculated letting $A = x$, $n = 20$, $i = 0.10$, and $P = \text{SFC}$.

$$A = P\frac{(1+i)^n - 1}{i}$$

$$x = \text{SFC}\frac{(1+0.10)^{20} - 1}{0.10}$$

$$x = \text{SFC}(57.274999)$$

$$\text{SFC} = 0.0174596x$$

The required annual ROI plus the annual SFC equals the annual net income of $20,000.

$$0.15x + 0.0174506x = \$20{,}000$$

$$0.1674596x = \$20{,}000$$

$$x = \$119{,}431.77$$

Mr. Graf should pay $119,432 for the gold mine.

35. Mr. Doody wants to receive $n = 12 \cdot 15 = 180$ monthly payments of $300. He can invest the money now at an interest rate $i = \frac{0.12}{12} = 0.01$ per month. Mr. Doody needs

$$V = P\frac{1-(1+i)^{-n}}{i} = \$300\frac{1-(1.01)^{-180}}{0.01} = \$24{,}996.50$$

to insure his retirement goals are met.

37. Mr. Jones is saving $500.00 every six months at an interest rate $i = \frac{0.06}{2} = 0.03$ per payment period. After 8 years, $n = 2 \cdot 8 = 16$ payments, he will have

$$A = P\frac{(1+i)^n - 1}{i} = \$500\frac{(1+0.03)^{16} - 1}{0.03} = \$10{,}078.44$$

39. If Bill is to pay his debt and interest in 7 years he will need

$$A = P(1 + i)^n = \$1000(1.05^7) = \$1407.10$$

at that time. Bill makes $n = 4 \cdot 7 = 28$ quarterly payments into the sinking fund that earns interest $i = \frac{0.08}{4} = 0.02$ per quarter. Each payment is

$$A = P\frac{(1+i)^n - 1}{i}$$

$$\$1407.10 = P\frac{(1+0.02)^{28} - 1}{0.02}$$

$$\$1407.10 = P(37.0512103)$$

$$P = \$37.98$$

41. A loan of \$3000 at an interest rate per payment period of $i = \frac{0.12}{12} = 0.01$ is to be amortized in 2 years with $n = 12 \cdot 2 = 24$ equal monthly payments. Each payment is

$$P = V\frac{i}{1-(1+i)^{-n}} = \$3000\frac{0.01}{1-(1.01)^{-24}} = \$141.22$$

43. The effective rate of interest is the simple interest rate that yields the same amount as the actual compounded rate. The annual interest rate of 9% compounded monthly gives an interest rate of $i = \frac{0.09}{12} = 0.0075$ per month.

$$I_{\text{effective}} = (1.0075)^{12} - 1 = 0.0938$$

The effective rate of interest is 9.38%.

45. If John's trust fund is earning an interest rate $i = \frac{0.08}{2} = 0.04$ every 6 months it will be amortized in 15 years with $n = 2 \cdot 15 = 30$ equal payments. Each semiannual payment will be

$$P = V\frac{i}{1-(1+i)^{-n}} = 20{,}000\frac{0.04}{1-(1+0.04)^{-30}} = \$1156.60$$

47. At the end of $n = 30$ months, there will be

$$A = P\frac{(1+i)^n - 1}{i} = \$60\frac{(1.01)^{30}-1}{0.01} = \$2087.09$$

in the employee's retirement fund.

49. The student is to amortize the \$4000 loan with $n = 4 \cdot 4 = 16$ quarterly payments. The interest rate per payment period of the loan is $i = \frac{0.14}{4} = 0.035$, and the quarterly payments are

$$P = V\frac{i}{1-(1+i)^{-n}} = \$4000\frac{0.035}{1-(1.035)^{-16}} = \$330.74$$

Chapter 5 Project

1. The monthly payment for the 30 year mortgage with no points is

$$P = V\frac{i}{1-(1+i)^{-n}} = \$120{,}000\frac{\frac{0.0650}{12}}{1-\left(1+\frac{0.0650}{12}\right)^{-360}} = \$758.48$$

3. The difference in the monthly payments between the loan with no points and the mortgage with 0.5 point is \$758.48 – \$749.04 = \$9.44

5. Answers will vary, but they should include justification.

7. If the loan is decreased by the amount that would have been spent on points and the larger interest rate of 5.50% is used, the monthly payments become:

Rate	Points	Fee Paid	Loan Amount	Monthly Payment
6.5%	0	\$0.00	\$120,000 − \$3240 = \$116,760	$116{,}760\dfrac{\frac{0.065}{12}}{1-\left(1+\frac{0.065}{12}\right)^{-360}} = \$738.00$
6.38%	0.50	\$600	\$120,000 − \$2640 = \$117,360	$117{,}360\dfrac{\frac{0.0638}{12}}{1-\left(1+\frac{0.0638}{12}\right)^{-360}} = \$732.56$
6.25%	1.10	\$1320	\$120,000 − \$1920 = \$118,080	$118{,}080\dfrac{\frac{0.0625}{12}}{1-\left(1+\frac{0.0625}{12}\right)^{-360}} = \$727.04$
5.90%	2.70	\$3240	\$120,000	$120{,}000\dfrac{\frac{0.590}{12}}{1-\left(1+\frac{0.590}{12}\right)^{-360}} = \$711.76$

Mathematical Questions from Professional Exams

1. (b)

3. (b)

5. (d)

7. (c)

Chapter 6

Sets; Counting Techniques

6.1 Sets

1. $\subseteq, =$ **3.** $\cap$

17. True **19.** False **21.** False **23.** True

25. True **15.** True

17. $\{1, 2, 3\} \cap \{2, 3, 4, 5\} = \{2, 3\}$

19. $\{1, 2, 3\} \cup \{2, 3, 4, 5\} = \{1, 2, 3, 4, 5\}$

21. $\{2, 4, 6, 8\} \cap \{1, 3, 5, 7\} = \varnothing$

23. $\{a, b, e\} \cup \{d, e, f, q\} = \{a, b, d, e, f, q\}$

25. (a) $A \cup B = \{0, 1, 2, 3, 5, 7, 8\}$

(b) $B \cap C = \{5\}$

(c) $A \cap B = \{5\}$

(d) $\overline{A \cap B} = \{0, 1, 2, 3, 4, 6, 7, 8, 9\}$

(e) $\overline{A} \cap \overline{B} = \{2, 3, 4, 6, 8, 9\} \cap \{0, 1, 4, 6, 7, 9\} = \{4, 6, 9\}$

(f) $A \cup (B \cap A) = \{0, 1, 5, 7\} \cup \{5\} = \{0, 1, 5, 7\} = A$

(g) $(C \cap A) \cap \left(\overline{A}\right) = \{5\} \cap \{2, 3, 4, 6, 8, 0\} = \varnothing$

(h) $(A \cap B) \cup (B \cap C) = \{5\} \cup \{5\} = \{5\}$

27. (a) $A \cup B = \{b, c, d, e, f, g\}$

(b) $A \cap B = \{c\}$

(c) $\overline{A} \cap \overline{B} = \overline{A \cup B} = \{a, h, i, j, k, l, m, n, o, p, q, r, s, t, u, v, w, x, y, z\}$

(d) $\overline{A} \cup \overline{B} = \overline{A \cap B} = \{a, b, d, e, f, g, h, i, j, k, l, m, n, o, p, q, r, s, t, u, v, w, x, y, z\}$

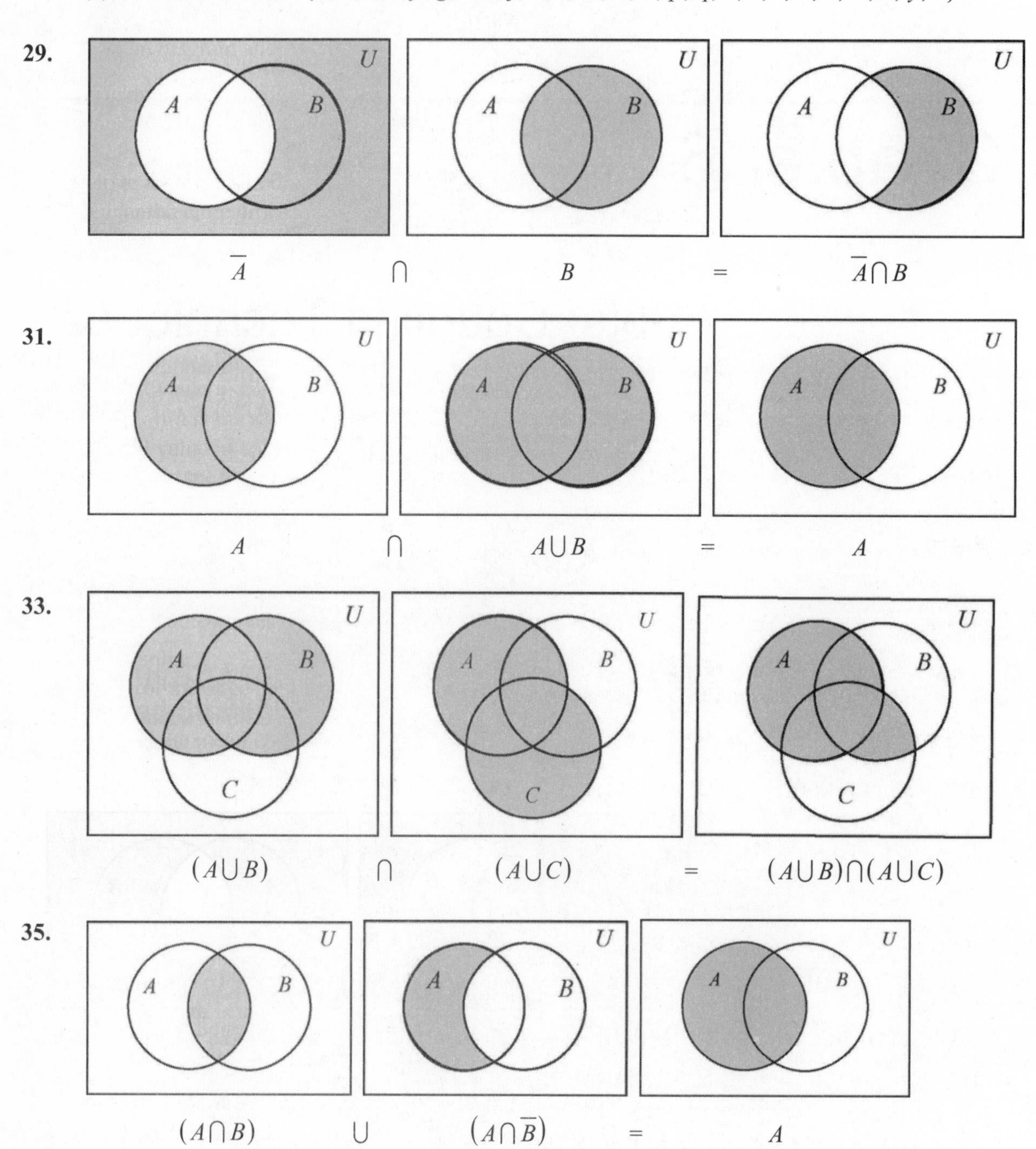

37. The Distributive Property: $A \cap (B \cup C) = (A \cap B) \cup (A \cap C)$

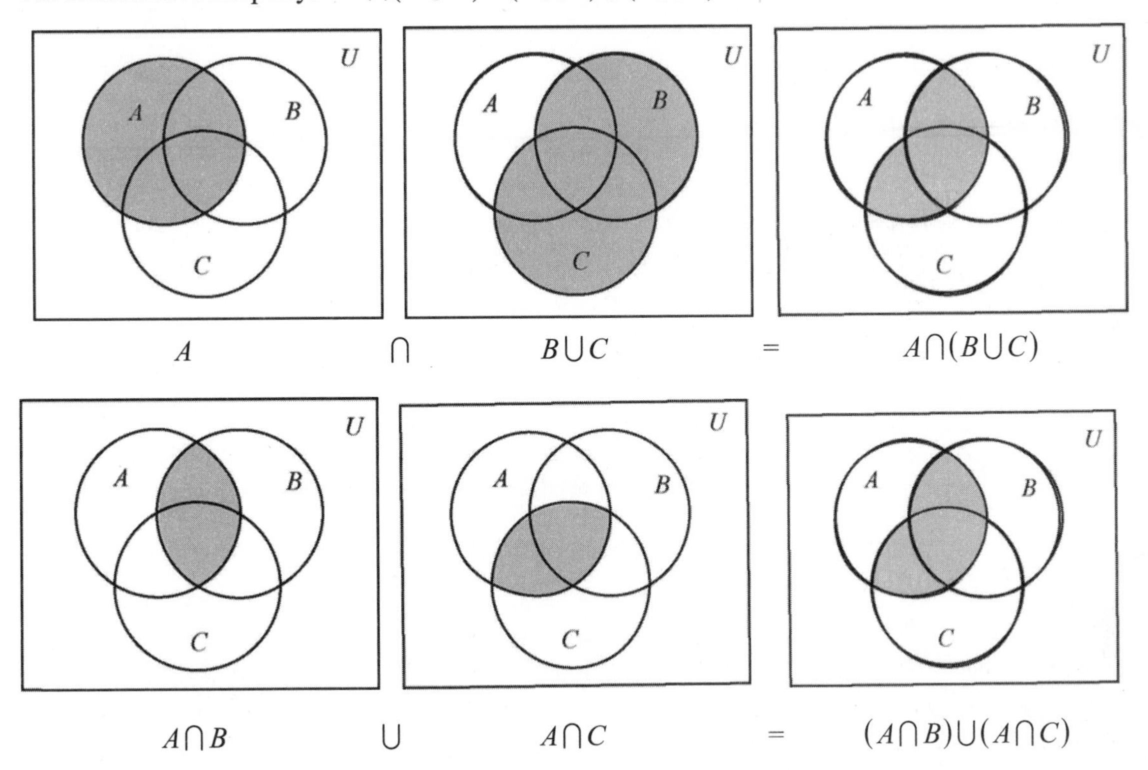

39. De Morgan's Property: $\overline{A \cap B} = \overline{A} \cup \overline{B}$

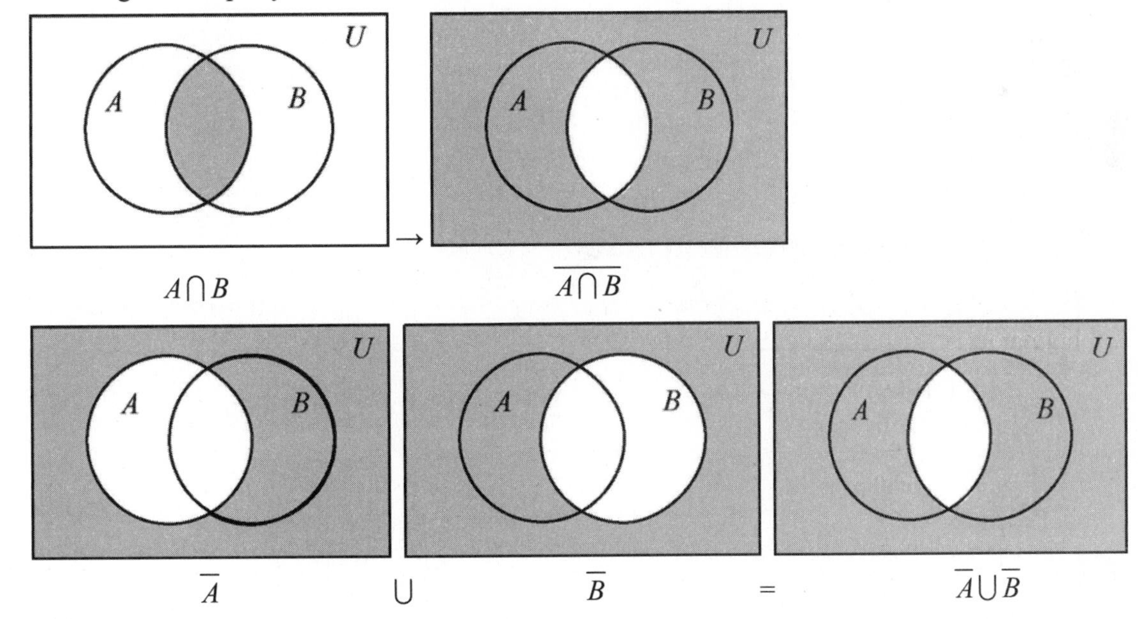

41. $A \cap E$ is the set of people who are both customers of IBM and are on the Board of Directors of IBM.

43. $A \cup D$ is the set of people who are customers of IBM or who are stockholders of IBM.

45. $\overline{A} \cap D$ is the set of people who are not customers of IBM, but who are stockholders of IBM.

47. $M \cap S$ is the set of all male students who smoke.

49. $\overline{M} \cup \overline{F}$ is the set of students who are female or who are not freshmen.

51. $F \cap S \cap M$ is the set of all male freshmen students who smoke.

53. (a) $O \cap A = \{$Schumer, Feingold, Grassley$\}$ This set represects the members of the Judiciary Committee who are on both the Administrative Oversight and the Antitrust, Competition Policy and Consumer Rights subcommittees.

(b) $\overline{O} = \{x | x$ is a member of the Judiciary Committee who is not on the Administrative Oversight subcommittee.$\}$

(c) $O \cup A = \{x | x$ is a member of the Judiciary Committee who is on the Administrative Oversight subcommittee or on the Antitrust, Competition Policy, and Consumer Rights subcommittee.$\}$

(d) $\overline{O \cup A} = \{x | x$ is a member of the Judiciary Committee who is not on the Administrative Oversight subcommittee and is not on the Antitrust, Competition Policy, and Consumer Rights subcommittee.$\}$

55. (a) $I \cup H = \{$Black, Bodden, Forbes, Gallagher, Murphy, Petevis, Russell, Smith, Stein, Sutton, Brown, Earnest, Johnson, Randolph, Rhodes$\}$. This set represents the clients who own either Intel Corp. stock or Hewlett-Packard stock.

(b) $I \cap H = \{$Bodden, Gallagher, Sutton$\}$ This set is the persons in the advisors data base who own both Intel Corp. and Hewlett-Packard stock.

57. The subsets of $\{a, b, c\}$ are Ø, $\{a\}$, $\{b\}$, $\{c\}$, $\{a, b\}$, $\{a, c\}$, $\{b, c\}$, and $\{a, b, c\}$.

6.2 The Number of Elements in a Set

1. False. If A and B are sets, then $n(A \cup B) = n(A) + n(B) - n(A \cap B)$.

3. $n(A) = 6$

5.
$$A \cap B = \{2, 4, 6\}$$
$$n(A \cap B) = 3$$

7.
$$\begin{aligned}(A \cap B) \cup A &= \{2, 4, 6\} \cup \{1, 2, 3, 4, 5, 6\} \\ &= \{1, 2, 3, 4, 5, 6\} \\ &= A\end{aligned}$$

$n[(A \cap B) \cup A] = n(A) = 6$

9.
$$\begin{aligned}n(A \cup B) &= n(A) + n(B) - n(A \cap B) \\ &= 4 + 3 - 2 \\ &= 5\end{aligned}$$

11.
$$\begin{aligned}n(A \cup B) &= n(A) + n(B) - n(A \cap B) \\ n(A \cap B) &= n(A) + n(B) - n(A \cup B) \\ &= 5 + 4 - 7 \\ &= 2\end{aligned}$$

13.
$$\begin{aligned}n(A \cup B) &= n(A) + n(B) - n(A \cap B) \\ n(A) &= n(A \cup B) - n(B) + n(A \cap B) \\ &= 14 - 8 + 4 \\ &= 10\end{aligned}$$

15. Here we are looking for the number of cars that have either GPS system or satellite radio. It is the same as finding $n(A \cup B)$. So the number of cars we are looking for is:

$$\begin{pmatrix}\text{the number of cars}\\ \text{with GPS systems}\end{pmatrix}+\begin{pmatrix}\text{the number of cars}\\ \text{with satellite radio}\end{pmatrix}-\begin{pmatrix}\text{the number of cars}\\ \text{with both options}\end{pmatrix}=325+216-89=452$$

The company manufactured 452 cars.

17. $n(A) = 10 + 6 + 3 + 5$
$= 24$

19. Before determining the number of elements in A or B, we need to find the number of elements in $A \cap B$.

$$n(A \cap B) = 6 + 3 = 9$$

So,

$$\begin{aligned} n(A \cup B) &= n(A) + n(B) - n(A \cap B) \\ &= 24 + 19 - 9 \\ &= 34 \end{aligned}$$

21. To find the number of elements in A but not in B, subtract $n(A \cap B)$ from $n(A)$.

$$n(A) - n(A \cap B) = 24 - 9 = 15$$

23. The number of elements in A or B or C is found by expanding the formula for finding the number of elements in A or C.

$$\begin{aligned} c(A \cup B \cup C) &= n(A) + n(B) + n(C) - n(A \cap B) - n(A \cap C) - n(B \cap C) + n(A \cap B \cap C) \\ &= 24 + 19 + 30 - 9 - 8 - 5 + 3 \\ &= 54 \end{aligned}$$

25. The number of elements in A and B and C is the number of elements common to all three sets.

$$n(A \cap B \cap C) = 3$$

27. The problem is made easier if we find some totals first. Adding the columns we get:

The number of voters under age 35:	195
The number of voters between ages 35 and 54:	265
The number of voters older than 54:	166

Adding the rows reveals how the voters of the two religions voted.

The number of Protestants who voted Republican:	345
The number of Protestants who voted Democrat:	90
The number of Catholics who voted Republican:	67
The number of Catholics who voted Democrat:	124

From these last four we can determine the number of voters who are Protestant and the number who are Catholic.

Protestant: $345 + 90 = 435$
Catholic: $67 + 124 = 191$

(a) To determine number of voters who are Catholic or Republican we need to find

$$\begin{pmatrix}\text{number of}\\ \text{Catholics}\end{pmatrix}+\begin{pmatrix}\text{number of}\\ \text{Republicans}\end{pmatrix}-\begin{pmatrix}\text{number of}\\ \text{Catholic Republicans}\end{pmatrix}=191+412-67=536$$

536 voters were either Catholic or Republican.

(b) To find the number of voters who are Catholic or over 54 we determine

$$\begin{pmatrix}\text{number of}\\ \text{Catholics}\end{pmatrix}+\begin{pmatrix}\text{number of}\\ \text{voters older than 54}\end{pmatrix}-\begin{pmatrix}\text{number of Catholics}\\ \text{older than 54}\end{pmatrix}=191+166-40=317$$

Of the voters polled 317 were either Catholic or over 54 years of age.

(c) Since a person cannot be both younger than 35 and older than 54, to find the number of persons who voted Democrat and are below 35 or over 54 simply add.

$$\begin{pmatrix}\text{Protestants voting}\\ \text{Democrat under 35}\end{pmatrix}+\begin{pmatrix}\text{Catholics voting}\\ \text{Democrat under 35}\end{pmatrix}+\begin{pmatrix}\text{Protestants voting}\\ \text{Democrat older than 54}\end{pmatrix}$$
$$+\begin{pmatrix}\text{Catholics voting}\\ \text{Democrat older than 54}\end{pmatrix}$$
$$=42+44+15+33=134$$

134 voters under the age of 35 or over the age of 54 voted Democrat.

29. (a) Number who are unemployed:

$$n(\text{unemployed males})+n(\text{unemployed females})=3226+2707=5933$$

5933 U.S. citizens 20 years or older were unemployed.

(b) Number who are unemployed or not in the labor force:

$$n(\text{unemployed})+n(\text{males not in labor force})+n(\text{females not in labor force})$$
$$=5933+24{,}572+43{,}442=76{,}947$$

76,947,000 U.S. citizens 20 years or older were unemployed or not in the labor force.

(c) Number who are female or employed:

$$n(\text{females})+n(\text{employed})-n(\text{females and employed})$$
$$=185{,}961$$

185,961,000 U.S. citizens 20 years or older were female or employed.

(d) There are 24,572,000 males not in the labor force.

31. Before doing this problem, it is suggested that you add the numbers in the rows and in the columns of the table.

(a) New hires who are executive managerial or faculty or professional nonfaculty is the sum of the first three rows of the table.

$$11+84+161=256$$

256 persons were hired as executive managerial or faculty or professional nonfaculty.

(b) New hires who are female is the sum of the female columns of the table.

$$134+12+8+14=168$$

168 females were hired.

(c) $n(\text{black female column})+n(\text{faculty row})-n(\text{black, female, faculty})=12+84-3=93$

93 persons were hired who were black female or faculty.

(d) $n(\text{white male column})+n(\text{executive managerial row})-n(\text{white male managerial hires})$

$$119+11-5=125$$

125 persons were hired who were white males or executive managerial.

33. (a)

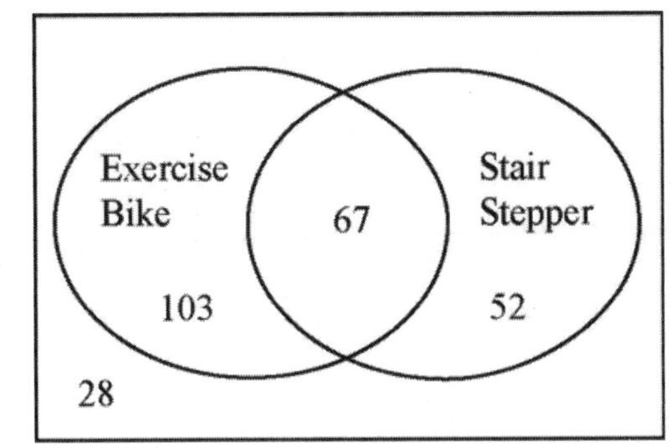

For parts (a) through (e) let E denote exercise bike and let S denote stair stepper.

(b) $n(E \cup S) = 103 + 67 + 52 = 222$ members use an exercise bike or a stair stepper regularly.

(c) $n(E \cap \overline{S}) = 103$ members use an exercise bike regularly, but not a stair stepper.

(d) $n(S \cap \overline{E}) = 52$ members use a stair stepper regularly, but not an exercise bike.

(e) $250 - n(E \cup S) = 250 - 222 = 28$ members use neither an exercise bike nor a stair stepper regularly.

35. (a)

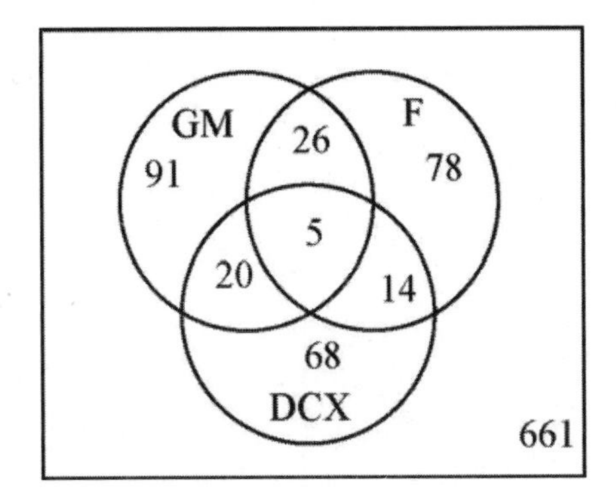

(b) 91 clients own only GM.

(c) 78 clients own only F.

(d) 68 clients own only DCX.

(e) $n\left[(\text{GM} \cup \text{DCX}) \cap \overline{\text{F}}\right] = 91 + 20 + 68 =$ 179 clients own GM or DCX but not F.

(f) $n\left[(\text{GM} \cap \text{DCS}) \cap \overline{\text{F}}\right] = 20$ clients own GM and DCX but not F.

(g) $n(\text{GM}) + n(\text{DCX}) + n(\text{F}) - n(\text{GM} \cap \text{DCX}) - n(\text{GM} \cap \text{F}) - n(\text{DCX} \cap \text{F}) + n(\text{GM} \cap \text{DCX} \cap \text{F})$
$= 142 + 107 + 123 - 25 - 31 - 19 + 5$
$= 302$ clients own at least one of the three stocks.

$$n\left(\overline{\text{GM} \cup \text{DCX} \cup \text{F}}\right) = 963 - 302 = 661 \text{ of the clients own none of the stocks.}$$

37. We denote the sets of females, seniors, and students on the dean's list by F, S, and D respectively. From the data given we determine that

$$n(F \cap S \cap D) = 31 \quad n\left(F \cap \overline{S} \cap D\right) = 62 \quad n\left(\overline{F} \cap S \cap D\right) = 45$$
$$n\left(F \cap S \cap \overline{D}\right) = 87 \quad n\left(\overline{F} \cap S \cap \overline{D}\right) = 96 \quad n\left(F \cap \overline{S} \cap \overline{D}\right) = 275$$
$$n\left(\overline{F} \cap \overline{S} \cap D\right) = 89 \quad n\left(\overline{F} \cap \overline{S} \cap \overline{D}\right) = 227$$

We next place the values on a Venn Diagram, so that we can answer the questions

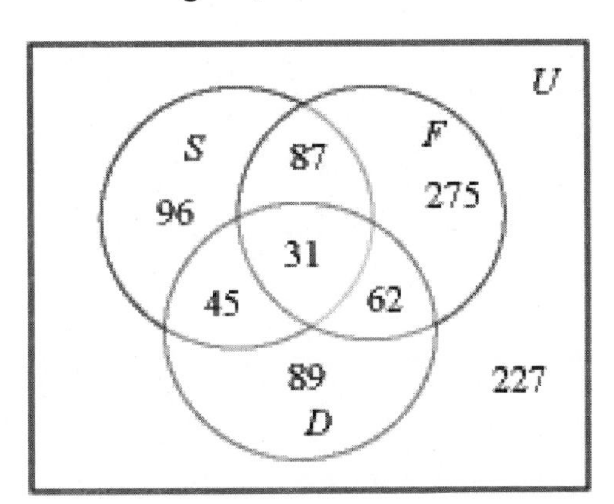

(a) There were 96 + 87 + 31 + 45 = 259 seniors at the college.
(b) There were 275 + 87 + 31 + 62 = 455 women at the college.
(c) There were 89 + 45 + 31 + 62 = 227 students on the dean's list.
(d) There were 45 + 31 = 76 seniors on the dean's list.
(e) There were 87 + 31 = 118 female seniors.
(f) There were 31 + 62 = 93 women on the dean's list.
(g) There were 96 + 87 + 275 + 45 + 31 + 62 + 89 + 227 = 912 students at the college.

39. We denote the sets of cars with heated seats, GPS, and satellite radio by H, G, and R respectively. From the data given we determine that

$$n(H)=90 \qquad n(G)=100 \qquad n(R)=75$$
$$n(H\cap G\cap R)=5 \qquad n\left(\overline{H\cup G\cup R}\right)=20 \qquad n\left(H\cap \overline{G}\cap \overline{R}\right)=20$$
$$n\left(H\cap G\cap \overline{R}\right)=60 \qquad n\left(\overline{H}\cap \overline{G}\cap \overline{R}\right)=30 \qquad n(G\cap R)=10$$

We next place the values on a Venn Diagram, so that we can answer the questions.

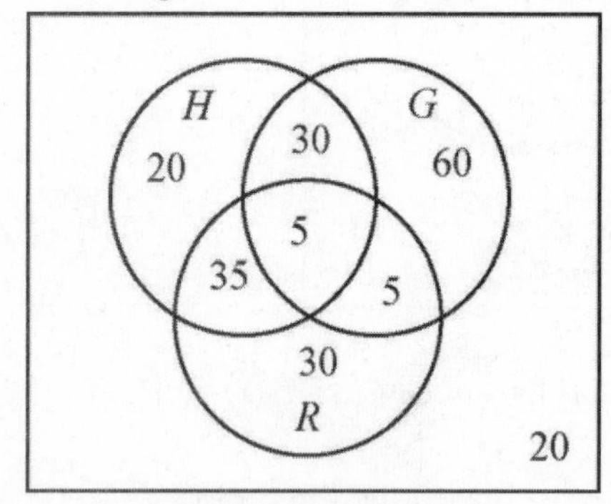

(a) 35 + 5 = 40 cars had satellite radio and heated seats.
(b) 30 + 5 = 35 cars had GPS and heated seats.
(c) 20 + 20 = 40 cars had neither satellite radio nor GPS.
(d) There were 20 + 20 + 30 + 5 + 35 + 30 + 5 + 60 = 205 cars sold in July.
(e) 100 + 90 – 35 = 155 cars were sold with GPS or heated seats or both.

41. There are 8 different possible blood types, as shown in the Venn diagram below. They can be listed as A^+, A^-, B^+, B^-, AB^+, AB^-, O^+, and O^-.

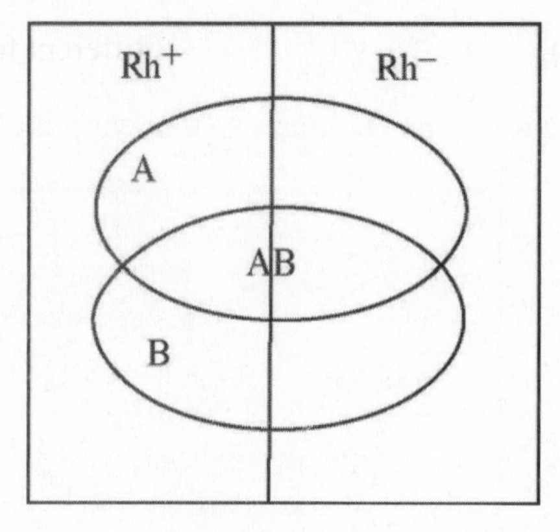

43. We denote the sets of users of HP, Mac, and Dell by H, M, and D respectively. From the information given, we determine that

$$n(H)=27, \quad n(M)=35, \quad n(D)=35, \quad n(H\cap M)=10, \quad n(H\cap D)=10, \quad n(M\cap D)=10$$

$$n(H\cap M\cap D)=3, \quad \text{and} \quad n\left(\overline{H\cup M\cup D}\right)=30$$

Putting this information onto a Venn diagram will help answer the question.

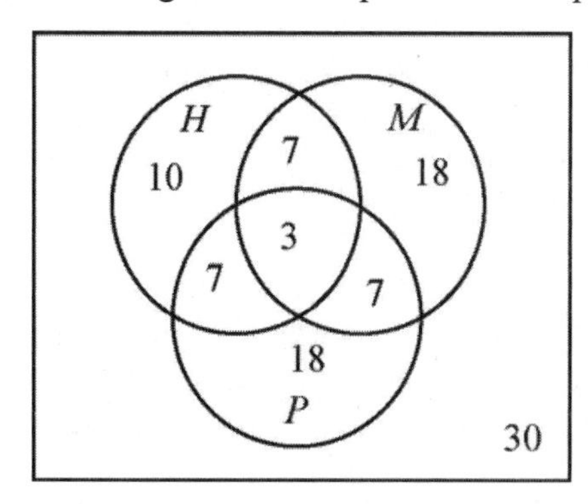

There are 10 + 18 + 18 = 46 users who exclusively use only one brand of computer.

45. The subsets of $\{a, b, c, d\}$ are $\emptyset$, $\{a\}$, $\{b\}$, $\{c\}$, $\{d\}$, $\{a, b\}$, $\{a, c\}$, $\{a, d\}$, $\{b, c\}$, $\{b, d\}$, $\{c, d\}$, $\{a, b, c\}$, $\{a, b, d\}$, $\{a, c, d\}$, $\{b, c, d\}$, and $\{a, b, c, d\}$.

There are $2^4 = 16$ subsets, where 4 is the number of elements in the original set.

6.3 The Multiplication Principle

1. By the Multiplication Principle the man can wear

$$5\cdot 3=15$$

different shirt and tie combinations.

3. There are 9 choices for the first digit and ten choices for each of the remaining 3 digits. By the Multiplication Principle

$$9\cdot 10\cdot 10\cdot 10=9000$$

different four digit numbers can be formed.

5. By the Multiplication Principle one can travel

$$2\cdot 4=8$$

different routes from town A to town C through town B.

7. If there are 3 different car models and 8 different color schemes, then by the Multiplication Principle, you must display

$$3\cdot 8=24$$

different cars to show all possibilities.

9. Using the Multiplication Principle, a person can enter the house through a window and exit through a door

$$12\cdot 3=36$$

different ways.

11. By the Multiplication Principle, there are

$$3\cdot 8\cdot 10\cdot 4=960$$

different lunch selections available.

13. There are six people. Once a person sits in a seat, he is no longer available to sit in another seat.
So by the Multiplication Principle, there are

$$6 \cdot 5 \cdot 4 \cdot 3 \cdot 2 \cdot 1 = 720$$

different ways 6 people can sit in a row of 6 seats.

15. Since there are 26 different letters and 10 different digits, by the Multiplication Principle, a user name formed by choosing 4 letters followed by four digits can be done

$$26 \cdot 26 \cdot 26 \cdot 26 \cdot 10 \cdot 10 \cdot 10 \cdot 10 = 26^4 \cdot 10^4 = 4{,}569{,}760{,}000$$

ways. So theoretically there are 4,569,760,000 different user names possible.

17. In this problem letters and digits can be repeated, but no two adjacent symbols can be the same. In the first spot we can choose any one of 26 letters, but in the second position there are only 25 letters available. (We cannot repeat the letter to its left.) In the third position there are again 25 possible letters because the letter in the first position can be used, but the one in the second position cannot. The fourth position also has 25 available letters.

Similar reasoning is used for choosing the digits. Any one of 10 digits can be chosen for the fifth position, but only 9 that can be used in the sixth position. The seventh and eighth positions each have 9 possible digits that can be used.

Using this reasoning and the Multiplication Principle, we find that there are

$$26 \cdot 25 \cdot 25 \cdot 25 \cdot 10 \cdot 9 \cdot 9 \cdot 9 = 2{,}961{,}562{,}500$$

user names for the system.

19. Since in the World Series one team from the National League plays one team from the American League, there are $16 \cdot 14 = 224$ different match ups possible.

21. By the Multiplication Principle, we see that Adam has

$$9 \cdot 9 \cdot 3 \cdot 2 = 486$$

different insurance options to choose from.

23. The pharmaceutical sales rep can have $5 \cdot 2 \cdot 3 = 30$ different sales portfolios.

25. David must choose among $4 \cdot 2 \cdot 18 = 144$ different one-topping pizzas.

27. Since a network doesn't show the same show twice in one week, there were

$$19 \cdot 18 \cdot 17 = 5814$$

different new-comedy lineups possible.

29. According to the Multiplication Principle, the audits can be scheduled in

$$10 \cdot 9 \cdot 8 \cdot 7 \cdot 6 \cdot 5 \cdot 4 \cdot 3 \cdot 2 \cdot 1 = 3{,}628{,}800$$

different ways.

31. (a) According to the Multiplication Principle there can be

$$10 \cdot 9 \cdot 8 \cdot 7 \cdot 6 \cdot 5 \cdot 4 = 604{,}800$$

different 7-digit telephone numbers if no digit is allowed to repeat.

(b) If the first number cannot be 0 and digits cannot be repeated, there can be only

$$9 \cdot 9 \cdot 8 \cdot 7 \cdot 6 \cdot 5 \cdot 4 = 544{,}320$$

different 7-digit telephone numbers.

(c) The number of 7-digit telephone numbers possible if digits can be repeated and if 0 is allowed in the first position, is

$$10 \cdot 10 \cdot 10 \cdot 10 \cdot 10 \cdot 10 \cdot 10 = 10^7 = 10{,}000{,}000$$

33. (a) The 7 letters of the word PROBLEM can be arranged $7 \cdot 6 \cdot 5 \cdot 4 \cdot 3 \cdot 2 \cdot 1 = 5040$ different ways.

(b) If the P must come first then there are $1 \cdot 6 \cdot 5 \cdot 4 \cdot 3 \cdot 2 \cdot 1 = 720$ ways to arrange the letters of the word PROBLEM.

(c) If the letter P must come first and the letter M must be last, then there are

$$1 \cdot 5 \cdot 4 \cdot 3 \cdot 2 \cdot 1 \cdot 1 = 120$$

ways to arrange the letters of the word PROBLEM.

35. The license plates have 2 letters (26 possibilities) followed by 4 digits (10 possibilities).

(a) If the letters and digits can be repeated, then using the Multiplication Principle we find that

$$26 \cdot 26 \cdot 10 \cdot 10 \cdot 10 \cdot 10 = 26^2 \cdot 10^4 = 6{,}760{,}000$$

different license plates can be made.

(b) If the letters can be repeated, but the digits cannot be repeated, then using the Multiplication Principle we find that

$$26 \cdot 26 \cdot 10 \cdot 9 \cdot 8 \cdot 7 = 3{,}407{,}040$$

different license plates can be made.

(c) If neither the letters nor the digits can be repeated, then using the Multiplication Principle we find that

$$26 \cdot 25 \cdot 10 \cdot 9 \cdot 8 \cdot 7 = 3{,}276{,}000$$

different license plates can be made.

37. This problem involves two steps since two models have 3 body styles and one model has only two body styles. It is easiest to see using a tree diagram.

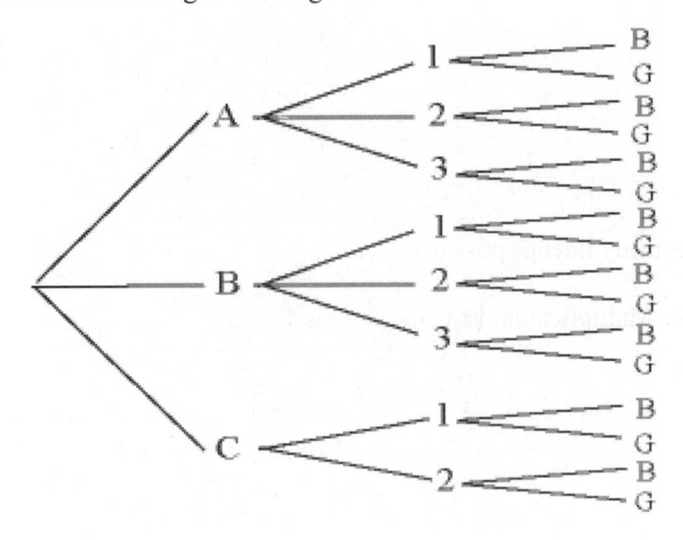

From the tree we see that there are 16 distinguishable car types.

39. By the Multiplication Principle, the contractor can build $5 \cdot 3 \cdot 4 = 60$ different types of homes.

41. By the Multiplication Principle, there can be $2^8 = 256$ different numbers (called bytes by computer scientists) possible.

43. Since there are 50 numbers on the lock and repetition is allowed, there are $50^3 = 125{,}000$ different lock combinations possible.

45. Think of the maze as if it were a tree diagram with each door of the maze as a branch on the tree. Then using the Multiplication Principle, we find that there are $4 \cdot 2 \cdot 1 = 8$ different paths from start to finish.

6.4 Permutations

1. 1; 6

3. False The number of ordered arrangements of r objects chosen from n objects, in which the n objects are distinct and repetition is allowed is n^r.

5. $\dfrac{5!}{2!} = \dfrac{5 \cdot 4 \cdot 3 \cdot 2!}{2!} = 60$

7. $\dfrac{6!}{3!} = \dfrac{6 \cdot 5 \cdot 4 \cdot 3!}{3!} = 120$

9. $\dfrac{10!}{8!} = \dfrac{10 \cdot 9 \cdot 8!}{8!} = 90$

11. $\dfrac{9!}{8!} = \dfrac{9 \cdot 8!}{8!} = 9$

13. $\dfrac{8!}{2!\ 6!} = \dfrac{8 \cdot 7 \cdot 6!}{2 \cdot 1 \cdot 6!} = 28$

15. $P(7,\ 2) = 7 \cdot 6 = 42$

17. $P(8,\ 7) = \dfrac{8!}{(8-7)!} = 8 \cdot 7 \cdot 6 \cdot 5 \cdot 4 \cdot 3 \cdot 2$
$= 40{,}320$

19. $P(6,\ 0) = \dfrac{6!}{(6-0)!} = \dfrac{6!}{6!} = 1$

21. $\dfrac{8!}{(8-3)!\ 3!} = \dfrac{8!}{5!\ 3!}$
$= \dfrac{8 \cdot 7 \cdot 6 \cdot 5!}{5! \cdot 3 \cdot 2 \cdot 1}$
$= 56$

23. $\dfrac{6!}{(6-6)!\ 6!} = \dfrac{6!}{0!\ 6!} = 1$

25. The ordered arrangements of length 3 formed from the letters a, b, c, d, and e are:

abc, abd, abe, acb, acd, ace, adb, adc, ade, aeb, aec, aed,
bac, bad, bae, bca, bcd, bce, bda, bdc, bde, bea, bec, bed,
cab, cad, cae, cba, cbd, cbe, cda, cdb, cde, cea, ceb, ced,
dab, dac, dae, dba, dbc, dbe, dca, dcb, dce, dea, deb, dec,
eab, eac, ead, eba, ebc, ebd, eca, ecb, ecd, eda, edb, edc

$P(5, 3) = 5 \cdot 4 \cdot 3 = 60$

27. The ordered arrangements of length 3 formed from the objects 1, 2, 3, and 4 are:

123, 124, 132, 134, 142, 143, 213, 214, 231, 234, 241, 243
312, 314, 321, 324, 341, 342, 412, 413, 421, 423, 431, 432

$P(4, 3) = 4 \cdot 3 \cdot 2 = 24$

29. Each letter of the code has 4 possible symbols, so there are $4^2 = 16$ possible two-letter codes using the letters *A, B, C,* and *D.*

31. Each digit of the number has two possible values, so there are $2^3 = 8$ possible three-digit numbers formed using the digits 0 and 1.

33. Four people can be lined up in $4! = 4 \cdot 3 \cdot 2 \cdot 1 = 24$ ways.

35. There are $5 \cdot 4 \cdot 3 = 60$ different three-letter codes that can be formed from the 5 letters *A, B, C, D,* and *E* if no letter can be used more than once.

This is the same as finding $P(5, 3) = 60$.

37. To find how many ways there are to seat 5 people in 8 chairs, find the number of ways to arrange 8 objects taken 5 at a time. This is $P(8, 5) = 8 \cdot 7 \cdot 6 \cdot 5 \cdot 4 = 6720$.

So there are 6720 ways to seat 5 people in 8 chairs.

39. Since the sets of symbols one-letter long, two-letters long, and three-letters long are disjoint, we can add the number of elements in each set.

$$\begin{aligned}\begin{pmatrix}\text{maximum number}\\ \text{of companies}\\ \text{in NYSE}\end{pmatrix} &= n\begin{pmatrix}\text{companies with}\\ \text{1-letter symbol}\end{pmatrix} + n\begin{pmatrix}\text{companies with}\\ \text{2- letter symbol}\end{pmatrix} + n\begin{pmatrix}\text{companies with}\\ \text{3 -letter symbol}\end{pmatrix}\\ &= 26 + 26^2 + 26^3\\ &= 18{,}278\end{aligned}$$

So, there can be 18,278 different companies listed on the New York Stock Exchange.

41. This is an ordered arrangement where repetition is allowed, so the number of ways the firm can purchase the printers is $7^3 = 343$.

43. This is the number of permutations of 14 shows taken 4 at a time.

$$P(14, 4) = \frac{14!}{(14-4)!} = \frac{14!}{10!} = 14 \cdot 13 \cdot 12 \cdot 11 = 24{,}024$$

The network can create 24,024 different comedy line-ups.

45. This is the number of permutations of 11 businesses taken 6 at a time.

$$P(11, 6) = \frac{11!}{(11-6)!} = \frac{11!}{5!} = 332{,}640$$

There are 33,640 different promotion schedules possible.

47. She needs the number of permutations of 21 facilities taken 8 at a time.

$$P(21, 8) = \frac{21!}{(21-8)!} = \frac{21!}{13!} = 8{,}204{,}716{,}800$$

There are 8,204,716,800 ways that she can schedule the first month's itinerary.

49. Since the 13th digit is determined by the other 12, there is no choice in picking it. The other 12 places are free to vary and repetitions are allowed. So there are 10^{12} possible ISBN numbers.

51. This is a permutation in which 2 different days are selected from a possible 365 days without a repetition. It is given by

$$P(365,\ 2)=\frac{365!}{(365-2)!}=\frac{365\cdot 364\cdot 363!}{363!}=365\cdot 364=132{,}860$$

So, 2 people can each have a different birthday 132,860 different ways.

53. (a) SUNDAY has 6 letters, and none of them are repeated. They can be arranged $P(6, 6) = 6! = 720$ different ways.

(b) If the letter S must come first, then we are really only arranging 5 letters. This can be done $P(5, 5) = 5! = 120$ different ways.

(c) If the letter S must come first and the letter Y must come last, then only four letters are being arranged. It is the permutation of 4 objects which can be done $P(4, 4) = 4! = 24$ ways.

55. Here we have to select 8 of the 12 children to whom we will distribute the books and then distribute the books among them. This is the number of permutations

$$P(12,\ 8)=\frac{12!}{(12-8)!}=\frac{12\cdot 11\cdot 10\cdot 9\cdot 8\cdot 7\cdot 6\cdot 5\cdot 4!}{4!}=19{,}958{,}400$$

The 8 books can be distributed to the 12 children 19,958,400 different ways.

57. We need to choose the three winning lottery tickets from the 1500 tickets sold because the prizes differ the order the tickets is chosen is important. The winning tickets can be chosen

$$P(1500,\ 3)=\frac{1500!}{(1500-3)!}=\frac{1500\cdot 1499\cdot 1498\cdot 1497!}{1497!}=3{,}368{,}253{,}000$$

different ways.

59. The number of ways to choose the president, vice-president, secretary, and treasurer from a club with 15 members (assuming no one individual holds more than one position) is given by $P(15, 4)$.

$$P(15,\ 4)=\frac{15!}{(15-4)!}=\frac{15\cdot 14\cdot 13\cdot 12\cdot 11!}{11!}=32{,}760$$

So, the 4 officers can be chosen 32,760 different ways.

6.5 Combinations

1. Combination

3. False. If $r \le n$, then $P(n, r) = r!C(n, r)$.

5. $C(6,\ 4)=\dfrac{6!}{(6-4)!\ 4!}=\dfrac{6\cdot 5\cdot 4!}{2!\ 4!}=\dfrac{6\cdot 5}{2}=15$

7. $C(7,\ 2)=\dfrac{7!}{(7-2)!\ 2!}=\dfrac{7\cdot 6\cdot 5!}{5!\ 2!}=\dfrac{7\cdot 6}{2\cdot 1}=21$

9. $C(5,\ 1)=\dfrac{5!}{(5-1)!\ 1!}=\dfrac{5\cdot 4!}{4!\ 1!}=5$

11. $C(8,\ 6)=\dfrac{8!}{(8-6)!\ 6!}=\dfrac{8\cdot 7\cdot 6!}{2!\ 6!}=\dfrac{8\cdot 7}{2\cdot 1}=28$

13. The combinations of 5 objects a, b, c, d, and e taken three at a time are:

abc, abd, abe, acd, ace, ade, bcd, bce, bde, cde

$$C(5,\ 3) = \frac{5!}{(5-3)!\ 3!} = \frac{5 \cdot 4 \cdot 3!}{2!\ 3!} = \frac{5 \cdot 4}{2 \cdot 1} = 10$$

15. The combinations of 4 objects 1, 2, 3, and 4 taken 3 at a time are:

123, 124, 134, 234

$$C(4,\ 3) = \frac{4!}{(4-3)!\ 3!} = \frac{4 \cdot 3!}{1!\ 3!} = 4$$

17. A committee of 4 students chosen from a pool of 7 students can be formed $C(7, 4) = 35$ ways.

$$C(7,\ 4) = \frac{7!}{(7-4)!\ 4!} = \frac{7 \cdot 6 \cdot 5 \cdot 4!}{3!\ 4!} = \frac{7 \cdot 6 \cdot 5}{3 \cdot 2 \cdot 1} = 35$$

19. The math department needs to select 4 professors without regard to order from a department of 17 eligible teachers. This can be done

$$C(17,\ 4) = \frac{17!}{(17-4)!\ 4!} = \frac{17 \cdot 16 \cdot 15 \cdot 14 \cdot 13!}{13!\ 4!} = \frac{17 \cdot 16 \cdot 15 \cdot 14}{4 \cdot 3 \cdot 2 \cdot 1} = 2380 \text{ ways.}$$

21. A committee is an unordered selection of people. We are interested in a committee of 3 members chosen from the 20 members of the Math Club. The committee can be selected

$$C(20,\ 3) = \frac{20!}{(20-3)!\ 3!} = \frac{20 \cdot 19 \cdot 18 \cdot 17!}{17!\ 3!} = \frac{20 \cdot 19 \cdot 18}{3 \cdot 2 \cdot 1} = 1140 \text{ ways.}$$

23. A bit is either a 0 or a 1. An 8-bit string is a list of 8 digits, all of which are either 0s or 1s. To determine how many 8-bit strings have exactly three 1s we need to select 3 positions to place the 1s. This can be done

$$C(8,\ 3) = \frac{8!}{(8-3)!\ 3!} = \frac{8 \cdot 7 \cdot 6 \cdot 5!}{5!\ 3!} = \frac{8 \cdot 7 \cdot 6}{3 \cdot 2 \cdot 1} = 56 \text{ ways.}$$

25. The historians are interested in choosing an unordered collection of 5 presidencies out of 43 presidencies. This can be done

$$C(43,\ 5) = \frac{43!}{(43-5)!\ 5!} = \frac{43 \cdot 42 \cdot 41 \cdot 40 \cdot 39 \cdot 38!}{38!\ 5!} = \frac{43 \cdot 42 \cdot 41 \cdot 40 \cdot 39}{5 \cdot 4 \cdot 3 \cdot 2 \cdot 1} = 962{,}598 \text{ ways.}$$

27. In the word ECONOMICS there are 9 letters, but they are not all distinct. There are 2 C's, 2 O's, and one each of the remaining 5 letters. We want the number of permutations of 9 objects, not all of which are distinct.

The number of 9-letter words that can be formed is given by

$$\frac{9!}{2!\ 2!\ 1!\ 1!\ 1!\ 1!\ 1!} = 90{,}720$$

29. There are 12 colored lights to be arranged in a string, but the colors are not all distinct. There are 3 reds, 4 yellows, and 5 blues. We want the number of permutations of 12 objects, not all of which are distinct.

The number of different colored light arrangements possible is

$$\frac{12!}{3!\ 4!\ 5!}=\frac{12\cdot 11\cdot 10\cdot 9\cdot 8\cdot 7\cdot 6\cdot 5!}{3\cdot 2\cdot 1\cdot 4\cdot 3\cdot 2\cdot 1\cdot 5!}=27,720$$

31. This problem consists of two tasks: selecting the boys for the committee which can be done C(4, 2) ways and selecting the girls for the committee which can be done $C(8, 3)$ ways. Then by the Multiplication Principle, we find that the committee can be formed in

$$C(4,\ 2)\cdot C(8,\ 3)=\frac{4!}{(4-2)!\ 2!}\cdot\frac{8!}{(8-3)!\ 3!}=\frac{4\cdot 3\cdot 2!}{2\cdot 1\cdot 2!}\cdot\frac{8\cdot 7\cdot 6\cdot 5!}{5!\cdot 3\cdot 2\cdot 1}=6\cdot 56=336 \text{ ways.}$$

33. Once on a team the children are no longer distinct, so placing people on teams is much like forming words in which all the letters are not distinct.

The 12 children can be placed on 3 teams, a first having 3 players, a second having 5 players and a third having 4 players in

$$\frac{12!}{3!\cdot 5!\cdot 4!}=\frac{12\cdot 11\cdot 10\cdot 9\cdot 8\cdot 7\cdot 6\cdot 5!}{3\cdot 2\cdot 1\cdot 5!\cdot 4\cdot 3\cdot 2\cdot 1}=1\cdot 11\cdot 10\cdot 9\cdot 4\cdot 7\cdot 1=27,720 \text{ ways}$$

35. There are two tasks here. The first to choose the 5 basic numbers. The number of ways of doing choosing these numbers is $C(55, 5)$. Then the power number is chosen. By the Multiplication Principle, the number of distinct Powerball tickets is

$$C(55,\ 5)\cdot 42=\frac{55!}{50!\cdot 5!}\cdot 42=3,478,761\cdot 42=146,107,962.$$

37. We first select one member from each division, then we pool the remaining faculty to choose the 2 members at large. Finally we use the Multiplication Principle to find the number of negotiating teams possible.

$$\begin{aligned}&C(25,1)\cdot C(23,1)\cdot C(15,1)\cdot C(8,1)\cdot C(19,1)\cdot C(85,2)\\&=25\cdot 23\cdot 15\cdot 8\cdot 19\cdot\frac{85!}{83!\cdot 2!}\\&=4,680,270,000\end{aligned}$$

There are 4,680,270,000 different negotiating teams possible.

39. We want to select 8 accounts from 58 accounts without regard to the order of selection. This can be done $C(58, 8)$ = 1,916,797,311 different ways.

41. Since the prizes are different, the order in which the winners are chosen is important. This is a permutation in which not all the elements are distinct.

The number of ways prizes can be distributed among the 80 employees is

$$\frac{80!}{2!\,3!\,75!}=240,400,160$$

43. The 100 senators can be placed on the committees in

$$\frac{100!}{22!\cdot 13!\cdot 10!\cdot 5!\cdot 16!\cdot 17!\cdot 17!}=1.157\times 10^{76} \text{ ways.}$$

45. This problem consists of four separate tasks: choosing the winner in each of the three divisions which can be done $C(5, 1)$, $C(6, 1)$, and $C(5, 1)$ ways respectively, and then choosing the wild card team from the remaining 13 teams. This can be done $C(13, 1)$ ways. By the Multiplication Principle the playoff participants can be chosen

$$C(5, 1)\cdot C(6, 1)\cdot C(5, 1)\cdot C(13, 1)=5\cdot 6\cdot 5\cdot 13=1950 \text{ ways.}$$

47. Since we do not care whether the sample includes smokers or non-smokers, we add the two groups together. Then we select the sample. This can be done $C(55, 8)$ ways.

$$C(55, 8)=\frac{55!}{(55-8)!\ 8!}=\frac{55!}{47!\ 8!}=1{,}217{,}566{,}350 \text{ ways.}$$

49. Three people from a group of 5 people can be chosen to participate in one of the distinct tests in

$$P(5, 3)=\frac{5!}{(5-3)!}=\frac{5!}{2!}=60 \text{ ways.}$$

51. In choosing a sample, order is not important. Four light bulbs can be selected from a box of 24 bulbs in

$$C(24, 4)=\frac{24!}{(24-4)!\cdot 4!}=\frac{24!}{20!\cdot 4!}=10{,}626 \text{ ways.}$$

53. Since the sportswriters rating is in order, the top 15 teams can be chosen from among 50 teams in

$$P(50, 15)=\frac{50!}{(50-15)!}=\frac{50!}{35!}=2.943\times 10^{24}$$

different ways.

55. Here the order in which the cards are drawn is important, because different orders will result in different numbers. Also since the cards are not replaced, there is no repetition of the digits in the number formed. So there are $P(10, 4) = 10\cdot 9\cdot 8\cdot 7 = 5040$ different numbers possible.

6.6 The Binomial Theorem

1. Pascal Triangle

3. False. $\binom{n}{j}=\dfrac{n!}{(n-j)!\,j!}$

5. $\binom{5}{3}=\dfrac{5!}{(5-3)!\,3!}=\dfrac{5!}{2!\,3!}=10$

7. $\binom{7}{5}=\dfrac{7!}{(7-5)!\,5!}=\dfrac{7!}{2!\,5!}=21$

9. $\binom{50}{49}=\dfrac{50!}{(50-49)!\,49!}=\dfrac{50!}{1!\,49!}=50$

11. $\binom{1000}{1000}=\dfrac{1000!}{(1000-1000)!\,1000!}=\dfrac{1}{0!}=1$

13. $\binom{55}{23} = \dfrac{55!}{(55-23)!\,23!} = 1.86644 \times 10^{15}$

15. $\binom{47}{25} = \dfrac{47!}{(47-25)!\,25!} = 1.48339 \times 10^{13}$

17.
$$(x+y)^5 = \binom{5}{0}x^5y^0 + \binom{5}{1}x^4y^1 + \binom{5}{2}x^3y^2 + \binom{5}{3}x^2y^3 + \binom{5}{4}x^1y^4 + \binom{5}{5}x^0y^5$$
$$= x^5 + 5x^4y + 10x^3y^2 + 10x^2y^3 + 5xy^4 + y^5$$

19.
$$(x+3y)^3 = \binom{3}{0}x^3(3y)^0 + \binom{3}{1}x^2(3y)^1 + \binom{3}{2}x^1(3y)^2 + \binom{3}{3}x^0(3y)^3$$
$$= x^3 + 3x^2 \cdot 3y + 3x \cdot 9y^2 + 27y^3$$
$$= x^3 + 9x^2y + 27xy^2 + 27y^3$$

21.
$$(2x-y)^4 = \binom{4}{0}(2x)^4(-y)^0 + \binom{4}{1}(2x)^3(-y)^1 + \binom{4}{2}(2x)^2(-y)^2 + \binom{4}{3}(2x)^1(-y)^3 + \binom{4}{4}(2x)^0(-y)^4$$
$$= 16x^4 - 4 \cdot 8x^3y + 6 \cdot 4x^2y^2 - 4 \cdot 2xy^3 + y^4$$
$$= 16x^4 - 32x^3y + 24x^2y^2 - 8xy^3 + y^4$$

23.
$$(x+y)^5 = \binom{5}{0}x^5 + \binom{5}{1}x^4y + \binom{5}{2}x^3y^2 + \binom{5}{3}x^2y^3 + \binom{5}{4}xy^4 + \binom{5}{5}y^5$$

The coefficient of x^2y^3 is $\binom{5}{3} = 10$.

25.
$$(x+3)^{10} = \binom{10}{0}x^{10} + \binom{10}{1}x^9 \cdot 3 + \binom{10}{2}x^8 \cdot 3^2 + \binom{10}{3}x^7 \cdot 3^3 + \binom{10}{4}x^6 \cdot 3^4$$
$$+ \binom{10}{5}x^5 \cdot 3^5 + \binom{10}{6}x^4 \cdot 3^6 + \binom{10}{7}x^3 \cdot 3^7 + \binom{10}{8}x^2 \cdot 3^8 + \binom{10}{9}x \cdot 3^9 + \binom{10}{10}3^{10}$$

The coefficient of x^8 is $\binom{10}{2} \cdot 3^2 = 405$.

27. A set with 8 elements has $2^8 = 256$ subsets.

29.
$$(1.001)^5 = (1+0.001)^5$$
$$= \binom{5}{0}1^5 \cdot 0.001^0 + \binom{5}{1}1^4 \cdot 0.001^1 + \binom{5}{2}1^3 \cdot 0.001^2 + \binom{5}{3}1^2 \cdot 0.001^3 + \binom{5}{4}1^1 \cdot 0.001^4 + \binom{5}{5}1^0 \cdot 0.001^5$$
$$= 1 + 0.005 + 0.00001 + 0.00000001 + 0.000000000005 + 0.000000000000001$$
$$= 1.005010010005001$$

$1.001^5 = 1.00501$ correct to 5 decimal places.

31. A set with 10 elements has $2^{10} - 1 = 1023$ non-empty subsets.

33. For any set A half of the subsets of A have an even number of elements, and half of the subsets have an odd number of elements. So, a set with 10 elements has $\frac{2^{10}}{2}=\frac{1024}{2}=512=2^9$ subsets with an odd number of elements.

35. To show that $\binom{10}{7}=\binom{6}{6}+\binom{7}{6}+\binom{8}{6}+\binom{9}{6}$, we repeatedly use the identity $\binom{n}{k}=\binom{n-1}{k}+\binom{n-1}{k-1}$.

So

$$\begin{aligned}\binom{10}{7}&=\binom{9}{7}+\binom{9}{6}\\&=\left[\binom{8}{7}+\binom{8}{6}\right]+\binom{9}{6}\\&=\left[\binom{7}{7}+\binom{7}{6}\right]+\binom{8}{6}+\binom{9}{6}\end{aligned}$$

Now since $\binom{7}{7}=\binom{6}{6}=1$, we substitute $\binom{6}{6}$ for $\binom{7}{7}$ in the line above and prove the statement.

$$\binom{10}{7}=\binom{6}{6}+\binom{7}{6}+\binom{8}{6}+\binom{9}{6}$$

37. We use the identity $\binom{n}{k}=\binom{n-1}{k}+\binom{n-1}{k-1}$ to determine that $\binom{11}{6}+\binom{11}{5}=\binom{12}{6}$.

39. The equation given is the binomial expansion of $\left(\frac{1}{4}+\frac{3}{4}\right)^5$. Since $\frac{1}{4}+\frac{3}{4}=1$, the expression on the left is equivalent to 1^5 or 1.

Chapter 6 Review

1. $\subset$, $\subseteq$ **3.** none of these **5.** none of these **7.** $\subset$, $\subseteq$

9. $\subset$, $=$ **11.** $\subseteq$, $=$ **13.** $\subset$, $\subseteq$ **15.** $\subseteq$, $=$

17. (a) $(A\cap B)\cup C=\{3, 6\}\cup\{6, 8, 9\}=\{3, 6, 8, 9\}$

(b) $(A\cap B)\cap C=\{3, 6\}\cap\{6, 8, 9\}=\{6\}$

(c) $(A\cup B)\cap B=\{1,2,3,5,6,7,8\}\cap\{2,3,6,7\}=\{2,3,6,7\}=B$

(d) $B\cup\varnothing=B$

(e) $A\cap\varnothing=\varnothing$

(f) $(A\cup B)\cup C=\{1, 2, 3, 5, 6, 7, 8\}\cup\{6, 8, 9\}=\{1, 2, 3, 5, 6, 7, 8, 9\}$

19. (a) (b) (c) (d) (e)

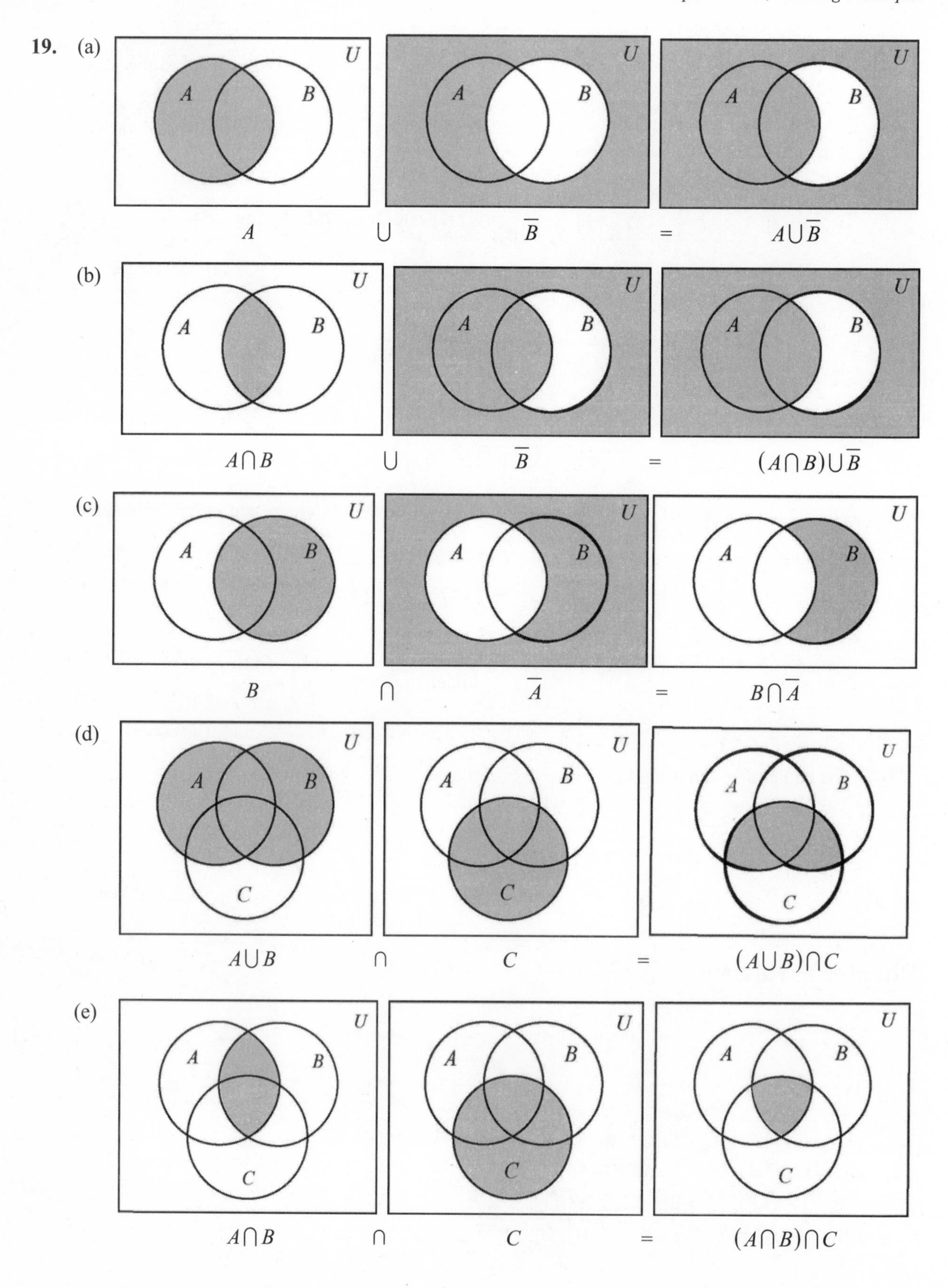
(a)
U
A
B
A $\cup$ $\overline{B}$ = $A \cup \overline{B}$
(b)
$A \cap B$ $\cup$ $\overline{B}$ = $(A \cap B) \cup \overline{B}$
(c)
B $\cap$ $\overline{A}$ = $B \cap \overline{A}$
(d)
C
$A \cup B$ $\cap$ C = $(A \cup B) \cap C$
(e)
$A \cap B$ $\cap$ C = $(A \cap B) \cap C$

(f)

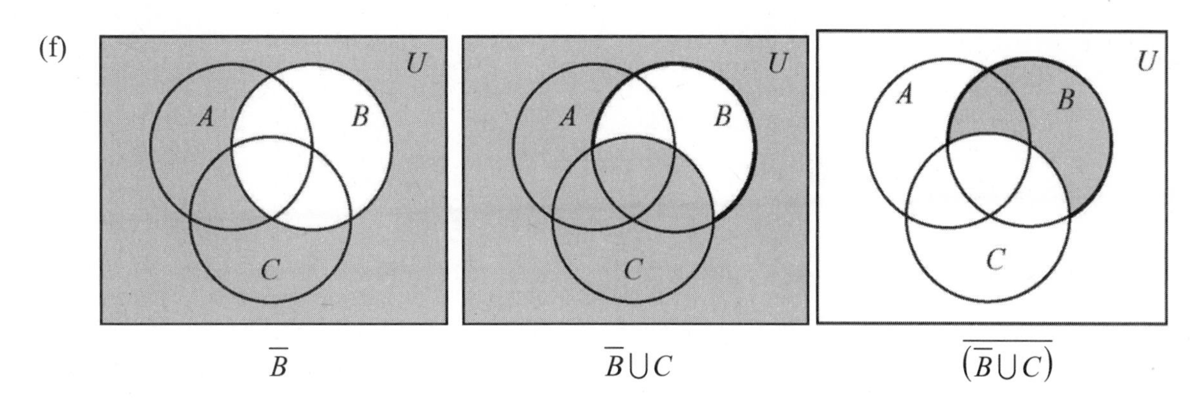

$\overline{B}$ $\qquad$ $\overline{B} \cup C$ $\qquad$ $\overline{(\overline{B} \cup C)}$

21. $A \cap B = \varnothing$

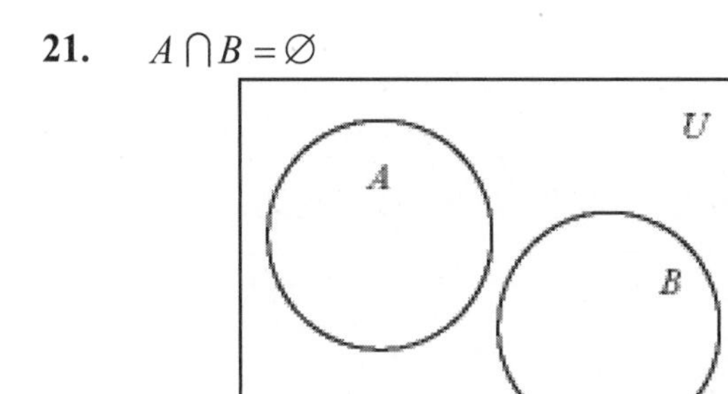

23. $A \cup V$ is the set of all states whose names begin with an A or end with a vowel.

25. $V \cap E$ is the set of all states whose names end with a vowel and which lie east of the Mississippi River.

27. $(A \cup V) \cap E$ is the set of all states whose names start with an A or end with a vowel and which lie east of the Mississippi.

29. $n(A \cup B) = n(A) + n(B) - n(A \cap B)$, so

$$\begin{aligned} n(A \cap B) &= n(A) + n(B) - n(A \cup B) \\ &= 24 + 12 - 33 \\ &= 3 \end{aligned}$$

31. (a) $$\begin{aligned} n(A \cup B) &= n(A) + n(B) - n(A \cap B) \\ &= 3 + 17 - 0 \\ &= 20 \end{aligned}$$

(b) Sets A and B are disjoint since their intersection has no elements.

33. $A = \{2, 4, 6\}$; $B = \{1, 2, 3\}$

$A \cap B = \{2\}$; $n(A \cap B) = 1$

$A \cup B = \{1, 2, 3, 4, 6\}$; $n(A \cup B) = 5$

35. We denote the sets of cars with heated seats, GPS, and a sun roof by S, G, and R respectively. From the data given we determine that

$$n(S) = 75 \qquad n(R) = 95 \qquad n(G) = 100$$
$$n(S \cap R \cap G) = 20 \quad n\left(\overline{S \cap P \cap G}\right) = 10 \quad n\left(S \cap \bar{R} \cap \bar{G}\right) = 10$$
$$n(R \cap G) = 50 \qquad n(S \cap R) = 60$$

We next place the values on a Venn Diagram, so that we can answer the questions.

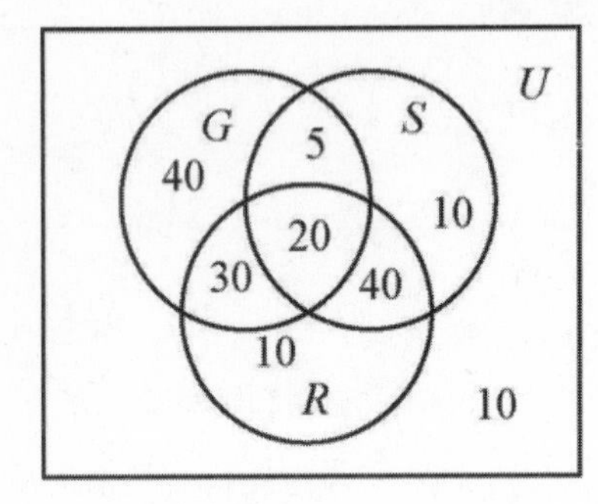

(a) There were
$10 + 5 + 20 + 40 + 10 + 30 + 40 + 10 = 165$
cars sold in June.

(b) 40 of the cars had only GPS.

37. If $U = \{1, 2, 3, 4, 5\}$; $B = \{1, 4, 5\}$, and $A \cap B = \{1\}$, then A could equal any of the following sets:

$$\{1\}, \{1, 2\}, \{1, 3\}, \text{ or } \{1, 2, 3\}$$

39. $0! = 1$

41. $\dfrac{7!}{4!} = \dfrac{7 \cdot 6 \cdot 5 \cdot 4!}{4!} = 210$

43. $\dfrac{12!}{11!} = \dfrac{12 \cdot 11!}{11!} = 12$

45. $P(5, 2) = 5 \cdot 4 = 20$

47. $P(12, 1) = 12$

49. $P(100, 2) = 100 \cdot 99 = 9900$

51. $C(10, 2) = \dfrac{10!}{(10-2)!\ 2!} = \dfrac{10!}{8!\ 2!} = \dfrac{10 \cdot 9 \cdot 8!}{8! \cdot 2 \cdot 1} = 45$

53. $C(6, 6) = \dfrac{6!}{(6-6)!\ 6!} = \dfrac{6!}{0!\ 6!} = 1$

55. $\dbinom{7}{4} = \dfrac{7!}{(7-4)!\ 4!} = \dfrac{7!}{3!\ 4!} = \dfrac{7 \cdot 6 \cdot 5 \cdot 4!}{3 \cdot 2 \cdot 1 \cdot 4!} = 35$

57. $\dbinom{9}{1} = \dfrac{9!}{(9-1)!\ 1!} = \dfrac{9!}{8!\ 1!} = \dfrac{9 \cdot 8!}{8! \cdot 1} = 9$

59. When choosing committees order is not important so a committee of 3 people can be chosen from 5 people in

$$C(5, 3) = \binom{5}{3} = \frac{5!}{(5-3)!\ 3!} = 10 \text{ ways.}$$

61. When books are arranged on a shelf different orderings mean different arrangements. So 3 books can be arranged

$$P(3, 3) = 3! = 6 \text{ ways.}$$

63. To count the number of different house styles that are available we use the Multiplication Principle. There are

$$3 \cdot 4 \cdot 6 = 72$$

different styles of houses possible.

65. Each question on a true-false test has 2 possible answers, and answers can be used more than once. So on a 10 question test, there are $2^{10} = 1024$ different ways the test can be answered.

67. (a) The number of 3-digit words that can be formed without repeating a digit from the symbols 1, 2, 3, 4, 5, 6 is given by

$$P(6, 3) = 6 \cdot 5 \cdot 4 = 120 \text{ words.}$$

So there are 120 words in this language.

(b) If the same letters in different a order indicate the identical word, then there are only

$$C(6, 3) = 20$$

possible words in this language.

69. (a) A committee of 3 boys and 4 girls can be formed from 7 boys and 6 girls in

$$C(7, 3) \cdot C(6, 4) = 35 \cdot 15 = 525$$

different ways.

(b) If the committee must have at least one person of each sex, the committee could consist of 1 boy and 6 girls or 2 boys and 5 girls or 3 boys and 4 girls or 4 boys and 3 girls or 5 boys and 2 girls or 6 boys and 1 girl. Since each configuration is mutually exclusive, we will determine the number of ways the committee can be formed by using the Multiplication Principle 6 times and adding the results. The number of different committees possible is

$$\begin{aligned} &C(7, 1) \cdot C(6, 6) + C(7, 2) \cdot C(6, 5) + C(7, 3) \cdot C(6, 4) + C(7, 4) \cdot C(6, 3) \\ &\quad + C(7, 5) \cdot C(6, 2) + C(7, 6) \cdot C(6, 1) = 7 \cdot 6 + 21 \cdot 15 + 35 \cdot 20 + 35 \cdot 15 + 21 \cdot 6 + 7 \cdot 1 \\ &\quad = 42 + 315 + 700 + 525 + 126 + 7 \\ &\quad = 1715 \end{aligned}$$

71. Since the books of the same subject must be kept together, there are 4! ways to arrange the history books, 5! ways to arrange the English books, and 6! ways to arrange the mathematics books. However, we must also decide the order the three subjects should be arranged on the shelf. This can be done 3! ways. Then by the Multiplication Principle there are $3! \cdot 4! \cdot 5! \cdot 6! = 12{,}441{,}600$ ways to arrange the books on the shelf.

73. In choosing a committee order is not important. So a committee of 8 boys and 5 girls can be formed form a group of 10 boys and 11 girls in

$$C(10, 8) \cdot C(11, 5) = 45 \cdot 462 = 20{,}790 \text{ ways.}$$

75. Three words, one each from five 3-letter words, six 4-letter words, and eight 5-letter words, can be chosen in $C(5, 1) \cdot C(6, 1) \cdot C(8, 1) = 5 \cdot 6 \cdot 8 = 240$ ways.

77. Although we are asked to order 5 speakers, we are told speaker B must come first. So in actuality we are only ordering 4 speakers which can be done $4! = 24$ ways.

79. Although we are told to order 5 speakers, we are told that B must go immediately after A. Think of tying them together. Then we need to order only 4 speakers which can be done $4! = 24$ ways.

81. We want to choose, without regard to order, 4 plums from a box of 25. Five of the plums are rotten, so we can partition the box into 20 good plums and 5 rotten plums.

(a) We can choose only good plums in $C(20, 4) = 4845$ ways.

(b) To determine how many ways there are to choose 3 good plums and 1 rotten plum, we use the Multiplication Principle. Three good plums and 1 rotten plum can be chosen in

$$C(20, 3) \cdot C(5, 1) = 1140 \cdot 5 = 5700 \text{ ways}$$

(c) The easiest way to find how many ways there are to get at least 1 rotten plum when choosing 4 plums is to subtract the number of ways we can get all good plums (part a) from the total number of ways we can choose 4 plums from the box of 25. This can be done

$$C(25, 4) - C(20, 4) = 12{,}650 - 4845 = 7805 \text{ ways.}$$

83. We need to find the number of permutations of 6 items where 2 of them are not distinct. The number of 6 letter words that can be made from the word FINITE is

$$\frac{6!}{1!\ 2!\ 1!\ 1!\ 1!} = \frac{720}{2} = 360$$

85. We need to find the number of permutations of 10 items 5 of which are not distinct. The number of ways of arranging the 10 books on the shelf when there are 3 copies of one book and 2 copies of another is

$$\frac{10!}{3!\ 2!\ 1!\ 1!\ 1!\ 1!\ 1!} = \frac{3{,}628{,}800}{12} = 302{,}400$$

87. Using the binomial theorem, we get

$$(x+2)^4 = \binom{4}{0}x^4 + \binom{4}{1}x^3 \cdot 2^1 + \binom{4}{2}x^2 \cdot 2^2 + \binom{4}{3}x^1 \cdot 2^3 + \binom{4}{4}2^4$$
$$= x^4 + 8x^3 + 24x^2 + 32x + 16$$

89. To find the coefficient of x^3 in the expansion of $(x + 2)^7$, simplify the term of the expansion,

$$\binom{7}{4}x^3 \cdot 2^4 = 35 \cdot 2^4 x^3 = 560x^3$$

The coefficient of x^3 is 560.

91. To show that $k\binom{n}{k} = n\binom{n-1}{k-1}$ we use the definition,

$$\binom{n}{k} = \frac{n!}{(n-k)!\ k!} \quad \text{and} \quad \binom{n-1}{k-1} = \frac{(n-1)!}{(n-1-(k-1))!\ (k-1)!} = \frac{(n-1)!}{(n-k)!\ (k-1)!}$$

Multiplying the equation on the left by k and the equation on the right by n, we get

$$k\binom{n}{k} = k \cdot \frac{n!}{(n-k)!\ k!} = \frac{n!}{(n-k)!\ (k-1)!} \quad \text{and} \quad n\binom{n-1}{k-1} = n \cdot \frac{(n-1)!}{(n-k)!\ (k-1)!} = \frac{n!}{(n-k)!\ (k-1)!}$$

Comparing the two expressions shows that they are equal and that

$$k\binom{n}{k} = n\binom{n-1}{k-1}.$$

Chapter 6 Project

1. Under the present system there are $7 \cdot 10^5 = 700{,}000$ different producer numbers available. The first digit can be chosen from the set $\{0, 2, 3, 4, 5, 6, 7\}$, but each of the other digits have 10 possible choices.

3. If all possible producer numbers are allowed there are 10^{11} possible correct UPC codes. The last digit is the check digit.

5. In Mauritius there are only 4 digits to be split between the publisher block and the title block. So the following 3 possibilities exist. There can be:

One digit for the publisher number and 3 digits for the title. This can be done $10 \cdot 10^3$ ways, which allows for $10^3 = 1000$ titles from each of 10 publishers or 10,000 titles.

Two digits for the publisher number and 2 digits for the title. This can be done $10^2 \cdot 10^2$ ways, which allows for $10^2 = 100$ titles from each of 100 publishers or 10,000 titles.

Three digits for the publisher number and 1 digit for the title. This can be done $10^3 \cdot 10$ ways, which allows for 10 titles from each of 1000 publishers or 10,000 titles.

Since only one ISBN number is assigned per title, we can use the counting formula.

$$10{,}000 + 10{,}000 + 10{,}000 = 30{,}000$$

titles can be published in Mauritius.

7. The restrictions placed on the publisher number limits the number of titles from English speaking areas. We consider the following 4 possibilities.

	Number of Country Codes	Number of Publisher Codes	Number of Titles	Total Number of Titles
1.	2	20	10^6	$2 \cdot 20 \cdot 10^6 = 40{,}000{,}000$
2.	2	500	10^5	$2 \cdot 500 \cdot 10^5 = 100{,}000{,}000$
3.	2	1500	10^4	$2 \cdot 1500 \cdot 10^4 = 30{,}000{,}000$
4.	2	500	10^3	$2 \cdot 500 \cdot 10^3 = 1{,}000{,}000$

Using the counting formula, we find that there are

$$40{,}000{,}000 + 100{,}000{,}000 + 30{,}000{,}000 + 1{,}000{,}000 = 171{,}000{,}000$$

possible titles from English speaking areas.

Chapter 7

Probability

7.1 Sample Spaces and the Assignment of Probabilities

3. Sample space

5. Event

7. Let G stand for green and R stand for red.
$S = \{GA, GB, GC, RA, RB, RC\}$

9. $S = \{AA, AB, AC, BA, BB, BC, CA, CB, CC\}$

11. $S = \{$ AA1, AA2, AA3, AA4, AB1, AB2, AB3, AB4, AC1, AC2, AC3, AC4, BA1, BA2, BA3, BA4, BB1, BB2, BB3, BB4, BC1, BC2, BC3, BC4, CA1, CA2, CA3, CA4, CB1, CB2, CB3, CB4, CC1, CC2, CC3, CC4$\}$

13. Let G stand for an outcome of green and R stand for an outcome of red on spinner 1.

$S = \{$GA1, GA2, GA3, GA4, GB1, GB2, GB3, GB4, GC1, GC2, GC3, GC4, RA1, RA2, RA3, RA4, RB1, RB2, RB3, RB4, RC1, RC2, RC3, RC4$\}$

15. Each time a coin is tossed there are 2 possible outcomes. So using the Multiplication Principle, when a coin is tossed 4 times there are $2 \cdot 2 \cdot 2 \cdot 2 = 2^4 = 16$ outcomes in the sample space.

17. Each time a die is tossed there are 6 possible outcomes. Using the Multiplication Principle there are $6 \cdot 6 \cdot 6 = 6^3 = 216$ outcomes in the sample space when 3 dice are tossed.

19. A regular deck contains 52 cards. There are 52 possibilities for the first outcome and 51 for the second outcome. So using the Multiplication Principle there are $52 \cdot 51 = 2652$ outcomes in the sample space.

21. There are 26 possible outcomes for the first letter chosen and 26 possible outcomes for the second letter chosen. Using the Multiplication Principle, there are $26 \cdot 26 = 26^2 = 676$ outcomes in the sample space.

23. Valid assignments must be nonnegative, and the sum of all probabilities in the sample space must equal one.
Assignments A, B, C, and F are valid.

25. If the coin always comes up tails HH, HT, and TH are impossible events and have probabilities of 0. Assignment B should be used.

27. S is the sample space. $n(S) = 23$

Define event E: A white ball is picked. There are 3 ways E can occur, so $n(E) = 3$, and

$$P(E) = \frac{n(E)}{n(S)} = \frac{3}{23}$$

29. S is the sample space. $n(S) = 23$

Define event E: A green ball is picked. There are 7 ways E can occur, so $n(E) = 7$, and

$$P(E) = \frac{n(E)}{n(S)} = \frac{7}{23}$$

31. S is the sample space. $n(S) = 23$

Define event E: A white or a red ball is picked. There are 8 ways E can occur, so $n(E) = 8$, and

$$P(E) = \frac{n(E)}{n(S)} = \frac{8}{23}$$

33. S is the sample space. $n(S) = 23$

Define event E: A white or a blue ball is picked. If neither red nor green is picked, then either white or blue is chosen. There are 11 ways E can occur, so $n(E) = 11$, and

$$P(E) = \frac{n(E)}{n(S)} = \frac{11}{23}$$

35. A regular deck of cards has 52 cards, so $n(S) = 52$.

Define event E: The ace of hearts is drawn. There is one ace of hearts so $n(E) = 1$, and

$$P(E) = \frac{n(E)}{n(S)} = \frac{1}{52}$$

37. A regular deck of cards has 52 cards, so $n(S) = 52$.

Define event E: A spade is drawn. There are 13 spades, $n(E) = 13$, and

$$P(E) = \frac{n(E)}{n(S)} = \frac{13}{52} = \frac{1}{4}$$

39. A regular deck of cards has 52 cards, so $n(S) = 52$.

Define event E: A picture card is drawn. There are 12 picture cards, making $n(E) = 12$, and

$$P(E) = \frac{n(E)}{n(S)} = \frac{12}{52} = \frac{3}{13}$$

41. A regular deck of cards has 52 cards, so $n(S) = 52$.

Define event E: A card with a number less than 6 is drawn. There are 5 numbers less than 6 and 4 of each number, making $n(E) = 20$, and

$$P(E) = \frac{n(E)}{n(S)} = \frac{20}{52} = \frac{5}{13}$$

43. A regular deck of cards has 52 cards, so $n(S) = 52$.

Define event E: A card that is not an ace is drawn. There are 48 cards that are not aces, making $n(E) = 48$, and

$$P(E) = \frac{n(E)}{n(S)} = \frac{48}{52} = \frac{12}{13}$$

45. Let F denote female and M denote male, and let D denote dry, O denote oil, and R denote regular. A sample space describing this experiment is S = {FD, FO, FR, MD, MO, MR}.

There are 6 outcomes in the sample space. That is $n(S) = 6$.

47. (a) Let G represent girl and B represent boy. We form sample space below. There are $n(S) = 16$ outcomes in the sample space. We assume the probability of a boy child equals the probability of a girl child. So we assign a probability of $\frac{1}{16}$ to each outcome.

e_i	e_1	e_2	e_3	e_4	e_5	e_6	e_7	e_8	e_9	e_{10}	e_{11}	e_{12}	e_{13}	e_{14}	e_{15}	e_{16}
S	GGGG	GGGB	GGBG	GBGG	BGGG	GGBB	GBBG	BBGG	GBGB	BGGB	BGBG	GBBB	BGBB	BBGB	BBBG	BBBB
$P(e_i)$	$\frac{1}{16}$	$\frac{1}{16}$	$\frac{1}{16}$	$\frac{1}{16}$	$\frac{1}{16}$	$\frac{1}{16}$	$\frac{1}{16}$	$\frac{1}{16}$	$\frac{1}{16}$	$\frac{1}{16}$	$\frac{1}{16}$	$\frac{1}{16}$	$\frac{1}{16}$	$\frac{1}{16}$	$\frac{1}{16}$	$\frac{1}{16}$

(b) (i) If E is the event "the first two children are girls, then E = {$GGGG$, $GGGB$, $GGBG$, $GGBB$}, and $n(E) = 4$. So, the probability the first two children are girls is

$$P(E) = \frac{n(E)}{n(S)} = \frac{4}{16} = \frac{1}{4}$$

(ii) If F is the event all children are boys, then F = {$BBBB$} and $n(F) = 1$. So the probability that all the children are boys is

$$P(F) = \frac{n(F)}{n(S)} = \frac{1}{16}$$

(iii) If G is the event at least one child is a girl then $n(G) = 15$ and the probability that at least one child is a girl is

$$P(G) = \frac{n(G)}{n(S)} = \frac{15}{16}$$

(iv) If H is the event the first child and the last child are girls, then H = {$GGGG$, $GGBH$, $GBGG$, $GBBG$} and $n(G) = 4$. So the probability the first and last children are girls is

$$P(H) = \frac{n(H)}{n(S)} = \frac{4}{16} = \frac{1}{4}$$

49. To form the probability model, we assign a probability to each income level that is proportional to the number of families in the category. First we add the right hand column to find $n(S) = 114{,}498.9$

$$P(<\$25{,}000) = \frac{30{,}998.1}{114{,}498.9} = 0.271 \qquad P(\$25{,}000 - \$49{,}999) = \frac{30083.0}{114{,}498.9} = 0.263$$

$$P(\$50{,}000 - \$74{,}999) = \frac{21046.7}{114{,}498.9} = 0.184 \qquad P(\$75{,}000 - \$99{,}999) = \frac{12{,}697.0}{114{,}498.9} = 0.111$$

$$P(\geq \$100{,}000) = \frac{19{,}674.0}{114{,}498.9} = 0.172$$

51. Using the information from the probability model in Problem 49, $P(\$50{,}000 - \$74{,}999) = 0.184$.

53. Using the information from the probability model in Problem 49,

$$P(< \$50{,}000) = 0.271 + 0.263 = 0.534$$

To do Problems 55-57 we need a probability model. We will use the following model in each of the four problems.

$$S = \{\text{Visa, MasterCard, American Express, Discover}\}$$

e_i	Visa	MasterCard	American Express	Discover
$P(e_i)$	0.521	0.380	0.046	0.053

55. Using the probability model above, $P(\text{MasterCard}) = 0.380$.

57. Using the probability model before Problem 56,

$$P(\text{American Express or Discover card}) = 0.046 + 0.053 = 0.099.$$

59. Using the information in the table on page 418 of the text, we find 11 stocks have prices that are greater than \$50 per share. So,

$$P(\text{Dow stock has a price greater than } \$50) = \frac{11}{30}.$$

61. Using the information in the table on page 418 of the text, we count 5 stocks with prices between \$40 and \$50 per share. So,

$$P(\text{Dow stock has a price between } \$40 \text{ and } \$50) = \frac{5}{30} = \frac{1}{6}.$$

63. The sample space is the set of all issues that advanced, declined, or that remained unchanged.

$$n(S) = 2422 + 1810 + 166 = 4398$$

$$P(\text{a randomly chosen issue advanced}) = \frac{2422}{4398} = 0.551$$

65. If the event E is "an issue made a 52-week high," then $n(E) = 253$.

$$P(\text{a randomly chosen issue made a 52-week high}) = \frac{253}{4398} = 0.058$$

67. There were 534 high yield issues that declined,

$$P(\text{a randomly chosen issue declined and was high yield}) = \frac{534}{4398} = 0.121$$

69. There were 8 issues that were at a 52-week low and high yield,

$$P(\text{a randomly chosen issue was at a 52-week low and high yield}) = \frac{8}{4398} = 0.002$$

71. Define the sample space as "all Americans. $n(S) = 247{,}257{,}000 + 46{,}577{,}000 = 293{,}834{,}000$

(a) If the event E is "an American was covered by health insurance," then $n(E) = 247{,}257{,}000$.

$$P(E) = \frac{n(E)}{n(S)} = \frac{247{,}257{,}000}{293{,}834{,}000} = 0.841$$

(b) If the event F is "an American was not covered by health insurance,) then $n(F) = 46{,}577{,}000$.

$$P(F) = \frac{n(F)}{n(S)} = \frac{46{,}577{,}000}{293{,}834{,}000} = 0.159$$

73. Simulation: Answers will vary.

Actual Probabilities: $n(S) = 30$, $n(R) = 18$, and $n(W) = 12$

$$P(R) = \frac{n(R)}{n(S)} = \frac{18}{30} = \frac{3}{5} = 0.6 \qquad P(W) = \frac{n(W)}{n(S)} = \frac{12}{30} = \frac{2}{5} = 0.4$$

75. Simulation: Answers will vary.

Actual Probabilities: $n(S) = 50$, $n(R) = 15$, and $n(W) = 35$

$$P(R) = \frac{n(R)}{n(S)} = \frac{15}{50} = \frac{3}{10} = 0.3 \qquad P(W) = \frac{n(W)}{n(S)} = \frac{35}{50} = \frac{7}{10} = 0.7$$

77. Simulation: Answers will vary.

Actual Probabilities: $P(\text{Adam wins}) = 0.22$
$P(\text{ Beatrice wins}) = 0.60$
$P(\text{Cathy wins}) = 0.18$

7.2 Properties of the Probability of an Event

5. Event

7. False. If E is an event, then $P(E) = 1 - P(\overline{E})$.

9.
$$\begin{aligned} P(E \cup F) &= P(E) + P(F) - P(E \cap F) \\ &= 0.4 + 0.5 - 0.2 \\ &= 0.7 \end{aligned}$$

11. $P(E \cup F) = P(E) + P(F) - P(E \cap F)$

$$P(E \cap F) = P(E) + P(F) - P(E \cup F) = 0.7 + 0.5 - 0.8 = 0.4$$

13. $P(E \cup F) = P(E) + P(F) - P(E \cap F)$

$$P(F) = P(E \cup F) + P(E \cap F) - P(E) = 0.6 + 0.1 - 0.4 = 0.3$$

15. $P(\overline{E}) = 1 - P(E) = 1 - 0.4 = 0.6$

17. $P(A) = 0.5$, $P(B) = 0.4$ and $P(A \cap B) = 0.2$

(a) $P(A \text{ or } B) = P(A) + P(B) - P(A \cap B) = 0.5 + 0.4 - 0.2 = 0.7$

(b) $P(A \text{ but not } B) = P(A) - P(A \cap B) = 0.5 - 0.2 = 0.3$

(c) $P(B \text{ but not } A) = P(B) - P(A \cap B) = 0.4 - 0.2 = 0.2$

(d) $P(\text{neither } A \text{ nor } B) = P(\overline{A \cup B}) = 1 - P(A \cup B) = 1 - 0.7 = 0.3$

19. $P(A \cap B) = \varnothing$, $P(A) = 0.6$, and $P(B) = 0.2$

(a) $P(A \cap B) = \varnothing$

(b) $P(A \cup B) = P(A) + P(B) = 0.6 + 0.2 = 0.8$

(c) $P(\overline{A \cup B}) = 1 - P(A \cup B) = 1 - 0.8 = 0.2$

(d) $P(\overline{B}) = 1 - P(B) = 1 - 0.2 = 0.8$

(e) $P(\overline{A}) = 1 - P(A) = 1 - 0.6 = 0.4$

(f) $P(\overline{A \cap B}) = 1 - P(A \cap B) = 1 - 0 = 1$

21. $P(E) = \dfrac{3}{3+1} = \dfrac{3}{4}$

23. $P(E) = \dfrac{5}{5+7} = \dfrac{5}{12}$

25. $P(E) = \dfrac{1}{1+1} = \dfrac{1}{2}$

27. $P(E) = 0.6$ $\quad P(\overline{E}) = 1 - P(E) = 1 - 0.6 = 0.4$

Odds for E: $\dfrac{P(E)}{P(\overline{E})} = \dfrac{6}{4}$ or 3 to 2

Odds against E: $\dfrac{P(\overline{E})}{P(E)} = \dfrac{4}{6}$ or 4 to 6 or 2 to 3

29. $P(F) = \dfrac{3}{4}$ $\quad P(\overline{F}) = 1 - P(F) = \dfrac{1}{4}$

Odds for F: $\dfrac{P(F)}{P(\overline{F})} = \dfrac{\frac{3}{4}}{\frac{1}{4}} = \dfrac{3}{1}$ or 3 to 1

Odds against F: $\dfrac{P(\overline{F})}{P(F)} = \dfrac{\frac{1}{4}}{\frac{3}{4}} = \dfrac{1}{3}$ or 1 to 3

31. Let E and F be the events: E: The Bears win; F: The Bears tie.
E and F are mutually exclusive, so the probability that the Bears either win or tie is

$$P(E \cup F) = P(E) + P(F) = 0.65 + 0.05 = 0.70$$

The probability that the Bears lose is the complement of their winning or tying.

$$P(\overline{E \cup F}) = 1 - P(E \cup F) = 1 - 0.70 = 0.30$$

33. Let M and E be the events Anne passes mathematics and Anne passes English respectively. We are given $P(M) = 0.4$; $P(E) = 0.6$, and $P(M \cup E) = 0.8$. The probability that Anne passes both courses is

$$P(M \cap E) = P(M) + P(E) - P(M \cup E) = 0.4 + 0.6 - 0.8 = 0.2$$

35. Define the events T: Car needs a tune-up.
B: Car needs a brake job.

$P(T) = 0.6$ $\quad P(B) = 0.1$ $\quad P(T \cap B) = 0.02$

(a) The probability the car needs either a tune-up or a brake job is

$$P(T \cup B) = P(T) + P(B) - P(T \cap B) = 0.6 + 0.1 - 0.02 = 0.68$$

(b) Probability a car needs a tune-up but not a brake job is

$$P(T \cap \overline{B}) = P(T) - P(T \cap B) = 0.6 - 0.02 = 0.58$$

(c) The probability it needs neither repair is

$$P(\overline{T \cup B}) = 1 - P(T \cup B) = 1 - 0.68 = 0.32$$

37. (a) $P(1 \text{ or } 2 \text{ TVs}) = P(1 \text{ TV}) + P(2 \text{ TVs}) = 0.24 + 0.33 = 0.57$

(b) $P(1 \text{ or more TVs}) = 1 - P(0 \text{ TVs}) = 1 - 0.05 = 0.95$

(c) $P(3 \text{ or fewer TVs}) = 1 - P(4 \text{ or more}) = 1 - 0.17 = 0.83$

(d) $P(3 \text{ or more TVs}) = P(3 \text{ TVs}) + P(4 \text{ or more}) = 0.21 + 0.17 = 0.38$

(e) $P(\text{fewer than } 2 \text{ TVs}) = P(0 \text{ TV}) + P(1 \text{ TV}) = 0.05 + 0.24 = 0.29$

(f) $P(\text{not even } 1 \text{ TV}) = P(0 \text{ TV}) = 0.05$

(g) $P(1, 2, \text{ or } 3 \text{ TVs}) = 0.24 + 0.33 + 0.21 = 0.78$

(h) $P(2 \text{ or more TVs}) = 0.33 + 0.21 + 0.17 = 0.71$

39. Define event E: A person has Rh-positive blood.

$$\begin{aligned} P(E) &= P(\text{O-Positive}) + P(\text{A-Positive}) + P(\text{B-Positive}) + P(\text{AB-Positive}) \\ &= 0.38 + 0.34 + 0.09 + 0.03 \\ &= 0.84 \end{aligned}$$

The probability that a person selected at random has a Rh-positive blood type is 0.84.

41. Define event A: A person selected at random has blood that contains the A antigen.

$$\begin{aligned} P(A) &= P(\text{A-Negative}) + P(\text{A-Positive}) + P(\text{AB-Negative}) + P(\text{AB-Positive}) \\ &= 0.06 + 0.34 + 0.01 + 0.03 \\ &= 0.44 \end{aligned}$$

The probability that a person selected at random has blood containing the A antigen is 0.44.

43. Define E to be the event, a person has Rh-positive blood. From Problem 39, $P(E) = 0.84$. Define B to be the event, a person has type B blood. $P(B) = 0.09 + 0.02 = 0.11$.

$$\begin{aligned} P(B \cup E) &= P(B) + P(E) - P(B \cap E) \\ &= 0.11 + 0.84 - 0.02 \\ &= 0.93 \end{aligned}$$

The probability that a person selected at random has type B blood or is Rh-positive is 0.93.

45. Define E to be the event, a woman giving birth in 2003 gave birth to her first child. $P(E) = 0.393$ The probability that a woman who gave birth in 2003 had previously given birth is

$$\begin{aligned} P(\overline{E}) &= 1 - P(E) \\ &= 1 - 0.393 \\ &= 0.607 \end{aligned}$$

47. If 156,000 were rated deficient or obsolete, then $595{,}000 - 156{,}000 = 439{,}000$ were rated as neither deficient nor obsolete. The odds that a randomly selected bridge is either deficient or obsolete is

$$\frac{156{,}000}{439{,}000} = \frac{156}{439} \text{ or 156 to 439.}$$

49. Let E define the event, a homeowner with a subprime loan faces a higher mortgage payment. If $P(E) = 0.59$, then the probability a homeowner with a subprime loan does not face a higher mortgage payment is $P(\overline{E})$.

$$P(\overline{E}) = 1 - P(E) = 1 - 0.59 = 0.41$$

51. Assuming every type of patent is mutually exclusive with each of the other types, the probability that a randomly selected patent was either for a design or a botanical plant is

$$P(\text{design or botanical plant}) = P(\text{design}) + P(\text{botanical plant})$$

$$P(\text{design} \cup \text{botanical plant}) = \frac{n(\text{design})}{n(\text{patents})} + \frac{n(\text{botanical plants})}{n(\text{patents})}$$

$$= \frac{12{,}950}{157{,}717} + \frac{716}{157{,}717}$$

$$= \frac{13{,}666}{157{,}717} = 0.0866$$

53. The probability the fund will outperform the market in at least one of the next two years is the probability it outperforms the market in the first year plus the probability it outperforms the market in the second year minus the probability it outperforms the market in both years.

$$P(\text{fund outperforms in at least one of the next 2 years}) = 0.12 + 0.12 - 0.05 = 0.19$$

55. If the odds A wins are 1 to 2, the probability A wins is $P(A) = \frac{1}{3}$. If the odds B wins are 2 to 3, then the probability B wins is $P(B) = \frac{2}{5}$. Since only one person can win the race the events winning are mutually exclusive and

$$P(A \text{ or } B \text{ wins}) = P(A \cup B) = P(A) + P(B)$$

$$= \frac{1}{3} + \frac{2}{5} = \frac{11}{15}$$

$$\text{The odds for } A \text{ or } B \text{ winning} = \frac{P(A \cup B)}{P(\overline{A \cup B})} = \frac{P(A \cup B)}{1 - P(A \cup B)} = \frac{\frac{11}{15}}{\frac{4}{15}} = \frac{11}{4} \text{ or 11 to 4.}$$

57. Prove: $P(E \cup F) = P(E) + P(F) - P(E \cap F)$

Proof: $E \cup F$ can be written as the union of disjoint sets in the following way

$$E \cup F = (E \cap \overline{F}) \cup (E \cap F) \cup (\overline{E} \cap F)$$

then,

$$P(E \cup F) = P[(E \cap \overline{F}) \cup (E \cap F) \cup (\overline{E} \cap F)]$$

Since the sets are disjoint,

$$P(E \cup F) = P(E \cap \overline{F}) + P(E \cap F) + P(\overline{E} \cap F) \quad (1)$$

Next write E and F as the union of disjoint sets

$$E = (E \cap \overline{F}) \cup (E \cap F) \quad \text{and} \quad F = (\overline{E} \cap F) \cup (E \cap F)$$

So this means

$$P(E) = P(E \cap \overline{F}) + P(E \cap F) \quad \text{and} \quad P(F) = P(\overline{E} \cap F) + P(E \cap F)$$

or

$$P(E \cap \overline{F}) = P(E) - P(E \cap F) \quad \text{and} \quad P(\overline{E} \cap F) = P(F) - P(E \cap F)$$

Substituting the above two equations into (1) we have

$$P(E \cup F) = [P(E) - P(E \cap F)] + P(E \cap F) + [P(F) - P(E \cap F)]$$

which simplifies to

$$P(E \cup F) = P(E) + P(F) - P(E \cap F)$$

59. Show: $P(A \cup B \cup C) = P(A) + P(B) + P(C) - P(A \cap B) - P(A \cap C) - P(B \cap C) + P(A \cap B \cap C)$

Proof: Let $E = A \cup B$. The addition rule states

$$P(E \cup C) = P(E) + P(C) - P(E \cap C)$$

Substitute $A \cup B$ for E in the addition rule.

$$P[(A \cup B) \cup C] = P(A \cup B) + P(C) - P[(A \cup B) \cap C)]$$

Apply the addition rule to $P(A \cup B)$.

$$P[(A \cup B) \cup C] = [P(A) + P(B) - P(A \cap B)] + P(C) - P[(A \cup B) \cap C)]$$

Apply the distributive property to the sets in the last bracket.

$$P[(A \cup B) \cup C] = [P(A) + P(B) - P(A \cap B)] + P(C) - P[(A \cap C) \cup (B \cap C)]$$

Apply the addition rule to the union of sets in the last bracket.

$$P(A \cup B \cup C) = P(A) + P(B) - P(A \cap B) + P(C) - [P(A \cap C) + P(B \cap C) - P(A \cap B \cap C)]$$

Remove the brackets and rearrange terms.

$$P(A \cup B \cup C) = P(A) + P(B) + P(C) - P(A \cap B) - P(A \cap C) - P(B \cap C) + P(A \cap B \cap C)$$

7.3 Probability Problems Using Counting Techniques

5. The number of elements in the sample space S is found by using the Multiplication Principle. Since the coin is tossed 5 times, there are

$$n(S) = 2^5 = 32 \text{ elements.}$$

(a) Define E as the event, "Exactly 3 heads appear."

$$P(E) = \frac{n(E)}{n(S)} = \frac{C(5, 3)}{32} = \frac{10}{32} = \frac{5}{16}$$

(b) Define F as the event, "No heads appear."

$$P(F) = \frac{n(F)}{n(S)} = \frac{C(5, 0)}{32} = \frac{1}{32}$$

7. The number of elements in the sample space S is found by using the Multiplication Principle. Since the dice are thrown 3 times,

$$n(S) = 36^3 = 46{,}656$$

(a) Define E as the event, "The sum of 7 appears 3 times."
When two dice are thrown, a sum of 7 can appear 6 ways, $\{(1, 6), (2, 5), (3, 4), (4, 3), (5, 2), (6, 1)\}$. Using the Multiplication Principle we find that on 3 throws a sum of 7 can appear 6^3 ways.

$$P(E) = \frac{n(E)}{n(S)} = \frac{6^3}{36^3} = \frac{1}{6^3} = \frac{1}{216} \approx 0.005$$

(b) Define F as the event, "The sum of 7 or 11 appears at least twice."
When 2 dice are thrown a sum of 7 can appear 6 ways and a sum of 11 can appear 2 ways. Using the addition rule, the sum of 7 or 11 can appear 8 ways.

Event F is equivalent to the event: Exactly two throws result in a 7 or 11, or all three throws result in a 7 or 11. These are mutually exclusive, so we are interested in finding

$$P(F) = P(7 \text{ or } 11 \text{ appears twice}) + P(7 \text{ or } 11 \text{ appear } 3 \text{ times})$$

Using the Multiplication Principle 7 or 11 can appear 3 times in $8^3 = 512$ ways.

7 or 11 can appear 2 times in $8^2 = 64$ ways, and the third number can appear $36 - 8 = 28$ ways. By the Multiplication Principle 2 throws of 7 or 11 and 1 throw of another number can appear $8^2 \cdot 28 = 1792$ ways. However, we still need to choose which of the two throws will result in the 7 or 11. We choose 2 out of 3 tries $C(3, 2) = 3$ ways.

$$P(F) = \frac{n(F)}{n(S)} = \frac{C(3, 2) \cdot 8^2 \cdot 28}{36^3} + \frac{8^3}{36^3}$$
$$= \frac{5376}{36^3} + \frac{512}{36^3} = \frac{5888}{46656} = 0.126$$

9. The number of elements in the sample space, S, is found by using the Multiplication Principle. Since there are 7 digits in the phone number, and there are no restrictions,

$$n(S) = 10^7$$

Define event E: A phone number has one or more repeated digits.

It is easier to do this problem looking at the complement of E. $\overline{E}$: A phone number has no repeated digits. $n(\overline{E}) = P(10, 7)$. So we get

$$P(E) = 1 - P(\overline{E})$$
$$= 1 - \frac{n(\overline{E})}{n(S)}$$
$$= 1 - \frac{P(10, 7)}{10^7} = 0.940$$

11. The number of elements in the sample space, S, is found by using the Multiplication Principle. Since there are 26 letters in the alphabet and we are going to select 5, allowing repetitions

$$n(S) = 26^5$$

Define event E: No letters are repeated. $n(E) = C(26, 5)$

$$P(E) = \frac{n(E)}{n(S)} = \frac{C(26, 5)}{26^5} = 0.0055$$

13. The number of elements in the sample space, S, is found by using the Multiplication Principle. Since there are 12 months in the year, and we are going to select 3,

$$n(S) = 12^3 = 1728$$

Define event E: At least 2 were born in the same month. It is easier to do this problem by looking at the complement of E,

$\overline{E}$: No two people were born in the same month.

$$n(\overline{E}) = P(12, 3) = 12 \cdot 11 \cdot 10 = 1320$$

The probability that at least 2 were born in the same month is

$$P(E) = 1 - P(\overline{E}) = 1 - \frac{n(\overline{E})}{n(S)} = 1 - \frac{1320}{1728} = 0.236$$

15. The number of elements in the sample space, *S*, is found by using the Multiplication Principle. Since there are 365 days in a year, and 100 senators, $n(S) = 365^{100}$.

Define event *E*: At least 2 senators have the same birthday. It is easier to do this problem by looking at the complement of *E*,

$\overline{E}$: No two senators have the same birthday.

$$n(\overline{E}) = P(365, 100) = 365 \cdot 364 \cdot 363 \cdot \ldots \cdot 266$$

The probability that at least 2 senators have the same birthday is

$$P(E) = 1 - P(\overline{E}) = 1 - \frac{n(\overline{E})}{n(S)} = 1 - \frac{P(365, 100)}{365^{100}} = 0.99999 \approx 1$$

17. The experiment consists of choosing 2 stocks from 10 stocks which is done $C(10, 2) = 45$ ways.

(a) Define the event *E* as both stock prices are greater than \$50.

$$P(E) = \frac{C(3, 2)}{C(10, 2)} = \frac{3}{45} = \frac{1}{15}$$

(b) Define the event *F* as neither stock price is above \$50. This is equivalent to choosing 2 stocks from the 7 with prices less than \$50.

$$P(F) = \frac{C(7, 2)}{C(10, 2)} = \frac{\frac{7!}{2! \cdot 5!}}{45} = \frac{21}{45} = \frac{7}{15}$$

(c) Define the event *G* as at least one stock is priced above \$50.

$$P(G) = 1 - P(F) = 1 - \frac{7}{15} = \frac{8}{15}$$

19. (a) The Puppies of the Dow are Pfizer, Verizon, AT&T, General Electric, and General Motors.

(b) Define event *E* as both have yields above 4%. Two of the five Puppies of the Dow have yields above 4%.

$$P(E) = \frac{C(2, 2)}{C(5, 2)} = \frac{1}{10}$$

(c) Define event *F* as at one stock has a yield above 4%.

$$P(F) = \frac{C(2, 1) \cdot C(3, 1)}{C(5, 2)} = \frac{2 \cdot 3}{10} = \frac{6}{10} = \frac{3}{5}$$

(d) Define event G as neither have a yield above 4%.

$$P(G)=\frac{C(3,2)}{C(5,2)}=\frac{3}{10}$$

(e) Define event H as at least one will have a yield above 4%.

$$P(H)=P(\overline{G})=1-P(G)=1-\frac{3}{10}=\frac{7}{10}$$

21. The experiment consists of choosing 5 accounts from a list of 35 accounts. The accounts can be partitioned into two groups: 32 that are correct and 3 that contain errors. The number of elements in the sample space is $C(35, 5) = 324{,}632$.

(a) The probability of choosing 5 accounts without errors:

$$\frac{C(32,5)\cdot C(3,0)}{C(35,5)}=\frac{\frac{32!}{5!\cdot 27!}\cdot 1}{\frac{35!}{5!\cdot 30!}}=\frac{116}{187}\approx 0.620$$

(b) The probability of choosing exactly 1 account with errors is

$$\frac{C(32,4)\cdot C(3,1)}{C(35,5)}=\frac{\frac{32!}{4!\cdot 28!}\cdot 3}{324{,}632}=\frac{435}{1309}\approx 0.332$$

(c) The probability of choosing at least 2 accounts containing errors is easiest done by using the complement. The probability of choosing at least 2 accounts containing errors is 1 minus the probability of choosing fewer than 2 accounts with errors.

1 – [probability of choosing 5 accounts with no errors + probability of choosing 4 accounts with no errors] = 1 – 0.620 – 0.332 = 0.048

23. The codes are formed by a letter, 5 numbers, and 2 final letters. Repetition is allowed and order is important.

(a) There are $26 \cdot 10 \cdot 10 \cdot 10 \cdot 10 \cdot 10 \cdot 26 \cdot 26 = 26^3 \cdot 10^5 = 1{,}757{,}600{,}000$ different SIM card codes.

(b) The number of codes that end in AA is $26 \cdot 10^5 = 2{,}600{,}000$. So the probability a code ends in AA is

$$\frac{n(\text{codes ending in AA})}{n(\text{codes})}=\frac{2{,}600{,}000}{1{,}757{,}600{,}000}=\frac{26}{17576}=\frac{1}{676}\approx 0.00148$$

(c) The number of codes that begin with A and end with A is the same as the number that end in AA. So the probability a code begins and ends in A is 0.00148.

(d) The number of codes with no repeated letter and no repeated number is given by

$$P(26, 3) \cdot P(10, 5) = 15{,}600 \cdot 30{,}240 = 471{,}744{,}000$$

The probability a code has no repeated letter and no repeated number is

$$\frac{P(26,3)\cdot P(10,5)}{26^3\cdot 10^5}=\frac{471{,}744{,}000}{1{,}757{,}600{,}000}=\frac{1134}{4225}\approx 0.2684$$

(e) The number of codes that have no repeated letter and all five of the same numbers is

$$P(26, 3) \cdot 10 = 15{,}600 \cdot 10 = 156{,}000$$

The probability a code has no repeated letter and all five of the same number is

$$\frac{156{,}000}{1{,}757{,}600{,}000} \approx 0.0001$$

25. The number of elements in the sample space S is equal to the number of combinations of 50 refrigerators taken 5 at a time,

$$C(50, 5) = \frac{50!}{5!\ 45!} = 2{,}118{,}760$$

Define E as the event, "5 refrigerators are defective."

$$P(E) = \frac{C(6,\ 5)}{C(50,\ 5)} = \frac{6}{2{,}118{,}760} = \frac{3}{1{,}059{,}380} = 2.83 \times 10^{-6}$$

Define F as the event, "at least 2 refrigerators are defective"

$$\begin{aligned} P(F) &= 1 - P(0 \text{ refrigerators are defective or } 1 \text{ refrigerator is defective}) \\ &= 1 - \left[\frac{C(6, 0) \cdot C(44, 5)}{2{,}118{,}760} + \frac{C(6,1) \cdot C(44,4)}{2{,}118{,}760}\right] \\ &= 1 - \left[\frac{1{,}086{,}008}{2{,}118{,}760} + \frac{6 \cdot (135{,}751)}{2{,}118{,}760}\right] \\ &= 0.103 \end{aligned}$$

27. The sample space consists of all four digit permutations (with repetition). $n(S) = 10^4 = 10{,}000$

(a) Define event E as the last two digits are 0s. $n(E) = 10^2 = 100$. The probability of E is

$$P(E) = \frac{n(E)}{n(S)} = \frac{100}{10{,}000} = \frac{1}{100} = 0.01$$

(b) Define event F as the last two digits are 0s, but the first digit is not 0. $n(F) = 9 \cdot 10 = 90$. The probability of F is

$$P(F) = \frac{n(F)}{n(S)} = \frac{90}{10{,}000} = \frac{9}{1000} = 0.009$$

29. The sample space consists of the 60 numbers. $n(S) = 60$

Define the event E as at least two students choose the same number. The easiest way to approach this problem is to consider its complement: no students choose the same number.

$$P(\overline{E}) = \frac{P(60, 30)}{60^{30}} \approx 0$$

Then $P(E) \approx 1$.

31. The experiment consists of choosing a committee of 7 senators from the 100 in the Senate. The number of elements in the sample space is $n(S) = C(100, 7)$.

(a) Define the event E as the committee has only Democratic members. (There are 51 Democrats.)

$$P(E) = \frac{n(E)}{n(S)} = \frac{C(51,7)}{C(100,7)} \approx 0.0072$$

(b) Define the event F as the committee has only Republican members. (There are 49 Republicans.)

$$P(F)=\frac{n(F)}{n(S)}=\frac{C(49,7)}{C(100,7)}\approx 0.0054$$

(c) Define the event G as the committee has 4 Democrats and 3 Republicans.

$$P(G)=\frac{n(G)}{n(S)}=\frac{C(51,4)\cdot C(49,3)}{C(100,7)}\approx 0.2876$$

33. The experiment consists of choosing 2 printers from 100. The sample space is all combinations of the 2 printers that can be chosen. $n(S) = C(100, 2)$. Define event E as the two printers chosen work. $n(E) = C(95, 2)$.
The probability the shipment is rejected is $P(\overline{E})$.

$$P(\overline{E})=1-P(E)=1-\frac{C(95,2)}{C(100,2)}=1-\frac{893}{990}=\frac{97}{990}\approx 0.0980$$

35. The sample space, S, consist of all the possible combinations of 5 cards; $n(S) = C(52, 5)$

(a) Define event E: All are hearts; $n(E) = C(13, 5)$

$$P(E)=\frac{n(E)}{n(S)}=\frac{C(13,\ 5)}{C(52,\ 5)}\approx 0.0005$$

(b) Define event F: Exactly 4 are spades.
The number of elements in F is determined as follows: Exactly 4 spades means 4 spades and 1 non-spade. The 4 spades can be chosen $C(13, 4)$ ways. The non-spade can be chosen $C(39, 1)$ ways. Using the Multiplication Principle, $c(F) = C(13, 4) \cdot C(39, 1)$

$$P(F)=\frac{n(F)}{n(S)}=\frac{C(13,\ 4)\cdot C(39,\ 1)}{C(52,\ 5)}=0.0107$$

(c) Define event G: Exactly 2 are clubs.
The number of elements in E is determined as follows: Exactly 2 clubs means 2 clubs and 3 non-clubs. The 2 clubs can be chosen $C(13, 2)$ ways and the 3 non-clubs can be chosen $C(39, 3)$ ways. Using the Multiplication Principle, $c(G) = C(13, 2) \cdot C(39, 3)$

$$P(G)=\frac{n(G)}{n(S)}=\frac{C(13,\ 2)\cdot C(39,\ 3)}{C(52,\ 5)}=0.2743$$

37. The sample space, S, consist of all the possible combinations of 5 cards; $n(S) = C(52, 5)$

(a) Define event E: The hand is a royal flush; $n(E) = 4$ (One from each of 4 suits.)

$$P(E)=\frac{n(E)}{n(S)}=\frac{4}{C(52,\ 5)}=1.539\times 10^{-6}$$

(b) Define event F: A hand consists of a straight flush.

The number of elements in F is determined as follows: Each suit has 9 straight flushes, (cards A,2,3,4,5; 2,3,4,5,6; 3,4,5,6,7; ... ; 9,10,J,Q,K) and there are 4 suits. Using the Multiplication Principle, $n(F) = 4 \cdot 9 = 36$

$$P(F) = \frac{n(F)}{n(S)} = \frac{36}{C(52,\ 5)} = 1.385 \times 10^{-5}$$

(c) Define the event G: A hand contains 4 of a kind.

G is equivalent to 4 cards of one face value and 1 of another;

$$n(G) = C(13,\ 1) \cdot C(4,\ 4) \cdot C(48,\ 1) = 624$$

$$P(G) = \frac{n(G)}{n(S)} = \frac{624}{C(52,\ 5)} = 2.401 \times 10^{-4}$$

(d) Define event H: A hand contains one pair and one triple of the same face values.

The number of elements in H can be determined as follows:

1. Choose two suits for the pair, $C(4,\ 2) = 6$ ways, and then choose a face value for the pair, $C(13,\ 1) = 13$.
2. Choose three suits for the triple, $C(4,\ 3) = 4$, and then choose a face value for the triple, $C(12,\ 1) = 12$. (One face value had been used to make the pair.)
3. Use the Multiplication Principle to find

$$n(H) = C(4,\ 2) \cdot C(13,\ 1) \cdot C(4,\ 3) \cdot C(12,\ 1) = 6 \cdot 13 \cdot 4 \cdot 12 = 3744$$

$$P(H) = \frac{n(H)}{n(S)} = \frac{3744}{C(52,\ 5)} = 0.0014$$

(e) Define event K: The hand contains 5 nonconsecutive cards from a single suit.

The number of elements in K is determined as follows:

1. Choose the suit to be used, $C(4,\ 1) = 4$, and then choose 5 cards from the suit, $C(13,\ 5)$.
2. Eliminate the cards that are in order, that is, the royal flushes from part (a) and the straight flushes from part (b).

$$n(K) = 4 \cdot C(13,\ 5) - (4 + 36) = 5108$$

$$P(K) = \frac{n(K)}{n(S)} = \frac{5108}{C(52,\ 5)} = 0.0020$$

(f) Define event L: The hand consists of 5 cards in order.

The number of elements in L is determined as follows:

1. There are 10 straights. (A,2,3,4,5; 2,3,4,5,6; 3,4,5,6,7; ... ; 10,J,Q,K,A) Each card in the straight can come from any one of 4 suits, so can be chosen 4^5 ways.
2. Eliminate the cards that are from the same suits, that is, the royal flushes from part (a) and the straight flushes from part (b).

$$n(K) = 10 \cdot 4^5 - (4 + 36) = 10{,}200$$

$$P(L) = \frac{n(L)}{n(S)} = \frac{10{,}200}{C(52,\ 5)} = 0.0039$$

39. The number of elements in the sample space, S, is the number of ways the playoff teams can be chosen. Since the selection is made in several steps we use the Multiplication Principle.

$$n(S) = C(5, 1) \cdot C(5, 1) \cdot C(4, 1) \cdot C(11, 1) = 1100$$

Define event E: The Yankees and the Red Sox are in the playoffs.

The number of elements in E is determined as follows: One of the two teams is the division winner, this can happen 1 way; and the other team is the wild card, this can happen 1 way. Allowing for each team to be division winner doubles this number. The division winners from the West and Central Divisions have $C(4, 1)$ and $C(5, 1)$ ways of occurring. Using the Multiplication Principle, we find

$$n(E) = 2(1 \cdot 4 \cdot 5 \cdot 1) = 40$$

The probability that both the Yankees and the Red Sox are in the playoffs is

$$P(E) = \frac{n(E)}{n(S)} = \frac{40}{1100} = \frac{2}{55} \approx 0.036$$

41. The number of elements in the sample space, S, is the number of ways the playoff teams can be chosen from the three divisions. We use the Multiplication Principle.

$$n(S) = C(5, 1) \cdot C(5, 1) \cdot C(4, 1) \cdot C(11, 1) = 1100$$

Define event E: The wild card team is from the East Division.

$$n(E) = C(5, 1) \cdot C(5, 1) \cdot C(4, 1) \cdot C(4, 1) = 400$$

The probability the wild card team is from the East Division is

$$P(E) = \frac{n(E)}{n(S)} = \frac{400}{1100} = \frac{4}{11} \approx 0.364$$

43. The number of elements in the sample space is the number of ways of choosing 2 out of 4 teams from each of 8 groups. $n(S) = [C(4, 2)]^8 = 6^8 = 1{,}679{,}616$

Define event E as the United States, Mexico and England (different sections) proceed to the second round. $n(E) = 1 \cdot C(3, 1) \cdot 1 \cdot C(3, 1) \cdot 1 \cdot C(3, 1) \cdot 6^5 = 27 \cdot 6^5 = 209{,}952$

$$P(E) = \frac{n(E)}{n(S)} = \frac{27 \cdot 6^5}{6^8} = \frac{3^3}{2^3 \cdot 3^3} = \frac{1}{8} = 0.125$$

7.4 Conditional Probability

1. False. The conditional probability of the event E given the event F is denoted by $P(E|F)$.

3. $P(E) = 0.2 + 0.3 = 0.5$

5. $P(E|F) = \dfrac{P(E \cap F)}{P(F)} = \dfrac{0.3}{0.7} = \dfrac{3}{7} \approx 0.429$

7. $P(E \cap F) = 0.3$

9. $P(\overline{E}) = 1 - P(E) = 1 - 0.5 = 0.5$

11. $P(E|F) = \dfrac{P(E \cap F)}{P(F)} = \dfrac{0.1}{0.4} = \dfrac{1}{4} = 0.25$

$P(F|E) = \dfrac{P(E \cap F)}{P(E)} = \dfrac{0.1}{0.2} = \dfrac{1}{2} = 0.5$

13. $P(E \mid F) = \dfrac{P(E \cap F)}{P(F)}$

$$P(F) = \frac{P(E \cap F)}{P(E \mid F)} = \frac{0.2}{0.4} = \frac{1}{2} = 0.5$$

15. $P(E \mid F) = \dfrac{P(E \cap F)}{P(F)}$

$$P(E \cap F) = P(F) \cdot P(E \mid F) = \frac{5}{13} \cdot \frac{4}{5} = \frac{4}{13}$$

17. (a) $P(F \mid E) = \dfrac{P(E \cap F)}{P(E)}$

$$P(E) = \frac{P(E \cap F)}{P(F \mid E)} = \frac{\frac{1}{3}}{\frac{2}{3}} = \frac{1}{2}$$

(b) $P(E \mid F) = \dfrac{P(E \cap F)}{P(F)}$

$$P(F) = \frac{P(E \cap F)}{P(E \mid F)} = \frac{\frac{1}{3}}{\frac{1}{2}} = \frac{2}{3}$$

19. $P(C) = (0.7)(0.9) + (0.3)(0.2)$
$= 0.63 + 0.06 = 0.69$

21. $P(C \mid A) = 0.9$

23. $P(C \mid B) = 0.2$

25. $P(E \cap F) = P(E) + P(F) - P(E \cup F)$
$= 0.5 + 0.4 - 0.8 = 0.1$

27. $P(F \mid E) = \dfrac{P(E \cap F)}{P(E)} = \dfrac{0.1}{0.5} = \dfrac{1}{5} = 0.2$

29. A tree diagram helps to see this problem.

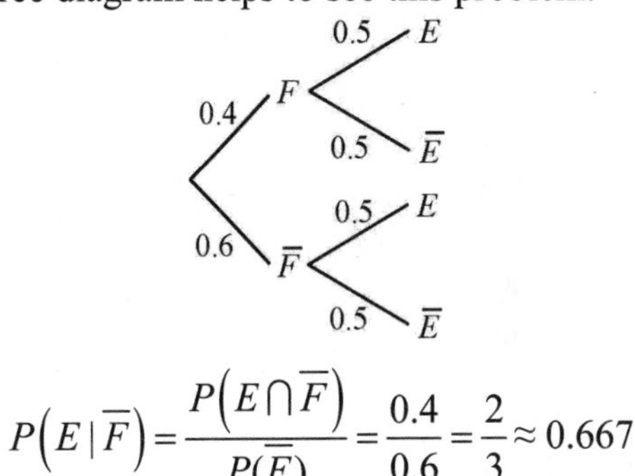

$$P(E \mid \overline{F}) = \frac{P(E \cap \overline{F})}{P(\overline{F})} = \frac{0.4}{0.6} = \frac{2}{3} \approx 0.667$$

31. $S = \{$ BBB, BBG, BGB, GBB, GGB, GBG, BGG, GGG$\}$; $n(S) = 8$

(a) Let E be the event, "The family has 2 girls." $E = \{$GGB, GBG, BGG$\}$
Let F be the event, "The first child is a girl." $F = \{$ GBB, GGB, GBG, GGG$\}$

$P(F) = \dfrac{4}{8} = \dfrac{1}{2}$ $\quad E \cap F = \{$GGB, GBG$\}$ $\quad P(E \cap F) = \dfrac{1}{4}$

So the probability a family with 3 children has exactly 2 girls, given the first child is a girl is

$$P(E \mid F) = \frac{P(E \cap F)}{P(F)} = \frac{\frac{1}{4}}{\frac{1}{2}} = \frac{1}{2}$$

(b) Let E be the event, "The family has 1 girl." $E = \{\text{GBB, BGB, BBG}\}$

Let F be the event, "The first child is a boy." $F = \{\text{BBB, BBG, BGB, BGG}\}$; $P(F) = \frac{4}{8} = \frac{1}{2}$

$$E \cap F = \{\text{BGB, BBG}\} \qquad P(E \cap F) = \frac{2}{8} = \frac{1}{4}$$

So the probability a family with 3 children has exactly 1 girl, given the first child is a boy, is

$$P(E \mid F) = \frac{P(E \cap F)}{P(F)} = \frac{\frac{1}{4}}{\frac{1}{2}} = \frac{1}{2}$$

33. The experiment consists of drawing two cards without replacement from a deck of 52 cards.

(a) Define the event E: The first card is a heart. Define the event F: The second card is red.

$$P(\text{E}) = \frac{13}{52} = \frac{1}{4} \qquad P(F \mid E) = \frac{25}{51}$$

The probability that when two cards are drawn without replacement, the first is a heart and the second is red is

$$P(E \cap F) = P(E) \cdot P(F \mid E) = \frac{1}{4} \cdot \frac{25}{51} = \frac{25}{204}$$

(b) Define the event G: The first card is a red. Define the event H: The second card is a heart.

$$P(G) = \frac{26}{52} = \frac{1}{2} \qquad P(H \mid G) = \frac{13}{51} \cdot \frac{1}{2} + \frac{12}{51} \cdot \frac{1}{2} = \frac{25}{102}$$

The probability that when two cards are drawn without replacement, the first is a red and the second is a heart is

$$P(G \cap H) = P(G) \cdot P(H \mid G) = \frac{1}{2} \cdot \frac{25}{102} = \frac{25}{204}$$

35. The experiment consists of drawing two balls without replacement from an box containing 8 balls. Define event E: A white ball and a yellow are drawn.

The probability of choosing a white and a yellow ball without replacement can be considered as the union of two mutually exclusive events.

$$P(E) = P(\text{W on first}) \cdot P(\text{Y} \mid \text{W on first}) + P(\text{Y on first}) \cdot P(\text{W} \mid \text{Y on first})$$

$$= \frac{3}{6} \cdot \frac{1}{5} + \frac{1}{6} \cdot \frac{3}{5} = \frac{1}{10} + \frac{1}{10} = \frac{1}{5}$$

37. The experiment consists of drawing a card from a deck of 52 cards; $n(S) = 52$.

(a) Define event E to be "A red ace is drawn." $n(E) = 2$

$$P(E) = \frac{n(E)}{n(S)} = \frac{2}{52} = \frac{1}{26}$$

(b) Define event F to be "An ace is drawn." $n(F) = 4$

$$P(E \mid F) = \frac{P(E \cap F)}{P(F)} = \frac{\frac{2}{52}}{\frac{4}{52}} = \frac{1}{2}$$

(c) Define event G to be "A red card is picked." $n(G) = 26$

$$P(E|G) = \frac{P(E \cap G)}{P(G)} = \frac{\frac{2}{52}}{\frac{26}{52}} = \frac{2}{26} = \frac{1}{13}$$

39. $P(E) = 0.4$

41. $P(H) = 0.24$

43. $P(E \cap H) = 0.10$

45. $P(G \cap H) = 0.08$

47. $P(E|H) = \frac{P(E \cap H)}{P(H)} = \frac{0.10}{0.24} = \frac{5}{12}$

49. $P(G|H) = \frac{P(G \cap H)}{P(H)} = \frac{0.08}{0.24} = \frac{1}{3}$

51. $P(E|G) = \frac{P(E \cap G)}{P(G)} = \frac{180}{180+60+20} = \frac{180}{260} = \frac{9}{16}$

The probability the customer likes the deodorant given he/she is from group *I* is $\frac{9}{13}$ or 0.692.

53. $P(H|E) = \frac{P(H \cap E)}{P(E)} = \frac{110}{180+110+55} = \frac{110}{345} = \frac{22}{69}$

The probability a customer is from group *II* given he/she likes the deodorant is 0.319.

55. $P(F|G) = \frac{P(F \cap G)}{P(G)} = \frac{60}{180+60+20} = \frac{60}{260} = \frac{3}{13} \approx 0.231$

The probability a customer does not like the deodorant given he/she is from group *I* is $\frac{3}{13}$ or 0.231.

57. $P(H|F) = \frac{P(H \cap F)}{P(F)} = \frac{85}{60+85+65} = \frac{85}{210} = \frac{17}{42} \approx 0.405$

The probability a customer is from group *II* given he/she does not like the deodorant is $\frac{17}{42}$ or 0.405.

59. $P(F|I) = \frac{P(F \cap I)}{P(I)} = \frac{25}{30+25} = \frac{25}{55} = \frac{5}{11} = 0.4545$

The probability the resident is female given he/she is an Independent is approximately 0.455.

61. $P(M|D) = \frac{P(M \cap D)}{P(D)} = \frac{50}{50+60} = \frac{50}{110} = \frac{5}{11} = 0.4545$

The probability the resident is male given he/she is a Democrat is approximately 0.455.

63. $$P(M\mid R\cup I)=\frac{P(M\cap(R\cup I))}{P(R\cup I)}=\frac{P[(M\cap R)\cup(M\cap I)]}{P(R\cup I)}$$
$$=\frac{P(M\cap R)+P(M\cap I)}{P(R)+P(I)}=\frac{40+30}{40+30+30+25}=\frac{70}{125}=\frac{14}{25}=0.56$$

The probability the resident is male given he/she is either a Republican or an Independent is 0.56.

65. (a) $P(M)=\frac{1448}{2018}=\frac{724}{1009}\approx 0.7175$ (b) $P(A)=\frac{666}{2018}=\frac{333}{1009}\approx 0.3300$

(c) $P(F\cap B)=\frac{144}{2018}=\frac{72}{1009}\approx 0.0714$ (d) $P(F\mid E)=\frac{P(F\cap E)}{P(E)}=\frac{102}{526}=\frac{51}{263}\approx 0.194$

(e) $P(A\mid M)=\frac{P(A\cap M)}{P(M)}=\frac{342}{1448}=0.2362$

(f) $P(F\mid A\cup E)=\frac{P[F\cap(A\cup E)]}{P(A\cup E)}=\frac{P(F\cap A)+P(F\cap E)}{P(A)+P(E)}=\frac{324+102}{666+526}=\frac{426}{1192}=\frac{213}{593}\approx 0.357$

(g) $P(M\cap\overline{B})=P[M\cap(A\cup E)]=P(M\cap A)+P(M\cap E)=\frac{342}{2018}+\frac{424}{2016}=\frac{766}{2018}=\frac{383}{1009}\approx 0.380$

(h) $P(F\mid\overline{E})=P(F\mid A\cup B)=\frac{P[F\cap(A\cup B)]}{P(A\cup B)}=\frac{P(F\cap A)+P(F\cap B)}{P(A)+P(B)}=\frac{324+144}{666+826}=\frac{468}{1492}=\frac{117}{373}$

67. $$P(B\mid \text{Rh}+)=\frac{P(B\cap\text{Rh+})}{P(\text{Rh+})}=\frac{0.09}{0.38+0.34+0.09+0.03}=\frac{0.09}{0.84}=\frac{3}{28}\approx 0.107$$

69. $$P(\text{Rh+}\mid O)=\frac{P(\text{Rh+}\cap O)}{P(O)}=\frac{0.38}{0.38+0.07}=\frac{0.38}{0.45}=\frac{38}{45}\approx 0.844$$

71. Define event E as a household receives the mailing, and define event F as a household responds.
$$P(E\cap F)=P(E)\cdot P(F\mid E)=0.20\cdot 0.07=0.014$$

73. (a) $P(<30)=0.20+0.07+0.18=0.45$

(b) $P(>50\mid\geq \$500)=\frac{P(>50\cap\geq \$500)}{P(\geq \$500)}=\frac{0.02}{0.18+0.10+0.02}=\frac{0.02}{0.30}=\frac{1}{15}\approx 0.067$

(c) $P(\text{30-50}\mid \$100\text{-}\$499.99)=\frac{P(\text{30-50}\cap\$100\text{-}\$499.99)}{P(\$100\text{-}\$499.99)}=\frac{0.21}{0.07+0.21+0.04}=\frac{0.21}{0.32}=\frac{21}{32}\approx 0.656$

75. $$P(\text{Issue declined}\mid\text{high yield})=\frac{519}{751+519+90}=\frac{519}{1360}\approx 0.382$$

77. $P(\text{Issue made 52-week high} \mid \text{investment grade}) = \dfrac{134}{1791+991+68} = \dfrac{134}{2850} \approx 0.047$

79. Define event H as an individual has health insurance.

(a) $P(\overline{H} \mid <18) = \dfrac{P(\overline{H} \cap <18)}{P(<18)} = \dfrac{8310}{65{,}675+8310} = \dfrac{8310}{73{,}985} \approx 0.112$

(b) $P(<18 \mid \overline{H}) = \dfrac{P(<18 \cap \overline{H})}{P(\overline{H})} = \dfrac{8310}{8310+27{,}068+10{,}740+459} = \dfrac{8310}{46{,}577} \approx 0.178$

(c) $P(<18 \mid H) = \dfrac{P(<18 \cap H)}{P(H)} = \dfrac{65{,}675}{65{,}675+83{,}549+63{,}038+35{,}046} = \dfrac{65{,}675}{247{,}308} = 0.266$

81. Define event E as the student went to private school, and define event F as the student went to public school.

$P(A \cap E) = P(E) \cdot P(A \mid E) = 0.40 \cdot 0.30 = 0.12$

$$P(E \mid A) = \frac{P(A \cap E)}{P(A)} = \frac{0.12}{0.24} = \frac{1}{2} = 0.5$$

The probability the student attended a private school given the student has an A is 0.5.

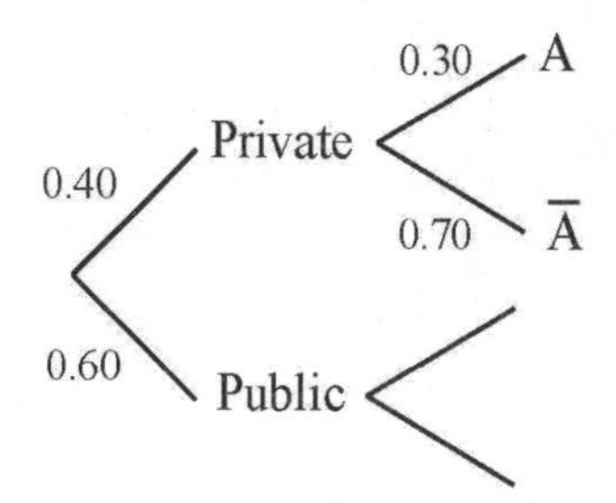

83. The sample space, S, is the set of possible outcomes when 2 dice are rolled. $n(S) = 36$

We define 3 events: W: The player wins; $n(W) = 5$ (number of ways to roll 8)
L: The player loses; $n(L) = 6$ (number of ways to roll 7)
R: The player rolls again; $n(R) = 25$ (number of ways to roll a number other than 7 or 8)

The player has already rolled an 8, so we can think of the next roll as the first roll. We want $P(W)$. This can happen on the first roll or it can happen on 2nd roll or it can happen on the 3rd roll, and so on. The probability of each individual outcome is determined using the Multiplication Principle. See the table below.

Player wins on roll	Outcome	Probability
1	W	$\frac{5}{36}$
2	RW	$\frac{25}{36} \cdot \frac{5}{36}$
3	RRW	$\frac{25}{36} \cdot \frac{25}{36} \cdot \frac{5}{36} = \left(\frac{25}{36}\right)^2 \cdot \frac{5}{36}$
4	$RRRW$	$\left(\frac{25}{36}\right)^3 \cdot \frac{5}{36}$
⋮	⋮	⋮

Each of these outcomes is mutually exclusive so we can use the addition rule.

$$P(W) = P(W) + P(R)\,P(W) + P(R)\,P(R)\,P(W) + P(R)\,P(R)\,P(R)\,P(W) + \ldots$$

This is a geometric series with $a = \frac{5}{36}$ and $r = \frac{25}{36}$. In Appendix A we learned that the sum of a geometric series is $S = a\left(\frac{1-r^n}{1-r}\right)$. If n gets larger $r^n = \left(\frac{25}{36}\right)^n$ becomes closer to 0. (Try it!) In an infinite series, as we have here, one that can continue forever, and $S = a\left(\frac{1}{1-r}\right)$. The probability of the player winning this game of craps is

$$S = \left(\frac{5}{36}\right)\left(\frac{1}{1-\frac{25}{36}}\right) = \left(\frac{5}{36}\right)\left(\frac{36}{11}\right) = \frac{5}{11} \approx 0.455$$

85. We use a tree diagram to determine the probability a person successfully completes training if no pre-test was administered.
Let events E, F, and S be defined as follow: E: A person passes the test; F: A person fails the test; and S: A person successfully completes the training.

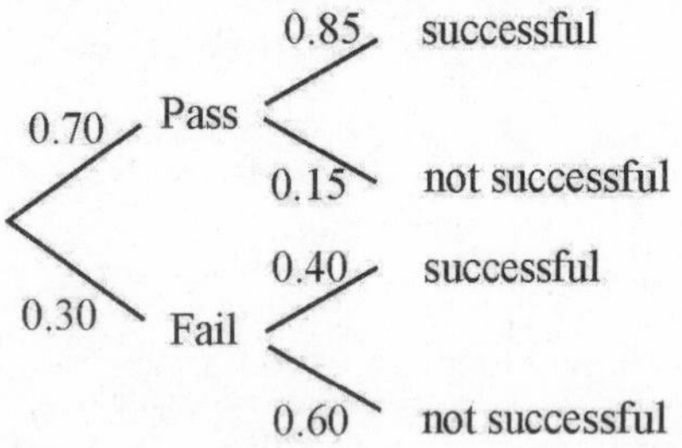

0.70 Pass
- 0.85 successful
- 0.15 not successful

0.30 Fail
- 0.40 successful
- 0.60 not successful

$$P(S) = P(S \mid E)\cdot P(E) + P(S \mid F)\cdot P(F)$$
$$= (0.85)(0.70) + (0.40)(0.30)$$
$$= 0.715$$

87. Define the events, M, F, and S as follows. M: A person is male; F: A person is female; S: A person smokes.

First we must determine the probability that a randomly selected person is male (or female).

$$P(M) = \frac{19.913}{19.913 + 21.987} = 0.475 \qquad P(F) = 1 - P(M) = 1 - 0.475 = 0.525$$

The probability a person 16 years of age or older randomly selected in Great Britain smokes is

$$P(S) = P(S \mid M)\cdot P(M) + P(S \mid F)\cdot P(F) = (0.28)(0.475) + (0.26)(0.525) = 0.270$$

89. If $P(E) \neq 0$, then $P(E \mid E) = \frac{P(E \cap E)}{P(E)}$

Since $E \cap E = E$, $P(E \cap E) = P(E)$, and $P(E \mid E) = \frac{P(E)}{P(E)} = 1$

91. $P(E \mid S) = \frac{P(E \cap S)}{P(S)}$ where S is the sample space.

From set theory, we know that $E \cap S = E$, and from the properties of probability, we know that $P(S) = 1$. So, $P(E \mid S) = P(E)$.

7.5 Independent Events

1. False. Mutually exclusive events are independent only if at least one event is the impossible event.

3.
$$\begin{aligned}P(F\cap E)&=P(E)\cdot P(F)\\&=(0.4)\cdot(0.6)\\&=0.24\end{aligned}$$

5.
$$\begin{aligned}P(E\cup F)&=P(E)+P(F)-P(E\cap F)\\&=P(E)+P(F)-P(E)P(F)\\&=P(E)+P(F)(1-P(E))\end{aligned}$$
$$\begin{aligned}P(F)&=\frac{P(E\cup F)-P(E)}{1-P(E)}\\&=\frac{0.3-0.2}{1-0.2}=\frac{0.1}{0.8}=\frac{1}{8}=0.125\end{aligned}$$

7. E and F are independent if
$$P(F\cap E)=P(E)\cdot P(F)$$
$$\frac{2}{9}\neq\left(\frac{4}{21}\right)\left(\frac{7}{12}\right)=\frac{1}{9}$$
So, the events are not independent.

9. (a) $P(E\mid F)=P(E)=0.2$

(b) $P(F\mid E)=P(F)=0.4$

(c)
$$\begin{aligned}P(E\cap F)&=P(E)\cdot P(F)\\&=(0.2)(0.4)\\&=0.08\end{aligned}$$

(d)
$$\begin{aligned}P(E\cup F)&=P(E)+P(F)-P(E\cap F)\\&=0.2+0.4-0.08\\&=0.52\end{aligned}$$

11.
$$\begin{aligned}P(E\cap F\cap G)&=P(E)\cdot P(F)\cdot P(G)\\&=\frac{2}{3}\cdot\frac{3}{7}\cdot\frac{2}{21}=\frac{4}{147}\end{aligned}$$

13.
$$\begin{aligned}P(E\cap F)&=P(E)+P(F)-P(E\cup F)\\&=0.3+0.2-0.4=0.1\end{aligned}$$
$$P(E\mid F)=\frac{P(F\cap E)}{P(F)}=\frac{0.1}{0.2}=\frac{1}{2}$$

$P(E)\cdot P(F)=(0.3)(0.2)=0.06\neq P(E\cap F)$

E and F are not independent.

15. The sample space is the set of all possible outcomes. Let R stand for the mouse goes right and L stand for the mouse goes left. $S = \{RRR, RRL, RLR, LRR, RLL, LRL, LLR, LLL\}$

On the first two runs $P(L) = P(R) = \frac{1}{2}$, and we are told that on the third try the mouse is twice as likely to choose L. From this we get $P(L) = \frac{2}{3}$ and $P(R) = \frac{1}{3}$.

A Venn diagram and the Multiplication Principle will help to assign the probabilities.

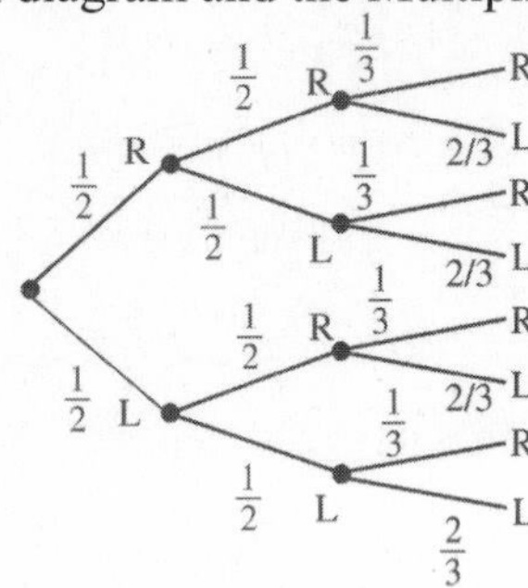

$P(RRR) = \frac{1}{12}$ $P(RLL) = \frac{2}{12}$

$P(RRL) = \frac{2}{12}$ $P(LRL) = \frac{2}{12}$

$P(RLR) = \frac{1}{12}$ $P(LLR) = \frac{1}{12}$

$P(LRR) = \frac{1}{12}$ $P(LLL) = \frac{2}{12}$

(a) $P(E) = P(RRL \cup LRR)$
$= P\,P(RRL) + P(LRR)$
$= \frac{2}{12} + \frac{1}{12} = \frac{3}{12} = \frac{1}{4}$

(b) $P(F) = P(LLL) = \frac{2}{12} = \frac{1}{6}$

(c) $P(G) = P(LRR \cup LRL \cup LLR \cup LLL)$
$= P(LRR) + P(LRL) + P(LLR) + P(LLL)$
$= \frac{1}{12} + \frac{2}{12} + \frac{1}{12} + \frac{2}{12} = \frac{6}{12} = \frac{1}{2}$

(d) $P(H) = P(RRR \cup RRL \cup LRR \cup LRL)$
$= P(RRR) + P(RRL) + P(LRR) + P(LRL)$
$= \frac{1}{12} + \frac{2}{12} + \frac{1}{12} + \frac{2}{12} = \frac{6}{12} = \frac{1}{2}$

17. Two marbles are chosen with replacement.
Define the events R: The marble chosen is red; $P(R) = 0.60$
W: The marble chosen is white; $P(W) = 0.40$
The events are independent.

(a) Probability both marbles are red is
$P(R \cap R) = P(R) \cdot P(R) = 0.60 \cdot 0.60 = 0.36$

(b) Probability exactly 1 of the marbles is red is
$P(R \cap W) + P(W \cap R) = P(R) \cdot P(W) + P(W) \cdot P(R) = 0.60 \cdot 0.40 + 0.40 \cdot 0.60 = 0.48$

19. Let H denote a child with heart disease, and $\overline{H}$ denote a child with no heart disease. The sample space is the set of all possible outcomes. The couple has two children. $S = \{HH, H\overline{H}, \overline{H}H, \overline{HH}\}$

$$P(H) = \frac{3}{4} \qquad P(\overline{H}) = \frac{1}{4}$$

(a) $P(HH) = P(H \cap H) = P(H) \cdot P(H) = \frac{3}{4} \cdot \frac{3}{4} = \frac{9}{16}$

(b) $P(\overline{HH}) = P(\overline{H} \cap \overline{H}) = P(\overline{H}) \cdot P(\overline{H}) = \frac{1}{4} \cdot \frac{1}{4} = \frac{1}{16}$

(c) $P(H\overline{H} \cup \overline{H}H) = P(H \cap \overline{H}) + P(\overline{H} \cap H) = P(H) \cdot P(\overline{H}) + P(\overline{H}) \cdot P(H) = \frac{3}{4} \cdot \frac{1}{4} + \frac{1}{4} \cdot \frac{3}{4} = \frac{6}{16} = \frac{3}{8}$

21. Define the events G: Baby born is a girl; $P(G) = 0.49$.
The events are independent, so the probability of all 4 babies born being girls is

$$P(GGGG) = P(G) \cdot P(G) \cdot P(G) \cdot P(G) = 0.49^4 = 0.0576$$

23. Define the events G: The seed germinates and V: The seed chosen is a violet seed.

$$P(G) = 0.60 \qquad P(V) = \frac{1}{3}$$

The events are independent, so the probability a planted seed will grow into a violet is

$$P(G \cap V) = P(G) \cdot P(V) = 0.60 \cdot \frac{1}{3} = 0.20$$

25. (a) Because the inspectors are independent, the probability both inspectors miss a defective piece of furniture is $0.20^2 = 0.04$.

(b) To find the number of inspectors needed to assure that the probability of missing a defect is less than 0.01, solve the equation

$$0.20^n = 0.01$$

Write the exponential equation in logarithmic form and then use a change of base formula.

$$n = \log_{0.20} 0.01 = \frac{\log 0.01}{\log 0.20} = 2.861$$

Efraim Furniture Company should hire 3 inspectors to insure the probability of a missed defect is less than 0.01

27. Since the probability that the two stocks increase in value is independent, the probability both stocks improve is $0.60^2 = 0.36$.
The probability that at least one stock will not increase in value is

$$1 - \text{probability both stocks increase in value} = 1 - 0.36 = 0.64$$

29. (a) $P(< 18 \text{ years old}) = \dfrac{65{,}675 + 8310}{65{,}675 + 83{,}549 + 63{,}038 + 8{,}310 + 27{,}068 + 10{,}740 + 459}$

$= \dfrac{73{,}985}{293{,}885} \approx 0.252$

$P(< 18 \text{ years old} \mid \text{no health insurance}) = \dfrac{8310}{8310 + 27{,}068 + 10{,}740 + 459} = \dfrac{8310}{46{,}577} \approx 0.178$

(b) The events are not independent. If they were independent then P(< 18 years old | no health insurance) and P(< 18 years old) would be equal.

31. (a) The failures are not mutually exclusive because $P(\text{both pumps fail}) \neq 0$.

(b) The probability at least one pump fails is

$$1 - P(\text{neither pump fails}) = 1 - 0.95^2 = 0.0975$$

(c) The failures in the pumps are independent because

$$P(\text{one pump fails}) \cdot P(\text{other pump fails}) = P(\text{both pumps fail})$$

$$0.05 \cdot 0.05 = 0.0025$$

33. (a) $P(\text{burglarized} \mid \text{suburban}) = \dfrac{30}{146} = \dfrac{15}{73} \approx 0.205$

(b) $P(\text{rural} \mid \text{vehicle theft}) = \dfrac{4}{28} = \dfrac{1}{7} \approx 0.143$

(c) If rural and vehicle theft are independent then P(rural | vehicle theft) = P(rural).

$$P(\text{rural}) = \frac{131}{500} = 0.262 \neq P(\text{rural} \mid \text{vehicle theft})$$

The events are not independent.

(d) The events "Urban" and "Burglary" are independent if

$$P(\text{urban}) \cdot P(\text{burglary}) = P(\text{urban} \cap \text{burglary})$$

$P(\text{urban}) \cdot P(\text{burglary}) = \dfrac{220}{500} \cdot \dfrac{100}{500} = \dfrac{11}{125} = 0.088$; $P(\text{urban} \cap \text{burglary}) = \dfrac{44}{500} = \dfrac{11}{125} = 0.088$

The events are independent.

35. Let V: The first voter votes for the candidate, and W: The second voter votes for the candidate.

$$P(V) = P(W) = \frac{2}{3}$$

(a) $P(V \cap W) = P(V)P(W) = \dfrac{2}{3} \cdot \dfrac{2}{3} = \dfrac{4}{9}$ (b) $P(\overline{V} \cap \overline{W}) = P(\overline{V})P(\overline{W}) = \dfrac{1}{3} \cdot \dfrac{1}{3} = \dfrac{1}{9}$

(c) $P(V\overline{W} \cup \overline{V}W) = P(V\overline{W}) + P(\overline{V}W)$

$= \dfrac{2}{3} \cdot \dfrac{1}{3} + \dfrac{1}{3} \cdot \dfrac{2}{3} = \dfrac{4}{9}$

37. (a) The probability of obtaining at least one 1 in four throws of a die is the complement of throwing no 1s. Define event E as a one is thrown.

$$P(\overline{E}) = \frac{5}{6}; \quad \text{The probability of no 1s is } P(\overline{E}\ \overline{E}\ \overline{E}\ \overline{E}) = \left(\frac{5}{6}\right)^4$$

The probability at least one 1 is $= 1 - \left(\frac{5}{6}\right)^4 = 0.518$

(b) The probability of throwing a pair of 1s in a throw of 2 dice is $\frac{1}{36}$.

The probability of obtaining at least one pair of 1s in 24 throws of a pair of dice is the complement of obtaining no pairs.

$P(\overline{E}) = \frac{35}{36}$; The probability of no pairs of 1s is $P(\overline{E}_1\ \overline{E}_2\ \overline{E}_3 \ldots \overline{E}_{24}) = \left(\frac{35}{36}\right)^{24}$

The probability of at least one pair of 1s in 24 throws is $= 1 - \left(\frac{35}{36}\right)^{24} = 0.491$

The first event is more likely.

39. E is any event; F is the impossible event. That means $P(F) = 0$ and that $F = \emptyset$. Two events are independent if $P(E \cap F) = P(E)P(F)$.

$$P(E \cap F) = P(E \cap \emptyset) = P(\emptyset) = 0$$

We also have $P(E)P(F) = P(E) \cdot 0 = 0$.
Since both $P(E \cap F) = 0$ and $P(E)P(F) = 0$, they are equal to each other, showing that the events are independent.

41. E and F are independent events, and $P(E) \neq 0$ and $P(F) \neq 0$.

Assume E and F mutually exclusive, which makes $P(E \cap F) = 0$.
Also,

$P(E \cap F) = P(E)P(F)$ (The events are independent.)

and $P(E)P(F) > 0$ (The product of two non-negative, non-zero numbers is positive.)

but this is a contradiction, so E and F cannot be mutually exclusive.

43. To prove formula (1) we must look at 2 parts:
1. we assume E and F are independent events and show $P(E \cap F) = P(E) \cdot P(F)$, and
2. we assume $P(E \cap F) = P(E) \cdot P(F)$ and show E and F are independent events.

1. If E and F are independent events, then $P(E|F) = P(E)$.

   By definition of conditional probability, $P(E|F) = \frac{P(E \cap F)}{P(F)}$, so $\frac{P(E \cap F)}{P(F)} = P(E)$

   or $P(E \cap F) = P(E) \cdot P(F)$.

2. If we know $P(E \cap F) = P(E) \cdot P(F)$, then $P(E|F) = \frac{P(E) \cdot P(F)}{P(F)}$ or $P(E|F) = P(E)$ which is the definition of independent events.

Chapter 7 Review

1. $S = \{0, 1, 2, 3, 4, 5\}$

3. $S = \{BB, BG, GB, GG\}$

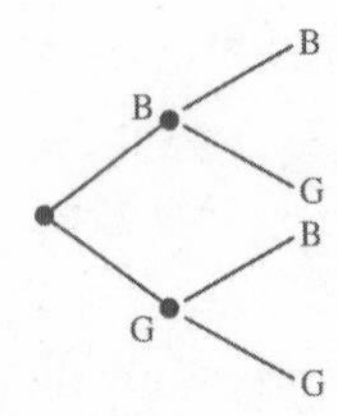

5. Let P represent penny, Q represent quarter, and D represent dime. $S = \{P, Q, D\}$

$$P(P) = \frac{4}{15} \quad P(Q) = \frac{2}{5} \quad P(D) = \frac{1}{3}$$

7. $S = \{1, 2, 3, 4, 5, 6\}$

Let x denote the probability a 1 occurs.

$$P(1) = P(3) = P(4) = P(6) = x$$
$$P(2) = P(5) = 2x$$

$$P(1) + P(2) + P(3) + P(4) + P(5) + P(6) = 1$$
$$x + 2x + x + x + 2x + x = 1$$
$$8x = 1$$
$$x = \frac{1}{8}$$

Assign the probabilities

$$P(2) = P(5) = \frac{2}{8} = \frac{1}{4}$$
$$P(1) = P(3) = P(4) = P(6) = \frac{1}{8}$$

9. (a) The sample space is the number of girls in the family. $S = \{0, 1, 2, 3, 4\}$
The simple event
0 occurs when there is no girl, BBBB.
1 occurs when there is 1 girl, GBBB, BGBB, BBGB, BBBG.
2 occurs when there are 2 girls, GGBB, GBGB, GBBG, BGGB, BGBG, BBGG.
3 occurs when there are 3 girls, GGGB, GGBG, GBGG, BGGG.
4 occurs when all 4 children are girls, GGGG.

We assign valid probabilities to each of these events, assuming it is equally likely for a child to be born G or B.

$$P(0) = \frac{1}{16}, \quad P(1) = \frac{4}{16} = \frac{1}{4}, \quad P(2) = \frac{6}{16} = \frac{3}{8}, \quad P(3) = \frac{4}{16} = \frac{1}{4}, \quad P(4) = \frac{1}{16}$$

(b) *i.* $P(0) = \frac{1}{16}$ *ii.* $P(2) = \frac{3}{8}$

iii. $P(1) + P(2) + P(3) = \frac{1}{4} + \frac{3}{8} + \frac{1}{4} = \frac{7}{8}$

iv. $1 - P(4) = 1 - \frac{1}{16} = \frac{15}{16}$

11. Let W denote a white marble is chosen, Y denote a yellow marble is chosen, R denote a red marble is chosen, and B denote a blue marble is chosen.

(a) $P(\text{BB}) = P(\text{B on } 1^{\text{st}})P(\text{B on } 2^{\text{nd}}) = \left(\frac{5}{14}\right)\left(\frac{4}{13}\right) = \frac{10}{91}$

(b) The probability exactly one is blue is the union of two mutually exclusive events. The probability is

$$P(\text{B}\overline{\text{B}} \cup \overline{\text{B}}\text{B}) = \left(\frac{5}{14}\right)\left(\frac{9}{13}\right) + \left(\frac{9}{14}\right)\left(\frac{5}{13}\right) = \frac{45}{91}$$

(c) The probability at least one is blue is the union of both are blue and exactly one is blue. Using the results from parts a and b,

$$P(\text{at least one marble is blue}) = \frac{10}{91} + \frac{45}{91} = \frac{55}{91}$$

13. (a) $P(\text{A} \cup B) = P(A) + P(B) - P(A \cap B)$
$= 0.3 + 0.5 - 0.2$
$= 0.6$

(b) $P(\overline{A}) = 1 - A$
$= 1 - 0.3$
$= 0.7$

(c) $P(\overline{A \cup B}) = 1 - P(\text{A} \cup B)$
$= 1 - 0.6$
$= 0.4$

(d) $P(\overline{A} \cup \overline{B}) = 1 - P(A \cap B)$
$= 1 - 0.2$
$= 0.8$

15. (a) $P(\text{head}) = \frac{243}{300} = \frac{81}{100} = 0.81$

(b) $P(\text{tail}) = \frac{57}{300} = \frac{19}{100} = 0.19$

17. (a) $P(\overline{F}) = 1 - P(F) = 1 - 0.5 = 0.5$

(b) $P(E \cup F) = P(E) + P(F) - P(E \cap F)$
$= 0.63 + 0.50 - 0.35 = 0.78$

(c) E and F are not mutually exclusive. If they were, $P(E \cap F)$ would equal 0, but in this problem we are told that $P(E \cap F) = 0.35$.

19. (a) $P(E \cup F) = P(E) + P(F) - P(E \cap F)$
$= 0.2 + 0.6 - 0.1 = 0.7$

(b) $P(\overline{E}) = 1 - P(E) = 1 - 0.2 = 0.8$

(c) $P(\overline{E \cap F}) = 1 - P(E \cap F) = 1 - 0.1 = 0.9$

21. (a) $P(\overline{E}) = 1 - P(E) = 1 - 0.25 = 0.75$

(b) $P(\overline{F}) = 1 - P(F) = 1 - 0.3 = 0.7$

(c) $P(E \cap F) = P(E) + P(F) - P(E \cup F)$
$= 0.25 + 0.3 - 0.55 = 0$

(d) $P(\overline{E \cap F}) = 1 - P(E \cap F) = 1 - 0 = 1$

(e) $P(\overline{E} \cap \overline{F}) = P(\overline{E \cup F}) = 1 - P(E \cup F)$
$= 1 - 0.55 = 0.45$

(f) $P(\overline{E} \cup \overline{F}) = P(\overline{E \cap F}) = 1 - P(E \cap F)$
$= 1 - 0 = 1$

23. (a) The events are not equally likely.
(b) The outcome 0 has the highest probability.
(c) $P(F) = P(3, 3) \cdot \left(\frac{4}{8}\right) \cdot \left(\frac{3}{8}\right) \cdot \left(\frac{1}{8}\right) = 6 \cdot \left(\frac{4}{8}\right) \cdot \left(\frac{3}{8}\right) \cdot \left(\frac{1}{8}\right) = \frac{9}{64}$

25. Define the event E: A family has 4 boys. $P(E)=\dfrac{1}{16}$

The odds in favor of E are defined as the ratio $\dfrac{P(E)}{P(\overline{E})}$.

The odds in favor of a family with 4 children to have 4 boys is $\dfrac{1}{15}$ or 1 to 15.

27. The odds for the Giants winning are 5:3.

The probability the Giants win is $\dfrac{5}{5+3}=\dfrac{5}{8}$.

The probability the Giants lose is $\dfrac{3}{5+3}=\dfrac{3}{8}$.

29. Since a 6 is 3 times more likely to appear than any other number, the probabilities assigned to each of the outcomes in the sample space are

$$P(1)=P(2)=P(3)=P(4)=P(5)=\frac{1}{8};\ P(6)=\frac{3}{8}$$

$E=\{(3,1),(3,2),(3,3),(3,4),(3,5),(3,6)\}$
$F=\{(1,6),(2,6),(3,6),(4,6),(5,6),(6,6)\}$
$E\cap F=\{(3,6)\};\ P(E\cap F)=\dfrac{3}{64}$.

$P(E)\cdot P(F)=\dfrac{3}{64}$ the events are independent.

31. We define events E: a client exercises regularly, and F: a client eats a healthy diet.

$$P(E)=0.42;\ P(F)=0.51;\text{ and } P(E\cap F)=0.30.$$

(a) $P(F\mid E)=\dfrac{P(E\cap F)}{P(E)}=\dfrac{0.30}{0.42}=\dfrac{5}{7}\approx 0.714$ (b) $P(E\mid F)=\dfrac{P(E\cap F)}{P(F)}=\dfrac{0.30}{0.51}=\dfrac{10}{17}\approx 0.588$

(c) $P(E\cup F)=P(E)+P(F)-P(E\cap F)=0.42+0.51-0.30=0.63$
The probability a client exercises regularly or eats a healthy diet is 0.63.

33. Define events E: a person has blue eyes, F: a person has brown eyes, and G: a person is left handed.

$$P(E)=0.25 \text{ and } P(F)=0.75 \text{ Also, } P(G\mid E)=0.10 \text{ and } P(G\mid F)=0.05.$$

(a) $P(E\cap G)=P(G\mid E)\cdot P(E)=(0.10)\cdot(0.25)=0.025$
The probability a person is blue-eyed and left handed is 0.025.

(b) Notice that $P(E)+P(F)=0.25+0.75=1.00$, this indicates that no other eye-color is possible.

$$\begin{aligned}P(G)&=P(G\cap E)+P(G\cap F)\\&=0.025+P(G\mid F)\cdot P(F)\\&=0.025+0.05\cdot 0.75\\&=0.0625\end{aligned}$$

The probability a person is left handed is 0.0625.

(c) $P(E\mid G)=\dfrac{P(E\cap G)}{P(G)}=\dfrac{0.025}{0.0625}=\dfrac{2}{5}$.
The probability a person is blue-eyed given the person is left handed is 0.40.

35. Define events E: a student took Form A, F: a student took Form B, G: a student scored over 80%, H: a student scored under 80%.

(a) $P(E \mid G) = \dfrac{P(E \cap G)}{P(G)} = \dfrac{8}{20} = \dfrac{2}{5} = 0.40$

The probability that a student who scored over 80% took form A is $\dfrac{2}{5}$.

(b) $P(G \mid E) = \dfrac{P(E \cap G)}{P(E)} = \dfrac{8}{40} = \dfrac{1}{5} = 0.20$

The probability that a student who took form A scored over 80% is $\dfrac{1}{5}$.

(c) To show G and E are independent we need to show $P(E \cap G) = P(E) \cdot P(G)$.

$$P(E \cap G) = \frac{8}{100} = 0.8; \quad P(E) = \frac{40}{100} = 0.4; \quad \text{and } P(G) = \frac{20}{100} = 0.2$$

$P(E) \cdot P(G) = \dfrac{40}{100} \cdot \dfrac{20}{100} = \dfrac{8}{100} = 0.8 = P(E \cap G)$ So G and E are independent events.

(d) To determine if G and F are independent events we must check if $P(F \cap G) = P(F) \cdot P(G)$.

$$P(F \cap G) = \frac{12}{100} = 0.12; \quad (F) = \frac{60}{100} = 0.6; \quad \text{and } \; P(G) = \frac{20}{100} = 0.2$$

$P(F) \cdot P(G) = \dfrac{60}{100} \cdot \dfrac{20}{100} = \dfrac{12}{100} = 0.12 = P(F \cap G)$ So G and F are independent events.

37. The probability at least one person gets the correct letter is the complement of the probability everyone gets an incorrect letter.

Define event E: everyone get an incorrect letter.

$$P(E) = \frac{n(E)}{n(S)} = \frac{2}{3!} = \frac{1}{3}.$$

So the probability at least one person gets the correct letter is $1 - P(E) = 1 - \dfrac{1}{3} = \dfrac{2}{3}$.

39. The number of elements in the sample space S is equal to the number of ways 4 jars can be chosen from 72 jars of jam, or $C(72, 4) = 1{,}028{,}790$.

(a) Define E as the event, "All 4 jars are underweight." E can occur $C(10, 4) = 210$ ways.

$$P(E) = \frac{C(10,\ 4)}{C(72,\ 4)} = \frac{210}{1{,}028{,}790} = 0.0002$$

(b) Define F as the event, "2 jars are underweight." F can occur $C(10, 2) \cdot C(62, 2) = 85{,}095$ ways.

$$P(F) = \frac{C(10,\ 2) \cdot C(62,\ 2)}{C(72,\ 4)} = \frac{85{,}095}{1{,}028{,}790} = 0.083$$

(c) Define G as the event, "At most 1 jar is underweight." The event G is equivalent to selecting either 0 or 1 underweight jars. Since the events are mutually exclusive, the sum of their probabilities will give the probability of G.

$$\begin{aligned} P(G) &= P(\text{No underweight jars}) + P(1 \text{ underweight jar}) \\ &= \frac{C(10,\ 0)\cdot C(62,\ 4)}{C(72,\ 4)} + \frac{C(10,\ 1)\cdot C(62,\ 3)}{C(72,\ 4)} \\ &= 0.5422 + 0.3676 = 0.910 \end{aligned}$$

41. Define E as the event, "Each person has a different birth day of the month."

$n(S) = 31^{15}$ and $n(E) = P(31,\ 15)$, so $P(E) = \dfrac{P(31,\ 15)}{31^{15}} = 0.017$

The probability that each person in a room of 15 people have birthdays on different days of the month is 0.017.

43. E and F are independent events,

$$P(E \mid F) = \frac{P(E \cap F)}{P(F)} = \frac{0.2}{0.4} = \frac{1}{2} = 0.5$$

45. Define E: the basketball player makes a free throw. $P(E) = 0.7$

The probability the player misses a free throw, $P(\overline{E}) = 1 - P(E) = 1 - 0.7 = 0.3$.

(a) The probability the player misses a free throw and then makes 3 in a row is

$$P(\overline{E} \cap E \cap E \cap E) = P(\overline{E})\cdot P(E)\cdot P(E)\cdot P(E) = 0.3 \cdot 0.7^3 = 0.1029$$

(b) The probability of probability the player makes 10 free throws in a row is

$$[P(E)]^{10} = (0.7)^{10} = 0.0282$$

47. Define E: the car is black, F: the car is red, G: the interior is tan.
We make a table to organize the data.

	Black Car	Red Car	Total
Tan Interior	2	6	8
Other Interior	6	6	12
Total	8	12	20

$$P(E \mid G) = \frac{P(E \cap G)}{P(G)} = \frac{n(E \cap G)}{n(G)} = \frac{2}{8} = \frac{1}{4}$$

Chapter 7 Project

1. $P(E \mid H)$ is the probability a person who tossed a head is truly overweight. This probability will give the proportion of overweight students in the school.

3. $P(E \mid T) = 1$ Since respondent was instructed to answer "Yes" if a coin toss resulted in a tail. (No one who flipped a tail answered, "No.")

$P(E) = P(H) \cdot P(E \mid H) + P(T) \cdot P(E \mid T)$

$$P(E \mid H) = \frac{P(E) - P(T)\cdot P(E \mid T)}{P(H)}$$

5. $P(A \mid H) = \dfrac{P(A \cap H)}{P(H)}$

7. $P(C \mid H) = \dfrac{P(C \cap H)}{P(H)}$

Mathematical Questions from Professional Exams

1. (b) $P \cap Q = \varnothing$ means $P(P \cap Q) = 0$, but we were told $P(P) > 0$ and $P(Q) > 0$ so $P(P) \cdot P(Q) > 0$ which indicates P and Q are not independent.

3. (b) $\left(\dfrac{26}{52}\right)\left(\dfrac{25}{51}\right)\left(\dfrac{24}{50}\right) = \dfrac{2}{17}$

5. (b) The probability that all are not the same is the complement of the probability that the same face shows each time.

$$\begin{aligned} P(\text{ at least one different}) = 1 - P(\text{all faces the same}) &> 0.999 \\ P(\text{all faces the same}) &< 0.001 \\ \left(\frac{1}{6}\right)^n &< 0.001 \\ n \log\left(\frac{1}{6}\right) &< \log(.001) \\ n &> \frac{\log(0.001)}{\log\left(\frac{1}{6}\right)} > 3.9 \end{aligned}$$

7. (b) Define E: the first roll is an even number
F: the sum of the rolls is 8

$$P(F \mid E) = \frac{P(E \cap F)}{P(E)} = \frac{\frac{3}{36}}{\frac{1}{2}} = \frac{1}{6}$$

9. (c) $P(S \cap T) = P\left(S \cap \overline{T}\right) = P\left(\overline{S} \cap T\right) = p$

$$\begin{aligned} P(S \cup T) &= P(S) + P(T) - P(S \cap T) \\ &= [P\left(S \cap \overline{T}\right) + P(S \cap T)] + [P\left(\overline{S} \cap T\right) + P(S \cap T)] - P(S \cap T) \\ &= p + p + p + p - p = 3p \end{aligned}$$

Chapter 8

Additional Probability Topics

8.1 Bayes' Theorem

3. Partition

5. $P(E \mid A) = 0.4$

5. $P(E \mid A) = 0.4$

9. $P(E \mid C) = 0.7$

11.
$$\begin{aligned} P(E) &= P(E \cap A) + P(E \cap B) + P(E \cap C) \\ &= (0.4)(0.3) + (0.2)(0.6) + (0.7)(0.1) \\ &= 0.31 \end{aligned}$$

13. $$P(A \mid E) = \frac{P(A \cap E)}{P(E)} = \frac{P(E \mid A) \cdot P(A)}{P(E)} = \frac{(0.4)(0.3)}{0.31} = \frac{12}{31} = 0.387$$

15. $$P(C \mid E) = \frac{P(C \cap E)}{P(E)} = \frac{P(E \mid C) \cdot P(C)}{P(E)} = \frac{(0.7)(0.1)}{0.31} = \frac{7}{31} = 0.226$$

17. $$P(B \mid E) = \frac{P(B \cap E)}{P(E)} = \frac{P(E \mid B) \cdot P(B)}{P(E)} = \frac{(0.2)(0.6)}{0.31} = \frac{12}{31} = 0.387$$

19.

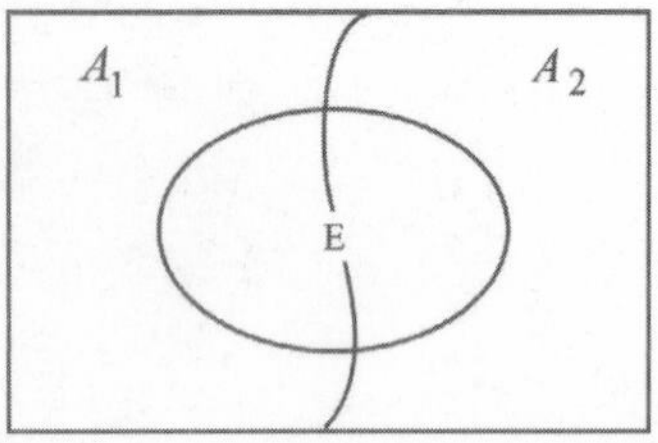

$$\begin{aligned} P(E) &= P(E \cap A_1) + P(E \cap A_2) \\ &= P(A_1) \cdot P(E \mid A_1) + P(A_2) \cdot P(E \mid A_2) \\ &= (0.4) \cdot (0.03) + (0.6) \cdot (0.02) \\ &= 0.024 \end{aligned}$$

21. $$\begin{aligned}P(E) &= P(E \cap A_1) + P(E \cap A_2) + P(E \cap A_3)\\ &= P(A_1) \cdot P(E \mid A_1) + P(A_2) \cdot P(E \mid A_2) + P(A_3) \cdot P(E \mid A_3)\\ &= (0.6) \cdot (0.01) + (0.2) \cdot (0.03) + (0.2) \cdot (0.02)\\ &= 0.016\end{aligned}$$

23. $$P(A_1 \mid E) = \frac{P(A_1 \cap E)}{P(E)} = \frac{P(A_1) \cdot P(E \mid A_1)}{P(E)} = \frac{(0.4)(0.03)}{0.024} = \frac{12}{24} = \frac{1}{2} = 0.5$$

$$P(A_2 \mid E) = \frac{P(A_2 \cap E)}{P(E)} = \frac{P(A_2) \cdot P(E \mid A_2)}{P(E)} = \frac{(0.6)(0.02)}{0.024} = \frac{12}{24} = \frac{1}{2} = 0.5$$

25. $$P(A_1 \mid E) = \frac{P(A_1 \cap E)}{P(E)} = \frac{P(A_1) \cdot P(E \mid A_1)}{P(E)} = \frac{(0.6)(0.01)}{0.016} = \frac{6}{16} = \frac{3}{8} = 0.375$$

$$P(A_2 \mid E) = \frac{P(A_2 \cap E)}{P(E)} = \frac{P(A_2) \cdot P(E \mid A_2)}{P(E)} = \frac{(0.2)(0.03)}{0.016} = \frac{6}{16} = \frac{3}{8} = 0.375$$

$$P(A_3 \mid E) = \frac{P(A_3 \cap E)}{P(E)} = \frac{P(A_3) \cdot P(E \mid A_3)}{P(E)} = \frac{(0.2)(0.02)}{0.016} = \frac{4}{16} = \frac{1}{4} = 0.250$$

27. Define the events: U_1: Jar 1 is selected, U_2: Jar 2 is selected, U_3: Jar 3 is selected

Since the jar is selected at random, $P(U_1) = P(U_2) = P(U_3) = \frac{1}{3}$.

(a) The probability the ball is red, $P(R)$, is

$$\begin{aligned}P(R) &= P(R \cap U_1) + P(R \cap U_2) + P(R \cap U_3)\\ &= P(U_1) \cdot P(R \mid U_1) + P(U_2) \cdot P(R \mid U_2) + P(U_3) \cdot P(R \mid U_3)\\ &= \left(\frac{1}{3}\right) \cdot \left(\frac{5}{16}\right) + \left(\frac{1}{3}\right) \cdot \left(\frac{3}{16}\right) + \left(\frac{1}{3}\right) \cdot \left(\frac{7}{16}\right) = \frac{15}{48} = \frac{5}{16} = 0.3125\end{aligned}$$

(b) The probability the ball is white, $P(W)$, is

$$\begin{aligned}P(W) &= P(W \cap U_1) + P(W \cap U_2) + P(W \cap U_3)\\ &= P(U_1) \cdot P(W \mid U_1) + P(U_2) \cdot P(W \mid U_2) + P(U_3) \cdot P(W \mid U_3)\\ &= \left(\frac{1}{3}\right) \cdot \left(\frac{6}{16}\right) + \left(\frac{1}{3}\right) \cdot \left(\frac{4}{16}\right) + \left(\frac{1}{3}\right) \cdot \left(\frac{5}{16}\right) = \frac{15}{48} = \frac{5}{16} = 0.3125\end{aligned}$$

(c) The probability the ball is blue, $P(B)$, is

$$\begin{aligned}P(B) &= P(B \cap U_1) + P(B \cap U_2) + P(B \cap U_3)\\ &= P(U_1) \cdot P(B \mid U_1) + P(U_2) \cdot P(B \mid U_2) + P(U_3) \cdot P(B \mid U_3)\\ &= \left(\frac{1}{3}\right) \cdot \left(\frac{5}{16}\right) + \left(\frac{1}{3}\right) \cdot \left(\frac{9}{16}\right) + \left(\frac{1}{3}\right) \cdot \left(\frac{4}{16}\right) = \frac{18}{48} = \frac{3}{8} = 0.375\end{aligned}$$

(d) $$P(U_1 \mid R) = \frac{P(U_1 \cap R)}{P(R)} = \frac{P(U_1) \cdot P(R \mid U_1)}{P(R)} = \frac{\left(\frac{1}{3}\right)\left(\frac{5}{16}\right)}{\left(\frac{5}{16}\right)} = \frac{1}{3} = 0.3333$$

(e) $P(U_2 \mid B) = \dfrac{P(U_2 \cap B)}{P(B)} = \dfrac{P(U_2)\cdot P(B \mid U_2)}{P(B)} = \dfrac{\left(\frac{1}{3}\right)\left(\frac{9}{16}\right)}{\left(\frac{3}{8}\right)} = \dfrac{1}{2} = 0.50$

(f) $P(U_3 \mid W) = \dfrac{P(U_3 \cap W)}{P(W)} = \dfrac{P(U_3)\cdot P(W \mid U_3)}{P(W)} = \dfrac{\left(\frac{1}{3}\right)\left(\frac{5}{16}\right)}{\left(\frac{5}{16}\right)} = \dfrac{1}{3} = 0.3333$

29. Define the events A_1: a person is male, A_2: a person is female, and E: a person is color-blind.

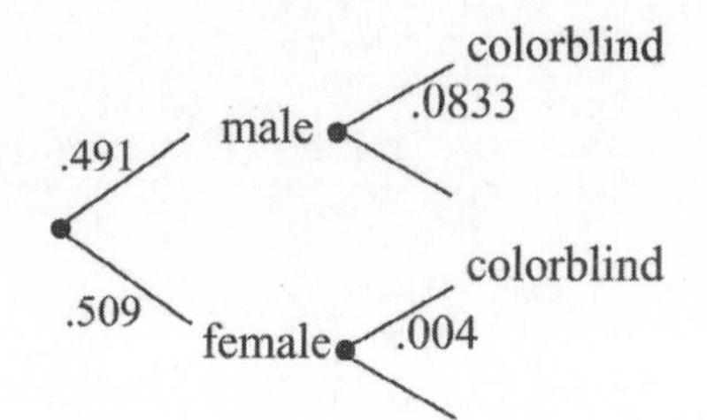

$P(A_1) = 0.491 \quad P(A_2) = 0.509 \quad P(E \mid A_1) = \dfrac{1}{12} = 0.0833$

$P(E \mid A_2) = 0.004$

$$\begin{aligned} P(E) &= P(E \cap A_1) + P(E \cap A_2) \\ &= P(A_1)\cdot P(E \mid A_1) + P(A_2)\cdot P(E \mid A_2) \\ &= 0.491 \cdot (0.0833) + 0.509 \cdot (0.004) \\ &= 0.043 \end{aligned}$$

$$P(A_1 \mid E) = \frac{P(A_1 \cap E)}{P(E)} = \frac{P(A_1)\cdot P(E \mid A_1)}{P(E)} = \frac{(0.491)(0.08333)}{0.04294} = 0.953$$

The probability that a color-blind person is male is 0.953.

31. Define the events E: the recipient sailed on Castaway before, M: the recipient is male, F: the recipient is female.

$$P(M) = \frac{6500}{10{,}000} = 0.65 \quad P(F) = \frac{3500}{10{,}000} = 0.35 \quad P(E \mid M) = 0.8 \quad P(E \mid F) = 0.4$$

(a) Since we are looking for the probability a recipient who books a cruise had never sailed before, we need $\overline{E}$. Before finding $P(\overline{E})$, we note

$$P(\overline{E} \mid M) = 1 - P(E \mid M) = 1 - 0.8 = 0.2 \text{ and } P(\overline{E} \mid F) = 1 - P(E \mid F) = 1 - 0.4 = 0.6$$

$$\begin{aligned} P(\overline{E}) = P(\overline{E} \cap M) + P(\overline{E} \cap F) &= P(M)\cdot P(\overline{E} \mid M) + P(F)\cdot P(\overline{E} \mid F) \\ &= 0.65 \cdot 0.2 + 0.35 \cdot 0.6 = 0.34 \end{aligned}$$

(b) $P(M \mid \overline{E}) = \dfrac{P(M \cap \overline{E})}{P(\overline{E})} = \dfrac{P(M)\cdot P(\overline{E} \mid M)}{P(\overline{E})} = \dfrac{0.65 \cdot 0.2}{0.34} = \dfrac{13}{34} \approx 0.38$

33. Define the events *E*: the plane is late; *S*: she is on Southwest.

$$P(S)=0.7 \qquad P(\overline{S})=0.3$$

$$P(E\mid S)=1-P(\overline{E}\mid S)=1-0.77=0.23 \qquad P(E\mid \overline{S})=1-P(\overline{E}\mid \overline{S})=1-0.72=0.28$$

$$P(E)=P(E\cap S)+P(E\cap \overline{S})=P(S)\cdot P(E\mid S)+P(\overline{S})\cdot P(E\mid \overline{S})$$
$$=0.7\cdot 0.23+0.3\cdot 0.28=0.245$$

So, the probability she is on Southwest given the plane is late is

$$P(S\mid E)=\frac{P(S\cap E)}{P(E)}=\frac{P(S)\cdot P(E\mid S)}{P(E)}=\frac{0.70\cdot 0.23}{0.245}\approx 0.657$$

35. Define the events *E*: there is negative feedback, *R*: the survey is from Riverside, *S*: the survey is from Springfield, *C*: the survey is from Centerville.

$$P(R)=0.4 \quad P(S)=0.25 \quad P(C)=0.35 \quad P(E\mid R)=0.08 \quad P(E\mid S)=0.12 \quad P(E\mid C)=0.09$$

(a) $P(E)=P(R)\cdot P(E\mid R)+P(S)\cdot P(E\mid S)+P(C)\cdot P(E\mid C)$

$$=0.4\cdot 0.08+0.25\cdot 0.12+0.35\cdot 0.09=0.0935$$

(b) $P(R\mid E)=\dfrac{P(R\cap E)}{P(E)}=\dfrac{P(R)\cdot P(E\mid R)}{P(E)}=\dfrac{0.4\cdot 0.08}{0.0935}\approx 0.342$

(c) $P(S\mid E)=\dfrac{P(S\cap E)}{P(E)}=\dfrac{P(S)\cdot P(E\mid S)}{P(E)}=\dfrac{0.25\cdot 0.12}{0.0935}\approx 0.321$

37. Define the events *E*: a tax return is audited, *M*: an individual's adjusted gross income ≥ million.

$$P(E\mid \overline{M})=\frac{1}{100} \qquad P(E\mid M)=\frac{1}{16} \qquad P(M)=\frac{1}{250}$$

$$P(M\mid E)=\frac{P(M\cap E)}{P(E)}=\frac{P(M)\cdot P(E\mid M)}{P(M)\cdot P(E\mid M)+P(\overline{M})\cdot P(E\mid \overline{M})}$$

$$\frac{\frac{1}{250}\cdot\frac{1}{16}}{\frac{1}{250}\cdot\frac{1}{16}+\frac{249}{250}\cdot\frac{1}{100}}=\frac{25}{1021}\approx 0.025$$

39. Define the events *E*: an employee is female, *M*: occupation is classified management and related; *S*: occupation is classified service; *O*: occupation is classified sales or office; *N*: occupation is classified as natural resources, etc.; *P*: occupation is classified production, etc.

$$P(M)=0.349 \quad P(S)=0.165 \quad P(O)=0.25 \quad P(N)=0.11 \quad P(P)=0.126$$

$$P(E\mid M)=0.506,\ P(E\mid S)=0.573,\ P(E\mid O)=0.633,\ P(E\mid N)=0.047,\ P(E\mid P)=0.228$$

(a) $P(E)=P(M)\cdot P(E\mid M)+P(S)\cdot P(E\mid S)+P(O)\cdot P(E\mid O)$

$$+P(N)\cdot P(E\mid N)+P(P)\cdot P(E\mid P)$$
$$=0.349\cdot 0.506+0.165\cdot 0.573+0.25\cdot 0.633+0.11\cdot 0.047+0.126\cdot 0.228\approx 0.463$$

(b) $P(M\mid E)=\dfrac{P(M\cap E)}{P(E)}=\dfrac{P(M)\cdot P(E\mid M)}{P(E)}=\dfrac{0.349\cdot 0.506}{0.463}\approx 0.381$

(c) $P(S|E) = \dfrac{P(S \cap E)}{P(E)} = \dfrac{P(S) \cdot P(E|S)}{P(E)} = \dfrac{0.165 \cdot 0.573}{0.463} \approx 0.204$

(d) $P(O|E) = \dfrac{P(O \cap E)}{P(E)} = \dfrac{P(O) \cdot P(E|O)}{P(E)} = \dfrac{0.25 \cdot 0.633}{0.463} \approx 0.342$

41. Define events D: a person is a Democrat, R: a person is a Republican, I: a person is an Independent, E: a person voted.

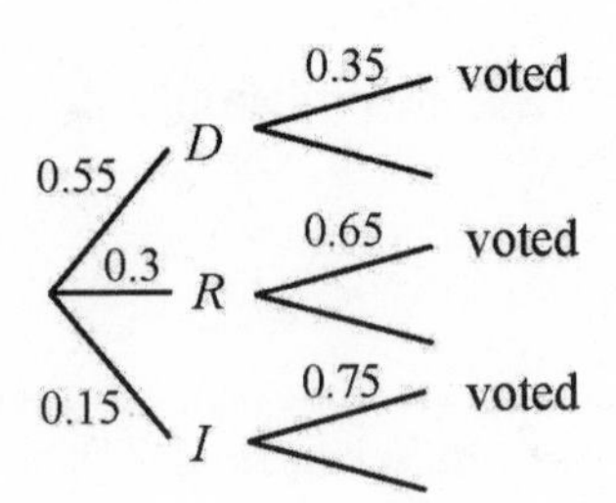

$P(E) = P(E \cap D) + P(E \cap R) + P(E \cap I)$
$= P(D) \cdot P(E \mid D) + P(R) \cdot P(E \mid R) + P(I) \cdot P(E \mid I)$
$= (0.55) \cdot (0.35) + (0.30) \cdot (0.65) + (0.15) \cdot (0.75)$
$= 0.5$

(a) $P(D \mid E) = \dfrac{P(D \cap E)}{P(E)} = \dfrac{P(D) \cdot P(E|D)}{P(E)} = \dfrac{(0.55)(0.35)}{0.5} = 0.385$

The probability that a voter was a Democrat was 0.385.

(b) $P(R \mid E) = \dfrac{P(R \cap E)}{P(E)} = \dfrac{P(R) \cdot P(E|R)}{P(E)} = \dfrac{(0.30)(0.65)}{0.5} = 0.39$

The probability that a voter was a Republican was 0.390.

(c) $P(I \mid E) = \dfrac{P(I \cap E)}{P(E)} = \dfrac{P(I) \cdot P(E|I)}{P(E)} = \dfrac{(0.15)(0.75)}{0.5} = 0.225$

The probability that a voter was an Independent was 0.225.

43. Define the events A_1: the soil is rock, A_2: the soil is clay, A_3: the is soil is sand, and E: the geological test is positive.

$P(A_1) = 0.53 \qquad P(A_2) = 0.21 \qquad P(A_3) = 0.26$

$P(E \mid A_1) = 0.35 \qquad P(E \mid A_2) = 0.48 \qquad P(E \mid A_3) = 0.75$

$P(E) = P(A_1) \cdot P(E \mid A_1) + P(A_2) \cdot P(E \mid A_2) + P(A_3) \cdot P(E \mid A_3)$
$= (0.53) \cdot (0.35) + (0.21) \cdot (0.48) + (0.26) \cdot (0.75) = 0.4813$

(a) $P(A_1 \mid E) = \dfrac{P(A_1 \cap E)}{P(E)} = \dfrac{P(A_1) \cdot P(E|A_1)}{P(E)} = \dfrac{(0.53)(0.35)}{0.4813} = 0.385$

The probability that the soil is rock given the test is positive is 0.385.

(b) $P(A_2 \mid E) = \dfrac{P(A_2 \cap E)}{P(E)} = \dfrac{P(A_2) \cdot P(E|A_2)}{P(E)} = \dfrac{(0.21)(0.48)}{0.4813} = 0.209$

There is a 0.209 probability that the soil is clay given that the test is positive.

(c) $P(A_3 \mid E) = \dfrac{P(A_3 \cap E)}{P(E)} = \dfrac{P(A_3) \cdot P(E|A_3)}{P(E)} = \dfrac{(0.26)(0.75)}{0.4813} = 0.405$

The probability is 0.405 that the soil is sand given that the test came out positive.

45. Define the additional event E: a person votes Republican.

$$P(N) = 0.40 \qquad P(S) = 0.10 \qquad P(M) = 0.25 \qquad P(W) = 0.25$$
$$P(E \mid N) = 0.40 \qquad P(E \mid S) = 0.56 \qquad P(E \mid M) = 0.48 \qquad P(E \mid W) = 0.52$$

(a)
$$\begin{aligned} P(E) &= P(E \cap N) + P(E \cap S) + P(E \cap M) + P(E \cap W) \\ &= P(N) \cdot P(E \mid N) + P(S) \cdot P(E \mid S) + P(M) \cdot P(E \mid M) + P(W) \cdot P(E \mid W) \\ &= (0.40) \cdot (0.40) + (0.10) \cdot (0.56) + (0.25) \cdot (0.48) + (0.25) \cdot (0.52) = 0.466 \end{aligned}$$

The probability that a person chosen at random votes Republican is 0.466.

(b)
$$P(N \mid E) = \frac{P(N \cap E)}{P(E)} = \frac{P(N) \cdot P(E \mid N)}{P(E)} = \frac{(0.40)(0.40)}{0.466} = 0.343$$

If a person has voted Republican, there is a 34.3% probability the person is from the Northeast.

47. Define the events A_1: the nurse forgets to give Mr. Brown his pill, A_2: the nurse remembers to give Mr. Brown his pill, and E: Mr. Brown dies.

$$P(A_1) = \frac{2}{3} \qquad P(A_2) = \frac{1}{3} \qquad P(E \mid A_1) = \frac{3}{4} \qquad P(E \mid A_2) = \frac{1}{3}$$

$$P(E) = P(A_1) \cdot P(E \mid A_1) + P(A_2) \cdot P(E \mid A_2) = \frac{2}{3} \cdot \frac{3}{4} + \frac{1}{3} \cdot \frac{1}{3} = \frac{11}{18}$$

$$P(A_1 \mid E) = \frac{P(A_1 \cap E)}{P(E)} = \frac{P(A_1) \cdot P(E \mid A_1)}{P(E)} = \frac{\frac{2}{3} \cdot \frac{3}{4}}{\frac{11}{18}} = \frac{9}{11} = 0.8182$$

The probability the nurse forgot to give Mr. Brown his pill given that he died is 0.818.

49. Define the events E: an accident occurs, A: a driver is young driver

$$P(A) = 0.132 \qquad P(\overline{A}) = 1 - P(A) = 1 - 0.132 = 0.868$$
$$P(E \mid \overline{A}) = 0.024$$

(a) Since young drivers have 2.5 times more accidents than older drivers, the probability a young driver is in an accident is $P(E \mid A) = 2.5 \cdot P(E \mid \overline{A}) = 2.5 \cdot 0.024 = 0.06$.

$$\begin{aligned} P(A \mid E) &= \frac{P(A \cap E)}{P(E)} = \frac{P(A) \cdot P(E \mid A)}{P(A) \cdot P(E \mid A) + P(\overline{A}) \cdot P(E \mid \overline{A})} \\ &= \frac{0.132 \cdot 0.06}{0.132 \cdot 0.06 + 0.868 \cdot 0.024} = 0.275 \end{aligned}$$

(b) If 20% of the insureds are young drivers, then 80% are older drivers. For this insurer,

$$P(\overline{A} \mid E) = \frac{P(\overline{A}) \cdot P(E \mid \overline{A})}{P(A) \cdot P(E \mid A) + P(\overline{A}) \cdot P(E \mid \overline{A})} = \frac{0.80 \cdot 0.024}{0.20 \cdot 0.06 + 0.80 \cdot 0.024} = 0.615$$

51. Define the events A_1: a student has the HIV virus, A_2: a student does not have the HIV virus, and E: the Elias test is positive.

(a) $P(A_1) = 0.002 \quad P(A_2) = 0.998 \quad P(E \mid A_1) = 0.998 \quad P(E \mid A_2) = 0.002$

$P(E) = P(A_1) \cdot P(E \mid A_1) + P(A_2) \cdot P(E \mid A_2) = (0.002) \cdot (0.998) + (0.998) \cdot (0.002) = 0.004$

$$P(A_1 \mid E) = \frac{P(A_1 \cap E)}{P(E)} = \frac{P(A_1) \cdot P(E \mid A_1)}{P(E)} = \frac{(0.002)(0.998)}{0.004} = 0.499$$

A student with a positive Elias test has a probability of 0.5 of having the HIV virus.

(b) $P(A_1) = 0.05 \quad P(A_2) = 0.95 \quad P(E \mid A_1) = 0.998 \quad P(E \mid A_2) = 0.002$

$P(E) = P(A_1) \cdot P(E \mid A_1) + P(A_2) \cdot P(E \mid A_2) = (0.05) \cdot (0.998) + (0.95) \cdot (0.002) = 0.0518$

$$P(A_1 \mid E) = \frac{P(A_1 \cap E)}{P(E)} = \frac{P(A_1) \cdot P(E \mid A_1)}{P(E)} = \frac{(0.05)(0.998)}{0.0518} = 0.963$$

In a high risk area a person with a positive Elias test has a 0.963 probability of having the HIV virus.

(c) Define event F: A person has 2 positive Elias tests.
Assuming, from part (a) that $P(A_1) = 0.002$ and that the tests are independent, we get

$P(F \mid A_1) = [P(E \mid A_1) \cdot P(E \mid A_1)]^2 = 0.998^2 = 0.996$ and
$P(F \mid A_2) = [P(E \mid A_2) \cdot P(E \mid A_2)]^2 = 0.002^2 = 4.0 \times 10^{-6}$

$$\begin{aligned} P(F) &= P(A_1) \cdot P(F \mid A_1) + P(A_2) \cdot P(F \mid A_2) \\ &= (0.002) \cdot (0.996) + (0.998) \cdot (4.0 \times 10^{-6}) = 0.002 \end{aligned}$$

$$P(A_1 \mid F) = \frac{P(A_1 \cap F)}{P(F)} = \frac{P(A_1) \cdot P(F \mid A_1)}{P(F)} = \frac{(0.002)(0.996)}{0.002} = 0.996$$

A student with 2 positive Elias tests has a 99.6% chance of having the HIV virus.

53. When F is a subset of E, $E \cap F = F$, and $P(E \cap F) = P(F)$.

By the definition of conditional probability, $P(E \mid F) = \dfrac{P(E \cap F)}{P(F)}$ provided $P(F) \neq 0$.

So $P(E \mid F) = \dfrac{P(F)}{P(F)} = 1$.

8.2 Binomial Probability Model

5. False. In a Bernoulli trial the probability of each outcome remains the same for each trial.

7. $b(7, 4; 0.20) = \binom{7}{4}(0.2)^4(0.8)^3 = (35)(0.0016)(0.512) = 0.0287$

9. $b(15, 8; 0.80) = \binom{15}{8}(0.80)^8(0.2)^7 = (6435)(0.16777)(1.28\times10^{-5}) = 0.01382$

11. $b(15, 10; \frac{1}{2}) = \binom{15}{10}\left(\frac{1}{2}\right)^{10}\left(\frac{1}{2}\right)^5 = 0.09164$

13. $b(15, 3; 0.3) + b(15, 2; 0.3) + b(15, 1; 0.3) + b(15, 0; 0.3)$

$$= \binom{15}{3}(0.3)^3(0.7)^{12} + \binom{15}{2}(0.3)^2(0.7)^{13} + \binom{15}{1}(0.3)^1(0.7)^{14} + \binom{15}{0}(0.3)^0(0.7)^{15}$$

$$= 0.17004 + 0.09156 + 0.03052 + 0.00475 = 0.29687$$

15. $b\left(3, 2; \frac{1}{3}\right) = \binom{3}{2}\left(\frac{1}{3}\right)^2\left(\frac{2}{3}\right)^1 = \frac{2}{9}$

17. $b\left(3, 0; \frac{1}{6}\right) = \binom{3}{0}\left(\frac{1}{6}\right)^0\left(\frac{5}{6}\right)^3 = \frac{125}{216}$

19. $b\left(5, 3; \frac{2}{3}\right) = \binom{5}{3}\left(\frac{2}{3}\right)^3\left(\frac{1}{3}\right)^2 = \frac{80}{243}$

21. $n = 10, k = 6, p = 0.3 \quad b(10, 6; 0.3) = \binom{10}{6}(0.3)^6(0.7)^4 = 0.0368$

23. $n = 12, k = 9, p = 0.8 \quad b(12, 9; 0.8) = \binom{12}{9}(0.8)^9(0.2)^3 = 0.2362$

25. $n = 8,\ p = 0.30$

The probability P of at least 5 successes is the probability of 5 or 6 or 7 or 8 successes. Since the events are mutually exclusive we can add the probabilities.

$$\begin{aligned} P &= b(8, 5; 0.30) + b(8, 6; 0.30) + b(8, 7; 0.30) + b(8, 8; 0.30) \\ &= \binom{8}{5}(0.30)^5(0.70)^3 + \binom{8}{6}(0.30)^6(0.70)^2 + \binom{8}{7}(0.30)^7(0.70)^1 + \binom{8}{8}(0.30)^8(0.70)^0 \\ &= 0.0467 + 0.0100 + 0.0012 + 0.0001 = 0.058 \end{aligned}$$

27. $n = 8,\ k = 1;\ p = 0.5$

$$P(\text{exactly 1 head}) = b(8, 1; 0.5) = \binom{8}{1}(0.5)^1(0.5)^7 = 0.03125$$

29. $n = 8, p = 0.5$

$$\begin{aligned} P(\text{at least 5 tails}) &= P(\text{exactly 5 tails}) + P(\text{exactly 6 tails}) + P(\text{exactly 7 tails}) + P(\text{exactly 8 tails}) \\ &= b(8, 5; 0.5) + b(8, 6; 0.5) + b(8, 7; 0.5) + b(8, 8; 0.5) = 0.3633 \end{aligned}$$

31. $n = 8, p = 0.5$

$$\begin{aligned} P(\text{at least 1 head}) &= P(\text{exactly 1 head}) + P(\text{exactly 2 heads}) + \ldots P(\text{exactly 8 heads}) \\ &= 1 - P(0 \text{ heads}) \end{aligned}$$

The intersection of “at least 1 head” and “exactly 2 heads” is “exactly 2 heads”

$$P(\text{at exactly 2 heads} \mid \text{at least 1 head}) = \frac{P(\text{exactly 2 heads})}{1 - P(\text{no heads})} = \frac{b(8,2;0.5)}{1 - b(8,0;0.5)} = 0.1098$$

33. The probability of rolling a sum of 7 with two dice is $\frac{1}{6}$. So we have $n = 5,\ k = 2,\ p = \frac{1}{6}$.

$$P(\text{exactly 2 sums of 7}) = b(5, 2; \tfrac{1}{6}) = 0.1608$$

35. (a)

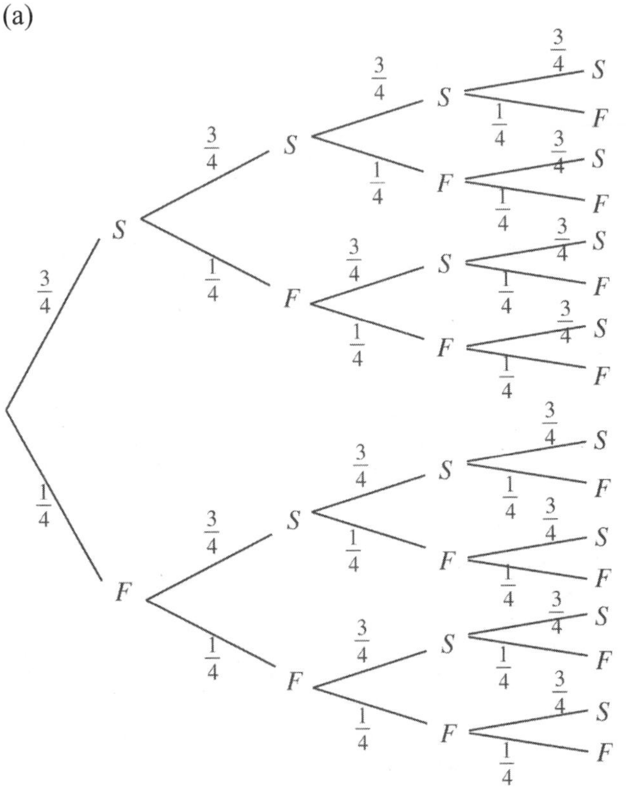

(b) P(exactly 2 success, 2 failures)

$P(SSFF) + P(SFSF) + P(SFFS) + P(FSSF) + P(FSFS) + P(FFSS)$

$$= \left(\frac{3}{4}\right)\left(\frac{3}{4}\right)\left(\frac{1}{4}\right)\left(\frac{1}{4}\right) + \left(\frac{3}{4}\right)\left(\frac{1}{4}\right)\left(\frac{3}{4}\right)\left(\frac{1}{4}\right) + \left(\frac{3}{4}\right)\left(\frac{1}{4}\right)\left(\frac{1}{4}\right)\left(\frac{3}{4}\right) + \left(\frac{1}{4}\right)\left(\frac{3}{4}\right)\left(\frac{3}{4}\right)\left(\frac{1}{4}\right) + \left(\frac{1}{4}\right)\left(\frac{3}{4}\right)\left(\frac{1}{4}\right)\left(\frac{3}{4}\right) + \left(\frac{1}{4}\right)\left(\frac{1}{4}\right)\left(\frac{3}{4}\right)\left(\frac{3}{4}\right)$$

$$= 6 \cdot \left(\frac{9}{256}\right) = \frac{27}{128}$$

(c) P(exactly 2 success, 2 failures)

$$= b\left(4, 2; \frac{3}{4}\right) = \binom{4}{2}\left(\frac{3}{4}\right)^2\left(\frac{1}{4}\right)^2 = \frac{27}{128}$$

37. $n = 8, p = 0.05$

(a) P(exactly 1 is defective) $= b(8, 1; 0.05) = 0.2793$

(b) P(exactly 2 are defective) $= b(8, 2; 0.05) = 0.0515$

(c) P(at least 1 is defective) $= 1 - P$(none are defective)
$= 1 - b(8, 0; 0.05)$
$= 1 - 0.6634 = 0.3366$

(d) P(fewer than 3 defective) $= P$(no defective) $+ P$(1 defective) $+ P$(2 defective)
$= b(8, 0; 0.05) + b(8, 1; 0.05) + b(8, 2; 0.05) = 0.9942$

39. (a) When exactly 3 are boys, $n = 6, k = 3, p = 0.5$, and $b(6, 3; 0.5) = 0.3125$

The probability that a family with 6 children has exactly 3 boys and 3 girls is 0.3125.

(b) The probability at least 5 are boys means that either or 5 or 6 of the children are boys.

$$P(\text{at least 5 are boys}) = b(6,5;\ 0.5) + b(6,6;\ 0.5) = 0.09375 + 0.015625 = 0.1094$$

The probability a family with 6 children has at least 5 boys is 0.1094.

(c) We now consider having a girl a success. The probability of a girl is 0.5, the probability at least 2 but fewer than 4 are girls is

$$P(2 \le \text{girls} < 4) = b(6,2;0.5) + b(6,3;\ 0.5) = 0.2344 + 0.3125 = 0.5469$$

The probability a family with 6 children has 2 or 3 girls is 0.5469.

41. $n = 20, p = 0.5$

(a) $P(\text{student guesses all correct}) = b(20, 20; 0.5) = 9.537 \times 10^{-7}$

(b) The student needs 12 correct to pass, but passes if he/she gets at least 12 correct. So the probability a student passes is the probability of getting 12, 13, 14, …, 20 correct.

$$P(\text{passing}) = b(20,12;0.5)+b(20,13;0.5)+b(20,14;0.5)+b(20,15;0.5)+b(20,16;0.5)$$
$$+b(20,17;0.5)+b(20,18;0.5)+b(20,19;0.5)+b(20,20;0.5)$$

$$= 0.1201+0.0739+0.0370+0.0148+0.0046+0.0011+0.0001+0.0000+0.0000$$
$$= 0.2516$$

The student's probability of passing is 0.252.

(c) The odds in favor of passing are $\dfrac{P(\text{passing})}{P(\text{failing})} = \dfrac{0.252}{0.748} = \dfrac{252}{748} = \dfrac{63}{187}$ or about 1 to 3.

43. $n = 15; p = 0.12$

(a) It is easier to find the probability of $k \geq 5$, by considering the complement.

$$P(\text{at least } 5) = 1 - P(k \leq 4) = 1 - [P(k=0)+P(k=1)+P(k=2)+P(k=3)+P(k=4)]$$
$$= 1 - b(15, 0; 0.12) - b(15, 1; 0.12) - b(15, 2; 0.12) - b(15, 3; 0.12) - b(15, 4; 0.12)$$
$$= 1 - 0.1470 - 0.3006 - 0.2870 - 0.1696 - 0.0694$$
$$= 0.0264$$ (Answer may differ by one in the thousandths place due to rounding.)

Probability that at least 5 of those surveyed think it is acceptable to cheat is 0.0264.

(b) $P(\text{fewer than } 3) = P(k < 3) = P(k=0)+P(k=1)+P(k=2)$
$$= b(15, 0; 0.12) + b(15, 1; 0.12) + b(15, 2; 0.12)$$
$$= 0.1470 + 0.3006 + 0.2870$$
$$= 0.7346$$

Probability fewer than 3 people surveyed think it is acceptable to cheat is 0.7346.

(c) $P(0) = b(15, 0; 0.12) = 0.1470$

Probability that no one thinks it is acceptable to cheat is 0.1470.

45. $n = 8; p = 0.10$

(a) $P(k = 2) = b(8, 2; 0.1) = 0.1488$

Probability two of the returns were filed on April 15 is 0.1488.

(b) $P(k = 5) = b(8, 5; 0.1) = 0.0004$

Probability five of the returns were filed on April 15 is 0.0004.

(c) $P(k = 0) = b(8, 0; 0.1) = 0.4305$

Probability that none of the returns were filed on April 15 is 0.4305.

47. $n = 10; p = 0.8$

(a) $P(k = 10) = b(10, 10; 0.8) = 0.1074$

Probability that all like their jobs is 0.1074.

(b) $P(k = 9) = b(10, 9; 0.8) = 0.2684$

Probability that 9 persons liked their jobs is 0.2684.

(c) $P(\text{no more than } 7) = P(k \leq 7) = 1 - P(k \geq 8)$
$$= 1 - [b(10, 8; 0.8) + b(10, 9; 0.8) + b(10, 10; 0.8)]$$
$$= 1 - 0.3020 - 0.2684 - 0.1074$$
$$= 0.3222$$

Probability that no more than 7 persons liked their jobs is 0.3222.

49. $n = 18; p = 0.77$

(a) $P(k = 14) = b(18, 14; 0.77) = 0.2205$

Probability that exactly 14 of those selected use coupons is 0.2205.

(b) $P(\text{at least } 14) = P(k \geq 14) = P(k = 14) + P(k = 15) + P(k = 16) + P(k = 17) + P(k = 18)$
$= b(18, 14; 0.77) + b(18, 15; 0.77) + b(18, 16; 0.77) + b(18, 17; 0.77) + b(18, 14; 0.77)$
$= 0.2205 + 0.1969 + 0.1236 + 0.0487 + 0.0091 = 0.5988$

Probability that at least 14 of those selected use coupons is 0.5988.

(c) $P(\text{at least } 16) = P(k \geq 16) = P(k = 16) + P(k = 17) + P(k = 18)$
$= b(18, 16; 0.77) + b(18, 17; 0.77) + b(18, 14; 0.77)$
$= 0.1236 + 0.0487 + 0.0091 = 0.1814$

Probability that at least 16 of those selected use coupons is 0.1814.

51. $n = 12; p = 0.15$

(a) $P(k = 0) = b(12, 0; 0.15) = 0.1422$

Probability that none of the 12 claims involves fraud is 0.1422

(b) $P(k = 6) = b(12, 6; 0.15) = 0.0040$

Probability that 6 of the 12 claims involves fraud is 0.0040.

(c) $P(\text{at least } 1) = P(k \geq 1) = 1 - P(k = 0) = 1 - 0.1422 = 0.8578$

Probability that at least one of the 12 claims involves fraud is 0.8578.

53. $n = 44; p = 0.12$

(a) $P(k = 2) = b(44, 2; 0.12) = 0.0635$

Probability two passengers of the 44 do not show is 0.0635.

(b) $P(k = 0) = b(44, 0; 0.12) = 0.0036$

Probability all passengers of the 44 show is 0.0036.

(c) $P(\text{at least } 2) = P(k \geq 2) = 1 - [P(k = 0) + P(k = 1)] = 1 - [b(44, 0; 0.12) + b(44, 1; 0.12)]$
$= 1 - 0.0036 - 0.0216 = 0.9748$

Probability at least two passengers of the 44 do not show is 0.9748.

55. $n = 4, \ p = 0.250$

(a) $P(\text{at least 2 hits}) = P(\text{exactly 2 hits}) + P(\text{exactly 3 hits}) + P(\text{exactly 4 hits})$
$= b(4, 2; 0.250) + b(4, 3; 0.250) + b(4, 4; 0.250)$
$= 0.2109 + 0.0469 + 0.0039 = 0.2617$

A player with a 0.250 batting average has a 0.2617 probability of getting at least 2 hits in 4 at bats.

(b) $P(\text{at least 1 hit}) = 1 - P(\text{no hits})$
$= 1 - b(4, 0; 0.250) = 1 - 0.3164 = 0.6836$

The probability the batter gets at least 1 hit is 0.6836.

57. $n = 15, p = 0.03$

(a) $k = 0$; $P(\text{no defective tubes}) = b(15, 0; 0.03) = 0.6333$

There is a 0.633 probability that there are no defective tubes in the sample of 15.

(b) $P(\text{more than 2 defective tubes}) = 1 - P(\text{2 or fewer defective tubes})$

$$= 1 - [P(\text{0 defective}) + P(\text{1 defective}) + P(\text{ 2 defective})]$$
$$= 1 - [b(15, 0; 0.03) + b(15, 1; 0.03) + b(15, 2; 0.03)]$$
$$= 1 - 0.6333 - 0.2938 - 0.0636 = 1 - 0.9907 = 0.0093$$

The probability that there are more than 2 defective tubes in the sample is 0.009.

59. $n = 8, k = 8; p = 0.40$

$P(\text{all 8 voters prefer Ms. Moran}) = b(8, 8; 0.40) = 0.0007.$

The probability that all 8 voters prefer Ms. Moran is 0.07%.

61. $n = 10, k = 4, p = 0.23$

$P(\text{exactly 4 deaths are due to heart attack}) = b(10, 4; 0.23) = 0.1225$

There is a 0.1225 probability that 4 of the next 10 unexpected deaths are due to a heart attack.

63. (a) $n = 6, p = 0.5$

$P(\text{identifies at least 5 cups}) = P(\text{identifies exactly 5 cups}) + P(\text{identifies exactly 6 cups})$

$$= b(6, 5; 0.5) + b(6, 6; 0.5)$$
$$= 0.09375 + 0.015625 = 0.109375$$

The probability of correctly identifying at least 5 out of 6 cups of coffee if merely guessing is 0.1094.

(b) $n = 6, p = 0.80$

$P(\text{identifies fewer than 5 cups}) = 1 - P(\text{identifies at least 5 cups})$

$$= 1 - [P(\text{identifies exactly 5 cups}) + P(\text{identifies exactly 6 cups})]$$
$$= 1 - [b(6, 5; 0.80) + b(6, 6; 0.80)]$$
$$= 1 - 0.393216 - 0.262144 = 0.34464$$

The probability her claim is rejected when she really has the ability to identify the coffee is 0.3446.

65. $n = 10, p = 0.124$

(a) $P(\text{exactly 4 are 65 or older}) = b(10, 4; 0.124) = 0.0224$

There is a probability of 0.0224 that exactly 4 of the 10 people are 65 or older.

(b) $P(\text{no one is 65 or older}) = b(10, 0; 0.124) = 0.2661$

There is a probability of 0.2661 that none of the 10 people are 65 or older.

(c) $P(\text{at most } 5 \geq 65) = P(0 \geq 65) + P(1 \geq 65) + P(2 \geq 65) + P(3 \geq 65) + P(4 \geq 65) + P(5 \geq 65)$

$$= b(10, 0; 0.124) + b(10, 1; 0.124) + b(10, 2; 0.124) +$$
$$b(10, 3; 0.124) + b(10, 4; 0.124) + b(10, 5; 0.124)$$
$$= 0.2661 + 0.3767 + 0.2399 + 0.0906 + 0.0224 + 0.0038 = 0.9995$$

The probability that no more than 5 people selected are 65 years of age or older is 0.9995.

67. The code is of length 15, and corrects 1 error. The probability a digit is transmitted correctly is 0.98. So we have $n = 15, p = 0.98$.
The probability a message is received correctly will be the probability it was received without error plus the probability if was received with one corrected error.

$$b(15, 15; 0.98) + b(15, 14; 0.98) = 0.7386 + 0.2261 = 0.9647$$

The message will be received correctly 96.47% of the time using this Hamming code.

69. The code is of length 31, and corrects up to 2 errors. The probability a digit is transmitted correctly is 0.97. So we have $n = 31, p = 0.97$
The probability a message is received correctly will be the probability it was received without error or with one or two corrected errors.

$$\begin{aligned} &b(31, 31; 0.97) + b(31, 30; 0.97) + b(31, 29; 0.97) \\ &= 0.3890 + 0.3729 + 0.1730 = 0.9349 \end{aligned}$$

The probability the code is correctly received is 0.9349.

71. Estimates will vary, but all should be close to the actual theoretical probabilities.

The actual values of $P(k)$ are obtained from $b(4, k; 0.2)$ where k denotes the number of heads obtained in 4 tosses of the coin. Here since the probability of getting a tail is 0.80, the probability of getting a head is 0.20.

k	Estimate of $P(k)$	Actual $P(k)$
0		0.4096
1		0.4096
2		0.1536
3		0.0256
4		0.0016

73. Estimates will vary, but they should be close to the actual theoretical probability.
The actual value of $P(\text{Exactly 3 heads}) = b(8, 3; 0.2) = 0.1468$

8.3 Expected Value

3. True

5. $E = (2)(0.4) + (3)(0.2) + (-2)(0.1) + (0)(0.3) = 1.2$

7. $E = (30{,}000)(0.08) + (40{,}000)(0.42) + (60{,}000)(0.42) + (80{,}000)(0.08) = 50{,}800$

9. $E = (8)(0.1) + (0)(0.90) = 0.8$

Mary should pay \$0.80 for one draw.

11. $P(\text{double when throwing 2 dice}) = \dfrac{1}{6}$

$$E = (12)\left(\frac{1}{6}\right) + (0)\left(\frac{5}{6}\right) = \frac{12}{6} = 2$$

David should pay \$2 for a throw.

13. $E = (100)(0.001) + (50)(0.003) + (0)(0.996) = 0.25$

The price of a ticket exceeds the expected value by \$0.75.

15. $P(3 \text{ tails}) = \frac{1}{8}$ $\quad P(2 \text{ tails}) = \frac{3}{8}$ $\quad P(1 \text{ tail}) = \frac{3}{8}$ $\quad P(0 \text{ tails}) = \frac{1}{8}$

(a) $E = (3)\left(\frac{1}{8}\right) + (2)\left(\frac{3}{8}\right) + (0)\left(\frac{3}{8}\right) + (-3)\left(\frac{1}{8}\right) = \frac{6}{8} = 0.75$

The expected value of the game is \$0.75.

(b) The game is not fair. Fair games have an expected value of 0.

(c) Let x represent the payoff for tossing 1 tail.

$$x\left(\frac{3}{8}\right) = -\frac{6}{8} \text{ or } x = -2.$$

To make the game fair the player should lose \$2.00 if one tail is thrown.

17. $P(\text{team A wins}) = \frac{9}{14}$ $\quad P(\text{team B wins}) = \frac{5}{14}$

If team A wins you lose \$4; if team B wins you win \$6. The expected value of the game is

$$E = (-4)\left(\frac{9}{14}\right) + (6)\left(\frac{5}{14}\right) = -\frac{6}{14} = -0.4286$$

The bet is not fair to you. You should expect to lose 43 cents.

19. $P(\text{selecting a heart other than the ace}) = \frac{12}{52}$; $P(\text{selecting an ace other than the heart}) = \frac{3}{52}$

$$P(\text{selecting the ace of hearts}) = \frac{1}{52}$$

$$E = (40)\left(\frac{12}{52}\right) + (50)\left(\frac{3}{52}\right) + (90)\left(\frac{1}{52}\right) = \frac{720}{52} = 13.8462$$

Sarah's expected winnings are 13.8 – 15 = –1.2 cents.

21. First we find the distribution of the corporations by size.

Size	<\$25,000	\$25,000-\$49,999	\$50,000-\$99,000	\$100,000-\$499,000	>\$500,000	Total
Number	1280	346	522	1649	623	4420
Proportion	0.290	0.078	0.118	0.373	0.141	1.000

$E = (12{,}500)(0.29) + (37{,}500)(0.078) + (75{,}000)(0.118) + (250{,}000)(0.373) + (750{,}000)(0.141)$
$= \$214{,}400$

The expected receipts of the chosen corporation are \$214,400.

23. $E = (50)(0.627) + (150)(0.241) + (300)(0.091) + (600)(0.034) + (900)(0.007) = 121.5$

The expected gross rentable area of the selected shopping center is 121,500 square feet.

25. This is a Bernoulli process, with $n = 20$ and $p = 0.15$.

$$E = np = (20)(0.15) = 3$$

Three of the adults surveyed would be expected to say that health care costs are paramount.

27. (a) If the person dies the insurance company has a loss of \$250,000 – \$450 = \$249,550. If the person survives the company has a gain of \$450.

$$E = (450)(0.9986) - (249{,}550)(1 - 0.9986) = 100$$

The insurance company can expect a profit of \$100.

(b) To have a profit of \$250 per policy, the company's expected value must be 250. Let x be the premium necessary to obtain an expected value of 250.

$$\begin{aligned} E = x(0.9886) - (250{,}000 - x)(1 - 0.9886) &= 250 \\ 0.9886x - 250{,}000 + 249{,}605 + x - 0.9886x &= 250 \\ -350 + x &= 250 \\ x &= 600 \end{aligned}$$

The company should set the premium at \$600.

29. This is an example of 2000 Bernoulli trials, where $p = \frac{1}{6}$ is the probability of success.

$$E = np = 2000\left(\frac{1}{6}\right) = 333.333$$

We expect 333.333 fives in 2000 rolls of a fair die.

31. This can be considered an example of 500 Bernoulli trials, where $p = 0.02$ is the probability a light bulb is defective.

$$E = np = (500)(0.02) = 10$$

The shop owner expects 10 defective light bulbs in a shipment of 500.

33. (a) We compare the two expected profits. $P(\text{success}) = \frac{1}{2}$

Location 1: $E = (15{,}000)\left(\frac{1}{2}\right) + (-3000)\left(\frac{1}{2}\right) = 6000$

The first location will provide an expected profit of \$6000.

Location 2: $E = (20{,}000)\left(\frac{1}{2}\right) + (-6000)\left(\frac{1}{2}\right) = 7000$

The second location will provide an expected profit of \$7000.

The management should choose the Location 2. It has a higher expected profit.

(b) We compare the two expected profits.

Location 1: $P(\text{success}) = \left(\frac{2}{3}\right)$; $E = (15{,}000)\left(\frac{2}{3}\right) + (-3000)\left(\frac{1}{3}\right) = 9000$

The first location will provide an expected profit of \$9000.

Location 2: $P(\text{success}) = \left(\frac{1}{3}\right)$; $E = (20{,}000)\left(\frac{1}{3}\right) + (-6000)\left(\frac{2}{3}\right) = 2666.67$

The second location will provide an expected profit of \$2666.67.

The management should choose the Location 1. It has a higher expected profit.

35. This is an example of 500 Bernoulli trials, where $p = 0.002$ is the probability of having an unfavorable reaction to the drug.

$$E = np = (500)(0.002) = 1$$

The doctor can expect 1 patient to have an unfavorable reaction to the drug.

37. (a) $E = 7p_7 + 8p_8 + 9p_9 + 10p_{10} + 11p_{11}$
$= (7)(0.10) + (8)(0.20) + (9)(0.40) + (10)(0.20) + (11)(0.10) = 9$

The expected number of customers is 9.

(b) To decide the optimal number of trucks to have on hand, we need to find the expected profit for each possible number of trucks.

$$\text{Profit} = \text{Revenue} - \text{Cost}$$

Expected profit is the product of the profit and the probability of obtaining it.

1. If 7 trucks are on hand, all will be rented, and the expected profit will be

$$E(\text{profit}) = (7)(90 - 20) = \$490$$

2. If 8 trucks are on hand, 7 will be rented with probability 0.10 and 8 will be rented with probability 0.90 $(0.90 = 0.20 + 0.40 + 0.20 + 0.10)$

$$E(\text{profit}) = [7(90 - 20) - 20](0.10) + [8(90 - 20)](0.90) = \$551$$

3. If 9 trucks are on hand, 7 will be rented with probability 0.10, 8 will be rented with probability 0.20, and 9 will be rented with probability $0.40 + 0.20 + 0.10 = 0.70$.

$$E(\text{profit}) = [7(90 - 20) - 2(20)](0.10) + [8(90 - 20) - 20](0.20) + [9(90 - 20)](0.70) = \$594$$

4. If 10 trucks are on hand, 7 will be rented with probability 0.10, 8 will be rented with probability 0.20, 9 will be rented with probability 0.40, and 10 will be rented with probability $0.20 + .010 = 0.70$.

$$E(\text{profit}) = [7(90 - 20) - 3(20)](0.10) + [8(90 - 20) - 2(20)](0.20) + [9(90 - 20) - 20)](0.40) + [10(90 - 20)](0.30) = \$601$$

5. If 11 trucks are on hand, 7 will be rented with probability 0.10, 8 will be rented with probability 0.20, and 9 will be rented with probability 0.40, 10 will be rented with probability 0.20, and 11 will be rented with probability 0.10.

$$E(\text{profit}) = [7(90 - 20) - 4(20)](0.10) + [8(90 - 20) - 3(20)](0.20) + [9(90 - 20) - 2(20)](0.40) + [10(90 - 20) - 20](0.20) + [11(90 - 20)](0.10) = \$590$$

The largest expected profit is \$594, and the rental agency should keep 10 trucks on hand.

39. (a) If each of the two stocks have equal weight then $w_1 = w_2 = 0.5$. The expected return on the investor's portfolio is

$$E = 0.5 \cdot 0.10 + 0.5 \cdot 0.15 = 0.125$$

The investor expects a return of 12.5%.

(b) To obtain a return of 14%, we need $E = 0.14$. Let x represent w_1 and let $1 - x$ represent w_2.

$$E = x \cdot 0.10 + (1 - x) \cdot 0.15 = 0.14$$
$$0.1x + 0.15 - 0.15x = 0.14$$
$$-0.05x = -0.01$$
$$x = 0.2$$

To realize a return of 14%, the investor's portfolio should be split with 20% Wal-Mart stock and 80% Viacom stock.

41. Profit = Revenue – Cost. We will compare the profit from each aircraft.

Aircraft A
Expected number of passengers:

$$E = (150)(0.2) + (180)(0.3) + (200)(0.2 + 0.2 + 0.1) = 30 + 54 + 100 = 184$$

Expected revenue: ticket price times the number of tickets sold = (\$500)(184) = \$92,000
Expected cost: fixed cost plus passenger cost = \$16,000 + \$200(184) = \$52,800
Expected profit: Revenue – Cost = \$92,000 – \$52,800 = \$39,200

The company can expect a profit of \$39,200 if it uses aircraft A.

Aircraft B
Expected number of passengers:

$$E = (150)(.2)+(180)(.3)+(200)(.2)+(250)(.2)+(300)(.1) = 30 + 54 + 40 + 50 + 30 = 204$$

Expected revenue: ticket price times the number of tickets sold = (\$500)(204) = \$102,000
Expected cost: fixed cost plus passenger cost = \$18,000 + \$230(204) = \$64,920
Expected profit: Revenue – Cost = \$102,000 – \$64,920 = \$37,080

The company can expect a profit of \$37,080 if it uses aircraft B.

The company should use aircraft A; it generates a larger expected profit.

43. If components are tested individually n components need n tests. If they tested together, then we need 1 test for n components if they are all good, and $n + 1$ tests if the grouped test fails. The probability of needing only one test is p^n and the probability of needing $n + 1$ tests is $1 - p^n$. From this we can get the expected number of tests, E.

$$\begin{aligned} E &= (1)(p^n) + (n + 1)(1 - p^n) \\ &= p^n + n - np^n + 1 - p^n \\ &= (n + 1) - np^n \end{aligned}$$

and the expected number of tests saved is

$$n - [(n + 1) - np^n] = np^n - 1$$

If we are testing n components the number of tests saved per component becomes

$$\frac{np^n - 1}{n} = p^n - \frac{1}{n}$$

The optimal group size is that which saves the most tests per component.

(a) We are asked to find the optimal group size if the probability a component is good, $p = 0.8$, in Table 4 (p. 508 of the text).

Group Size	Expected Tests Saved per Component $p = 0.8$	Percent Saving
2	$0.8^2 - \frac{1}{2} = 0.14$	14%
3	$0.8^3 - \frac{1}{3} = 0.1787$	17.9%
4	$0.8^4 - \frac{1}{4} = 0.1596$	16.0%
5	$0.8^5 - \frac{1}{5} = 0.12768$	12.8%
6	$0.8^6 - \frac{1}{6} = 0.09547$	9.5%

The optimal group size is 3. Its percent saving is 17.9%.

(b) If the probability a component is good is $p = 0.95$

Group Size	Expected Tests Saved per Component $p = 0.95$	Percent Saving
2	$0.95^2 - \frac{1}{2} = 0.4025$	40.3%
3	$0.95^3 - \frac{1}{3} = 0.5240$	52.4%
4	$0.95^4 - \frac{1}{4} = 0.5645$	56.5%
5	$0.95^5 - \frac{1}{5} = 0.5738$	57.4%
6	$0.95^6 - \frac{1}{6} = 0.5684$	56.8%
7	$0.95^7 - \frac{1}{7} = 0.55548$	55.5%

The optimal group size is 5. Its percent saving is 57.4%

(c) If the probability a component is good is 0.99, then the savings incurred by grouping the components before testing is significant. The table below shows the savings

Group Size	Expected Tests Saved per Component $p = 0.99$	Percent Saving
10	$0.99^{10} - \frac{1}{10} = 0.80438$	80.438%
11	$0.99^{11} - \frac{1}{11} = 0.80443$	80.443%
12	$0.99^{12} - \frac{1}{12} = 0.80305$	80.305%

The optimal group size is 11. Its percent saving is 80.44%.

45. If the probability a test is positive is p, the probability the test is negative is $(1 - p)$.

(a) The test for a pooled sample will be positive if at least one person tests positive. So,

$$\begin{aligned} P(\text{pooled test is positive}) &= 1 - P(\text{all tests are negative}) \\ &= 1 - (1-p)^{20} \end{aligned}$$

(b) If the pooled test is negative then 1 test suffices for 20 allergens, but if the pooled test is positive then 21 tests must be done. The expected number of tests necessary using the pooled method is

$$\begin{aligned} E &= (1)(1-p)^{20} + 21\,[1-(1-p)^{20}] \\ &= (1-p)^{20} + 21 - 21(1-p)^{20} \\ &= 21 - 20(1-p)^{20} \end{aligned}$$

(c) A pooled sampling saves approximately

$$20 - (21 - 20(1-p)^{20}) = 20(1-p)^{20} - 1$$

tests per individual.

8.4 Random Variables

1. Let X denote the number of heads that appear when a fair coin is tossed twice.

X	$x = 0$	$x = 1$	$x = 2$
$P(X = x)$	$p(0) = b(2, 0; 0.5) = \frac{1}{4}$	$p(1) = b(2, 1; 0.5) = \frac{1}{2}$	$p(2) = b(2, 2; 0.5) = \frac{1}{4}$

3. Let X denote the number of female children in 3-child family assuming probability that a child is female is $\frac{1}{2}$.

X	$P(X = x)$
$x = 0$	$0) = b\left(3,\ 0;\ \frac{1}{2}\right) = \frac{1}{8}$
$x = 1$	$p(1) = b\left(3,\ 1;\ \frac{1}{2}\right) = \frac{3}{8}$
$x = 2$	$p(2) = b\left(3,\ 2;\ \frac{1}{2}\right) = \frac{3}{8}$
$x = 3$	$p(3) = b\left(3,\ 3;\ \frac{1}{2}\right) = \frac{1}{8}$

5. Let X denote the number of red balls that are drawn when 3 balls are drawn with replacement from an urn with 10 balls.

$P(\text{red ball chosen}) = 0.4$;
$P(\text{white ball chosen}) = 0.6$

X	$P(X = x)$
0	$p(0) = b(3,\ 0;\ 0.4) = 0.216$
1	$p(1) = b(3,\ 1;\ 0.4) = 0.432$
2	$p(2) = b(3,\ 2;\ 0.4) = 0.288$
3	$p(3) = b(3,\ 3;\ 0.4) = 0.064$

7. $E(X) = x_1p_1 + x_2p_2 + x_3p_3 + x_4p_4 = (2)(0.4) + (3)(0.2) + (-2)(0.1) + (0)(0.3) = 1.2$

9. (a) $X = \{0, 1, 2, 3, 4, 5\}$

(b)

$X = k$	$X = 0$	$X = 1$	$X = 2$	$X = 3$	$X = 4$	$X = 5$
$b(5, k; 0.6)$	$\binom{5}{0} 0.6^0 \cdot 0.4^5$	$\binom{5}{1} 0.6^1 \cdot 0.4^4$	$\binom{5}{2} 0.6^2 \cdot 0.4^3$	$\binom{5}{3} 0.6^3 \cdot 0.4^2$	$\binom{5}{4} 0.6^4 \cdot 0.4^1$	$\binom{5}{5} 0.6^5 \cdot 0.4^0$
$p(X = k)$	0.01024	0.0768	0.2304	0.3456	0.2592	0.07776

(c) $E(X) = np = 5 \cdot 0.6 = 3$

11. (a) Define the random variable X as the number of days late during a 1 month period.

$X = x_i$	$X = 0$	$X = 1$	$X = 2$	$X = 3$	$X = 4$	$X = 5$	$X = 6$
n	46	16	8	5	3	1	1
$p(x_i) = \frac{n}{80}$	0.5750	0.2	0.1	0.0625	0.0375	0.0125	0.0125

(b) $E(X) = x_1 p(x_1) + x_2 p(x_2) + x_3 p(x_3) + x_4 p(x_4) + x_5 p(x_5) + x_6 p(x_6)$
$= (0)(0.5750) + (1)(0.2) + (2)(0.1) + (3)(0.0625) + (4)(0.0375) + (5)(0.0125) + (6)(0.0125)$
$= 0.875$ day

13. (a) The shop is empty 20% of the time, so $P(0) = 0.2$, and the shop is full 30% of the time, so $P(2) = 0.3$. Then, since we have a distribution, the sum of the probabilities must be 1, and $P(X = 1) = 1 - (0.2 + 0.3) = 0.5$.

$X = x_i$	$X = 0$	$X = 1$	$X = 2$
$P(X = x_i)$	0.2	0.5	0.3

(b) $E(X) = x_1 p(x_1) + x_2 p(x_2) + x_3 p(x_3) = (0)(0.2) + (1)(0.5) + (2)(0.3) = 1.1$

There are 1.1 cars expected per 15 minute time slot.

15. (a) The random variable X is the number of persons in the sample who are 65 or older. X can take on the values, 0, 1, 2, 3, 4, and 5. X has a binomial distribution with $n = 5$ and $p = 0.124$. The distribution is

$p(0) = b(5, 0; 0.124) = 0.5158$ $\quad$ $p(1) = b(5, 1; 0.124) = 0.3651$
$p(2) = b(5, 2; 0.124) = 0.1034$ $\quad$ $p(3) = b(5, 3; 0.124) = 0.0146$
$p(4) = b(5, 4; 0.124) = 0.0010$ $\quad$ $p(5) = b(5, 5; 0.124) = 0.00002$

(b) The expected value of this binomial distribution is $E(X) = np = (5)(0.124) = 0.62$. We expect 62% of the sample to be at least 65 years of age.

17. (a) The random variable X has the values 0, 1, 2, 3, 4, 5.

(b) The probability distribution of X, the number of defective pens in the 5-pen sample is

$$p(0)=\frac{\binom{10}{0}\binom{90}{5}}{\binom{100}{5}}=0.5838 \quad p(1)=\frac{\binom{10}{1}\binom{90}{4}}{\binom{100}{5}}=0.3394 \quad p(2)=\frac{\binom{10}{2}\binom{90}{3}}{\binom{100}{5}}=0.0702$$

$$p(3)=\frac{\binom{10}{3}\binom{90}{2}}{\binom{100}{5}}=0.0064 \quad p(4)=\frac{\binom{10}{4}\binom{90}{1}}{\binom{100}{5}}=0.0003 \quad p(5)=\frac{\binom{10}{5}\binom{90}{0}}{\binom{100}{5}}=0.0000$$

(c) The expected value of X is

$$E(X)=x_1 p(x_1)+x_2 p(x_2)+x_3 p(x_3)+x_4 p(x_4)+x_5 p(x_5).$$

$$E(X)=(0)(0.5838)+(1)(0.3394)+(2)(0.0702)+(3)(0.0064)+(4)(0.0003)+(5)(0.0)$$
$$=0.5 \text{ defective pen}$$

19. Estimates will vary, but all should be close to the actual theoretical probability.

The actual probability $P(0.6\le X<0.9)=\dfrac{0.9-0.6}{1.0}=0.3$

Chapter 8 Review

1. $P(E\mid A)=0.82$

3. $P(E\mid B)=0.10$

5. $$P(E)=P(E\cap A)+P(E\cap B)=P(A)\cdot P(E\mid A)+P(B)\cdot P(E\mid B)$$
$$=(0.90)(0.82)+(0.10)(0.10)=0.748$$

$$P(A\mid E)=\frac{P(E\cap A)}{P(E)}=\frac{P(A)\cdot P(E\mid A)}{P(E)}=\frac{(0.90)(0.82)}{0.748}=0.9866$$

7. $$P(E)=P(E\cap A)+P(E\cap B)=P(A)\cdot P(E\mid A)+P(B)\cdot P(E\mid B)$$
$$=(0.90)(0.82)+(0.10)(0.10)=0.748$$

$$P(B\mid E)=\frac{P(E\cap B)}{P(E)}=\frac{P(B)\cdot P(E\mid B)}{P(E)}=\frac{(0.10)(0.10)}{0.748}=0.0134$$

9. $P(E\mid A)=0.5$

11. $P(E\mid B)=0.4$

13. $P(E\mid C)=0.3$

15. $$P(E)=P(E\cap A)+P(E\cap B)+P(E\cap C)=P(A)\cdot P(E\mid A)+P(B)\cdot P(E\mid B)+P(C)\cdot P(E\mid C)$$
$$=(0.4)(0.5)+(0.5)(0.4)+(0.1)(0.3)=0.43$$

17. $P(A \mid E) = \dfrac{P(E \cap A)}{P(E)} = \dfrac{P(A) \cdot P(E \mid A)}{P(E)} = \dfrac{(0.4)(0.5)}{0.43} = 0.4651$

19. $P(B \mid E) = \dfrac{P(E \cap B)}{P(E)} = \dfrac{P(B) \cdot P(E \mid B)}{P(E)} = \dfrac{(0.5)(0.4)}{0.43} = 0.4651$

21. $P(C \mid E) = \dfrac{P(E \cap C)}{P(E)} = \dfrac{P(C) \cdot P(E \mid C)}{P(E)} = \dfrac{(0.1)(0.3)}{0.43} = 0.0698$

23. We make a tree diagram to illustrate the problem.

Define the events

A_1: The item comes from factory 1.
A_2: The item comes from factory 2.
A_3: The item comes from factory 3.
E: The item is defective.

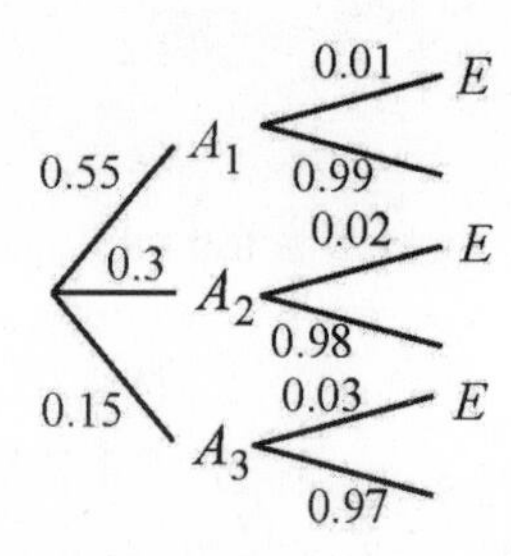

$P(E) = P(E \cap A_1) + P(E \cap A_2) + P(E \cap A_3)$
$= (0.55)(0.01) + (0.30)(0.02) + (0.15)(0.03) = 0.016$

(a) $P(A_1 \mid E) = \dfrac{P(E \cap A_1)}{P(E)} = \dfrac{P(A_1) \cdot P(E \mid A_1)}{P(E)} = \dfrac{(0.55)(0.01)}{0.016} = 0.3438$

(b) $P(A_2 \mid E) = \dfrac{P(E \cap A_2)}{P(E)} = \dfrac{P(A_2) \cdot P(E \mid A_2)}{P(E)} = \dfrac{(0.30)(0.02)}{0.016} = 0.375$

(c) $P(A_3 \mid E) = \dfrac{P(E \cap A_3)}{P(E)} = \dfrac{P(A_3) \cdot P(E \mid A_3)}{P(E)} = \dfrac{(0.15)(0.03)}{0.016} = 0.2813$

25. This is a Bernoulli experiment. We have $n = 5$ and $p = 0.2$.

(a) The probability that none of the people chosen will purchase the product is
$P(\text{exactly 0 purchases}) = b(5, 0; 0.2) = C(5, 0)(0.2^0)(0.8^5) = (1)(1)(0.32768) = 0.3277$

(b) The probability that exactly 3 will purchase the product is
$P(\text{exactly 3 purchases}) = b(5, 3; 0.2) = C(5, 3)(0.2^3)(0.8^2) = (10)(0.008)(0.64) = 0.0512$

27. This is a Bernoulli experiment with $n = 12$ and $p = 0.5$

(a) The probability the student will get all answers correct is
$P(\text{exactly 12 correct}) = b(12, 12; 0.5) = C(12, 12)(0.5^{12}) = 0.0002$

(b) The probability the student passes the test is

$P(\text{at least 7 correct})$
$= P(\text{exactly 7 correct}) + P(\text{exactly 8 correct}) + P(\text{exactly 9 correct}) + P(\text{exactly 10 correct}) + P(\text{exactly 11 correct}) + P(\text{exactly 12 correct})$
$= b(12, 7; 0.5) + b(12, 8; 0.5) + b(12, 9; 0.5) + b(12, 10; 0.5) + b(12, 11; 0.5) + b(12, 12; 0.5)$
$= 0.19336 + 0.12085 + 0.05371 + 0.01611 + 0.00293 + 0.00024 = 0.3872$

(c) The odds in favor of passing are 3872 to 6128 or 242 to 383.

29. This is a Bernoulli experiment with $n = 7$, the number of times the coin is tossed, and $p = 0.5$, the probability of tossing a head. The expected value of a Bernoulli experiment is given by $E = np$. The expected number of heads when a coin is tossed 7 times is

$$E = np = (7)(0.5) = 3.5$$

We expect 3.5 heads when we toss a coin 7 times.

31. Let the random variable X denote the number of red balls chosen. $X = 0$, 1, or 2.

$$p(X = 0) = \frac{C(2,0)\cdot C(4,2)}{C(6,2)} = \frac{6}{15} \qquad p(X = 1) = \frac{C(2,1)\cdot C(4,1)}{C(6,2)} = \frac{8}{15}$$

$$p(X = 2) = \frac{C(2,2)\cdot C(4,0)}{C(6,2)} = \frac{1}{15}$$

The expected value, E, of the game is

$$E = (0)(P(X = 0)) + (1)(P(X = 1)) + (2)(P(X = 2))$$
$$= 0\cdot\frac{6}{15} + 1\cdot\frac{8}{15} + 2\cdot\frac{1}{15} = \frac{10}{15} = \$0.6667$$

The expected value of the game is 66.67 cents, Frank paid 70 cents; so Frank paid $3\frac{1}{3}$ cents too much.

33. The game pictured in Problem 33 indicates that we can win \$0.80 with probability $\frac{1}{6}$, \$0.30 with probability $\frac{1}{3}$, and \$0.10 with probability $\frac{1}{2}$. The expected value of the game is

$$E = 0.80\cdot\frac{1}{6} + 0.30\cdot\frac{1}{3} + 0.10\cdot\frac{1}{2} = 0.2833$$

The game is not fair. Fair games have an expected value equal to 0.

35. In reality the player has paid the \$1.00 so if the color matches the bet, he wins only \$1.00. If the color does not match the bet he loses the \$1.00 that he bet.

$$E = (1)\left(\frac{18}{37}\right) + (-1)\left(\frac{19}{37}\right) = -\frac{1}{37} = -0.027$$

The game is not fair, a player can expect to lose 2.7 cents.

37. This is a Bernoulli experiment with $n = 100$. Since 1 out of 5 choices is correct, 4 of the 5 answer choices are incorrect. We are interested in the expected number of wrong answers, so $p = \frac{4}{5}$, and

$$E = np = (100)\left(\frac{4}{5}\right) = 80$$

A student who guesses on the 100 question test should expect to get 80 answers wrong.

39. (a) The random variable X has the values {0, 1, 2, 3} where X is the number of yellow flowers that grow.

(b) The probability distribution of X, the number of yellow flowers in the 3-bulb sample is

$$P(0) = \frac{\binom{3}{0}\binom{7}{3}}{\binom{10}{3}} = 0.2917 \qquad P(1) = \frac{\binom{3}{1}\binom{7}{2}}{\binom{10}{3}} = 0.525$$

$$P(2) = \frac{\binom{3}{2}\binom{7}{1}}{\binom{10}{3}} = 0.175 \qquad P(3) = \frac{\binom{3}{3}\binom{7}{0}}{\binom{10}{3}} = 0.0083$$

(c) The expected value of X is $E(X) = x_1 P(x_1) + x_2 P(x_2) + x_3 P(x_3)$.
Here $E(X) = (0)(0.2917) + (1)(0.525) + (2)(0.175) + (3)(0.0083) = 0.8999$.

We expect 0.8999 yellow flower to grow.

41. (a) The random variable X has the values {0, 1, 2, 3}.

(b) The probability distribution of X, the number of defective items in the sample, is Binomial with $n = 3$ and $p = 0.04$

$$P(0) = b(3, 0; 0.04) = 0.8847 \qquad P(1) = b(3, 1; 0.04) = 0.1106$$
$$P(2) = b(3, 2; 0.04) = 0.0046 \qquad P(3) = b(3, 3; 0.04) = 0.0001$$

(c) The expected value of X, $E(X) = np = (3)(0.04) = 0.12$ defective items per sample.

Chapter 8 Project

1. The probability that a defective spacecraft is detected using 4 or fewer tests means that a defect was found on the 1st test, or it passed the 1st test but failed the 2nd test, or it passed the tests 1 and 2 but failed the 3rd test, or it passed tests 1 through 3 but failed test 4. Since the tests are independent, $P(E \cap F) = P(E)P(F)$.

The probability a defect is found in 4 or fewer tests is

$$P(F) + P(E)P(F) + P(E)P(E)P(F) + P(E)P(E)P(E)P(F)$$
$$= (0.9) + (0.1)(0.9) + (0.1)(0.1)(0.9) + (0.1)(0.1)(0.1)(0.9) = 0.9999$$

3. We use G: The spacecraft is good, and D: The spacecraft is defective.

(a) $$P(G \mid E) = \frac{P(G)\cdot P(E \mid G)}{P(E)} = \frac{P(G)\cdot P(E \mid G)}{P(G)\cdot P(E \mid G) + P(D)\cdot P(E \mid D)} = \frac{(0.05)(1)}{(0.05)(1) + (0.95)(0.1)}$$
$$= 0.3448$$

(b) $$P(D \mid E) = \frac{P(D)\cdot P(E \mid D)}{P(E)} = \frac{P(D)\cdot P(E \mid D)}{P(G)\cdot P(E \mid G) + P(D)\cdot P(E \mid D)} = \frac{(0.95)(0.1)}{(0.05)(1) + (0.95)(0.1)}$$
$$= 0.6552$$

5. Define event K: The spacecraft passes 3 tests. Assuming the tests are independent, we get

$$P(K \mid G) = P(E \mid G) \cdot P(E \mid G) \cdot P(E \mid G) = 1 \text{ and}$$
$$P(K \mid D) = P(E \mid D) \cdot P(E \mid D) \cdot P(E \mid D) = 0.1^3 = 0.001$$

$$P(K) = P(G) \cdot P(K \mid G) + P(D) \cdot P(K \mid D)$$
$$= (0.05) \cdot (1) + (0.95) \cdot (0.001) = 0.05095$$

$$P(G \mid K) = \frac{P(G) \cdot P(K \mid G)}{P(K)} = \frac{(0.05)(1)}{0.05095} = 0.9814$$

7.

Number of Tests	0	1	2	3	4
$P(G)$	0.05	0.3448	0.8403	0.9814	0.9981

Mathematical Questions from the Professional Exams

1. (d) $P(\text{no fewer than 1 and no more than 9}) = 1 - [P(\text{no heads}) + P(\text{10 heads})]$

$$= 1 - \frac{1}{2^{10}} - \frac{1}{2^{10}} = 1 - 2\left(\frac{1}{2^{10}}\right) = 1 - \frac{1}{2^9}$$

3. (a) First we find the expected value of the investment if the investor holds it, and then we compare it to \$10,000, the return if the investor sells now.

$$E = (5000)(0.4) + (8000)(0.2) + (12{,}000)(0.3) + (30{,}000)(0.1) = \$10{,}200$$

5. (b) $E = (6000)(.2) + (8000)(.2) + (10{,}000)(.2) + (12{,}000)(.2) + (14{,}000)(.1) + (16{,}000)(.1) = 10{,}200$

7. (b) or (c)

$$E(\#1) = (100{,}000)(.7) + (70{,}000)(.3)$$
$$E(\#2) = (170{,}000 - 40{,}000)(.8) + (80{,}000 - 40{,}000)(.2)$$
$$= (130{,}000)(.8) + (40{,}000)(.2)$$
$$= (170{,}000)(.8) + (80{,}000)(.2) - (40{,}000)(.8 + .2)$$
$$= (170{,}000)(.8) + (80{,}000)(.2) - 40{,}000$$

Chapter 9

Statistics

9.1 Introduction to Statistics: Data and Sampling

1. Variable

3. False. A discrete random variable can assume only a finite set of values or as many values as there are whole numbers.

5. The variable is the number of heads thrown, and it is discrete.

7. The variable is the miles per gallon, and it is continuous.

9. The variable is the time of waiting in line, and it is continuous.

11. The variable is the number of flights, and it is discrete

13. The variable is the number of people crossing the intersection, and it is discrete.

15. The variable is length of time, and it is continuous.

17. Answers will vary. All answers should include a method to choose a group of viewers for which each viewer of the program has an equal chance of being chosen.

19. Answers will vary. All answers should include a method to choose a sample in which each member of the population has an equal chance of being selected.

21. Answers will vary. All answers should include a method to choose a sample in which each member of the population has an equal chance of being selected.

23–25. Answers will vary. All answers should give examples of possible bias.

9.2 Representing Data Graphically: Bar Graphs; Pie Charts

1. Bar graphs; pie charts

3. False. In a pie chart a circle is divided into sectors, one sector for each category represented by the data. The size of each sector is proportional to the size of the category.

5.

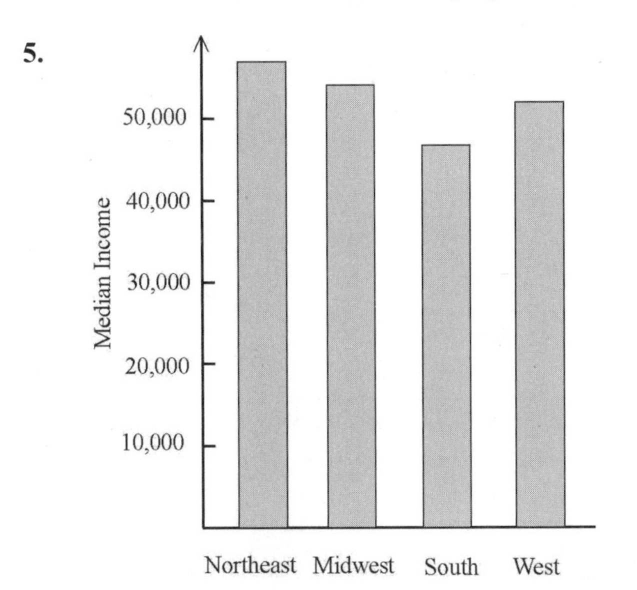

(b) The Northeast has the highest median income.

(c) The South has the lowest median income.

(d) Answers will vary. All discussions should conclude that pie chars are inappropriate.

7. (a)

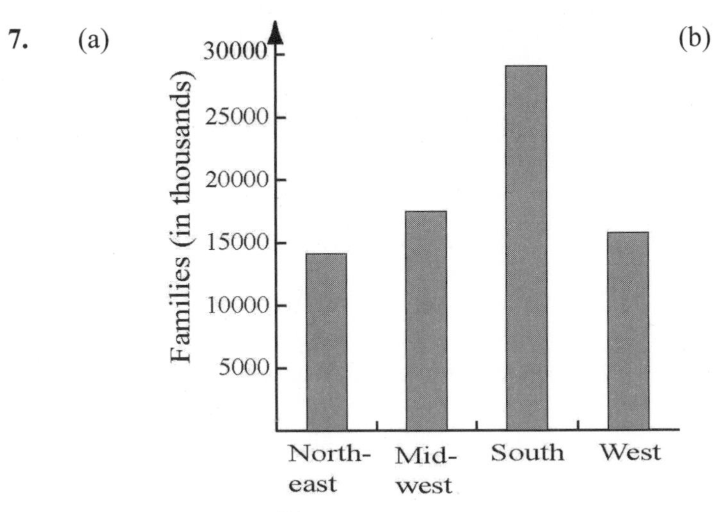

(b)

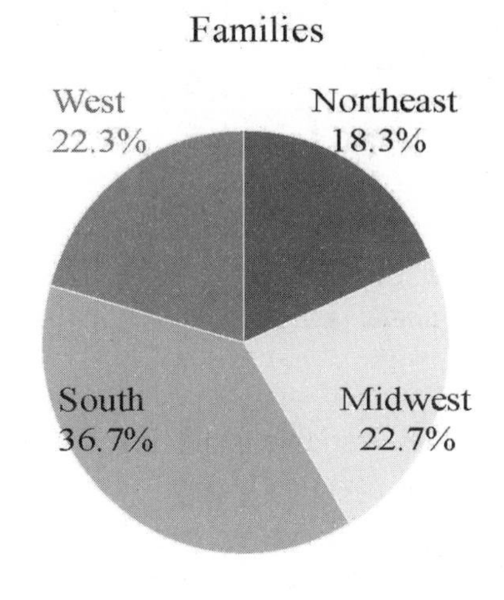

(c) Answers will vary.
(d) The South has the most families.
(e) The Northeast has the fewest families.
(f) Answers will vary.

9. (a)

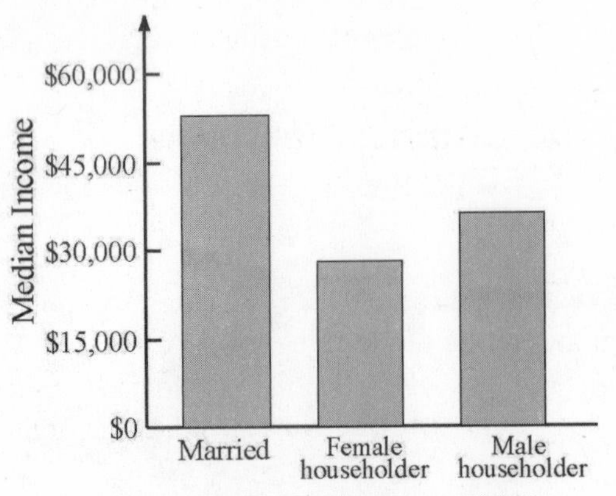

(b) Married – couple families have the highest median income.

(c) Female householder – no spouse families have the lowest median income.

(d) Answers will vary.

11. (a)

(b) A pie chart would be appropriate since all causes of death are represented.
(d) The leading cause of death among Americans in 2004 was heart disease.

13. (a) Hawaiian Airlines had the highest percentage of on-time flights with 91.9% of its flights on time.
(b) Sky West had the lowest percentage of on-time flights. Only 65.0% of its flights were on time.

(c) 73.2% of United Airlines' flights were on time in January 2007.

15. (a) The largest component of the CPI is housing, fuel, and utilities. It comprises 43% of the Consumer Price Index.
(b) The smallest component of the CPI is other goods and services. It accounts for only 3% of the CPI.

(c) Answers will vary.

9.3 Organizing and Displaying Quantitative Data

3. class intervals

5. (a)

Score	Frequency	Score	Frequency	Score	Frequency
25	1	36	2	47	1
26	1	37	4	48	3
27	0	38	1	49	1
28	1	39	1	50	1
29	1	40	1	51	1
30	3	41	5	52	3
31	2	42	3	53	2
32	1	43	1	54	2
33	2	44	2	55	1
34	2	45	1		
35	1	46	2		

(b)

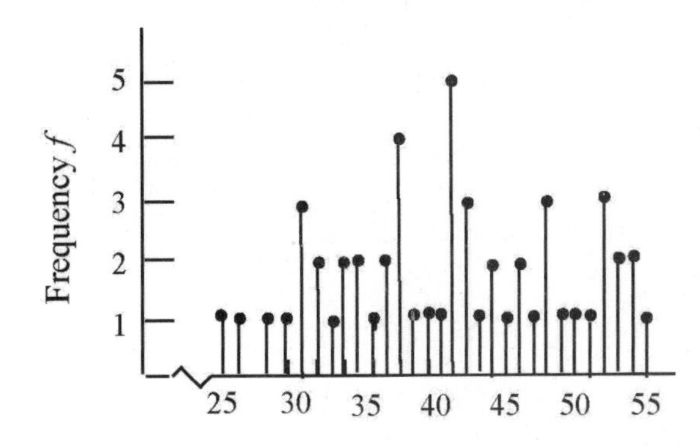

(c) and (f)

Class Interval	Frequency	Cumulative Frequency	Class Interval	Frequency	Cumulative Frequency
24 – 25.9	1	1	40 – 41.9	6	29
26 – 27.9	1	2	42 – 43.9	4	33
28 – 29.9	2	4	44 – 45.9	3	36
30 – 31.9	5	9	46 – 47.9	3	39
32 – 33.9	3	12	48 – 49.9	4	43
34 – 35.9	3	15	50 – 51.9	2	45
36 – 37.9	6	21	52 – 53.9	5	50
38 – 39.9	2	23	54 – 55.9	3	53

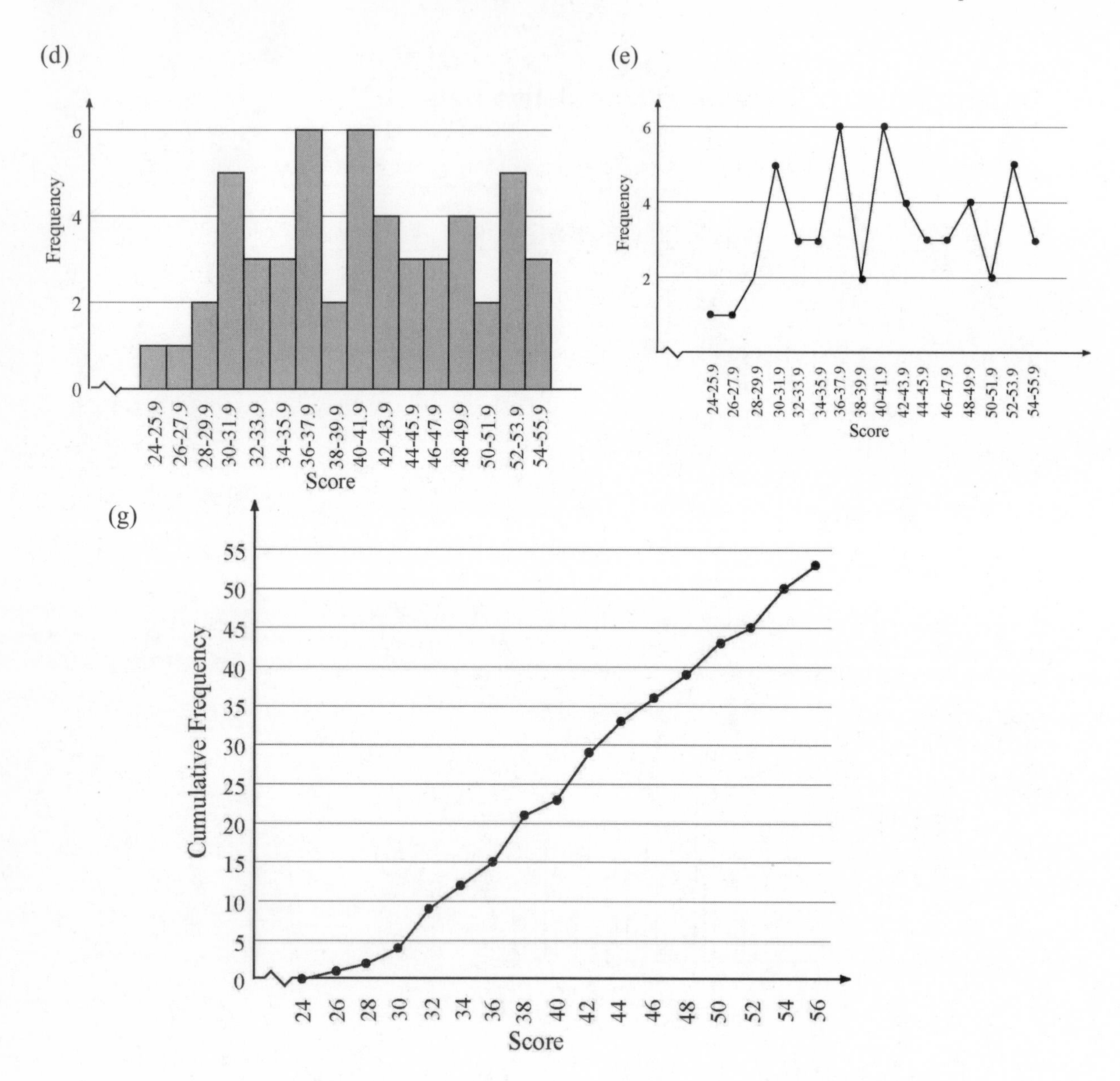

(d)
Frequency
0
2
4
6
24-25.9
26-27.9
28-29.9
30-31.9
32-33.9
34-35.9
36-37.9
38-39.9
40-41.9
42-43.9
44-45.9
46-47.9
48-49.9
50-51.9
52-53.9
54-55.9
Score
(e)
Frequency
2
4
6
24-25.9
26-27.9
28-29.9
30-31.9
32-33.9
34-35.9
36-37.9
38-39.9
40-41.9
42-43.9
44-45.9
46-47.9
48-49.9
50-51.9
52-53.9
54-55.9
Score
(g)
Cumulative Frequency
0
5
10
15
20
25
30
35
40
45
50
55
24
26
28
30
32
34
36
38
40
42
44
46
48
50
52
54
56
Score

7. (a) and (d)

Class Interval	Frequency	Cumulative Frequency
50 – 54.9	1	1
55 – 59.9	6	7
60 – 64.9	3	10
65 – 69.9	6	16
70 – 74.9	8	24
75 – 79.9	11	35
80 – 84.9	2	37
85 – 89.9	12	49
90 – 94.9	12	61
95 – 99.9	2	63
100 – 104.9	2	65
105 – 109.9	4	69
110 – 114.9	0	69
115 – 119.9	2	71

(b)

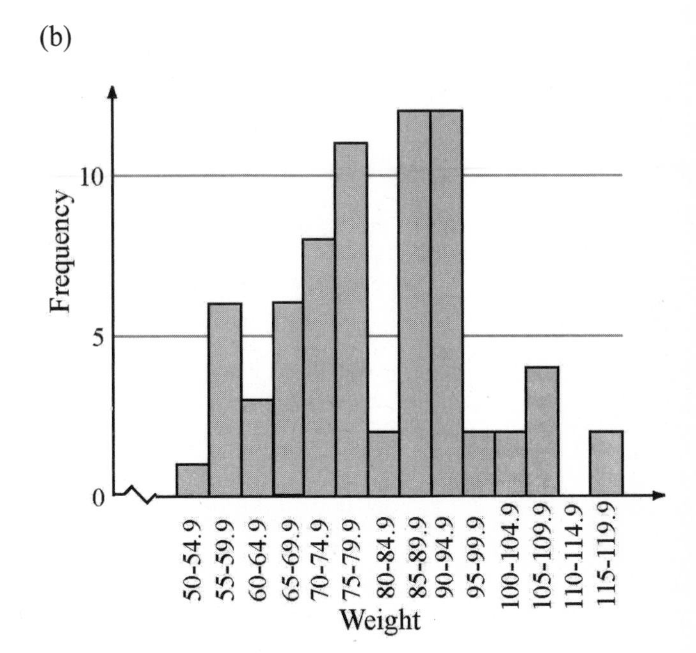

(c)

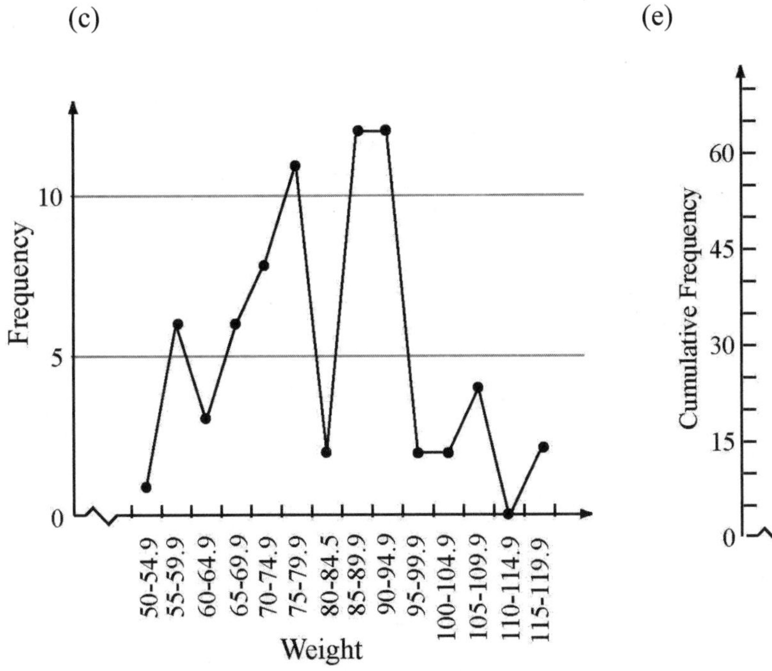

(e)

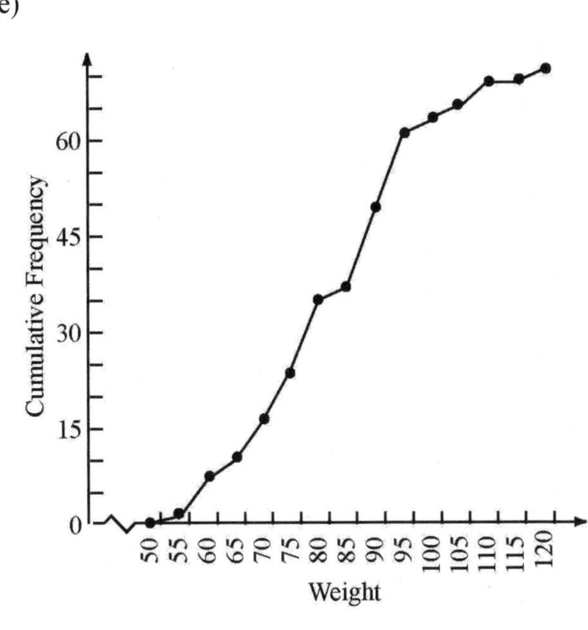

9. Since the data is clustered on the left giving the graph a tail on the right, this distribution is skewed right.

11. (a) There are 13 class intervals.

(b) The lower class limit of the first class interval is 20 and the upper class limit of the first class interval is 24.

(c) The class width is the difference between consecutive lower class limits. The class width is 5.

(d) To find the number of licensed drivers from 70 to 84 years old, add the drivers in the class intervals 70–74, 75–79, and 80–84. There are approximately 1,500,000 drivers between 70 and 84 years old.

(e) The 40–44 year old class interval has the most drivers.

(f) The 80–84 year old class interval has the fewest drivers.

(g) The distribution is clustered on the left and has a longer tail on the right. The distribution is skewed right.

(h)

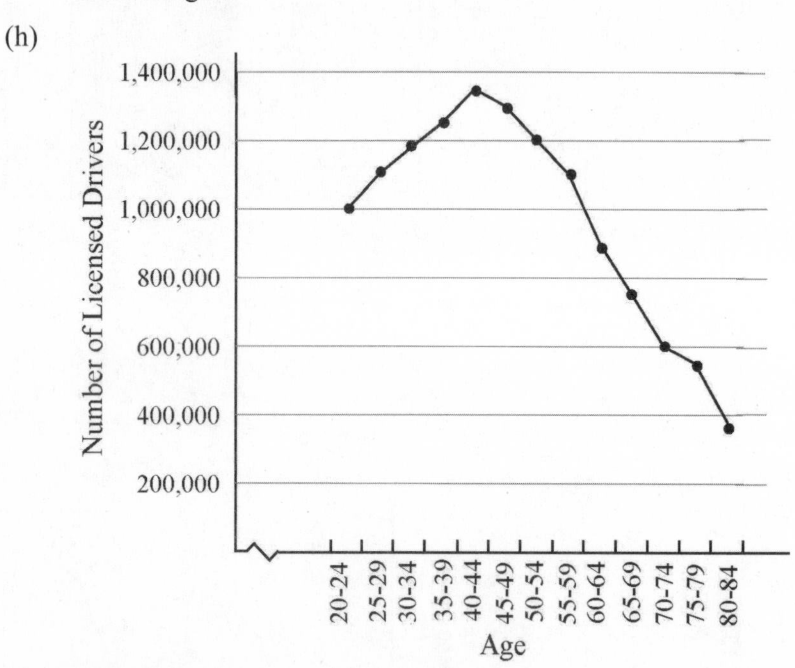

13. (a) There are 7 class intervals.

(b) The lower class limit of the last class interval is 80, and the upper class limit of the last class interval is 89.

(c) and (f)

Class Interval	Frequency	Cumulative Frequency
20 – 29	2,100,000	2,100,000
30 – 39	2,400,000	4,500,000
40 – 49	2,700,000	7,200,000
50 – 59	2,300,000	9,500,000
60 – 69	1,700,000	11,200,000
70 – 79	1,150,000	12,350,000
80 – 89	380,000	12,730,000

(d) The new class width makes the distribution less skewed.

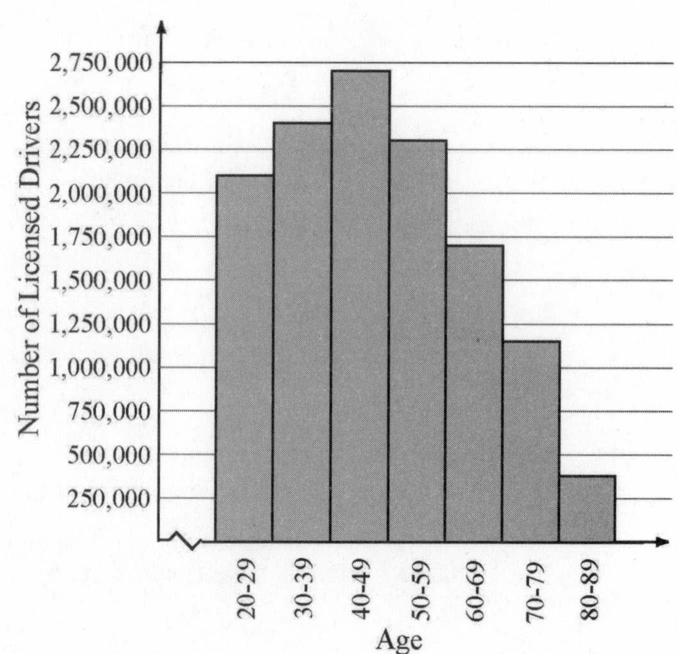

(e)

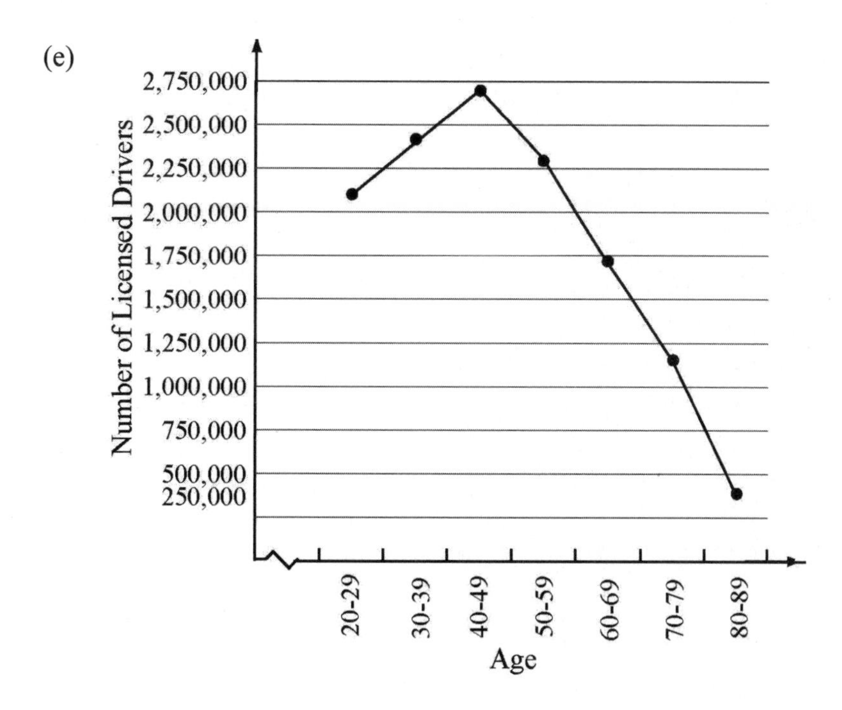

(g)

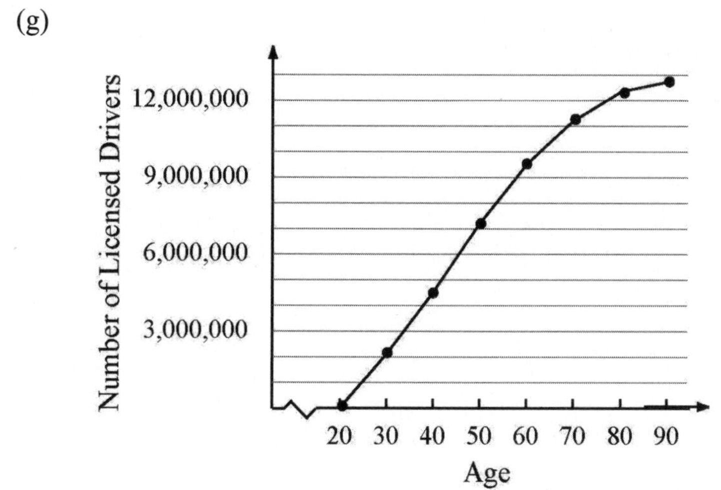

15. (a) There are 13 class intervals.

(b) The lower class limit of the first class interval is 20, and the upper class limit of the first class interval is 24.

(c) The class width is difference between consecutive lower class limits. Here the class width is 5 years.

(d)

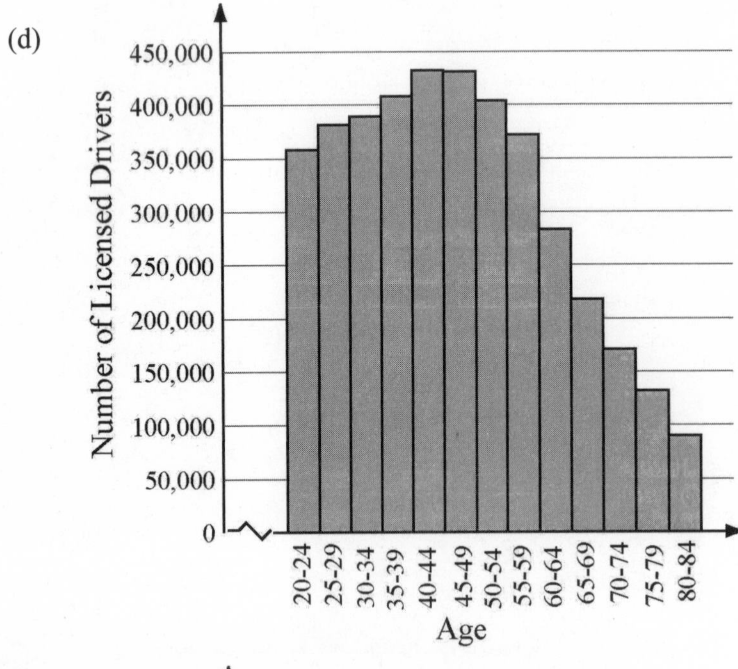

(e)

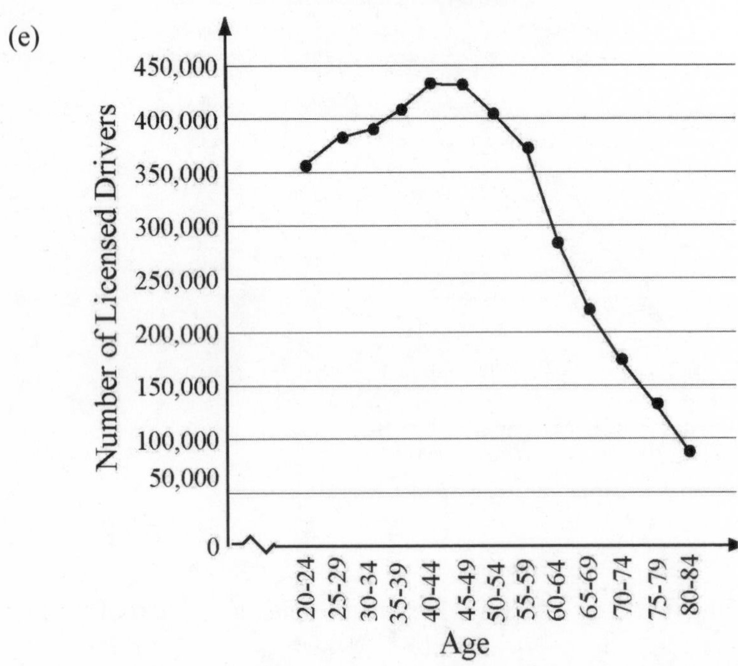

(f) Most licensed drivers are in the 40-44 year-old age group.

(g) The fewest licensed drivers are in the 80–84 year-old age group.

17. (a) There are 19 class intervals.

(b) The lower class limit of the first class interval is \$0, and the upper class limit of the first class interval is \$1999.

(c) The class width is difference between consecutive lower class limits. Here the class width is \$2000.

(d)

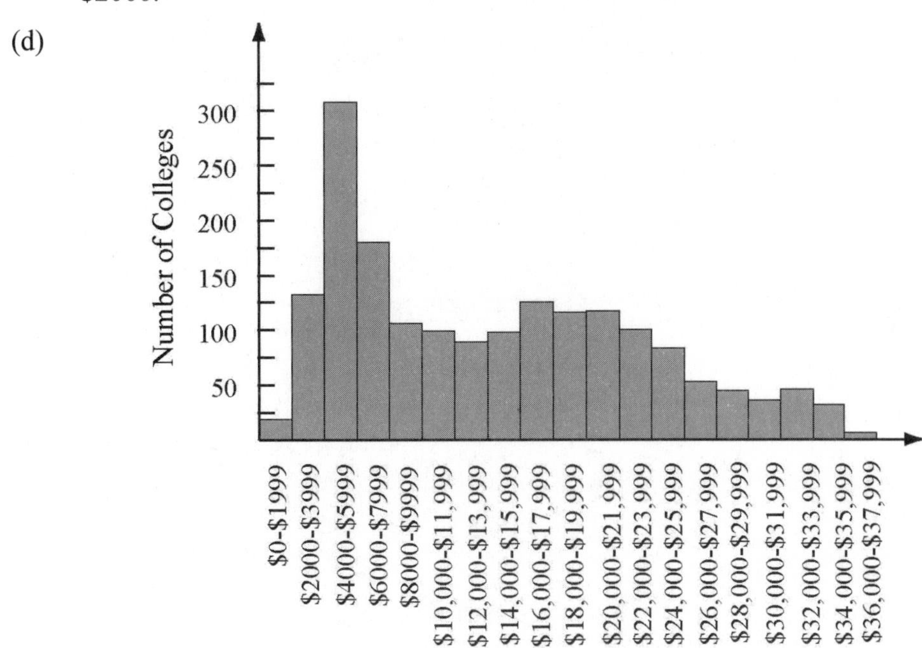

(e)

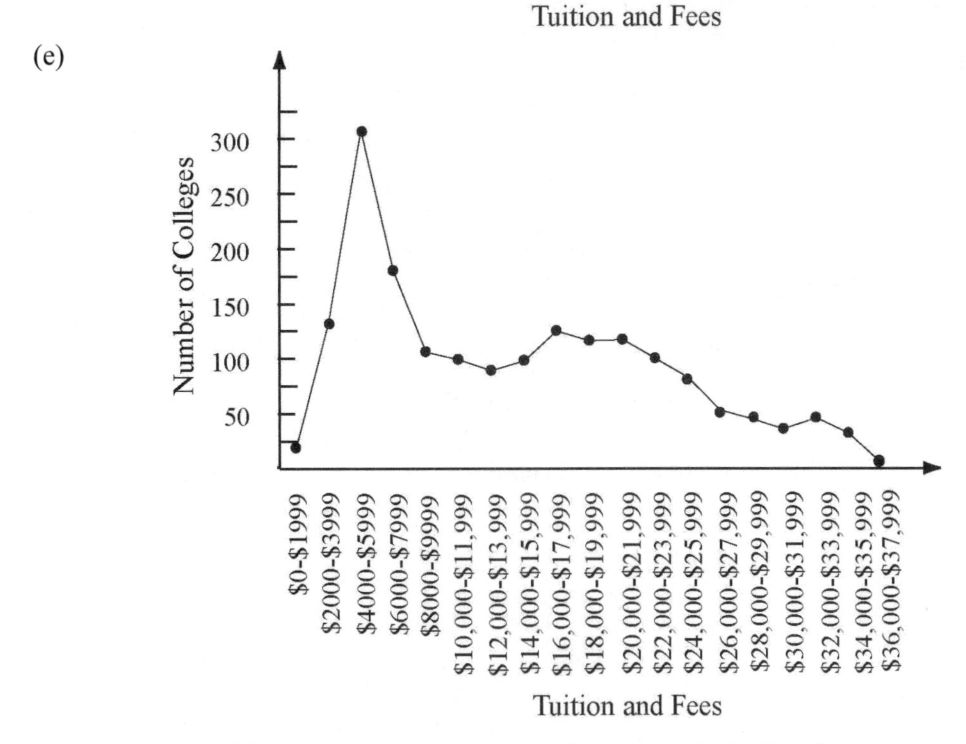

(f) Tuition in 2006-2007 was most frequently in the \$4000–\$5999 range.

19. (a)

	Class Interval	Frequency
1	11.7–12.2	2
2	12.3–12.8	4
3	12.9–13.4	2
4	13.5–14.0	6
5	14.1–14.6	1
6	14.7–15.2	1
7	15.3–15.8	3
8	15.9–16.4	1

(b)

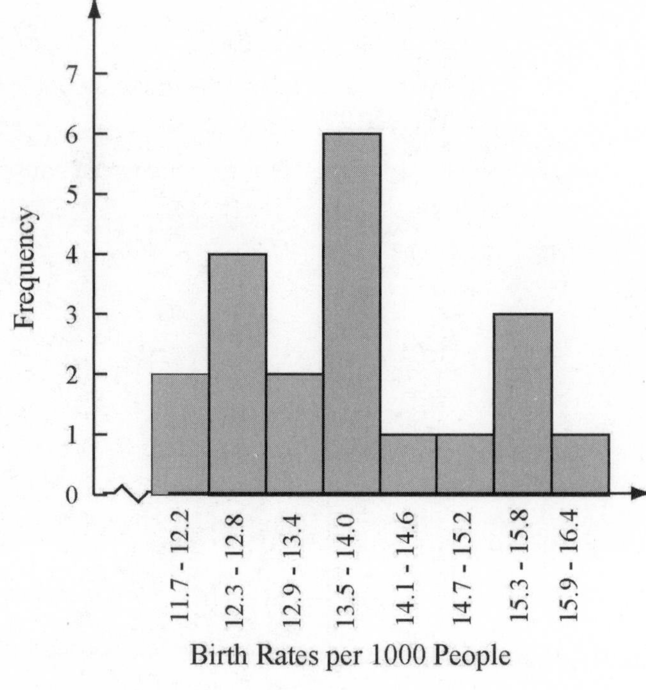

(c)

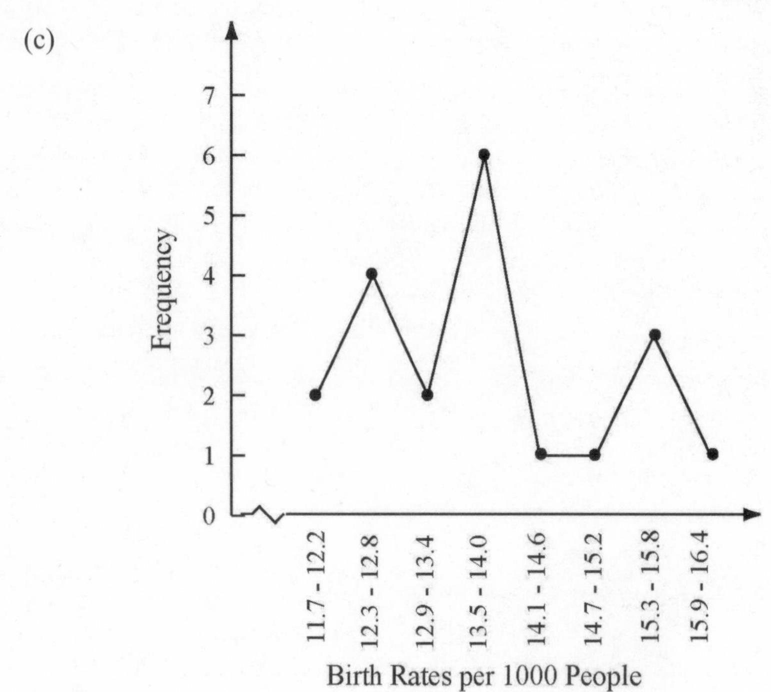

21. (a)

	Class Interval	Frequency
1	0.0 – 1.9	4
2	2.0 – 3.9	9
3	4.0 – 5.9	3
4	6.0 – 7.9	1
5	8.0 – 9.9	1
6	10.0 – 11.9	2

(b)

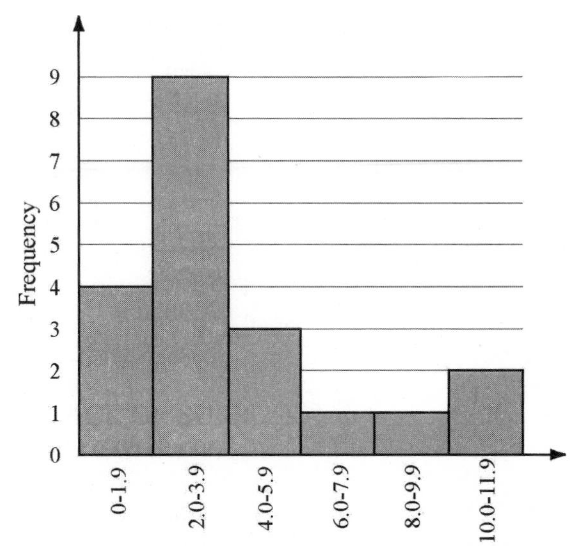

(c)

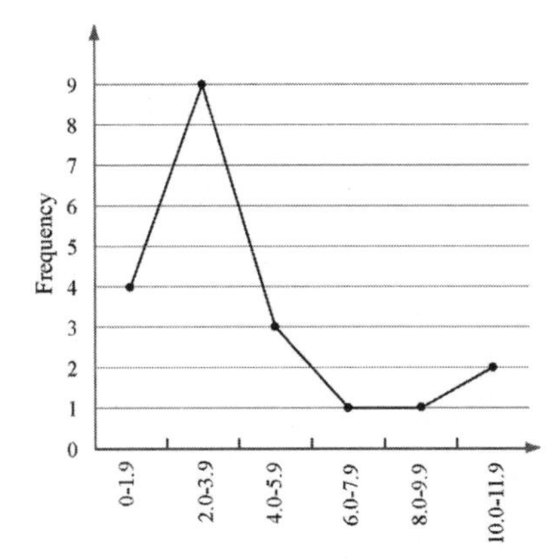

9.4 Measures of Central Tendency

1. Mean, median, mode

3. True

5. μ; $\bar{x}$

7. (a) Mean: $\bar{x} = \dfrac{21+25+43+36}{4} = 31.25$

(b) To find the median the data must be ranked from smallest to largest: 21, 25, 36, 43. Then since there is an even number of data points, the median is given by:

$$\frac{25+36}{2} = 30.5$$

(c) Since no point is repeated more than once, there is no mode.

9. (a) Mean: $\bar{x} = \dfrac{55+55+80+92+70}{5} = 70.4$

(b) To find the median, the data must be ranked from smallest to largest: 55, 55, 70, 80, 92 Now since there is an odd number of data points, the median is given by the middle point, 70.

(c) The mode is 55.

11. (a) Mean: $\bar{x} = \dfrac{65+82+82+95+70}{5} = 78.8$

(b) To find the median, the data must be ranked from smallest to largest: 65, 70, 82, 82, 95
There is an odd number of data points, so the median is given by the middle point, 82.

(c) The mode is 82.

13. (a) Mean: $\bar{x} = \dfrac{48+65+80+92+80+75}{6} = 73.333$

(b) To find the median, the data must be ranked from smallest to largest: 48, 65, 75, 80, 80, 92
Then since there is an even number of data points, the median is given by

$$\frac{75+80}{2} = 77.5$$

(c) The mode is 80.

15. (a) To find the mean age of the players on the Yankees, add all their ages and divide by 40, the number of players on the team.

$$\bar{x} = \frac{1153}{40} = 28.825 \text{ years of age.}$$

(b) To find the median, the data must be ranked from smallest to largest.

22, 23, 23, 23, 23, 24, 24, 24, 24, 25, 25, 25, 26, 26, 26, 26, 27, 27, 28, 29,
29, 29, 30, 30, 30, 31, 31, 32, 32, 32, 32, 32, 33, 33, 34, 35, 36, 37, 37, 38

Since there is an even number of players on the team, the median age is mean of the two middle values on the ordered list. The median age $= \dfrac{29+29}{2} = 29$ years.

(c) The mode is the value that occurs most often in the list. The mode is 32 years.

17. To find the mean cost per share, we use a method similar to finding the mean of grouped data. The result is often called the weighted average, or the weighted mean.

$$\bar{x} = \frac{\sum(\text{cost per share})(\text{number of shares purchased})}{\text{total number of shares purchased}}$$

$$\bar{x} = \frac{(\$85)(50) + (\$105)(90) + (\$110)(120) + (\$130)(75)}{50+90+120+75}$$

$$= \frac{36650}{335} = 109.40$$

The investor paid a mean price of \$109.40 per share of the IBM stock.

19. (a) To find the mean of grouped data we use the following steps:

Step 1: Multiply the midpoint, m_i, of each class interval by the frequency, f_i, for that class interval.

Step 2: Sum the products found in step 1.

Step 3: Compute the mean by dividing the sum of $f_i m_i$ by the number n of entries.

Class Interval	Midpoint, m_i	Frequency, f_i	$f_i m_i$
10–14	12.5	7	87.5
15–19	17.5	415	7262.5
20–24	22.5	1034	23,265
25–29	27.5	1104	30,360
30–34	32.5	966	31,284
35–39	37.5	476	17,850
40–44	42.5	104	4,420
45–49	47.5	6	285
Total		n = 4112	114,925

The mean age of a new mother in the United States in 2004 was $\bar{x} = \dfrac{114{,}925}{4112} = 27.95$ years.

(b) To find the median of grouped data we use the following steps.

Step 1: Find the interval containing the median. The median is the middle value when the 4112 items are in ascending order. So the median in this example is the mean of the 2056th and 2057th entries. The 2056th entry is in the interval 25–29.

Step 2: In the interval containing the median, count the number p of items remaining to reach the median. The first three intervals account for 1456 data points, there are $2056 - 1456 = 600$ points left, so $p = 600$.

Step 3: Calculate the interpolation factor.

q is the frequency for the interval containing the median; $q = 1104$

i is the size of the interval; $i = 5$

$$\text{interpolation factor } = \frac{p}{q} \cdot i = \frac{600}{1104} \cdot 5 = 2.72$$

Step 4: The median M is

$$M = \begin{bmatrix} \text{lower limit of interval} \\ \text{containing the median} \end{bmatrix} + [\text{interpolation factor}]$$

$$M = 25 + 2.717 = 27.72$$

21. (a) To find the mean of grouped data we use the following steps:

Step 1: Multiply the midpoint, m_i, of each class interval by the frequency, f_i, for that class interval.

Step 2: Sum the products found in step 1.

Step 3: Compute the mean by dividing the sum of $f_i m_i$ by the number n of entries.

Class Interval	Frequency, f	Midpoint, m_i	$f_i m_i$
20 – 24	357,986	22.5	8,054,685
25 – 29	381,703	27.5	10,496,832.5
30 – 34	389,501	32.5	12,658,782.5
35 – 39	408,043	37.5	15,301,612.5
40 – 44	432,601	42.5	18,385,542.5
45 – 49	431,431	47.5	20,492,972.5
50 – 54	403,996	52.5	21,209,790
55 – 59	371,684	57.5	21,371,830
60 – 64	283,136	62.5	17,696,000
65 – 69	217,832	67.5	14,703,660
70 – 74	171,004	72.5	12,397,790
75 – 79	132,012	77.5	10,230,930
80 – 84	89,812	82.5	7,409,490
Total	4,070,741		190,409,917.5

The mean age of a driver in Tennessee is $\bar{x} = \frac{\Sigma f_i m_i}{n} = \frac{190{,}409{,}917.5}{4{,}070{,}741} = 46.78$ years.

(b) To find the median of grouped data we use the following steps:

Step 1: Find the interval containing the median. The median is the 2,035,370[th] entry. It is in the interval. This entry is in the interval 45 – 49.

Step 2: In the interval, count the number p of items remaining to reach the median. The first five intervals account for 1,969,834 data points. We need 2,035,370 – 1,969,834 = 65,536 items, so $p = 65{,}536$.

Step 3: Calculate the interpolation factor.
q is the frequency for the interval containing the median; $q = 431{,}431$
i is the size of the interval; $i = 5$

$$\text{interpolation factor } = \frac{p}{q} \cdot i = \frac{65{,}536}{431{,}431} \cdot 5 = 0.760$$

Step 4: The median M is

$$M = \begin{bmatrix} \text{lower limit of interval} \\ \text{containing the median} \end{bmatrix} + [\text{interpolation factor}] = 45 + 0.760 = 45.8$$

The median age of a driver in Tennessee is 45.8 years.

23. To find the mean of grouped data we use the following steps:

Step 1: Multiply the midpoint, m_i, of each class interval by the frequency, f_i, for that class interval.

Step 2: Sum the products found in step 1.

Step 3: Compute the mean by dividing the sum of $f_i m_i$ by the number n of entries.

Interval	Midpoint, m_i	Frequency, f_i	$m_i \cdot f_i$
$0 – $1,999	$1000	19	19,000
$2000 – $3,999	$3000	132	396,000
$4000 – $5,999	$5000	308	1,540,000
$6,000 – $7,999	$7000	180	1,260,000
$8,000 – $9,999	$9000	106	954,000
$10,000 – $11,999	$11,000	99	1,089,000
$12,000 – $13,999	$13,000	89	1,157,000
$14,000 – $15,999	$15,000	98	1,470,000
$16,000 – $17,999	$17,000	125	2,125,000
$18,000 – $19,999	$19,000	116	2,204,000
$20,000 – $21,999	$21,000	117	2,457,000
$22,000 – $23,999	$23,000	100	2,300,000
$24,000 – $25,999	$25,000	83	2,075,000
$26,000 – $27,999	$27,000	53	1,431,000
$28,000 – $29,999	$29,000	45	1,305,000
$30,000 – $31,999	$31,000	36	1,116,000
$32,000 – $33,999	$33,000	46	1,518,000
$34,000 – $35,999	$35,000	32	1,120,000
$36,000 – $37,999	$37,000	6	222,000
Total		1790	25,758,000

The mean tuition at 4 year colleges in 2006-2007 is $\bar{x} = \dfrac{\Sigma f_i m_i}{n} = \dfrac{25{,}758{,}000}{1790} = \$14{,}390$.

25. (a) Mean:

$$\bar{x} = \frac{34{,}000 + 35{,}000 + 36{,}000 + 36{,}500 + 65{,}000}{5} = \frac{206{,}500}{5} = 41{,}300$$

The mean faculty salary is $41,300.

Median: The median is the middle entry once the items are placed in ascending order.

$$M = \$36000$$

(b) The median describes the situation more realistically because the data is skewed to the right. This means that 4 salaries are clustered, while the salary of $65,000 is much higher.

9.5 Measures of Dispersion

1. Dispersion or spread

3. 68

5. False. The standard deviation is preferred to the variance because the units are not squared.

7.

Score, x_i	Deviation from the Mean, $x - \bar{x}$	Deviation Squared $\left(x - \bar{x}\right)^2$
4	–6.85714	47.02041
5	–5.85714	34.30612
9	–1.85714	3.44898
9	–1.85714	3.44898
10	–0.85714	0.73469
14	3.142857	9.87755
25	14.14286	200.0204
Mean $= 10.86$ $n = 7$	Sum = 0	Sum = 298.8571

The standard deviation is $s = \sqrt{\dfrac{\sum (x_i - \bar{x})^2}{n-1}} = \sqrt{\dfrac{298.8571}{7-1}} = 7.058$.

9.

Score, x_i	Deviation from the Mean, $x - \bar{x}$	Deviation Squared $\left(x - \bar{x}\right)^2$
62	–3	9
58	–7	49
70	5	25
70	5	25
$n = 4$ $\bar{x} = 65$		sum = 108

The standard deviation is $s = \sqrt{\dfrac{\sum (x_i - \bar{x})^2}{n-1}} = \sqrt{\dfrac{108}{3}} = 6$

11.

Score, x_i	Deviation from the Mean, $x-\bar{x}$	Deviation Squared $\left(x-\bar{x}\right)^2$
85	5	25
75	−5	25
62	−18	324
78	−2	4
100	20	400
$\bar{x}=80$		sum = 778

The standard deviation is $s=\sqrt{\dfrac{\sum(x_i-\bar{x})^2}{n-1}}=\sqrt{\dfrac{778}{4}}=13.946$

13.

Class Interval	Class Midpoint m_i	Frequency f_i	$m_i\cdot f_i$	$m_i-\bar{x}$	$\left(m_i-\bar{x}\right)^2$	$\left(m_i-\bar{x}\right)^2\cdot f_i$
10–16	13	1	13	−18.879	356.409	356.409
17–23	20	3	60	−11.879	141.106	423.317
24–30	27	10	270	−4.879	23.803	238.026
31–37	34	12	408	2.121	4.500	53.994
38–44	41	5	205	9.121	83.197	415.983
45–51	48	2	96	16.121	259.893	519.787
Sum		33	1052			2007.515

The mean is $\bar{x}=\dfrac{\sum m_i\cdot f_i}{n}=\dfrac{1052}{33}=31.879$

The standard deviation is $s=\sqrt{\dfrac{\sum(x_i-\bar{x})^2}{n-1}}=\sqrt{\dfrac{2007.515}{32}}=7.921$

15. The mean is $\bar{x}=\dfrac{968+893+769+845+922+915}{6}=\dfrac{5312}{6}=885.333$

Time x	Deviation from the Mean $x-\bar{x}=x-885.333$	Deviation Squared $\left(x-\bar{x}\right)^2=\left(x-885.333\right)^2$
968	82.667	6833.778
893	7.667	58.778
769	−116.333	13533.444
845	−40.333	1626.778
922	36.667	1344.444
915	29.667	880.111
Sum = 5312		24277.333

The standard deviation is $s=\sqrt{\dfrac{\sum(x_i-\bar{x})^2}{n-1}}=\sqrt{\dfrac{24277.333}{5}}=69.681$

17. (a) The range is the difference between the largest value and the smallest value. The range for the Yankees' ages is 38 – 22 = 16 years.

For (b) and (c):

The mean (from 9.4 Problem 15) is 28.825 years of age.

Age x_i	Frequency f_i	$f_i \cdot x_i$	$x_i - \bar{x}$	$\left(x_i - \bar{x}\right)^2$	$\left(x_i - \bar{x}\right)^2 \cdot f_i$
22	1	22	-6.825	46.581	46.681
23	4	92	-5.825	33.931	135.723
24	4	96	-4.825	23.281	93.123
25	3	75	-3.825	14.631	43.892
26	4	104	-2.825	7.981	31.923
27	2	54	-1.825	3.331	6.661
28	1	28	-0.825	0.681	0.681
29	3	87	0.175	0.031	0.092
30	3	90	1.175	1.381	4.142
31	2	62	2.175	4.731	9.462
32	5	160	3.175	10.081	50.403
33	2	66	4.175	17.431	34.861
34	1	34	5.175	26.781	26.781
35	1	35	6.175	38.131	38.131
36	1	36	7.175	51.481	51.481
37	2	74	8.175	66.831	133.661
38	1	38	9.175	84.181	84.181
Total	40	1153			791.775

(b) The standard deviation assuming sample data is

$$s = \sqrt{\frac{\sum (x_i - \bar{x})^2 \cdot f_i}{n-1}} = \sqrt{\frac{791.775}{39}} = 4.506 \text{ years.}$$

(c) The standard deviation assuming population data is

$$\sigma = \sqrt{\frac{\sum (x_i - \mu)^2 \cdot f_i}{n}} = \sqrt{\frac{791.775}{40}} = 4.449 \text{ years.}$$

19. (a) These are population data. We have all the mothers in the United States represented.

(b) The mean age of the mother (from 9.4 Problem 19) is 27.949 years. This is $\bar{x}$ in the following table.

Class Midpoint m_i	Frequency f_i	$m_i - \mu$	$(m_i - \mu)^2$	$(m_i - \mu)^2 \cdot f_i$
12.5	7	−15.449	238.672	1,670.701
17.5	415	−10.449	109.182	45,310.36
22.5	1034	−5.449	29.692	30,701.12
27.5	1104	−0.449	0.202	222.568
32.5	966	4.551	20.712	20,007.41
37.5	476	9.551	91.222	43,421.48
42.5	104	14.551	211.732	22,020.09
47.5	6	19.551	382.242	2,293.45
Sum	4112			165,647.2

The standard deviation, assuming population data, is

$$\sigma = \sqrt{\frac{\sum (x - \mu)^2 \cdot f_i}{n}} = \sqrt{\frac{165{,}647.2}{4112}} = 6.347$$

21. (a) These are population data since all earthquakes are included.
To compute the mean and standard deviation, set up the table below. Use the first three columns to determine the mean. Then use the mean to complete the table and to calculate the standard deviation.

Class Midpoint m_i	Frequency f_i	$m_i \cdot f_i$	$m_i - \mu$	$(m_i - \mu)^2$	$(m_i - \mu)^2 \cdot f$
0.5	851	425.5	−3.334	11.116	9,459.333
1.5	19	28.5	−2.334	5.448	103.504
2.5	4016	10,040	−1.334	1.780	7,146.697
3.5	9953	34,835.5	−0.334	0.112	1,110.317
4.5	13,069	58,810.5	0.666	0.444	5,796.833
5.5	1483	8,156.5	1.666	2.776	4,116.15
6.5	132	858	2.666	1.108	938.197
7.5	10	75	3.666	13.440	134.396
Sum		113,229.5			28,805.43

(b) The mean magnitude of the earthquakes worldwide in 2006 is

$$\mu = \frac{\sum m_i \cdot f_i}{n} = \frac{113{,}229.5}{29{,}533} = 3.834 \text{ in magnitude.}$$

(c) The population standard deviation is

$$\sigma = \sqrt{\frac{\sum (x - \mu)^2 \cdot f_i}{n}} = \sqrt{\frac{28{,}805.43}{29{,}533}} = 0.988$$

23. In Problem 21 in 9.4 we found the mean age of the drivers in Tennessee to be 46.775 years of age. We will use the mean in the table below:

Class Midpoint m_i	Frequency f_i	$m_i - \bar{x}$	$(m_i - \bar{x})^2$	$(m_i - \bar{x})^2 \cdot f_i$
22.5	357986	−24.275	589.276	210,952,423.9
27.5	381703	−19.275	371.526	141,812,445.6
32.5	389501	−14.275	203.776	79,370,809.7
37.5	408,043	−9.275	86.026	35,102,154.1
42.5	432,601	−4.275	18.276	7,906,053.7
47.5	431,431	0.725	0.526	226,770.9
52.5	403,996	5.725	32.776	13,241,221.4
57.5	371,684	10.725	115.026	42,753,184.4
62.5	283,136	15.725	247.276	70,012,631.4
67.5	217,832	20.725	429.526	93,564,426.0
72.5	171,004	25.725	661.776	113,166,279.0
77.5	132,012	30.725	944.026	124,622,710.8
82.5	89,812	35.725	1276.276	114,624,866.4
Sum	4,070,741			1,047,355,977

(a) The standard deviation of the age of drivers in Tennessee, assuming that the data are from a sample is

$$s = \sqrt{\frac{\sum (x_i - \bar{x})^2 \cdot f_i}{n-1}} = \sqrt{\frac{1,047,355,977}{4,070,740}} = 16.05$$

(b) The standard deviation of the age of drivers in Tennessee, assuming that the data are the population is

$$\sigma = \sqrt{\frac{\sum (x_i - \bar{x})^2 \cdot f_i}{n}} = \sqrt{\frac{1,047,355,977}{4,070,741}} = 16.05$$

25. (a) These are population data; all 4-year colleges are represented.

(b) In Problem 23 of Section 9.4 we found that the mean tuition at 4-year colleges in 2006-2007 was \$14,390. To find the standard deviation use technology and find the standard deviation

$$\sigma = \sqrt{\frac{\sum (x_i - \mu)^2 \cdot f_i}{n}} = \sqrt{\frac{147,685,819,000}{1790}} = 9083.28 = \$9083.28$$

27. We know that the mean of the IQ test is 100, the standard deviation is 15, and the distribution is symmetric. So the Empirical Rule applies.

(a) Since 70 is two standard deviations below the mean and 130 is two standard deviations above the mean, the Empirical Rule states that approximately 95%of the IQ scores are between 70 and 130. So we conclude that approximately 95% of persons has an IQ score between 70 and 130.

(b) Since 95% of the scores are between 70 and 130, 1 – 0.95 = 0.05 = 5% of the scores are either below 70 or above 130.

(c) Since the Empirical Rule requires the distribution to be roughly symmetric, we assume that the percent of scores over 130 is equal to the percent below 70. So approximately 2.5% of IQ scores are above 130.

29. We are told the distribution of kidney weight is bell shaped with a mean of 325 grams and a standard deviation of 30 grams, so the Empirical Rule applies.

(a) About 95% of kidneys will be between two standard deviations of the mean. That means 95% of the kidneys will weigh between 325 – 2(30) = 325 – 60 = 265 grams and 325 + 2(30) = 325 + 60 = 385 grams.

(b) $\frac{325-235}{30}=\frac{90}{30}=3$ and $\frac{415-325}{30}=\frac{90}{30}=3$. According to the Empirical Rule approximately 99.7% of the data lie within 3 standard deviations of the mean. So approximately 99.7% of adult male kidneys weigh between 235 grams and 415 grams.

(c) Kidneys weighing less than 235 grams or more than 415 grams are more than 3 standard deviations from the mean. So, 1 – 0.997 = 0.003 or 0.3% of adult male kidneys weigh less than 235 grams or more than 415 grams.

(d) $\frac{325-295}{30}=\frac{30}{30}=1$ and $\frac{385-325}{30}=\frac{60}{30}=2$. Since 295 is one standard deviation below the mean, and 385 are two standard deviations above the mean from the Empirical Rule, we see that 0.34 + 0.34 + 0.135 = 0.815 or 81.5% of the adult male kidneys weight between 295 and 385 grams.

31. We are told that $\mu = 25$ and $\sigma = 3$

(a) We want the outcome to be between 19 and 31, so $k = \mu - 19 = 25 - 19 = 6$, and according to Chebychev's theorem, the probability is at least

$$1 - \frac{\sigma^2}{k^2} = 1 - \frac{3^2}{6^2} = 1 - \frac{9}{36} = \frac{3}{4} = 0.75$$

At least 75% of the outcomes are between 19 and 31.

(b) We want the outcome to be between 20 and 30, so $k = \mu - 20 = 25 - 20 = 5$, and according to Chebychev's theorem, the probability is at least

$$1 - \frac{\sigma^2}{k^2} = 1 - \frac{3^2}{5^2} = 1 - \frac{9}{25} = 0.64$$

At least 64% of the outcomes are between 20 and 30.

(c) We want the outcome to be between 16 and 34, so $k = \mu - 16 = 25 - 16 = 9$, and according to Chebychev's theorem, the probability is at least

$$1 - \frac{\sigma^2}{k^2} = 1 - \frac{3^2}{9^2} = 1 - \frac{9}{81} = \frac{8}{9} = 0.889$$

At least 88.9% of the outcomes are between 16 and 34.

(d) We want the outcome to be less than 19 or more than 31. This is the opposite event from part (a), so the probability is at most 1 – 0.75 = 0.25.

At most 25% of the outcomes are less than 19 or greater than 31.

(e) We want the outcome to be less than 16 or more than 34. This is the opposite event from part (c) so the probability is at most 1 – 0.889 = 0.111.

At most 11.1% of the outcomes are less than 16 or more than 34.

33. We are told that $\mu = 6$ and $\sigma = 2$. We want to estimate the number of boxes that have between 0 and 12 defective watches.

$$k = \mu - 0 = 6$$

According to Chebychev's theorem the probability is at least $1 - \frac{2^2}{6^2} = 0.8889$ that there are between 0 and 12 defective watches in a box. So we expect at least 66.7% of the 1000 boxes or at least 889 boxes to have between 0 and 12 defective watches.

35. (a) These are population data because they record all live births in the United States for the years listed.

(b) The mean number of live births is

$$\mu = \frac{4{,}112{,}052 + 4{,}089{,}950 + 4{,}021{,}726 + 4{,}025{,}933 + 4{,}058{,}814 + 3{,}959{,}417}{6} = 4{,}044{,}648.67$$

(c) To find the standard deviation, first complete the table below.

Year	Births, f_i	$f_i - \mu$	$(f_i - \mu)^2$
2004	4,112,052	67,403.33	4,543,208,895
2003	4,089,950	45,301.33	2,052,210,500
2002	4,021,726	-22,922.67	525,448,800
2001	4,025,933	-18,715.67	350,276,304
2000	4,058,814	14,165.33	200,656,574
1999	3,959,417	-85,231.67	7,264,437,571
Sum	24,267,892		14,936,238,643

The standard deviation of births over the 6-year period is

$$\sigma = \sqrt{\frac{(f_i - \mu)^2}{n}} = \sqrt{\frac{14{,}936{,}238{,}643}{6}} = 49{,}893.62$$

(d) The mean and standard deviations are exact because the data are not grouped.

9.6 The Normal Distribution

5. *Z-score*

7. 0.4

9. The mean is always at the center of the normal curve.

$$\mu = 8$$

34.135% of the area under the curve lies between μ and σ.

$$\sigma = 2$$

11. The mean is always at the center of the normal curve.

$$\mu = 18$$

68.27% of the area under the normal curve lies between $\mu - \sigma$ and $\mu + \sigma$.

$$\sigma = 1$$

13. $Z = \frac{x-\mu}{\sigma}$ Here we are told $\mu = 13.1$ and $\sigma = 9.3$.

(a) $x = 7$

$$Z = \frac{x-\mu}{\sigma} = \frac{7-13.1}{9.3} = -0.66$$

(b) $x = 9$

$$Z = \frac{x-\mu}{\sigma} = \frac{9-13.1}{9.3} = -0.44$$

(c) $x = 13$

$$Z = \frac{x-\mu}{\sigma} = \frac{13-13.1}{9.3} = -0.01$$

(d) $x = 29$

$$Z = \frac{x-\mu}{\sigma} = \frac{29-13.1}{9.3} = 1.71$$

(e) $x = 37$

$$Z = \frac{x-\mu}{\sigma} = \frac{37-13.1}{9.3} = 2.57$$

(f) $x = 41$

$$Z = \frac{x-\mu}{\sigma} = \frac{41-13.1}{9.3} = 3$$

15. Using the Standard Normal Curve Table on the inside back cover of the text, we can find the area under the standard normal curve between the standard score, Z, and the mean

(a) $Z = 0.89$
Read down the table under Z until you reach the row beginning 0.8. Then read across the row until you reach the entry under the column marked 0.09. The area under the standard normal curve between 0 and $Z = 0.89$ is 0.3133.

(b) $Z = 1.10$
Read down the table under Z until you reach the row beginning 1.1. The next entry in the row, 0.3642, represents the area under the standard normal curve between the mean and $Z =$ 1.10.

(c) $Z = 3.06$
Read down the table under Z until you reach the row beginning 3.0. Then read across the row until you reach the entry under the column marked 0.06. The area under the standard normal curve between 0 and $Z = 3.06$ is 0.4989.

(d) $Z = -1.22$
There are no negative Z-scores on this table, but we use the symmetry of the normal curve to find the area under the curve between $Z = -1.22$ and the mean.
Read down the table under Z until you reach the row beginning 1.2. Then read across the row until you reach the entry under the column marked 0.02. Because of symmetry, 0.3888 is area under the standard normal curve between $Z = -1.22$ and 0, as well as between 0 and $Z = 1.22$.

(e) $Z = 2.30$
Read down the table under Z until you reach the row beginning 2.3. The next entry in the row, 0.4893, represents the area under the standard normal curve between the mean and $Z =$ 2.30.

(f) $Z = -0.75$
There are no negative Z-scores on this table, but we use the symmetry of the normal curve to find the area under the curve between $Z = -0.75$ and the mean.
Read down the table under Z until you reach the row beginning 0.7. Then read across the row until you reach the entry under the column marked 0.05. Because of symmetry, 0.2734 is area under the standard normal curve between $Z = -0.75$ and 0, as well as between 0 and $Z = 0.75$.

17. $Z = -0.5$; A = 0.1915
Since we want the area to the left of Z, subtract A from 0.5000.

$$\text{Area} = 0.5000 - 0.1915$$
$$\text{Area} = 0.3085$$

19. $Z_1 = -1.2$; $A_1 = 0.3849$
$Z_2 = 1.5$; $A_2 = 0.4332$
Since the Z-scores are on opposite sides of the mean, add the areas.

$$\text{Area} = A_1 + A_2$$
$$\text{Area} = 0.3849 + 0.4332$$
$$\text{Area} = 0.8181$$

21. To approximate the probability of obtaining between 285 and 315 successes in the 750 trials, we find the area under a normal curve from $x = 284.5$ to $x = 315.5$. We convert to Z-scores,

$$x = 284.5{:}\ Z_1 = \frac{x-\mu}{\sigma} = \frac{284.5-300}{13.4} = -1.16 \quad A_1 = 0.3770$$

$$x = 315.5{:}\ Z_2 = \frac{x-\mu}{\sigma} = \frac{315.5-300}{13.4} = 1.16 \quad A_2 = 0.3770$$

Since the values are on opposite sides of the mean, we add the areas.

$$A = A_1 + A_2 = 0.3770 + 0.3770 = 0.7540 \text{ (if you use technology, } A = 0.7526$$

The approximate probability that there are between 285 and 315 successes is 0.754.

23. To approximate the probability of obtaining 300 or more successes in the 750 trials, we find the area under a normal curve to the right of $x = 299.5$.

$$Z = \frac{x-\mu}{\sigma} = \frac{299.5-300}{13.4} = -0.0373 \qquad A = 0.0149$$

So, the approximate probability of obtaining 300 or more successes is 0.5 + 0.0149 = 0.5149

25. To approximate the probability of obtaining 325 or more successes in the 750 trials, we find the area under a normal curve to the right of $x = 324.5$. We first convert 324.5 to a Z-score.

$$x = 324.5{:}\ Z = \frac{x-\mu}{\sigma} = \frac{324.5-300}{13.4} = 1.8284 \ A = 0.4664$$

We need the area to the right of $Z = 1.83$, so subtract A from 0.5000.

$$\text{Area} = 0.5000 - 0.466 = 0.0336$$

The approximate probability of 325 or more successes is 0.0336, (using technology, 0.0337).

27. We use the Standard Normal Curve Table and the interpretation of a Z-score as the number of standard deviations the original score is from its mean to solve this problem.

A – a score exceeds $\mu + 1.6\sigma$. Here $Z = 1.6$; the area under the standard normal curve between 0 and 1.6 is 0.4452. We need the area to the right of $Z = 1.6$. So we subtract 0.4452 from 0.5, the area under the curve to the right of 0.

$$0.5000 - 0.4452 = 0.0548$$

So, 5.48% of the class will get a grade of A.

B – a score is between $\mu + 0.6\,\sigma$ and $\mu + 1.6\sigma$. We know that the area under the curve from 0 to $Z =$ 1.6 is 0.4452. We find the area under the curve from 0 to $Z = 0.6$ is 0.2257. The area under the curve between the two Z-scores is the difference between the two areas, 0.4452 – 0.2257, which is 0.2195. So, 21.95% of the class will get a grade of B.

C – a score is between $\mu - 0.3\sigma$ and $\mu + 0.6\sigma$. We use symmetry and find the area under the curve between 0 and $Z = -0.3$ is 0.1179. We found in part (b) that the area under the curve from 0 to $Z = 0.6$ is 0.2257. Here since the Z-scores have opposite signs we add the two areas, $0.1179 + 0.2257$, and get 0.3436. So, 34.36% of the class will get a grade of C.

D – a score is between $\mu - 1.4\sigma$ and $\mu - 0.3\sigma$. We use the symmetry of the normal curve to determine the areas between both 0 and $Z = -1.4$, which is 0.4192, and 0 and $Z = -0.3$, which is 0.1179. Now since both Z-scores are negative, the area between them is the difference between 0.4192 and 0.1179, which is 0.3013. So, 30.13% of the class will get a grade of D.

F – a score is below $\mu - 1.4\sigma$. Use the symmetry of the curve to determine that the area between 0 and $Z = -1.4$ is 0.4192. But we need the area to the left of Z, so we subtract 0.4192 from 0.5, the total area under the curve to the left of 0.

$$0.5000 - 0.4192 = 0.0808$$

So, 8.08% of the class will get a grade of F.

(Note: When you add all the percents you should get 100%, the entire class.)

29. We are told that $\bar{x} = 64$ and $S = 2$. We will convert the given heights to Z-scores to determine the percent of women in the required intervals. Then we will calculate how many of the 2000 women are in the interval.

(a) between 62 and 66 inches: These women are within 1 standard deviation of the mean, $64 - 2 = 62$ and $64 + 2 = 66$. This gives

$$Z_1 = -1.0;\ A_1 = 0.3413 \text{ and } Z_2 = 1.0;\ A_2 = 0.3413$$

Since the two Z-scores are on opposite sides of the mean, add the corresponding areas

$$A = A_1 + A_2 = 0.3413 + 0.3413 = 0.6826$$

So, approximately 68.26% of the women or 1365 of the 2000 women sampled will be between 62 and 66 inches tall.

(b) between 60 and 68 inches tall: These women are within 2 standard deviations of the mean, $64 - 2(2) = 60$ and $64 + 2(2) = 68$. This gives

$$Z_1 = -2.0;\ A_1 = 0.4772 \text{ and } Z_2 = 2.0;\ A_2 = 0.4772$$

Since the two Z-scores are on opposite sides of the mean, add the corresponding areas

$$A = A_1 + A_2 = 0.4772 + 0.4772 = 0.9544$$

So, approximately 95.44% of the women or 1909 of the 2000 women sampled will be between 60 and 68 inches tall.

(c) between 58 and 70 inches tall: These women are within 3 standard deviations of the mean, $64 - 3(2) = 58$ and $64 + 3(2) = 70$. This gives

$$Z_1 = -3.0;\ A_1 = 0.4987 \text{ and } Z_2 = 3.0;\ A_2 = 0.4987$$

Since the two Z-scores are on opposite sides of the mean, add the corresponding areas

$$A = A_1 + A_2 = 0.4987 + 0.4987 = 0.9974$$

So, approximately 99.74% of the women or 1995 of the 2000 women sampled will be between 58 and 70 inches tall.

(d) more than 70 inches tall: These women are more than 3 standard deviations from the mean. We need to subtract $A = 0.4987$ from 0.5000 to find the percent of the population more than 3 standard deviations from the mean.

$$0.5000 - 0.4987 = 0.0013$$

So, approximately 0.13% of the women or 3 of the 2000 women sampled will be more than 70 inches tall.

(e) shorter than 58 inches: These women are more than 3 standard deviations below the mean. This gives $Z = -3$ and A = 0.4987. Since we want the area to the left of Z, we need to subtract A from 0.5000.

$$0.5000 - 0.4987 = 0.0013$$

So, 0.13% of the women or 3 of the 2000 women sampled will be shorter than 58 inches.

31. We are given that $\mu = 130$ and $\sigma = 5.2$ pounds.

(a) Convert 142 pounds to a Z-score.

$$Z = \frac{x-\mu}{\sigma} = \frac{142-130}{5.2} = 2.308\text{ ; A} = 0.4896$$

We are interested in the area under the normal curve that to the right of Z, so we subtract A from 0.5000.

$$0.5000 - 0.4896 = 0.0104$$

Approximately 1.04% of the students or 1 student weighs at least 142 pounds.

(b) To find the range of weights that includes the middle 70% of the students, we need to find the Z-score corresponding to A = 0.35. (Because of symmetry 35% of the students will weigh more than the mean, and 35% will weigh less than the mean.)

Looking in the body of the Standard Normal Curve Table, we find that 0.3508 is closest to A = 0.3500, and 0.3508 corresponds to Z = 1.04 and Z = −1.04.

Using $Z = \pm 1.04$, $\mu = 130$ and $\sigma = 5.2$, we solve the equations

$$Z = \frac{x-\mu}{\sigma} \qquad\qquad Z = \frac{x-\mu}{\sigma}$$

$$1.04 = \frac{x-130}{5.2} \qquad\qquad -1.04 = \frac{x-130}{5.2}$$

$$5.408 = x - 130 \qquad\qquad -5.408 = x - 130$$

$$x = 135.408 \qquad\qquad x = 124.592$$

So, we would expect 70% of the students to weight between 124.59 and 135.41 pounds.

33. We are given that $\mu = 40$ months and $\sigma = 7$ months.

Standardize 28 months and 42 months by finding the Z-scores corresponding to 28 and 40.

$$Z_1 = \frac{x_1-\mu}{\sigma} = \frac{28-40}{7} = -1.71\text{ ; } A_1 = 0.4564$$

$$Z_2 = \frac{x_2-\mu}{\sigma} = \frac{42-40}{7} = 0.29\text{ ; } A_2 = 0.1141$$

Now find the area under the standard normal curve between the two Z-scores. Since the Z-scores are on opposite sides of the mean, add the areas.

$$A = A_1 + A_2 = 0.4564 + 0.1141 = 0.5705$$

So, 57.05% of the clothing can be expected to last between 28 and 42 months.

35. We are given that $\mu = 10{,}000$ and $\sigma = 1000$ persons.

(a) The lowest 70% of the attendance figures includes the 50% that are less than the mean and the 20% that are between the mean and some positive value of Z. We look in the body of the Standard Normal Curve Table for the number closest to 0.2000. We find 0.2000 is almost half way between 0.1985 and 0.2019, which correspond to Z = 0.52 and Z = 0.53 respectively. We will use an approximate Z = 0.525.

Using $Z = 0.525$, $\mu = 10{,}000$ and $\sigma = 1000$, we will solve the equation:

$$Z = \frac{x - \mu}{\sigma}$$

$$0.525 = \frac{x - 10000}{1000}$$

$$525 = x - 10000$$
$$x = 10{,}525$$

Attendance lower than 10,525 will be in the lowest 70% of the figures.

(b) To find the percent of attendance figures that falls between 8500 and 11,000 persons, find the Z-score for each and determine the area under the standard normal curve between the two Z-scores.

$$Z_1 = \frac{x_1 - \mu}{\sigma} = \frac{8500 - 10000}{1000} = -1.5\,;\ A_1 = 0.4332$$

$$Z_2 = \frac{x_2 - \mu}{\sigma} = \frac{11000 - 10000}{1000} = 1.0\,;\ A_2 = 0.3413$$

Since the Z-scores are on opposite sides of the mean, add the areas.

$$A = A_1 + A_2 = 0.4332 + 0.3413 = 0.7745$$

Approximately 77.45% of the attendance figures are between 8500 and 11,000 persons.

(c) Here we are looking for the percent of attendance figures that are more than 11,500 or less than 8500. First find the Z-scores and areas corresponding to $x = 11{,}500$ and $x = 8500$.

$$Z_1 = \frac{x_1 - \mu}{\sigma} = \frac{11500 - 10000}{1000} = 1.5\,;\ A_1 = 0.4332$$

$$Z_2 = -1.5;\ A_2 = 0.4332 \text{ (from part (b))}$$

Now find the area under the standard normal curve outside the two Z-scores. Since the Z-scores are on opposite sides of the mean, add these areas.

$$A = (0.5 - A_1) + (0.5 - A_2) = (0.5 - 0.4332) + (0.5 - 0.4332) = 0.1336$$

Approximately 13.36% of the attendance figures differ from the mean by 1500 persons or more.

37. Transform each score to a standard score and then compare.

Colleen's score was 76; Colleen's standard score is $Z = \frac{x - \mu}{\sigma} = \frac{76 - 82}{7} = -0.857$

Mary's score was 89; Mary's standard score is $Z = \frac{x - \mu}{\sigma} = \frac{89 - 93}{2} = -2.0$

Kathleen's score was 21; Kathleen's standard score is $Z = \frac{x - \mu}{\sigma} = \frac{21 - 24}{9} = -0.33$

Kathleen has the highest relative standing.

39. (a) To find the line chart and the frequency curve, we need to first find the frequency distribution for the experiment. It is binomial with $n = 15$ and $p = .3$. The distribution is

Number of Heads, k	Probability $b(15,\ k,\ 0.3)$
0	.005
1	.031
2	.092
3	.170
4	.219
5	.206
6	.147
7	.081
8	.035
9	.012
10	.003
11	.0006
12	.0001
13	< 0.0001
14	< 0.0001
15	< 0.0001

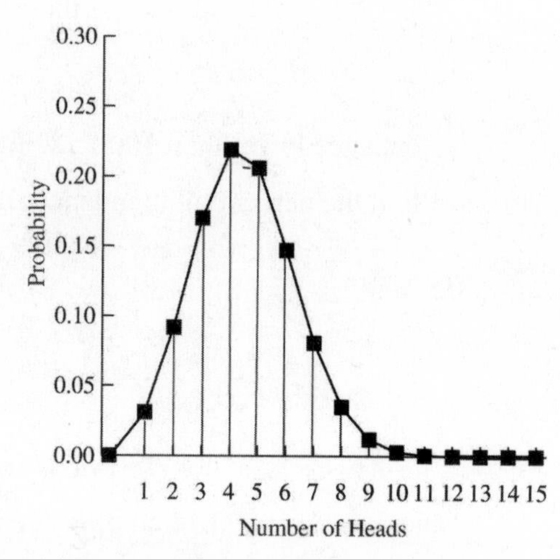

(b) The distribution is skewed right.

(c) Mean: $\mu = n \cdot p = 15 \cdot (0.3) = 4.5$
Standard Deviation: $\sigma = \sqrt{np(1-p)} = \sqrt{15 \cdot (0.3)(.7)} = 1.775$

41. This is a binomial distribution which we will approximate with a normal curve. We are told that the player's lifetime batting average is 0.250; this is his probability of success, and that he will bat 300 times; this is n.

We first find the mean and the standard deviation of the distribution.

mean: $\mu = np = (300)(0.250) = 75$ hits

standard deviation: $\sigma = \sqrt{np(1-p)} = \sqrt{300 \cdot (.250)(.750)} = 7.5$

(a) To approximate the probability that he gets at least 80 but no more than 90 hits, we find the area under a normal curve from $x = 79.5$ to $x = 90.5$. Convert 79.5 and 90.5 to Z-scores.

$$x = 79.5: \quad Z_1 = \frac{x-\mu}{\sigma} = \frac{79.5-75}{7.5} = 0.6 \quad A_1 = 0.2257$$

$$x = 90.5 \quad Z_2 = \frac{x-\mu}{\sigma} = \frac{90.5-75}{7.5} = 2.07 \quad A_2 = 0.4808$$

Since both 79.5 and 90.5 are greater than the mean, the area between the points is the difference between A_1 and A_2.

$$\text{Area} = A_2 - A_1 = 0.4808 - 0.2257 = 0.2551$$

The approximate probability of having at least 80, but no more than 90 hits is 0.2551 (technology answer is 0.2549).

(b) To approximate the probability that 85 or more hits occur, we find the area under a normal curve to the right of $x = 84.5$. We first convert 84.5 to a Z-score.

$$x = 85:\ Z = \frac{x-\mu}{\sigma} = \frac{84.5-75}{7.5} = 1.27 \qquad \text{A} = 0.3980$$

Since the area to the right of the mean is 0.5, we subtract A from 0.5 to obtain the area to the right of Z.

$$\text{Area} = 0.5000 - 0.3980 = 0.102$$

The approximate probability of 85 or more hits occurring is 0.102, (techmology 0.1026).

43. This is a binomial distribution which we will approximate with a normal curve. We are told that the 3% of the packages do not seal properly; so we will let $p = 0.03$. 500 packages will be selected; so $n = 500$. We first find the mean and the standard deviation of the distribution.

mean: $\mu = np = (500)(.03) = 15$ packages

standard deviation: $\sigma = \sqrt{np(1-p)} = \sqrt{500 \cdot (0.03)(0.97)} = 3.8$

To approximate the probability that at least 10 packages are not properly sealed, we find the area under a normal curve to the right of $x = 9.5$. Converting 9.5 to a Z-score gives

$$x = 9.5: \quad Z = \frac{x-\mu}{\sigma} = \frac{9.5-15}{3.8} = -1.45 \qquad \text{A} = 0.4265$$

Since we want the area to the right of $x = 9.5$, we add 0.5 to A.

$$\text{Area} = 0.5000 + 0.4265 = 0.9265$$

The approximate probability of selecting at least 10 unsealed packages is 0.9265.

45. $y = \frac{1}{\sqrt{2\pi}} e^{-(1/2)x^2}$

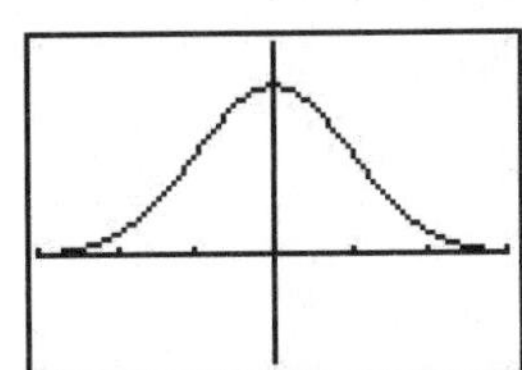

window: Xmin = –3 ; Xmax = 3

Ymin = –.25; Ymax = .5

The function assumes its maximum at $x = 0$.

47. From Problem 43, we have a binomial distribution with $n = 500$ and $p = 0.03$.
We want the probability that at least 10 bags of jelly beans are not sealed properly.
Graphing utilities often have a function, called the cumulative binomial distribution, that calculates the probability of being less than or equal to x. The keystrokes for the TI83/84 calculators are given. For other graphing utilities, consult your user's manual.

The probability at least 10 bags are improperly sealed will be given by 1 minus the opposite event, the probability that at most 9 bags are improperly sealed.

Enter 1 – 2nd DISTR, ALPHA A. You should see, 1 – binomcdf (
Enter 500, 0.03, 9). You should see, 1 – binomcdf (500, 0.03, 9)
Push ENTER. You should see .9330714394

The exact probability of selecting at least 10 unsealed packages is 0.933.

49. The probability of obtaining at least 25 bags of unsealed jelly beans in a sample of 500 is 0.010.

Chapter 9 Review

1. Variable: Circumference of head; it is continuous.

3. Variable: Number of people; it is discrete.

5. Variable: Number of defective products; it is discrete.

7. Answers will vary. All answers should include a method to choose a sample of 100 students from the population for which each student has an equal chance of being chosen.

9. Answers will vary. All answers should give examples of possible bias.

11. (a)

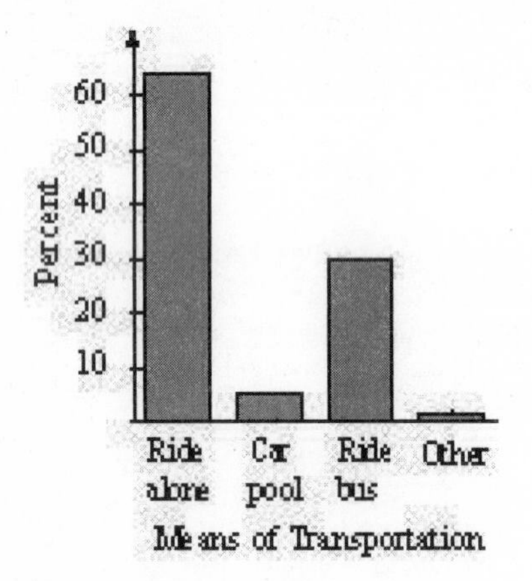

(b)

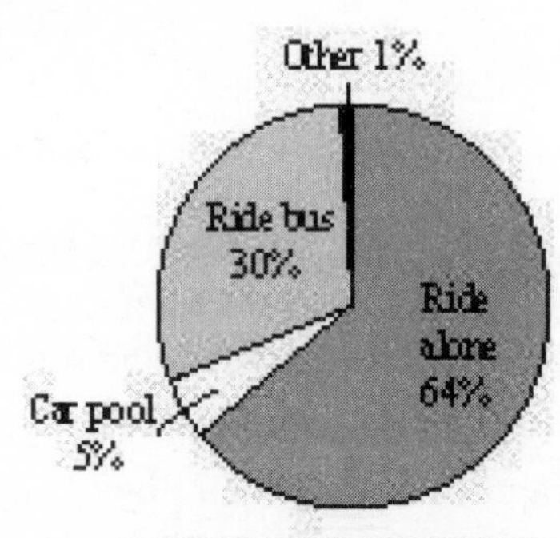

13. (a)

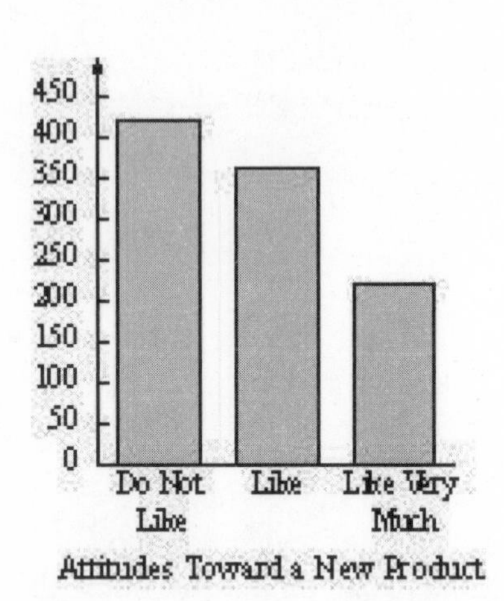

(b) To form the pie chart we express the results as percentages.

$$\frac{420}{1000} = 42\% \qquad \frac{360}{1000} = 36\%$$

$$\frac{220}{1000} = 22\%$$

Like Very Much 22%
Do Not Like 42%
Like 36%

Attitude Toward a New Product

15. (a) American Indians made up the smallest percentage of college enrollment with 1%.

(b) Asian-Americans were overrepresented in four-year colleges. They were 4% of the general population, but 6% of the college student population.

(c) Approximately, (0.08)(10,407,600) = 832,608 Hispanics were enrolled in four-year colleges in 2003.

17. (a) Most Americans' highest level of educational attainment is a high school diploma.

(b) About 52 million Americans have at least a bachelor's degree. (Add the numbers of people with bachelors' and graduate/ professional degrees.)

$$34{,}000 + 18{,}000 = 52{,}000 \text{ (thousands)} = 52{,}000{,}000$$

(c) About 28,000,000 Americans do not have a high school diploma.

(d) About 48 million Americans went to college but do not have a bachelor's degree. (Add the numbers with some college and with associates' degrees.)

$$32{,}000 + 16{,}000 = 48{,}000 \text{ (thousands)} = 48{,}000{,}000$$

19. (a)

Score	Frequency	Score	Frequency	Score	Frequency	Score	Frequency
21	2	62	1	74	1	87	2
33	1	63	2	75	1	89	1
41	2	66	2	77	1	90	2
42	1	68	1	78	2	91	1
44	1	69	1	80	4	92	1
48	1	70	2	82	1	95	1
52	2	71	1	83	1	99	1
55	1	72	2	85	2	100	2
60	1	73	2				

The range of scores is 79.

(b)

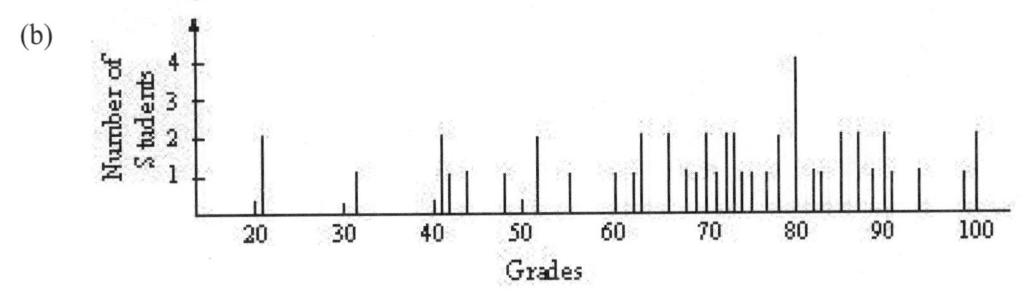

(c)

Class Interval	Frequency
20.0 – 29.9	2
30.0 – 39.9	1
40.0 – 49.9	5
50.0 – 59.9	3
60.0 – 69.9	8
70.0 – 79.9	12
80.0 – 89.9	11
90.0 – 99.9	6
100.0 – 109.9	2

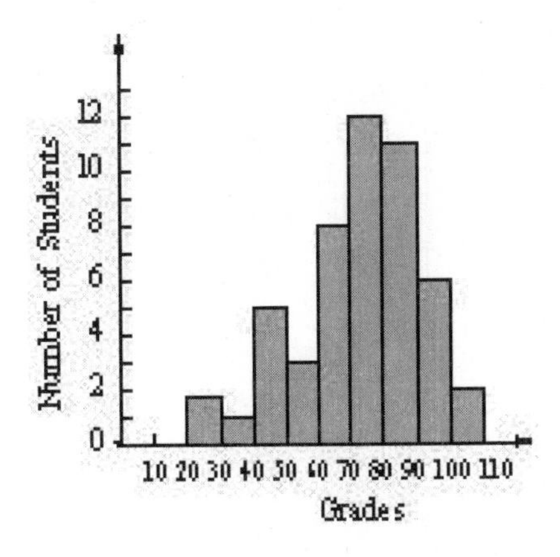

(d)

(e) Since there is a tail on the left, we say the distribution is skewed left.

(f)

Class Interval	Cumulative Frequency
20.0 – 29.9	2
30.0 – 39.9	3
40.0 – 49.9	8
50.0 – 59.9	11
60.0 – 69.9	19
70.0 – 79.9	31
80.0 – 89.9	42
90.0 – 99.9	48
100.0 – 109.9	50

(g)

21. (a)

Time	Freq.	Time	Freq.	Time	Freq.	Time	Freq.	Time	Freq
4'12"	1	4'46"	2	5'08"	1	5'43"	1	6'12"	1
4'15"	1	4'50"	1	5'12"	2	5'48"	1	6'30"	1
4'22"	1	4'52"	1	5'18"	1	5'50"	1	6'32"	1
4'30"	2	4'56"	1	5'20"	3	5'55"	1	6'40"	1
4'36"	1	5'01"	1	5'31"	2	6'01"	1	7'05"	1
4'39"	1	5'02"	1	5'37"	1	6'02"	1	7'15"	1
4'40"	2	5'06"	2	5'40"	2	6'10"	1		

The range is 3 minutes, 3 seconds.

(b)

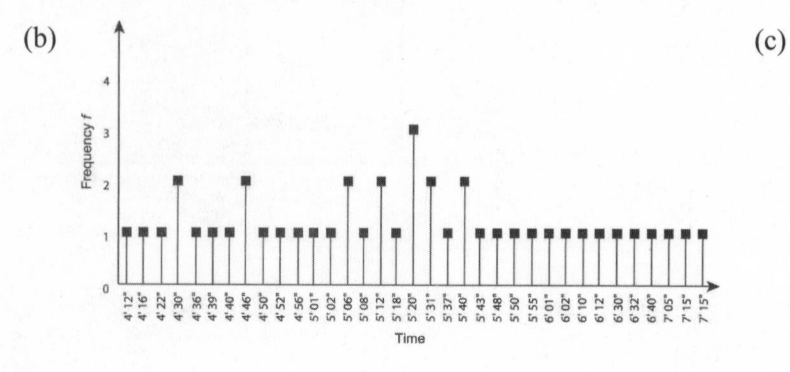

(c)

Class Interval	Frequency f_i
4'00"– 4'29"	3
4'30"– 4'59"	10
5'00"– 5'29"	11
5'30"– 5'59"	9
6'00"– 6'29"	4
6'30"– 6'59"	3
7'00"– 7'29"	2

(d)

(e)

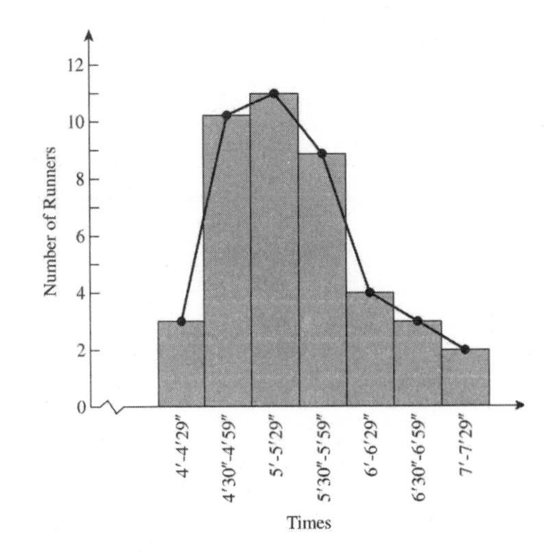

(f) Since the distribution has a tail that extends to the right, it is skewed right.

(g)

Class Interval	Cumulative Frequency
4'00"– 4'29"	3
4'30"– 4'59"	13
5'00"– 5'29"	24
5'30"– 5'59"	33
6'00"– 6'29"	37
6'30"– 6'59"	40
7'00"– 7'29"	42

(h)

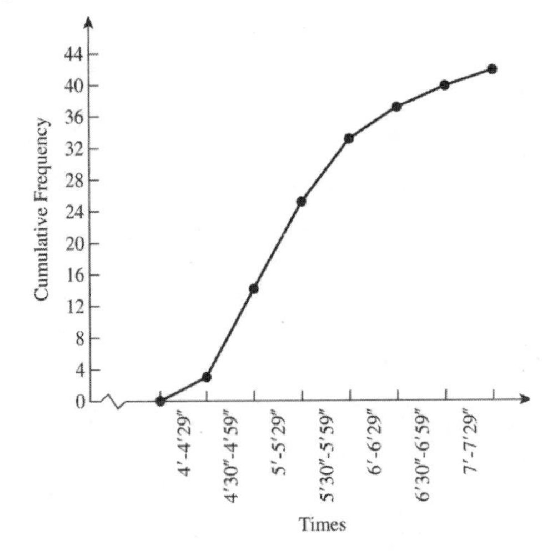

23. (a)

Age	Frequency	Age	Frequency	Age	Frequency
22	1	28	1	34	1
23	4	29	3	35	1
24	4	30	3	36	1
25	3	31	2	37	2
26	4	32	5	38	1
27	2	33	2		

(b)

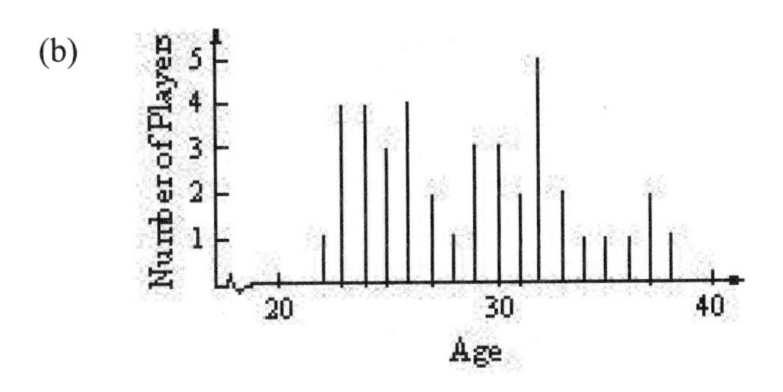

(c) and (g)

Class Interval	Frequency	Cumulative Frequency
20.0 – 24.9	9	9
25.0 – 29.9	13	22
30.0 – 34.9	13	35
35.0 – 39.9	5	40

There are 4 class intervals

(d)

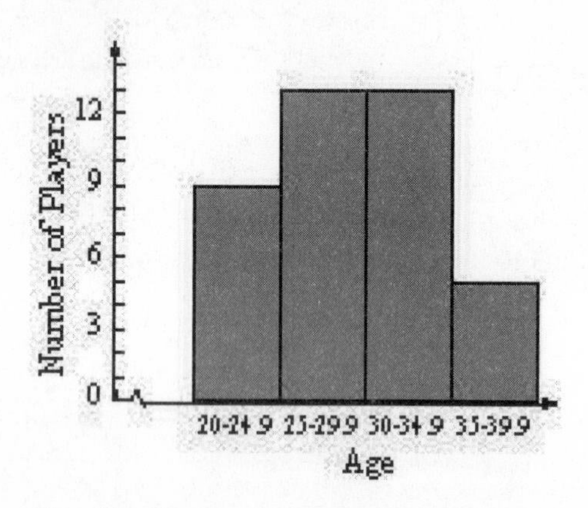

(e)

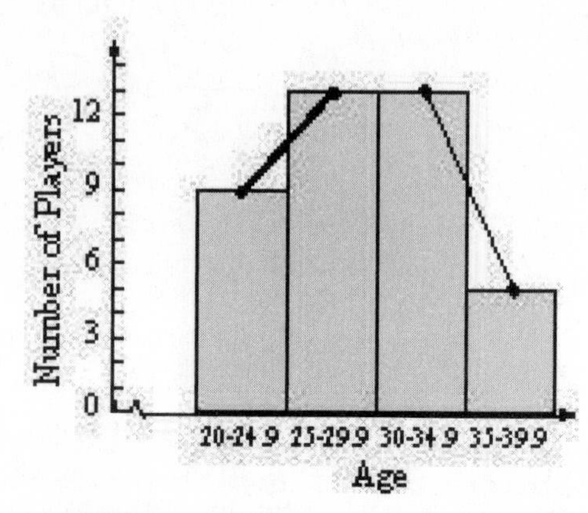

(f) The distribution is approximately symmetric.

(h)

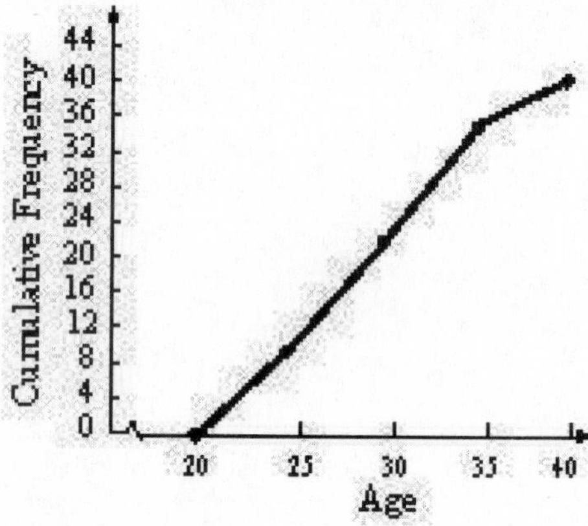

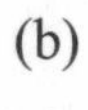

25. (a)

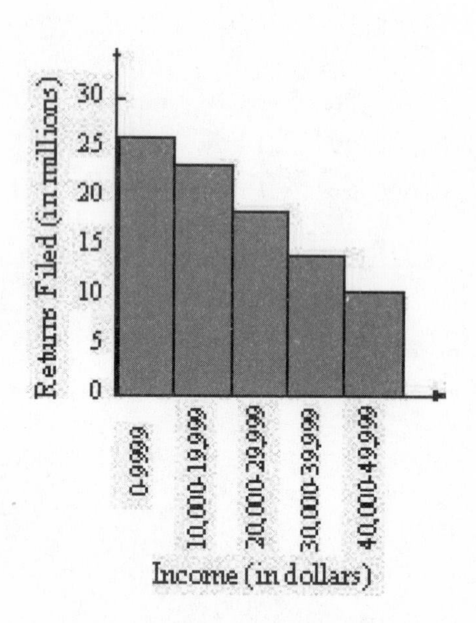

(b)

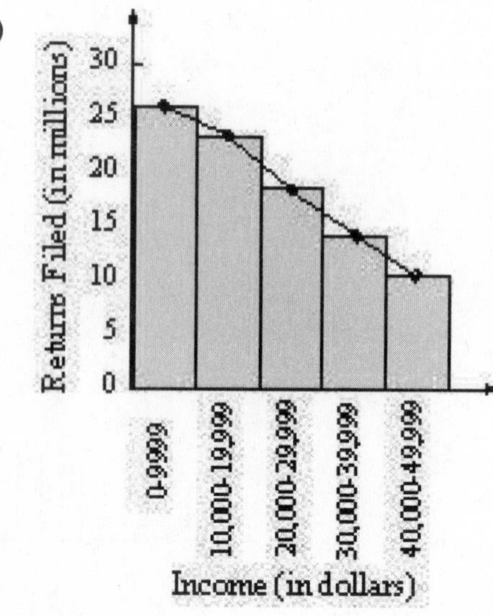

27. (a) Mean: $\bar{x} = \frac{63}{11} = 5.7273$

(b) Median: the middle value when the data are put in ascending order.

0, 0, 2, 4, 4, <u>5</u>, 8, 8, 10, 10, 12

Median = 5

(c) Mode: the value(s) with the greatest frequency exceeding a frequently of 1.

modes: 0, 4, 8, 10

(d) Range: the difference between the largest and smallest value.

Range = 12 – 0 = 12

(e) Standard Deviation:

value x	freq f	$x - \bar{x}$	$(x-\bar{x})^2$	$(x-\bar{x})^2 f$
0	2	–5.727	32.799	65.597
2	1	–3.727	13.891	13.891
4	2	–1.727	2.983	5.965
5	1	–0.727	0.529	0.529
8	2	2.273	5.167	10.333
10	2	4.273	18.259	36.517
12	1	6.273	39.351	39.351
Sum	11			172.182

$$s = \sqrt{\frac{\sum (x_i - \bar{x})^2 f}{n-1}} = \sqrt{\frac{172.182}{10}} = 4.149$$

29. (a) Mean: $\bar{x} = \frac{162}{10} = 16.2$

(b) Median: the middle value when the data are put in ascending order.

2, 5, 5, 7, <u>7, 7</u>, 9, 9, 11, 100

$$\text{Median} = \frac{7+7}{2} = 7$$

(c) Mode: the value(s) with the greatest frequency exceeding a frequently of 1.

mode = 7

(d) Range: the difference between the largest and smallest value.

Range = 100 – 2 = 98

(e) Standard Deviation

value x	freq f	$x - \bar{x}$	$(x-\bar{x})^2$	$(x-\bar{x})^2 f$
2	1	–14.2	201.64	201.64
5	2	–11.2	125.44	250.88
7	3	–9.2	84.64	253.92
9	2	–7.2	51.84	103.68
11	1	–5.2	27.04	27.04
100	1	83.8	7022.44	7022.44
162	10			7859.6

$$s = \sqrt{\frac{\sum (x_i - \bar{x})^2 f}{n-1}} = \sqrt{\frac{7859.6}{9}} = 29.5515$$

31. (a) Mean: $\bar{x} = \frac{70}{10} = 7$

(b) Median: the middle value when the data are put in ascending order.

1, 2, 5, 6, 7, 7, 9, 10, 11, 12

Median $= \frac{7+7}{2} = 7$

(c) Mode: the value(s) with the greatest frequency exceeding a frequently of 1.

mode = 7

(d) Range: the difference between the largest and smallest value.

Range = 12 – 1 = 11

(e) Standard Deviation.

value x	$x - \bar{x}$	$(x - \bar{x})^2$
5	–2	4
7	0	0
7	0	0
9	2	4
10	3	9
11	4	16
1	–6	36
6	–1	1
2	–5	25
12	5	25
Sum = 70		120

$$s = \sqrt{\frac{\sum (x_i - \bar{x})^2}{n-1}} = \sqrt{\frac{120}{9}} = 3.6515$$

33. (a) Answers may vary. For this problem, we will assume we have sample data; it is implied that Joe has played more than 7 rounds of golf.

(b) Mean:

$$\bar{x} = \frac{529}{7} = 75.57$$

(c) Standard deviation:

value x	$x - \bar{x}$	$(x - \bar{x})^2$
74	–1.571	2.469
72	–3.571	12.755
76	0.429	0.184
81	5.429	29.469
77	1.429	2.041
76	0.429	0.184
73	–2.571	6.612
Sum = 529		53.714

$$s = \sqrt{\frac{\sum (x_i - \bar{x})^2}{n-1}} = \sqrt{\frac{53.714}{6}} = 2.99$$

35. The size of this problem makes it perfect for technology. We put the midpoints and the frequencies into an Excel file and used it to produce the table.

Midpoint, *m*	Frequency, f	*fm*	(*m* - *mu*)	(*m* - *mu*)2	(*m* - *mu*)2 * *f*
2.5	9,922.0	24,805.0	-35.8	1281.64	12,716,432.08
7.5	9,545.0	71,587.5	-30.8	948.64	9,054,768.80
12.5	10,176.0	127,200.0	-25.8	665.64	6,773,552.64
17.5	10,249.0	179,357.5	-20.8	432.64	4,434,127.36
22.5	10,181.0	229,072.5	-15.8	249.64	2,541,584.84
27.5	9,798.0	269,445.0	-10.8	116.64	1,142,838.72
32.5	9,924.0	322,530.0	-5.8	33.64	333,843.36
37.5	10,439.0	391,462.5	-0.8	0.64	6,680.96
42.5	11,484.0	488,070.0	4.2	17.64	202,577.76
47.5	11,378.0	540,455.0	9.2	84.64	963,033.92
52.5	10,209.0	535,972.5	14.2	201.64	2,058,542.76
57.5	8,929.0	513,417.5	19.2	368.64	3,291,586.56
62.5	6,800.0	425,000.0	24.2	585.64	3,982,352.00
67.5	5,410.0	365,175.0	29.2	852.64	4,612,782.40
72.5	4,701.0	340,822.5	34.2	1169.64	5,498,477.64
77.5	4,294.0	332,785.0	39.2	1536.64	6,598,332.16
82.5	3,481.0	287,182.5	44.2	1953.64	6,800,620.84
87.5	2,118.0	185,325.0	49.2	2420.64	5,126,915.52
92.5	1,010.0	93,425.0	54.2	2937.64	2,967,016.40
97.5	307.0	29,932.5	59.2	3504.64	1,075,924.48
102.5	56.0	5,740.0	64.2	4121.64	230,811.84
Sum	150,411.0	5,758,762.5			80,412,803.04

(a) $\mu = \frac{\text{sum}(fm)}{\text{sum}(f)} = \frac{5,758,762.5}{150,411} = 38.3$

(b) To find the median, we need to determine where the middle value is located.

STEP 1: There are 150,411 entries, making the middle entry the 75,206th entry on the list. It is found in the class interval from 35 – 39. The first 7 intervals account for 69,795 entries.

STEP 2: In the 8th interval there are 75,206 – 69,795 = 5411 more entries to reach the median. We let $p = 5411$.

STEP 3: The interpolation factor will be

$$\frac{p}{q}i = \frac{5411}{10,439} \cdot 5 = 2.59$$

where q is the number of entries in the interval and i is the interval width.

STEP 4: The approximate median age of a female in the year 2005 was

M = 35 + 2.59 = 37.6 years of age.

(c) The approximate standard deviation of the age of females in the year 2005 was

$$\sigma = \sqrt{\frac{\sum (x_i - \mu)^2 f_i}{n}} = \sqrt{\frac{80,412,803.04}{150,411}} = 23.1 \text{ years.}$$

37. (a) 99.7% of the data lie between 3 standard deviations of the mean.

$$\mu - 3\sigma = 600 - 3\cdot 53 = 441$$
$$\mu + 3\sigma = 600 + 3\cdot 53 = 759$$

So, 99.7% of the light bulbs have lifetimes between 441 and 759 hours.

(b) We use the Empirical Rule and compute

$$\frac{494-600}{53} = -2 \text{ and } \frac{706-600}{53} = 2.$$

For a bell-shaped distribution, approximately 95% of the data fall between 2 standard deviations of the mean. So 95% of the light bulbs have a lifetime between 494 and 706 hours.

(c) We use the Empirical Rule and compute

$$\frac{547-600}{53} = -1 \text{ and } \frac{706-600}{53} = 2$$

Because of the symmetry of a bell-shaped distribution, $\frac{1}{2}\cdot 0.68 = 0.34$ of the data lie between the mean and one standard deviation and $\frac{1}{2}\cdot 0.95 - 0.475$ of the data lie between the mean and two standard deviations. So, 0.34 + 0.475 = 0.815 or 81.5% of the light bulbs will last between 547 and 706 hours.

(d) From part (a) 441 is three standard deviations. The manufacturer replaces all light bulbs that burn out before this time. Since 99.7% of the data lie between three standard deviations of the mean, and since the distribution is symmetric, the firm expects to replace

$$\frac{1}{2}(1-0.997) = \frac{1}{2}\cdot 0.003 = 0.0015 = 0.15\%$$

of the light bulbs.

39. Here $\mu = 12$ and $\sigma = 0.05$. We want to estimate the probability a jar contains between 11.9 and 12.1 ounces, so

$$k = 12.1 - \mu = 12.1 - 12 = 0.1 \text{ or}$$
$$k = \mu - 11.9 = 12 - 11.9 = 0.1$$

By Chebychev's theorem, the probability a jar has between 11.9 and 12.1 ounces is at least

$$1 - \frac{\sigma^2}{k^2} = 1 - \frac{0.05^2}{0.1^2} = 0.75$$

41. Here $\mu = 10$ and $\sigma = 0.25$ pounds. We want to estimate the probability a bag weighs less than 9.5 or more than 10.5 pounds.

$$k = 10.5 - \mu = 10.5 - 10 = 0.5, \text{ or}$$
$$k = \mu - 9.5 = 10 - 9.5 = 0.5$$

By Chebychev's theorem, the probability a bag weighs between 9.5 and 10.5 pounds is at least

$$1 - \frac{\sigma^2}{k^2} = 1 - \frac{0.25^2}{0.5^2} = 0.75$$

The probability a bag weighs less than 9.5 or more than 10.5 pounds is less than 1 – 0.75 = 0.25.

43. $Z = \frac{x-\mu}{\sigma} = \frac{8-10}{3}$

$Z = -\frac{2}{3} = -0.667$

45. $Z = \frac{x-\mu}{\sigma} = \frac{8-1}{5}$

$Z = \frac{7}{5} = 1.40$

47. $Z = \frac{x-\mu}{\sigma} = \frac{60-55}{3}$

$Z = \frac{5}{3} = 1.667$

49. $Z_1 = -1.35$ $\quad A_1 = 0.4115$
$Z_2 = -2.75$ $\quad A_2 = 0.4970$

Since both Z-scores are negative, they are both less than the mean. The area under the normal curve is the difference in A_1 and A_2.

$$\text{Area} = A_2 - A_1 = 0.4970 - 0.4115 = 0.0855$$

51. $Z_1 = -0.75$ $\quad A_1 = 0.2734$
$Z_2 = 2.1$ $\quad A_2 = 0.4821$

Since the Z-scores have opposite signs, one is less than and the other is greater than the mean. The area under the normal curve is the sum of A_1 and A_2.

$$\text{Area} = A_1 + A_2 = 0.2734 + 0.4821 = 0.7555$$

53. We are told that $\mu = 25$ and $\sigma = 5$.

(a) We need to find the area under a normal curve between $x = 20$ and $x = 30$.

$$x = 20: \quad Z_1 = \frac{x-\mu}{\sigma} = \frac{20-25}{5} = -1 \quad A_1 = 0.3413$$

$$x = 30: \quad Z_2 = \frac{x-\mu}{\sigma} = \frac{30-25}{5} = 1 \quad A_2 = 0.3413$$

Since the x-values are on opposite sides of the mean, the area under the normal curve is the sum of A_1 and A_2.

$$\text{Area} = A_1 + A_2 = 0.3413 + 0.3413 = 0.6826$$

So, 68.26% of the scores fall between 20 and 30.

(b) We need to find the area under a normal curve to the right of $x = 35$.

$$x = 35: \; Z = \frac{x-\mu}{\sigma} = \frac{35-25}{5} = 2 \quad A = 0.4772$$

Since the area to the right of the mean is 0.5, the area to the right of $x = 35$ will be

$$\text{Area} = 0.5000 - 0.4772 = 0.0228$$

So, 2.28% of the scores are above 35.

55. Here $\mu = 14$ and $\sigma = 1.25$ years. We need the area under a normal curve to the left of 10 years, 4 months which is equivalent to $10\frac{1}{3} = \frac{31}{3}$ years. We convert $x = \frac{31}{3}$ to a Z-score.

$$x = \frac{31}{3}: Z = \frac{x-\mu}{\sigma} = \frac{\frac{31}{3}-14}{1.25} = -2.93 \quad A = 0.4983$$

The area to the left of x is $0.500 - 0.4983 = 0.0017$. Approximately 0.17% of dogs will die before reaching the age of 10 years 4 months.

57. We need to convert both test grades to standard scores, and then compare the grades.

Mathematics: score: 89; $\mu = 79$; $\sigma = 5$

$$Z = \frac{x-\mu}{\sigma} = \frac{89-79}{5} = 2.0$$

Sociology: score: 79; $\mu = 72$; $\sigma = 3.5$

$$Z = \frac{x-\mu}{\sigma} = \frac{79-72}{3.5} = 2.0$$

Bob scored equally well on both exams.

59. We need to find the area under a normal curve between $x = 30$ and $x = 50$. We convert these values to Z-scores, using $\mu = 40$ and $\sigma = \sqrt{25} = 5$ units.

$$x = 30: \quad Z_1 = \frac{x-\mu}{\sigma} = \frac{30-40}{5} = -2 \qquad A_1 = 0.4772$$

$$x = 50: \quad Z_2 = \frac{x-\mu}{\sigma} = \frac{50-40}{5} = 2 \qquad A_2 = 0.4772$$

The area between the x-values is the sum of A_1 and A_2 which equals 0.9544. So, the probability that the week's production will be between 30 and 50 is 0.9544.

61. This is a binomial distribution which we will approximate using a normal curve with a correction for continuity. Since we are interested in the probability of more than 160 successes, we let $x = 160.5$, and convert x to a Z-score.
We have $p = 0.7$, the probability of a positive result, and $n = 200$, the number of blood samples.
We calculate

$$\mu = np = (200)(0.7) = 140, \text{ and } \sigma = \sqrt{np(1-p)} = \sqrt{200(0.7)(0.3)} = 6.48$$

$$x = 160.5: Z = \frac{x-\mu}{\sigma} = \frac{160.5-140}{6.48} = 3.16 \quad A = 0.4992$$

The probability of obtaining more than 160 positive results is $0.5000 - 0.4992 = 0.0008$.

63. This is a binomial distribution which we will approximate using a normal curve with corrections for continuity. We obtain the area under a normal curve between $x = 19.5$ and $x = 30.5$ by converting both to Z-scores.
We have $p = 0.05$, the probability the gate does not open, and $n = 500$, the number of times the gate is opened. We can calculate

$$\mu = np = (500)(0.05) = 25, \text{ and } \sigma = \sqrt{np(1-p)} = \sqrt{500(0.05)(0.95)} = 4.87$$

$$x = 19.5: \quad Z_1 = \frac{x-\mu}{\sigma} = \frac{19.5-25}{4.87} = -1.13 \qquad A_1 = 0.3708$$

$$x = 30.5: \quad Z_2 = \frac{x-\mu}{\sigma} = \frac{30.5-25}{4.87} = 1.13 \qquad A_2 = 0.3708$$

The area between the x-values is the sum of A_1 and A_2 which equals 0.7416. The probability that the toll gate fails to work between 20 and 30 times is 0.742, (technology answer is 0.7409).

Chapter 9 Project

1. Mean high temperature: $\frac{2342}{30} = 78.1°$ Fahrenheit.

Mean low temperature: $\frac{1937}{30} = 64.6°$ Fahrenheit.

Total rainfall: 7.32 inches

3. Mean monthly temperature $= \dfrac{\sum \frac{1}{2}(\text{daily high} + \text{daily low})}{30} = \dfrac{\frac{1}{2}\sum(\text{daily high} + \text{daily low})}{30}$

$$= \frac{\sum(\text{daily high} + \text{daily low})}{60}$$

$$= \frac{\sum(\text{daily high}) + \sum(\text{daily low})}{60}$$

Mean of mean high and mean low temperatures:

$$\frac{1}{2}\left[\left(\frac{\sum \text{daily high}}{30}\right) + \left(\frac{\sum \text{daily low}}{30}\right)\right]$$

$$= \frac{1}{2}\left[\frac{\sum(\text{daily high} + \text{daily low})}{30}\right]$$

$$= \frac{\sum(\text{daily high} + \text{daily low})}{60}$$

5. Modal high temperature was 83° and the modal low temperatures were 57° and 70°.

7. High temperatures:
Standard deviation:

$$\sigma = \sqrt{\frac{\sum(x - 78.1)^2}{30}} = \sqrt{\frac{1395.9}{30}} = 6.82$$

Low temperatures:
Standard deviation:

$$\sigma = \sqrt{\frac{\sum(x - 64.6)^2}{30}} = \sqrt{\frac{983.4}{30}} = 5.73$$

Since for the high temperatures, the mean < median < mode, the distribution is not symmetric, but is skewed to the left. This means that there are a few extremely low temperatures that are influencing the mean.

One might underestimate the probability of obtaining a temperature of between 90° and 92° using a normal distribution.

9.

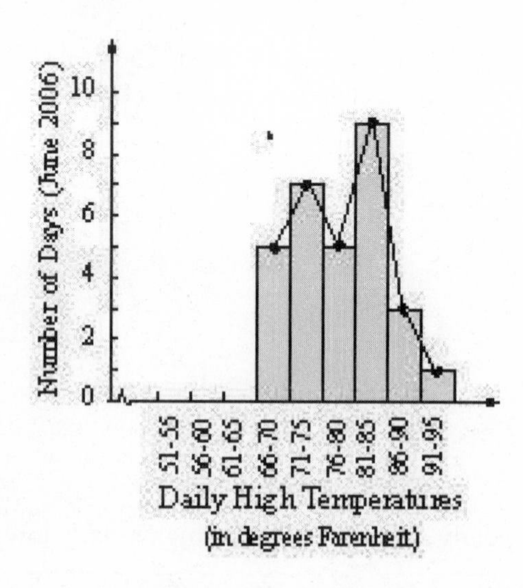

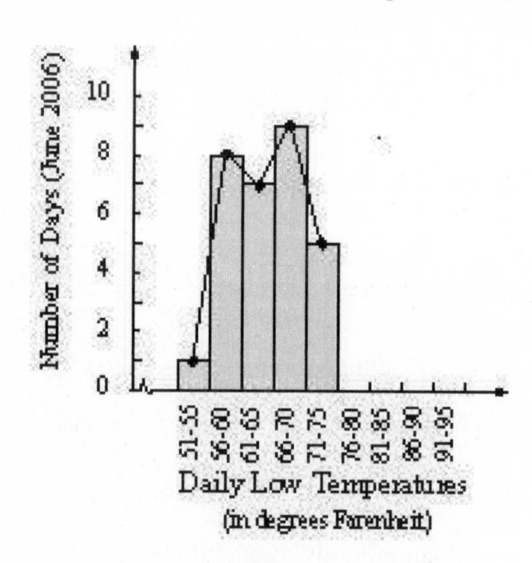

11.

	Midpoint, *m*	Frequency, *f*	*fm*	*m* - mean	(*m* - mean)2	*f*(*m* - mean)2
	52.5	0	0	-25.2	635.04	0
	57.5	0	0	-20.2	408.04	0
	62.5	0	0	-15.2	231.04	0
	67.5	5	337.5	-10.2	104.04	520.2
	72.5	7	507.5	-5.2	27.04	189.28
	77.5	5	387.5	-0.2	0.04	0.2
	82.5	9	742.5	4.8	23.04	207.36
	87.5	3	262.5	9.8	96.04	288.12
	92.5	1	92.5	14.8	219.04	219.04
Sum	652.5	30	2330			1424.2

$$\text{Approximate Mean: } \mu = \frac{\sum f \cdot m}{n} = \frac{2330}{30} = 77.7°$$

$$\text{Approximate Standard Deviation: } \sigma = \sqrt{\frac{\sum f \cdot (m-\mu)^2}{n}} = \sqrt{\frac{1424.2}{30}} = 6.9°$$

Answers and reasons will vary.

Mathematical Questions from Professional Exams

1. Answer: (e) 10

   In a normal curve, approximately 95% of the area under the curve lies within 2 standard deviations of the mean, so $K = 2\sigma$.

   The experiment is binomial, meaning $\sigma = \sqrt{np(1-p)}$.

   $p = \frac{1}{6}$, the probability of throwing a sum of 7

   $n = 180$ throws

   $$K = 2\sqrt{(180)\left(\frac{1}{6}\right)\left(\frac{5}{6}\right)} = 10$$

3. Answer (c) 0.68

   We are looking for the probability the balls are within 1 standard deviation of the mean weight of the 100 balls. The area under a normal curve between $\mu - \sigma$ and $\mu + \sigma$ is approximately 0.

Chapter 10

Markov Chains; Games

10.1 Markov Chains and Transition Matrices

5. False. In transition matrices all the entries are between 0 and 1 inclusive and the sum of the entries in every row is 1.

7. $1 \times m$

9. (a) The entry $\frac{1}{4}$ represents the probability that an object in state 2 will move to state 1.

(b) We use the tree diagram on the right. The probability distribution after one observation is

$$v^{(1)} = \begin{bmatrix} \frac{1}{3} & \frac{2}{3} \end{bmatrix}$$

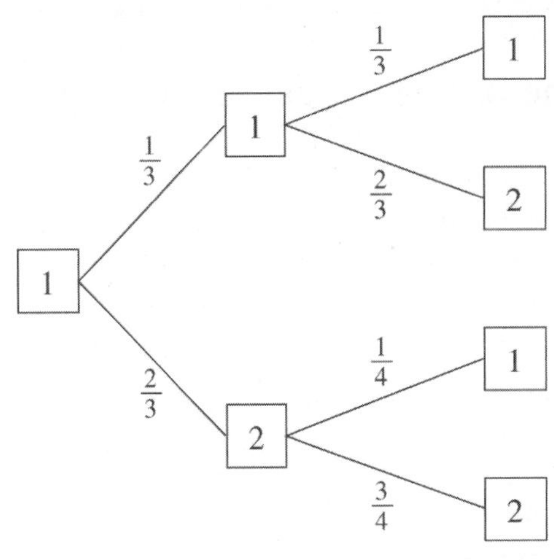

The probability distribution after two observations is

$$\text{from state 1 to 1} = \frac{1}{9} + \frac{1}{6} = \frac{5}{18}$$

$$\text{from state 1 to 2} = \frac{2}{9} + \frac{1}{2} = \frac{13}{18}$$

$$v^{(2)} = \begin{bmatrix} \frac{5}{18} & \frac{13}{18} \end{bmatrix}$$

(c) We use the tree diagram on the right. The probability distribution after one observation is

$$v^{(1)} = \begin{bmatrix} \frac{1}{4} & \frac{3}{4} \end{bmatrix}$$

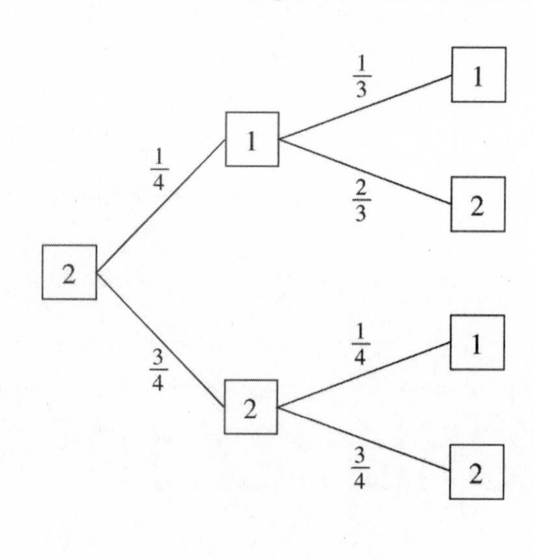

The probability distribution after two observations is

$$\text{from state 2 to 1} = \frac{1}{12} + \frac{3}{16} = \frac{13}{48}$$

$$\text{from state 2 to 2} = \frac{1}{6} + \frac{9}{16} = \frac{35}{48}$$

$$v^{(2)} = \begin{bmatrix} \frac{13}{48} & \frac{35}{48} \end{bmatrix}$$

11. Using Equation (2) (text p. 600), $v^{(k)} = v^{(0)}P^k$

$$v^{(2)} = v^{(0)} \begin{bmatrix} \frac{1}{3} & \frac{2}{3} \\ \frac{1}{4} & \frac{3}{4} \end{bmatrix}^2 = \begin{bmatrix} \frac{1}{4} & \frac{3}{4} \end{bmatrix} \begin{bmatrix} \frac{5}{18} & \frac{13}{18} \\ \frac{13}{48} & \frac{35}{48} \end{bmatrix} = \begin{bmatrix} \frac{157}{576} & \frac{419}{576} \end{bmatrix}$$

13. The probability distribution of the next observation is found by multiplying the initial distribution by the transition matrix.

$$v^{(1)} = v^{(0)}P = [0.25 \quad 0.25 \quad 0.5] \begin{bmatrix} 0.7 & 0.2 & 0.1 \\ 0.6 & 0.2 & 0.2 \\ 0.4 & 0.1 & 0.5 \end{bmatrix} = [0.525 \quad 0.15 \quad 0.325]$$

15. To be a transition matrix, all entries must be non-negative and the sum of the entries in a row must equal 1.

$$a = 1 - 0.2 - 0.4 = 0.4$$
$$b = 1 - 0.6 - 0.3 = 0.1$$
$$c = 1$$

$$\begin{bmatrix} 0.2 & 0.4 & 0.4 \\ 0.1 & 0.6 & 0.3 \\ 0 & 1 & 0 \end{bmatrix}$$

17. If in Example 3, $v^{(0)} = [0.7 \quad 0.3]$ then we can get the probability distribution after 5 years by using Equation (1) $v^{(k)} = v^{(k-1)}P$ (text p. 598), five times.

$$v^{(1)} = v^{(0)}P = [0.7 \quad 0.3] \begin{bmatrix} 0.93 & 0.07 \\ 0.01 & 0.99 \end{bmatrix} = [0.654 \quad 0.346]$$

$$v^{(2)} = v^{(1)}P = [0.654 \quad 0.346] \begin{bmatrix} 0.93 & 0.07 \\ 0.01 & 0.99 \end{bmatrix} = [0.6117 \quad 0.3883]$$

$$v^{(3)} = v^{(2)}P = [0.6117 \quad 0.3883] \begin{bmatrix} 0.93 & 0.07 \\ 0.01 & 0.99 \end{bmatrix} = [0.5727 \quad 0.4273]$$

$$v^{(4)} = v^{(3)}P = [0.5727 \quad 0.4273]\begin{bmatrix} 0.93 & 0.07 \\ 0.01 & 0.99 \end{bmatrix} = [0.5369 \quad 0.4631]$$

$$v^{(5)} = v^{(4)}P = [0.5369 \quad 0.4631]\begin{bmatrix} 0.93 & 0.07 \\ 0.01 & 0.99 \end{bmatrix} = [0.5040 \quad 0.4960]$$

After 5 years 50.4% of the residents live in the city and 49.6% live in the suburbs.

19. (a) This is an example of a Markov chain because it represents a sequence of experiments each of which results in one of two states, and the probability of being in a particular state depends only on the previous state.

(b) $P = \begin{array}{c} \\ \text{R} \\ \text{C} \end{array}\begin{array}{c} \begin{array}{cc} \text{R} & \text{C} \end{array} \\ \begin{bmatrix} 0.90 & 0.10 \\ 0.05 & 0.95 \end{bmatrix} \end{array}$ (c) $P^2 = \begin{array}{c} \\ \text{R} \\ \text{C} \end{array}\begin{array}{c} \begin{array}{cc} \text{R} & \text{C} \end{array} \\ \begin{bmatrix} 0.815 & 0.185 \\ 0.0925 & 0.9075 \end{bmatrix} \end{array}$ $P^3 = \begin{array}{c} \\ \text{R} \\ \text{C} \end{array}\begin{array}{c} \begin{array}{cc} \text{R} & \text{C} \end{array} \\ \begin{bmatrix} 0.74275 & 0.25725 \\ 0.128625 & 0.871375 \end{bmatrix} \end{array}$

21. (a) This is a Markov chain because it represents a sequence of experiments each of which results in one of two states, and the probability of being in a particular state depends only on the previous state.

(b)

$$P = \begin{bmatrix} \frac{1}{2} & 0 & 0 & \frac{1}{2} & 0 & 0 & 0 & 0 & 0 \\ 0 & \frac{1}{2} & 0 & 0 & \frac{1}{2} & 0 & 0 & 0 & 0 \\ 0 & 0 & \frac{1}{2} & 0 & 0 & \frac{1}{2} & 0 & 0 & 0 \\ \frac{1}{3} & 0 & 0 & \frac{1}{3} & 0 & 0 & \frac{1}{3} & 0 & 0 \\ 0 & \frac{1}{3} & 0 & 0 & \frac{1}{3} & 0 & 0 & \frac{1}{3} & 0 \\ 0 & 0 & 0 & 0 & 0 & 1 & 0 & 0 & 0 \\ 0 & 0 & 0 & 0 & 0 & 0 & 1 & 0 & 0 \\ 0 & 0 & 0 & 0 & 0 & 0 & 0 & 0 & 1 \\ 0 & 0 & 0 & 0 & 0 & 0 & 0 & 0 & 1 \end{bmatrix}$$

23. This is a Markov chain with a transition matrix,

$$P = \begin{array}{r} \\ \text{Pinot Grigio} \\ \text{Other White} \end{array}\begin{array}{c} \begin{array}{cc} \text{Pinot Grigio} & \text{Other White} \end{array} \\ \begin{bmatrix} 0.75 & 0.25 \\ 0.35 & 0.65 \end{bmatrix} \end{array}$$

If 50% of the people currently drink Pinot Grigio, the initial probability distribution is $v^{(0)} = [0.50 \quad 0.50]$. To find what percentage will drink Pinot Grigio after 2 months, we need to find $v^{(2)}$. We will use Equation (2) $v^{(2)} = v^{(0)}P^2$.

$$v^{(2)} = [0.50 \quad 0.50]\begin{bmatrix} 0.75 & 0.25 \\ 0.35 & 0.65 \end{bmatrix}^2 = [0.50 \quad 0.50]\begin{bmatrix} 0.65 & 0.35 \\ 0.49 & 0.51 \end{bmatrix} = [0.57 \quad 0.43]$$

After two months 57% of the people will drink Pinot Grigio.

25. We will use Equation (2) (text p. 600) to find the probability distribution after 10 observations.

$$v^{(10)} = v^{(0)}P^{10} = [0.4\ \ 0.3\ \ 0.1\ \ 0.2]\begin{bmatrix} 0.5 & 0.2 & 0.1 & 0.2 \\ 0.3 & 0.3 & 0.2 & 0.2 \\ 0.1 & 0.5 & 0.1 & 0.3 \\ 0.25 & 0.25 & 0.25 & 0.25 \end{bmatrix}^{10} = [0.3200\ \ 0.2892\ \ 0.1631\ 0.2277]$$

27. (a)

$$P = \begin{bmatrix} \frac{1}{2} & 0 & 0 & \frac{1}{2} & 0 & 0 \\ 0 & \frac{1}{2} & 0 & 0 & \frac{1}{2} & 0 \\ 0 & 0 & \frac{1}{2} & 0 & 0 & \frac{1}{2} \\ \frac{1}{2} & 0 & 0 & \frac{1}{2} & 0 & 0 \\ 0 & 0 & 0 & 0 & 0 & 1 \\ 0 & 0 & 0 & 0 & 0 & 1 \end{bmatrix}$$

(b) If the mouse is initially in room 2, the initial probability distribution is

$$v^{(0)} = [0\ \ 1\ \ 0\ \ 0\ \ 0\ \ 0]$$

(c) After 10 stages the probability distribution will be

$$v^{(10)} = v^{(0)}P^{10}$$

$$v^{(10)} = [0\ \ 0.001\ \ 0\ \ 0\ \ 0.001\ \ 0.998]$$

(d) The mouse will most likely be in room 6.

29. Since A is a transition matrix all of its entries are non-negative, and the sum of the rows is one.

$$a_{11} + a_{12} = 1 \qquad a_{21} + a_{22} = 1$$

Since $u = [u_1\ \ u_2]$ is a probability row vector its entries are non-negative and the sum of the entries is one.

$$uA = [u_1\ \ u_2]\begin{bmatrix} a_{11} & a_{12} \\ a_{21} & a_{22} \end{bmatrix} = [u_1a_{11} + u_2a_{21}\quad u_1a_{12} + u_2a_{22}]$$

If uA is a probability vector then its entries must be non-negative, which they are because they are the sums and products of non-negative numbers, and the entries must sum to 1.

$$u_1a_{11} + u_2a_{21} + u_1a_{12} + u_2a_{22} = 1$$

$$\begin{aligned} u_1a_{11} + u_1a_{12} + u_2a_{21} + u_2a_{22} &= u_1(a_{11} + a_{12}) + u_2(a_{21} + a_{22}) \\ &= u_1(1) + u_2(1) \qquad \text{(from above)} \\ &= 1 \qquad (u \text{ is a probability row vector.}) \end{aligned}$$

So, uA is a probability vector.

31. The matrix given cannot be the transition matrix for a Markov chain because it has a negative entry.

33. If A is a transition matrix then $A^{(2)}, A^{(3)}, \ldots, A^{(n)}$ are also transition matrices.

10.2 Regular Markov Chains

3. False. A Markov chain is regular if some power of the transition matrix P has positive entries.

5. P is not regular; the product of row 2 and column 1 is always 0. So $p_{21} = 0$ for every power of P.

7. P is not regular; the product of row 1 and column 2 is always 0. So $p_{12} = 0$ for every power of P.

9. P is regular.

$$P^2 = \begin{bmatrix} \frac{7}{36} & \frac{13}{36} & \frac{4}{9} \\ \frac{1}{8} & \frac{3}{8} & \frac{1}{2} \\ \frac{5}{24} & \frac{1}{3} & \frac{11}{24} \end{bmatrix} \text{ has all positive entries.}$$

11. P is not regular. The product of row 1 and column 3 is always 0. (So are the products of row 2 and column 3; row 3 and column 1; and row 3 and column 2). So $p_{13} = 0$, $p_{23} = 0$, $p_{31} = 0$, and $p_{32} = 0$ for every power of P.

13. P is regular.

$$P^2 = \begin{bmatrix} \frac{1}{3} & \frac{3}{16} & \frac{13}{48} & \frac{5}{24} \\ \frac{1}{3} & \frac{3}{16} & \frac{13}{48} & \frac{5}{24} \\ \frac{5}{18} & \frac{1}{6} & \frac{5}{18} & \frac{5}{18} \\ \frac{1}{3} & \frac{3}{16} & \frac{13}{48} & \frac{5}{24} \end{bmatrix} \text{ has all positive entries.}$$

15. Define $\mathbf{t} = [t_1 \quad t_2]$. We want to find $\mathbf{t}$ that makes $\mathbf{t}P = \mathbf{t}$ true.

$$[t_1 \quad t_2]\begin{bmatrix} \frac{1}{3} & \frac{2}{3} \\ \frac{1}{2} & \frac{1}{2} \end{bmatrix} = [t_1 \quad t_2]$$

These lead to the system of equations

$$\begin{cases} \frac{1}{3}t_1 + \frac{1}{2}t_2 = t_1 \\ \frac{2}{3}t_1 + \frac{1}{2}t_2 = t_2 \\ t_1 + t_2 = 1 \end{cases} \quad \text{or} \quad \begin{cases} -\frac{2}{3}t_1 + \frac{1}{2}t_2 = 0 \\ \frac{2}{3}t_1 - \frac{1}{2}t_2 = 0 \\ t_1 + t_2 = 0 \end{cases}$$

The first two equations are equivalent so we actually have a system of 2 equations,

$$\begin{cases} \frac{2}{3}t_1 - \frac{1}{2}t_2 = 0 & (1) \\ t_1 + t_2 = 1 & (2) \end{cases}$$

We will solve by substitution, substituting $t_2 = 1 - t_1$. Equation (1) becomes

$$\frac{2}{3}t_1 - \frac{1}{2}(1 - t_1) = 0$$

$$4t_1 - 3 + 3t_1 = 0$$

$$7t_1 = 3 \text{ or } t_1 = \frac{3}{7}$$

Back-substituting t_1 in equation (2) we get $t_2 = \frac{4}{7}$. The fixed probability vector is $\mathbf{t} = \begin{bmatrix} \frac{3}{7} & \frac{4}{7} \end{bmatrix}$.

17. Define $\mathbf{t} = [t_1 \quad t_2]$. We want to find $\mathbf{t}$ that makes $\mathbf{t}P = \mathbf{t}$ true.

$$[t_1 \quad t_2]\begin{bmatrix} \frac{1}{4} & \frac{3}{4} \\ \frac{1}{2} & \frac{1}{2} \end{bmatrix} = [t_1 \quad t_2]$$

These lead to the system of equations

$$\begin{cases} \frac{1}{4}t_1 + \frac{1}{2}t_2 = t_1 \\ \frac{3}{4}t_1 + \frac{1}{2}t_2 = t_2 \\ t_1 + \quad t_2 = 1 \end{cases} \quad \text{or} \quad \begin{cases} -\frac{3}{4}t_1 + \frac{1}{2}t_2 = 0 \\ \frac{3}{4}t_1 - \frac{1}{2}t_2 = 0 \\ t_1 + \quad t_2 = 1 \end{cases}$$

The first two equations are equivalent so we actually have a system of 2 equations,

$$\begin{cases} \frac{3}{4}t_1 - \frac{1}{2}t_2 = 0 & (1) \\ t_1 + \quad t_2 = 1 & (2) \end{cases}$$

We will solve by substitution, substituting $1 - t_1$ for t_2. Equation (1) becomes

$$\frac{3}{4}t_1 - \frac{1}{2}(1 - t_1) = 0$$

$$3t_1 - 2 + 2t_1 = 0$$

$$5t_1 = 2 \text{ or } t_1 = \frac{2}{5}.$$

Back-substituting t_1 in equation (2) we get $t_2 = \frac{3}{5}$. The fixed probability vector is $\mathbf{t} = \begin{bmatrix} \frac{2}{5} & \frac{3}{5} \end{bmatrix}$.

19. Define $\mathbf{t} = [t_1 \quad t_2 \quad t_3]$. We want to find $\mathbf{t}$ that makes $\mathbf{t}P = \mathbf{t}$ true.

$$[t_1 \quad t_2 \quad t_3]\begin{bmatrix} \frac{1}{3} & \frac{1}{3} & \frac{1}{3} \\ \frac{1}{2} & \frac{1}{4} & \frac{1}{4} \\ \frac{1}{4} & \frac{1}{4} & \frac{1}{2} \end{bmatrix} = [t_1 \quad t_2 \quad t_3]$$

These lead to the system of equations

$$\begin{cases} \frac{1}{3}t_1 + \frac{1}{2}t_2 + \frac{1}{4}t_3 = t_1 \\ \frac{1}{3}t_1 + \frac{1}{4}t_2 + \frac{1}{4}t_3 = t_2 \\ \frac{1}{3}t_1 + \frac{1}{4}t_2 + \frac{1}{2}t_3 = t_3 \\ t_1 + t_2 + t_3 = 1 \end{cases} \quad \text{or} \quad \begin{cases} -\frac{2}{3}t_1 + \frac{1}{2}t_2 + \frac{1}{4}t_3 = 0 \\ \frac{1}{3}t_1 - \frac{3}{4}t_2 + \frac{1}{4}t_3 = 0 \\ \frac{1}{3}t_1 + \frac{1}{4}t_2 - \frac{1}{2}t_3 = 0 \\ t_1 + t_2 + t_3 = 1 \end{cases}$$

The first equation is equivalent to the sum of the second two equations, so we have a system of 3 equations which we will solve by elimination.

$$\begin{cases} \frac{1}{3}t_1 - \frac{3}{4}t_2 + \frac{1}{4}t_3 = 0 & (1) \\ \frac{1}{3}t_1 + \frac{1}{4}t_2 - \frac{1}{2}t_3 = 0 & (2) \\ t_1 + t_2 + t_3 = 1 & (3) \end{cases} \quad \text{or multiplying (1) and (2) by 12,} \quad \begin{cases} 4t_1 - 9t_2 + 3t_3 = 0 & (1) \\ 4t_1 + 3t_2 - 6t_3 = 0 & (2) \\ t_1 + t_2 + t_3 = 1 & (3) \end{cases}$$

$$t_1 + t_2 + t_3 = 1 \quad (3) \quad \text{multiply by } (-3) \quad \begin{array}{rl} -3t_1 - 3t_2 - 3t_3 = -3 & (3) \\ 4t_1 - 9t_2 + 3t_3 = 0 & (1) \\ \hline t_1 - 12t_2 \quad = -3 & (1) \text{ add} \end{array}$$

$$t_1 + t_2 + t_3 = 1 \quad (3) \quad \text{multiply by } (6) \quad \begin{array}{rl} 6t_1 + 6t_2 + 6t_3 = 6 & (3) \\ 4t_1 + 3t_2 - 6t_3 = 0 & (2) \\ \hline 10t_1 + 9t_2 \quad = 6 & (2) \text{ add} \end{array}$$

We now have 2 equations $\begin{cases} t_1 = 12t_2 - 3 & (1) \\ 10t_1 + 9t_2 = 6 & (2) \end{cases}$ Substituting for t_1 in (2) we get

$10(12t_2 - 3) + 9t_2 = 6$ or $t_2 = \frac{36}{129} = \frac{12}{43}$. Back- substituting into (1), we get $t_1 = \frac{144}{43} - \frac{129}{43} = \frac{15}{43}$,

and putting both t_1 and t_2 into equation (3) we get $t_3 = 1 - \frac{15}{43} - \frac{12}{43} = \frac{16}{43}$.

The fixed probability vector is $\mathbf{t} = \begin{bmatrix} \frac{15}{43} & \frac{12}{43} & \frac{16}{43} \end{bmatrix}$.

21. (a) The transition matrix P is

$$P = \begin{array}{c} \\ A \\ B \\ C \end{array}\begin{array}{c} \begin{array}{ccc} A & B & C \end{array} \\ \begin{bmatrix} 0.7 & 0.15 & 0.15 \\ 0.1 & 0.8 & 0.1 \\ 0.2 & 0.2 & 0.6 \end{bmatrix} \end{array}$$

(b) We need to find the fixed probability vector **t** to learn her long term distribution of brands.

$$[t_1 \ \ t_2 \ \ t_3]\begin{bmatrix} 0.7 & 0.15 & 0.15 \\ 0.1 & 0.8 & 0.1 \\ 0.2 & 0.2 & 0.6 \end{bmatrix} = [t_1 \ \ t_2 \ \ t_3]$$

Multiplying we get the system of equations:

$$\begin{cases} 0.7t_1 + 0.1t_2 + 0.2t_3 = t_1 \\ 0.15t_1 + 0.8t_2 + 0.2t_3 = t_2 \\ 0.15t_1 + 0.1t_2 + 0.6t_3 = t_3 \\ t_1 + t_2 + t_3 = 1 \end{cases} \quad \text{or} \quad \begin{cases} -0.3t_1 + 0.1t_2 + 0.2t_3 = 0 \\ 0.15t_1 - 0.2t_2 + 0.2t_3 = 0 \\ 0.15t_1 + 0.1t_2 - 0.4t_3 = 0 \\ t_1 + t_2 + t_3 = 1 \end{cases}$$

The first equation is equivalent to the sum of the next two, so we have a system of three equations which we solve by elimination.

$$\begin{cases} 0.15t_1 - 0.2t_2 + 0.2t_3 = 0 & (1) \\ 0.15t_1 + 0.1t_2 - 0.4t_3 = 0 & (2) \\ t_1 + t_2 + t_3 = 1 & (3) \end{cases}$$

$$t_1 + t_2 + t_3 = 1 \ (3) \text{ multiply by } (-0.15) \quad \begin{array}{rl} -0.15t_1 - 0.15t_2 - 0.15t_3 = -0.15 & (3) \\ 0.15t_1 - 0.20t_2 + 0.20t_3 = 0 & (1) \\ \hline -0.35t_2 + 0.05t_3 = -0.15 & (1) \text{ add} \end{array}$$

$$t_1 + t_2 + t_3 = 1 \ (3) \text{ multiply by } (-0.15) \quad \begin{array}{rl} -0.15t_1 - 0.15t_2 - 0.15t_3 = -0.15 & (3) \\ 0.15t_1 + 0.10t_2 - 0.40t_3 = 0 & (2) \\ \hline -0.05t_2 - 0.55t_3 = -0.15 & (2) \text{ add} \end{array}$$

We now have two equations

$$\begin{cases} -.35t_2 + .05t_3 = -.15 & (1) \\ -.05t_2 - .55t_3 = -.15 & (2) \end{cases} \text{ which can be simplified to } \begin{cases} -7t_2 + t_3 = -3 & (1) \\ t_2 + 11t_3 = 3 & (2) \end{cases}$$

Multiplying equation (2) by 7 and adding, we get $78t_3 = 18$ or $t_3 = \frac{3}{13}$. Back-substituting in (2) gives $t_2 = -11\left(\frac{3}{13}\right) + \frac{39}{13} = \frac{6}{13}$ and finally, putting both t_2 and t_3 into equation (3), gives $t_1 = 1 - \frac{6}{13} - \frac{3}{13} = \frac{4}{13}$. The fixed probability vector is $\mathbf{t} = \left[\frac{4}{13} \ \ \frac{6}{13} \ \ \frac{3}{13}\right]$.

In the long run the grocer's stock will consist of $\frac{4}{13}$ or 30.8% brand A, $\frac{6}{13}$ or 46.2% brand B, and $\frac{3}{13}$ or 23.1% brand C detergent.

23. (a) First we form the transition matrix P, letting C, L, and S stand for Conservative, Labour, and Socialist respectively.

$$P = \begin{array}{c} \\ C \\ L \\ S \end{array}\begin{array}{c} \begin{array}{ccc} C & L & S \end{array} \\ \begin{bmatrix} 0.7 & 0.3 & 0 \\ 0.4 & 0.5 & 0.1 \\ 0.2 & 0.4 & 0.4 \end{bmatrix}\end{array}$$

Since we are interested in how the grandson of a Labourite votes, we need $v^{(2)}$ when the initial probability distribution is $v^{(0)} = [0 \quad 1 \quad 0]$.

$$v^{(2)} = v^{(0)}P^2 = [0 \quad 1 \quad 0]\begin{bmatrix} 0.7 & 0.3 & 0 \\ 0.4 & 0.5 & 0.1 \\ 0.2 & 0.4 & 0.4 \end{bmatrix}^2 = [0.5 \quad 0.41 \quad 0.09]$$

The probability the grandson of a Labourite votes Socialist is 0.09.

(b) The long term membership distribution is the fixed probability vector **t**. We find **t** that satisfies the equation **t**P = **t.**

$$[t_1 \quad t_2 \quad t_3]\begin{bmatrix} 0.7 & 0.3 & 0 \\ 0.4 & 0.5 & 0.1 \\ 0.2 & 0.4 & 0.4 \end{bmatrix} = [t_1 \quad t_2 \quad t_3]$$

$$\begin{cases} 0.7t_1 + 0.4t_2 + 0.2t_3 = t_1 \\ 0.3t_1 + 0.5t_2 + 0.4t_3 = t_2 \\ 0.1t_2 + 0.4t_3 = t_3 \\ t_1 + t_2 + t_3 = 1 \end{cases} \text{ or } \begin{cases} -0.3t_1 + 0.4t_2 + 0.2t_3 = 0 \\ 0.3t_1 - 0.5t_2 + 0.4t_3 = 0 \\ 0.1t_2 - 0.6t_3 = 0 \\ t_1 + t_2 + t_3 = 1 \end{cases}$$

The 1st equation is equivalent to the sum of the 2nd and 3rd equations, so we have a system of 3 equations which we will solve by elimination.

Multiply equation (1) by 10: $3t_1 - 5t_2 + 4t_3 = 0$ (1)

Multiply equation (3) by –3: $-3t_1 - 3t_2 - 3t_3 = -3$ (3)

Add: $-8t_2 + t_3 = -3$ or $t_3 = 8t_2 - 3$

Now substitute t_3 in equation (2): $0.1t_2 - 0.6(8t_2 - 3) = 0$

Simplify: $t_2 - 48t_2 + 18 = 0$ or $t_2 = \dfrac{18}{47}$

Back-substituting, we find $t_3 = 8\left(\dfrac{18}{47}\right) - 3 = \dfrac{3}{47}$.

Finally we substitute both t_2 and t_3 into equation (3) and solve for $t_1 = 1 - \dfrac{18}{47} - \dfrac{3}{47} = \dfrac{26}{47}$.

The fixed probability vector $\mathbf{t} = \begin{bmatrix} \dfrac{26}{47} & \dfrac{18}{47} & \dfrac{3}{47} \end{bmatrix}$.

In the long run $\dfrac{26}{47} = 55.3\%$ will vote Conservative.

25. (a) The transition matrix for this system is

$$P = \begin{array}{c} \\ \text{college} \\ \text{H.S.} \\ \text{E.S.} \end{array} \begin{array}{c} \begin{array}{ccc} \text{college} & \text{H.S.} & \text{E.S.} \end{array} \\ \begin{bmatrix} 0.80 & 0.18 & 0.02 \\ 0.40 & 0.50 & 0.10 \\ 0.20 & 0.60 & 0.20 \end{bmatrix} \end{array}$$

(b) To find the probability that a grandchild of an elementary school graduate completes college, we find $v^{(2)}$ when $v^{(0)} = [0 \quad 0 \quad 1]$.

$$v^{(2)} = [0 \quad 0 \quad 1] \begin{bmatrix} 0.80 & 0.18 & 0.02 \\ 0.40 & 0.50 & 0.10 \\ 0.20 & 0.60 & 0.20 \end{bmatrix}^2 = [0.44 \quad 0.456 \quad 0.104]$$

The probability that an elementary school graduate's grandson finishes college is 44%.

(c) To find the probability that a grandchild of a high school graduate also completes high school, we find $v^{(2)}$ when $v^{(0)} = [0 \quad 1 \quad 0]$.

$$v^{(2)} = [0 \quad 1 \quad 0] \begin{bmatrix} 0.80 & 0.18 & 0.02 \\ 0.40 & 0.50 & 0.10 \\ 0.20 & 0.60 & 0.20 \end{bmatrix}^2 = [0.54 \quad 0.382 \quad 0.078]$$

There is a 38.2% chance that a high school graduate's grandchild also finishes high school.

(d) The initial distribution for African Americans more than 25 years of age is $v^{(0)} = [0.13 \quad 0.62 \quad 0.25]$. To find the distribution of their grandchildren's educational attainment we need to find $v^{(2)}$.

$$v^{(2)} = [0.13 \quad 0.62 \quad 0.25] \begin{bmatrix} 0.80 & 0.18 & 0.02 \\ 0.40 & 0.50 & 0.10 \\ 0.20 & 0.60 & 0.20 \end{bmatrix}^2 = [0.538 \quad 0.383 \quad 0.079]$$

(e) The long-run distribution is the fixed probability vector **t**. **t** can be found from P^n

$$P^{20} = \begin{bmatrix} 0.649 & 0.298 & 0.053 \\ 0.649 & 0.298 & 0.053 \\ 0.649 & 0.298 & 0.053 \end{bmatrix}.$$

The long-run distribution is $\mathbf{t} = [0.649 \quad 0.298 \quad 0.053]$.

27. We enter the transition matrix P, into the graphing utility and calculate powers of the matrix until the rows are identical to 4 decimal places. The resulting row, rounded to three decimal places is the fixed probability vector **t**.

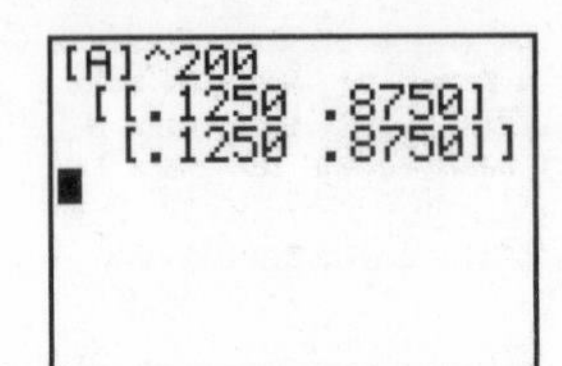

$\mathbf{t} = [0.125 \quad 0.875]$

29. We enter the transition matrix P, into the graphing utility and calculate powers of the matrix until the rows are identical to 5 decimal places. The resulting row, rounded to three decimal places is the fixed probability vector **t**.

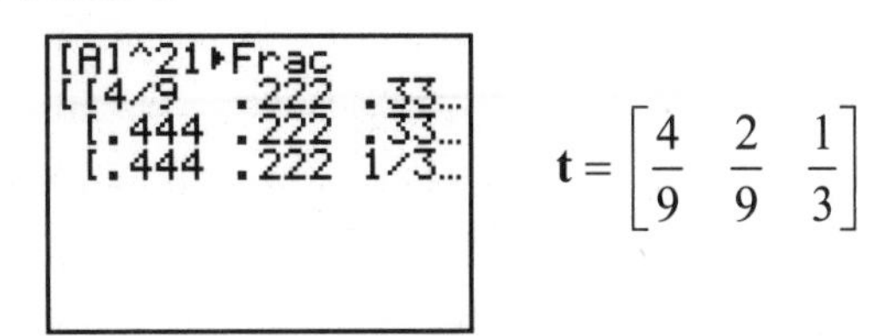

$$\mathbf{t} = \begin{bmatrix} \frac{4}{9} & \frac{2}{9} & \frac{1}{3} \end{bmatrix}$$

31. We enter the transition matrix P, into the graphing utility and calculate powers of the matrix until the rows are identical to 5 decimal places. The resulting row, rounded to three decimal places is the fixed probability vector **t**.

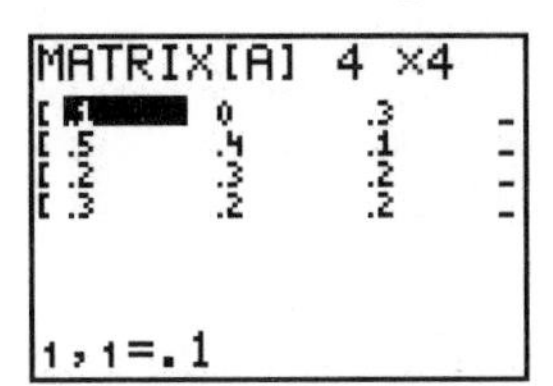

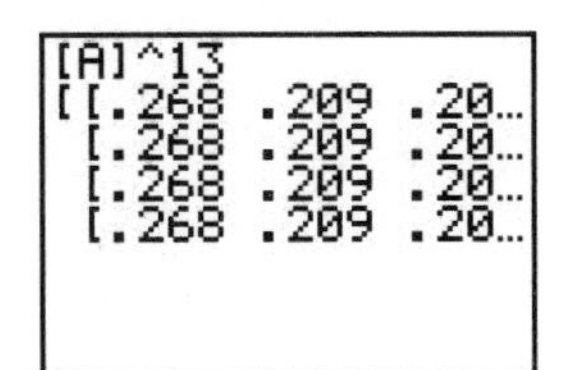

$$\mathbf{t} = [0.268 \quad 0.209 \quad 0.206 \quad 0.318]$$

33. A fixed probability vector of $\begin{bmatrix} \frac{1}{3} & 0 & \frac{1}{3} & \frac{1}{3} \end{bmatrix}$ cannot be from a transition matrix of a regular Markov chain. The definition of a regular Markov chain is that there is a power of P for which all entries are positive. After that power all higher powers will also be positive.

10.3 Absorbing Markov Chains

3. True

5. State 1 is an absorbing state, and it's possible to move from state 2 to state 1. So this is an absorbing Markov chain.

7. There are no absorbing states. So this transition matrix is not from an absorbing Markov chain.

9. States 1 and 3 are absorbing states, and it possible to move from state 2 to both absorbing states. So this is an absorbing Markov chain.

11. State 1 is an absorbing state, but it is impossible to move from either state 2 or state 3 to state 1. So this is not an absorbing Markov chain.

13. There are no absorbing states. So this transition matrix is not from an absorbing Markov chain.

15. States 2 and 3 are absorbing states, and it possible to move from both states 1 and 4 to the absorbing states. So this is an absorbing Markov chain.

17.

$$P = \begin{matrix} \\ 1 \\ 2 \\ 3 \end{matrix}\begin{matrix} \begin{matrix} 1 & 2 & 3 \end{matrix} \\ \begin{bmatrix} 1 & 0 & 0 \\ \frac{1}{3} & \frac{1}{3} & \frac{1}{3} \\ 0 & 0 & 1 \end{bmatrix} \end{matrix} = \begin{matrix} \\ 1 \\ 3 \\ 2 \end{matrix}\begin{matrix} \begin{matrix} 1 & 3 & 2 \end{matrix} \\ \left[\begin{array}{cc|c} 1 & 0 & 0 \\ 0 & 1 & 0 \\ \hline \frac{1}{3} & \frac{1}{3} & \frac{1}{3} \end{array}\right] \end{matrix}$$

19.

$$P = \begin{array}{c} \\ 1 \\ 2 \\ 3 \\ 4 \end{array}\begin{array}{c} \begin{array}{cccc} 1 & 2 & 3 & 4 \end{array} \\ \begin{bmatrix} \frac{1}{4} & \frac{1}{4} & \frac{1}{4} & \frac{1}{4} \\ 0 & 1 & 0 & 0 \\ \frac{1}{2} & 0 & \frac{1}{2} & 0 \\ 0 & 0 & 0 & 1 \end{bmatrix}\end{array} = \begin{array}{c} \\ 2 \\ 4 \\ 1 \\ 3 \end{array}\begin{array}{c} \begin{array}{cccc} 2 & 4 & 1 & 3 \end{array} \\ \left[\begin{array}{cc|cc} 1 & 0 & 0 & 0 \\ 0 & 1 & 0 & 0 \\ \hline \frac{1}{4} & \frac{1}{4} & \frac{1}{4} & \frac{1}{4} \\ 0 & 0 & \frac{1}{2} & \frac{1}{2} \end{array}\right]\end{array}$$

21.

$$P = \begin{array}{c} \\ 1 \\ 2 \\ 3 \\ 4 \\ 5 \end{array}\begin{array}{c} \begin{array}{ccccc} 1 & 2 & 3 & 4 & 5 \end{array} \\ \begin{bmatrix} 1 & 0 & 0 & 0 & 0 \\ \frac{1}{2} & 0 & \frac{1}{4} & 0 & \frac{1}{4} \\ 0 & 0 & 1 & 0 & 0 \\ 0 & \frac{1}{2} & 0 & \frac{1}{2} & 0 \\ 0 & 0 & \frac{1}{2} & 0 & \frac{1}{2} \end{bmatrix}\end{array} = \begin{array}{c} \\ 1 \\ 3 \\ 2 \\ 4 \\ 5 \end{array}\begin{array}{c} \begin{array}{ccccc} 1 & 3 & 2 & 4 & 5 \end{array} \\ \left[\begin{array}{cc|ccc} 1 & 0 & 0 & 0 & 0 \\ 0 & 1 & 0 & 0 & 0 \\ \hline \frac{1}{2} & \frac{1}{4} & 0 & 0 & \frac{1}{4} \\ 0 & 0 & \frac{1}{2} & \frac{1}{2} & 0 \\ 0 & \frac{1}{2} & 0 & 0 & \frac{1}{2} \end{array}\right]\end{array}$$

23. (a) The expected number of times the person will have \$3 given that she started with \$1 is the entry in row 1, column 3. If she started with \$1 we would expect her to have \$3 about 0.5 time.

The expected number of times the person will have \$3 given that she started with \$2 is the entry in row 2, column 3. If she started with \$2 we would expect her to have \$3 about 1 time.

(b) The expected number of games a person will play before absorption is the sum of the entries in the row corresponding to her starting value. A person who starts with \$3 can expect to play 0.5 + 1.0 + 1.5 = 3.0 games before absorption.

25. First we construct the transition matrix P and subdivide it so the absorbing states appear first.

$$P = \begin{array}{c} \\ 0 \\ 1 \\ 2 \\ 3 \end{array}\begin{array}{c} \begin{array}{cccc} 0 & 1 & 2 & 3 \end{array} \\ \begin{bmatrix} 1 & 0 & 0 & 0 \\ 0.6 & 0 & 0.4 & 0 \\ 0 & 0.6 & 0 & 0.4 \\ 0 & 0 & 0 & 1 \end{bmatrix}\end{array} = \begin{array}{c} \\ 0 \\ 3 \\ 1 \\ 2 \end{array}\begin{array}{c} \begin{array}{cccc} 0 & 3 & 1 & 2 \end{array} \\ \left[\begin{array}{cc|cc} 1 & 0 & 0 & 0 \\ 0 & 1 & 0 & 0 \\ \hline 0.6 & 0 & 0 & 0.4 \\ 0 & 0.4 & 0.6 & 0 \end{array}\right]\end{array}$$

Next we find the fundamental matrix T.

$$T = [I_2 - Q]^{-1} = \left[\begin{bmatrix} 1 & 0 \\ 0 & 1 \end{bmatrix} - \begin{bmatrix} 0 & 0.4 \\ 0.6 & 0 \end{bmatrix}\right]^{-1} = \begin{bmatrix} 1 & -0.4 \\ -0.6 & 1 \end{bmatrix}^{-1} = \begin{bmatrix} 1.3158 & 0.5263 \\ 0.7895 & 1.3158 \end{bmatrix} = \begin{bmatrix} \frac{25}{19} & \frac{10}{19} \\ \frac{15}{19} & \frac{25}{19} \end{bmatrix}$$

To find the probability of reaching an absorbing state we multiply $T \cdot S$

$$T \cdot S = \begin{bmatrix} 1.3158 & 0.5263 \\ 0.7895 & 1.3158 \end{bmatrix}\begin{bmatrix} .6 & 0 \\ 0 & .4 \end{bmatrix} = \begin{bmatrix} 0.7895 & 0.2105 \\ 0.4737 & 0.5263 \end{bmatrix} = \begin{bmatrix} \frac{15}{19} & \frac{4}{19} \\ \frac{9}{19} & \frac{10}{19} \end{bmatrix}$$

The probability of accumulating \$3 if starting with \$1 is the entry in row 1, column 2. The probability that a player wins \$3 if he started with \$1 is 0.2105 or $\frac{4}{19}$.

The probability of accumulating \$3 if starting with \$2 is 0.5263 or $\frac{10}{19}$.

27. We construct the transition matrix P and subdivide it so the absorbing states appear first.

$$P = \begin{array}{c} \$0 \\ \$1000 \\ \$2000 \\ \$3000 \\ \$4000 \end{array} \overset{\begin{array}{ccccc} \$0 & \$1000 & \$2000 & \$3000 & \$4000 \end{array}}{\begin{bmatrix} 1 & 0 & 0 & 0 & 0 \\ 0.6 & 0 & 0.4 & 0 & 0 \\ 0.6 & 0 & 0 & 0 & 0.4 \\ 0 & 0 & 0.6 & 0 & 0.4 \\ 0 & 0 & 0 & 0 & 1 \end{bmatrix}}$$

$$= \begin{array}{c} \$0 \\ \$4000 \\ \$1000 \\ \$2000 \\ \$3000 \end{array} \overset{\begin{array}{ccccc} \$0 & \$4000 & \$1000 & \$2000 & \$3000 \end{array}}{\left[\begin{array}{cc|ccc} 1 & 0 & 0 & 0 & 0 \\ 0 & 1 & 0 & 0 & 0 \\ \hline 0.6 & 0 & 0 & 0.4 & 0 \\ 0.6 & 0.4 & 0 & 0 & 0 \\ 0 & 0.4 & 0 & 0.6 & 0 \end{array}\right]}$$

Then we find the fundamental matrix T.

$$T = [I_3 - Q]^{-1} = \begin{bmatrix} 1 & -0.4 & 0 \\ 0 & 1 & 0 \\ 0 & -0.6 & 1 \end{bmatrix}^{-1} = \begin{bmatrix} 1 & 0.4 & 0 \\ 0 & 1 & 0 \\ 0 & 0.6 & 1 \end{bmatrix}$$

(a) Since Colleen started with \$1000 the expected number of wagers placed before the game ends is found by summing the entries in row 1 of T. She expects to place 1.4 wagers before the game ends.

(b) and (c) To answer parts (b) and (c) we need to the product of T and S.

$$T \cdot S = \begin{bmatrix} 1 & 0.4 & 0 \\ 0 & 1 & 0 \\ 0 & 0.6 & 1 \end{bmatrix} \begin{bmatrix} 0.6 & 0 \\ 0.6 & 0.4 \\ 0 & 0.4 \end{bmatrix} = \begin{bmatrix} 0.84 & 0.16 \\ 0.6 & 0.4 \\ 0.36 & 0.64 \end{bmatrix}$$

(b) The probability Colleen is wiped out is the entry in row 1, column 1 of matrix $T \cdot S$. She is wiped out with probability 0.84.

(c) The probability that Colleen wins enough to purchase the car is $1 - 0.84 = 0.16$. (It can also be found in row 1, column 2 of the matrix $T \cdot S$.)

29. Define the states of the system as

N:	No tank survives.	A:	Tank A alone survives.
B:	Tank B alone survives.	C:	Tank C alone survives.
AB:	Tanks A and B survive.	AC:	Tanks A and C survive.
BC:	Tanks B and C survive.	ABC:	All three tanks survive.

Since the game ends when either one tank or no tanks are left, states N, A, B, and C are absorbing states. We will set up the transition matrix P with this in mind and put those states first. We then need to determine the probabilities of moving from state to state in the nonabsorbing states. The rules of the game state that each tank aims at its strongest opponent. If strength is defined as the probability of hitting the target, then A has strength $\frac{1}{3}$, B has strength $\frac{1}{2}$, and C has strength $\frac{1}{6}$. At the start of the game all tanks are active. A tank survives a round if either it is not fired upon or it is missed.

Row ABC: A fires on B, B fires on A, C fires on B; possible resulting states and their probabilities: (Note that C must survive since no tank fires on it.)

$$\text{ABC: } P(\text{all three tanks miss their targets}) = P(\text{ABC} \mid \text{ABC}) = \frac{2}{3}\cdot\frac{1}{2}\cdot\frac{5}{6} = \frac{5}{18}$$

$$\text{AC: } P(\text{either A or C hit and B misses}) = \left(1-\frac{2}{3}\cdot\frac{5}{6}\right)\frac{1}{2} = \frac{2}{9}$$

$$\text{BC: } P(\text{B hits and both A and C miss}) = \frac{1}{2}\cdot\frac{2}{3}\cdot\frac{5}{6} = \frac{5}{18}$$

$$\text{C: } P(\text{either A or C hit and B hits}) = \left(1-\frac{2}{3}\cdot\frac{5}{6}\right)\frac{1}{2} = \frac{2}{9}$$

Row AB: A fires on B, B fires on A. Possible resulting states and their probabilities.

$$\text{AB: } P(\text{both tanks miss}) = \frac{2}{3}\cdot\frac{1}{2} = \frac{1}{3}$$

$$\text{A: } P(\text{A hits and B misses}) = \frac{1}{3}\cdot\frac{1}{2} = \frac{1}{6}$$

$$\text{B: } P(\text{A misses and B hits}) = \frac{2}{3}\cdot\frac{1}{2} = \frac{1}{3}$$

$$\text{N: } P(\text{both hit}) = \frac{1}{3}\cdot\frac{1}{2} = \frac{1}{6}$$

Row AC: A fires on C, C fires on A. Possible resulting states and their probabilities:

$$\text{AC: } P(\text{both tanks miss}) = \frac{2}{3}\cdot\frac{5}{6} = \frac{5}{9}$$

$$\text{A: } P(\text{A hits and C misses}) = \frac{1}{3}\cdot\frac{5}{6} = \frac{5}{18}$$

$$\text{C: } P(\text{A misses and C hits}) = \frac{2}{3}\cdot\frac{1}{6} = \frac{1}{9}$$

$$\text{N: } P(\text{both hit}) = \frac{1}{3}\cdot\frac{1}{6} = \frac{1}{18}$$

Row BC: B fires on C, C fires on B. Possible resulting states and their probabilities:

BC: $P(\text{both tanks miss}) = \frac{1}{2}\cdot\frac{5}{6} = \frac{5}{12}$

B: $P(\text{B hits and C misses}) = \frac{1}{2}\cdot\frac{5}{6} = \frac{5}{12}$

C: $P(\text{B misses and C hits}) = \frac{1}{2}\cdot\frac{1}{6} = \frac{1}{12}$

N: $P(\text{both tanks hit}) = \frac{1}{2}\cdot\frac{1}{6} = \frac{1}{12}$

$$P = \begin{array}{c} \\ N \\ A \\ B \\ C \\ ABC \\ AB \\ AC \\ BC \end{array}\begin{array}{c} \begin{array}{cccccccc} N & A & B & C & ABC & AB & AC & BC \end{array} \\ \left[\begin{array}{cccc|cccc} 1 & 0 & 0 & 0 & 0 & 0 & 0 & 0 \\ 0 & 1 & 0 & 0 & 0 & 0 & 0 & 0 \\ 0 & 0 & 1 & 0 & 0 & 0 & 0 & 0 \\ 0 & 0 & 0 & 1 & 0 & 0 & 0 & 0 \\ \hline 0 & 0 & 0 & \frac{2}{9} & \frac{5}{18} & 0 & \frac{2}{9} & \frac{5}{18} \\ \frac{1}{6} & \frac{1}{6} & \frac{1}{3} & 0 & 0 & \frac{1}{3} & 0 & 0 \\ \frac{1}{18} & \frac{5}{18} & 0 & \frac{1}{9} & 0 & 0 & \frac{5}{9} & 0 \\ \frac{1}{12} & 0 & \frac{5}{12} & \frac{1}{12} & 0 & 0 & 0 & \frac{5}{12} \end{array}\right] \end{array}$$

(a) There are 8 states in the chain.
(b) Four of the 8 states are absorbing.
(c) To find the expected number of rounds fired we need the fundamental matrix T.

$$T = [I_4 - Q]^{-1} = \begin{bmatrix} \frac{13}{18} & 0 & -\frac{2}{9} & -\frac{5}{18} \\ 0 & \frac{2}{3} & 0 & 0 \\ 0 & 0 & \frac{4}{9} & 0 \\ 0 & 0 & 0 & \frac{7}{12} \end{bmatrix}^{-1} = \begin{bmatrix} \frac{18}{13} & 0 & \frac{9}{13} & \frac{60}{91} \\ 0 & \frac{3}{2} & 0 & 0 \\ 0 & 0 & \frac{9}{4} & 0 \\ 0 & 0 & 0 & \frac{12}{7} \end{bmatrix}$$

Since all three tanks took part in the battle, the expected number of rounds fired is the sum of the entries in row 1 of T. We would expect about 2.74 rounds of fire.

(d) The product TS will give the probability A survives, or wins the game.

$$TS = \begin{bmatrix} \frac{17}{182} & \frac{5}{26} & \frac{25}{91} & \frac{40}{91} \\ \frac{1}{4} & \frac{1}{4} & \frac{1}{2} & 0 \\ \frac{1}{8} & \frac{5}{8} & 0 & \frac{1}{4} \\ \frac{1}{7} & 0 & \frac{5}{7} & \frac{1}{7} \end{bmatrix}$$

The probability that A survives is the entry in the first row (since all 3 tanks took part) and the second column (since A is the winner). We would expect A would win 19.2% of the time.

31. If I is an absorbing state, the transition matrix becomes

$$P = \begin{array}{c} \\ I \\ D \\ N \end{array}\begin{array}{c} \begin{array}{ccc} I & D & N \end{array} \\ \left[\begin{array}{c|cc} 1 & 0 & 0 \\ \hline 0.070 & 0.638 & 0.292 \\ 0.079 & 0.064 & 0.857 \end{array}\right] \end{array}$$

The fundamental matrix T will give the number of days until a decreasing stock begins to increase.

$$T = [I_2 - Q]^{-1} = \begin{bmatrix} 0.362 & -0.292 \\ -0.064 & 0.143 \end{bmatrix}^{-1} = \begin{bmatrix} 4.3231 & 8.8276 \\ 1.9348 & 10.9438 \end{bmatrix}$$

A decreasing stock should begin to increase in $4.3231 + 8.8276 = 13.1507$ days.

33. (a) and (d) The transition matrix is on the left. The subdivided matrix is on the right.

$$P = \begin{array}{c} \\ R1 \\ R2 \\ R3 \\ R4 \\ R5 \end{array}\begin{array}{c} \begin{array}{ccccc} R1 & R2 & R3 & R4 & R5 \end{array} \\ \begin{bmatrix} 1 & 0 & 0 & 0 & 0 \\ 0.25 & 0.25 & 0.25 & 0 & 0.25 \\ 0.20 & 0.20 & 0.20 & 0.20 & 0.20 \\ 0.25 & 0 & 0.25 & 0.25 & 0.25 \\ 0 & 0 & 0 & 0 & 1 \end{bmatrix} \end{array} = \begin{array}{c} \\ R1 \\ R5 \\ R2 \\ R3 \\ R4 \end{array}\begin{array}{c} \begin{array}{ccccc} R1 & R5 & R2 & R3 & R4 \end{array} \\ \left[\begin{array}{cc|ccc} 1 & 0 & 0 & 0 & 0 \\ 0 & 1 & 0 & 0 & 0 \\ \hline 0.25 & 0.25 & 0.25 & 0.25 & 0 \\ 0.20 & 0.20 & 0.20 & 0.20 & 0.20 \\ 0.25 & 0.25 & 0 & 0.25 & 0.25 \end{array}\right] \end{array}$$

(b) Two rooms, 1 and 5, represent absorbing states in this Markov chain.

(c) If the mouse starts in room 3, $v^{(0)} = [0\ \ 0\ \ 1\ \ 0\ \ 0]$. The probability the mouse is in room 5 after five steps is given by $v^{(5)}$.

$$v^{(5)} = v^{(0)}P^5 = [0\ \ 0\ \ 1\ \ 0\ \ 0]\begin{bmatrix} 1 & 0 & 0 & 0 & 0 \\ 0.25 & 0.25 & 0.25 & 0 & 0.25 \\ 0.20 & 0.20 & 0.20 & 0.20 & 0.20 \\ 0.25 & 0 & 0.25 & 0.25 & 0.25 \\ 0 & 0 & 0 & 0 & 1 \end{bmatrix}^5$$

$$= [0.474\ \ \ 0.015\ \ \ 0.022\ \ \ 0.015\ \ \ 0.474]$$

There is a probability of 0.474 that the mouse has found the cheese in exactly 5 steps.

(e) $I_2 = \begin{bmatrix} 1 & 0 \\ 0 & 1 \end{bmatrix}$ $\quad S = \begin{bmatrix} 0.25 & 0.25 \\ 0.20 & 0.20 \\ 0.25 & 0.25 \end{bmatrix}$ $\quad Q = \begin{bmatrix} 0.25 & 0.25 & 0 \\ 0.20 & 0.20 & 0.20 \\ 0 & 0.25 & 0.25 \end{bmatrix}$

(f) $T = [I_3 - Q]^{-1} = \begin{bmatrix} 1.4667 & 0.5 & 0.1333 \\ 0.4 & 1.5 & 0.4 \\ 0.1333 & 0.5 & 1.4667 \end{bmatrix}$

(g) $TS = \begin{bmatrix} 1.4667 & 0.5 & 0.1333 \\ 0.4 & 1.5 & 0.4 \\ 0.1333 & 0.5 & 1.4667 \end{bmatrix} \begin{bmatrix} 0.25 & 0.25 \\ 0.20 & 0.20 \\ 0.25 & 0.25 \end{bmatrix} = \begin{bmatrix} 0.5 & 0.5 \\ 0.5 & 0.5 \\ 0.5 & 0.5 \end{bmatrix}$

(h) Since the mouse started in room 3, the sum of the entries in row 2 of the fundamental matrix is the expected time until absorption. The mouse can stay out of the trap or away from the cheese for about 2.3 steps, but we do not know which absorbed state he will end.

10.4 Two Person Games

1. True

3. Katy's game matrix is

	one finger	two fingers
1 finger	−1	1
2 fingers	1	−1

where the entries are dimes.

5. Katy's game matrix is

	1	4	7
1	−2	5	−8
4	5	−8	11
7	−8	11	−14

where the entries stand for number of dimes.

7. A game is strictly determined when there is an entry that is the smallest element in its row and the largest element in its column. That entry, if it exists, is called the value of the game.

The game is strictly determined; –2 is the smallest element in row 1 and the largest element in column 1. The value of the game is –2.

9. A game is strictly determined when there is an entry that is the smallest element in its row and the largest element in its column. That entry, if it exists, is called the value of the game.

The game is strictly determined; 3 is the smallest element in row 1 and the largest element in column 2. The value of the game is 3.

11. A game is strictly determined when there is an entry that is the smallest element in its row and the largest element in its column. That entry, if it exists, is called the value of the game.

The game is not strictly determined. No entry satisfies the criteria.

13. A game is strictly determined when there is an entry that is the smallest element in its row and the largest element in its column. That entry, if it exists, is called the value of the game.

The game is strictly determined; 2 is the smallest element in row 3 and the largest element both in columns 1 and 2. The value of the game is 2.

15. A game is strictly determined when there is an entry that is the smallest element in its row and the largest element in its column. That entry, if it exists, is called the value of the game.

The game is not strictly determined. No entry satisfies the criteria.

17. A game is strictly determined when there is an entry that is the smallest element in its row and the largest element in its column. That entry, if it exists, is called the value of the game.

For the matrix to be strictly determined a must be the smallest value in row 1 and the largest value in column 1. To be the smallest value in row 1, a must be less than or equal to 3. To be the largest value in column 1 a must be greater than or equal to 0. There is no value of a that will make the entry in row 2, column 2 or in row 3, column 3 a saddle point.

So, for the matrix to be strictly determined, $0 \le a \le 3$.

19. The game is strictly determined if one of the following are true.

1. If a is the saddle point then $a \le 0$ to be the smallest value in row 1, and $a \ge 0$ to be the largest value in column 1, so $a = 0$.

2. If b is the saddle point then $b \le 0$ to be the smallest value in row 2, and $b \ge 0$ to be the largest value in column 2, so $b = 0$.

From the Zero Product Property, we can say that the game is strictly determined if and only if the product $ab = 0$.

10.5 Mixed Strategies

3. The expected payoff E of player I is given by PAQ where P is player I's strategy and Q is player II's strategy.

$$P = [0.30 \quad 0.70], \quad Q = \begin{bmatrix} 0.40 \\ 0.60 \end{bmatrix}. \quad E = PAQ = [0.30 \quad 0.70]\begin{bmatrix} 6 & 0 \\ -2 & 3 \end{bmatrix}\begin{bmatrix} 0.40 \\ 0.60 \end{bmatrix} = 1.42$$

Player I should expect to win \$1.42.

5. The expected payoff E of player I is given by PAQ where P is player I's strategy and Q is player II's strategy.

$$E = PAQ = \begin{bmatrix} \frac{1}{2} & \frac{1}{2} \end{bmatrix}\begin{bmatrix} 4 & 0 \\ 2 & 3 \end{bmatrix}\begin{bmatrix} \frac{1}{2} \\ \frac{1}{2} \end{bmatrix} = \begin{bmatrix} 3 & \frac{3}{2} \end{bmatrix}\begin{bmatrix} \frac{1}{2} \\ \frac{1}{2} \end{bmatrix} = \frac{9}{4} = 2.25$$

The game is biased in favor of player 1 and has an expected payoff of 2.25.

7. The expected payoff E of player I is given by PAQ where P is player I's strategy and Q is player II's strategy.

$$E = PAQ = \begin{bmatrix} \frac{1}{4} & \frac{3}{4} \end{bmatrix} \begin{bmatrix} 4 & 0 \\ 2 & 3 \end{bmatrix} \begin{bmatrix} \frac{1}{2} \\ \frac{1}{2} \end{bmatrix} = \begin{bmatrix} \frac{5}{2} & \frac{9}{4} \end{bmatrix} \begin{bmatrix} \frac{1}{2} \\ \frac{1}{2} \end{bmatrix} = \frac{19}{8} = 2.375$$

The game is biased in favor of player I and has an expected payoff of 2.375.

9. The expected payoff E of player I is given by PAQ where P is player I's strategy and Q is player II's strategy.

$$E = PAQ = \begin{bmatrix} \frac{2}{3} & \frac{1}{3} \end{bmatrix} \begin{bmatrix} 4 & 0 \\ -3 & 6 \end{bmatrix} \begin{bmatrix} \frac{1}{3} \\ \frac{2}{3} \end{bmatrix} = \begin{bmatrix} \frac{5}{3} & 2 \end{bmatrix} \begin{bmatrix} \frac{1}{3} \\ \frac{2}{3} \end{bmatrix} = \frac{17}{9}$$

The game is biased in favor of player I and has an expected payoff of $\frac{17}{9}$.

11. The expected payoff E of player I is given by PAQ where P is player I's strategy and Q is player II's strategy.

$$E = PAQ = \begin{bmatrix} \frac{1}{3} & \frac{1}{3} & \frac{1}{3} \end{bmatrix} \begin{bmatrix} 1 & 0 & 0 \\ 0 & 1 & 0 \\ 0 & 0 & 1 \end{bmatrix} \begin{bmatrix} \frac{1}{3} \\ \frac{1}{3} \\ \frac{1}{3} \end{bmatrix} = \begin{bmatrix} \frac{1}{3} & \frac{1}{3} & \frac{1}{3} \end{bmatrix} \begin{bmatrix} \frac{1}{3} \\ \frac{1}{3} \\ \frac{1}{3} \end{bmatrix} = \frac{1}{3}$$

The game is biased in favor of player I and has an expected payoff of $\frac{1}{3}$.

13. Player I's game matrix for matching pennies is $A = \begin{matrix} & \begin{matrix} H & T \end{matrix} \\ \begin{matrix} H \\ T \end{matrix} & \begin{bmatrix} 1 & -1 \\ -1 & 1 \end{bmatrix} \end{matrix}$

The expected payoff of player 1 is given by $E = PAQ$ where P is player I's strategy and Q is player II's strategy. In this game we are told each player chooses heads half the time, so

$$P = [0.5 \quad 0.5] \text{ and } Q = \begin{bmatrix} 0.5 \\ 0.5 \end{bmatrix} \qquad E = [0.5 \quad 0.5] \begin{bmatrix} 1 & -1 \\ -1 & 1 \end{bmatrix} \begin{bmatrix} 0.5 \\ 0.5 \end{bmatrix} = [0 \quad 0] \begin{bmatrix} 0.5 \\ 0.5 \end{bmatrix} = 0$$

The game has an expected payoff of 0. The game is fair.

15. Assume the matrix is not strictly determined. Then there is no element that is the smallest in its row and the largest in its column. That is, there is no saddle point.

There are 3 possible relations between a_{11} and a_{12} in row 1. They are

$$a_{11} < a_{12} \quad \text{or} \quad a_{11} > a_{12} \quad \text{or} \quad a_{11} = a_{12}$$

(1) If $a_{11} < a_{12}$ then $a_{11} < a_{21}$ otherwise it would be the smallest in the row, largest in the column – a saddle point. This forces $a_{21} > a_{22}$ so that a_{21} is not the saddle point, and $a_{12} > a_{22}$ preventing a_{22} from being the saddle point.

So if the matrix is not strictly determined and $a_{11} < a_{12}$ then $a_{11} < a_{21}$, $a_{21} > a_{22}$, and $a_{12} > a_{22}$ are also true (statement (b)).

(2) If $a_{11} > a_{12}$ then $a_{12} < a_{22}$ or a_{12} would be a saddle point. Since a_{22} is the largest in its column, it cannot be the smallest in row 2, so $a_{21} < a_{22}$. Finally, to prevent a_{21} from being the saddle point $a_{11} > a_{21}$.

So if the matrix is not strictly determined and $a_{11} > a_{12}$ then $a_{12} < a_{22}$, $a_{21} < a_{22}$, and $a_{11} > a_{21}$ are also true (statement (a)).

(3) If $a_{11} = a_{12}$ then either entry could be considered largest or smallest in row 1, and the matrix would be strictly determined.

10.6 Optimal Strategy in Two-Person Zero-Sum Games with 2 × 2 Matrices

3. $A = \begin{bmatrix} 1 & 2 \\ 4 & 1 \end{bmatrix}$

Player I's strategy is $[p \quad 1-p]$.

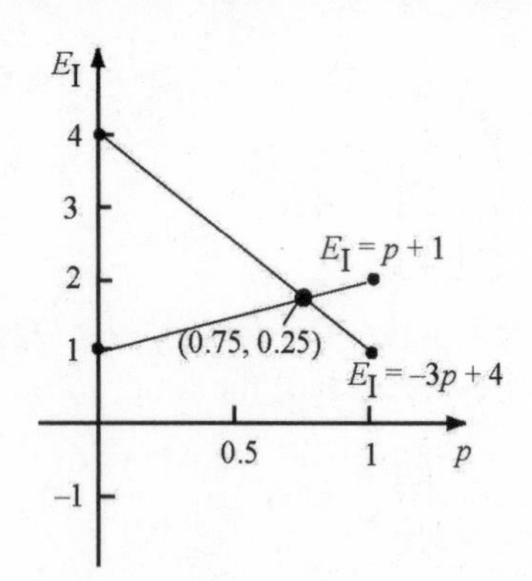

If player II chooses column 1, player I expects to win

$$E_I = p + 4(1-p) = -3p + 4 \quad (1)$$

If player II chooses column 2, player I expects to win

$$E_I = 2p + 1(1-p) = p + 1 \quad (2)$$

These 2 lines are graphed on the left. The intersection of the lines represents player I's maximum payoff and optimal strategy. We solve (1) and (2) simultaneously.

$$E_I = E_I$$
$$-3p + 4 = p + 1$$
$$p = \frac{3}{4} \quad \text{and} \quad 1 - p = \frac{1}{4}$$

Player I's optimal strategy is $P = \begin{bmatrix} \frac{3}{4} & \frac{1}{4} \end{bmatrix}$.

Using equation (5) to check the results, we get $P = [p_1 \quad p_2]$

$$p_1 = \frac{1-4}{1+1-2-4} = \frac{-3}{-4} = \frac{3}{4} \qquad p_2 = \frac{1-2}{1+1-2-4} = \frac{-1}{-4} = \frac{1}{4}$$

Player II's strategy is $\begin{bmatrix} q \\ 1-q \end{bmatrix}$.

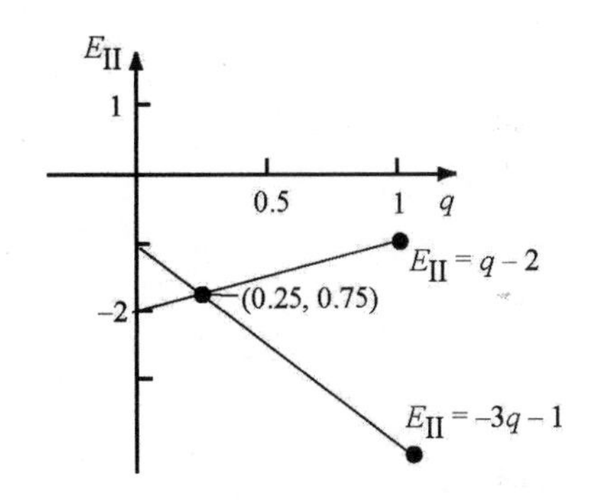

If player I chooses row 1, player II expects to win

$$E_{II} = -q - 2(1-q) = -q - 2 + 2q = q - 2 \quad (1)$$

If player I chooses row 2, player II expects to win

$$E_{II} = -4q - 1(1-q) = -3q - 1 \quad (2)$$

These 2 lines are graphed on the left. The intersection of the lines represents player II's maximum payoff and optimal strategy. We solve (1) and (2) simultaneously.

$$E_{II} = E_{II}$$
$$q - 2 = -3q - 1$$
$$4q = 1$$
$$q = \frac{1}{4} \quad \text{and} \quad 1 - q = \frac{3}{4}$$

Player II's optimal strategy is $Q = \begin{bmatrix} \frac{1}{4} \\ \frac{3}{4} \end{bmatrix}$

Using equations (6) to check the results we get $Q = \begin{bmatrix} q_1 \\ q_2 \end{bmatrix}$

$$q_1 = \frac{1-2}{1+1-2-4} = \frac{-1}{-4} = \frac{1}{4} \qquad q_2 = \frac{1-4}{1+1-2-4} = \frac{-3}{-4} = \frac{3}{4}$$

The expected payoff of the game is $E = PAQ = \begin{bmatrix} \frac{3}{4} & \frac{1}{4} \end{bmatrix} \begin{bmatrix} 1 & 2 \\ 4 & 1 \end{bmatrix} \begin{bmatrix} \frac{1}{4} \\ \frac{3}{4} \end{bmatrix} = \frac{7}{4} = 1.75$

5. $A = \begin{bmatrix} -3 & 2 \\ 1 & 0 \end{bmatrix}$

Player I's strategy is $[p \quad 1-p]$.

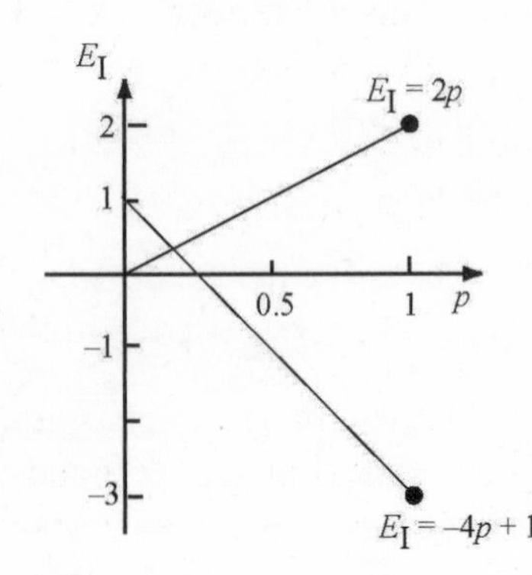

If player II chooses column 1, player I expects to win

$$E_I = -3p + 1(1-p) = -4p + 1 \quad (1)$$

If player II chooses column 2, player I expects to win

$$E_I = 2p + 0(1-p) = 2p \quad (2)$$

These 2 lines are graphed on the left. The intersection of the lines represents player I's maximum payoff and optimal strategy. We solve (1) and (2) simultaneously.

$$E_I = E_I$$
$$-4p + 1 = 2p$$
$$p = \frac{1}{6} \quad \text{and} \quad 1 - p = \frac{5}{6}$$

Player I's optimal strategy is $P = \begin{bmatrix} \frac{1}{6} & \frac{5}{6} \end{bmatrix}$.

Using equation (5) to check the results, we get $P = [p_1 \quad p_2]$

$$p_1 = \frac{0-1}{-3+0-2-1} = \frac{-1}{-6} = \frac{1}{6} \qquad p_2 = \frac{-3-2}{-3+0-2-1} = \frac{-5}{-6} = \frac{5}{6}$$

Player II's strategy is $\begin{bmatrix} q \\ 1-q \end{bmatrix}$.

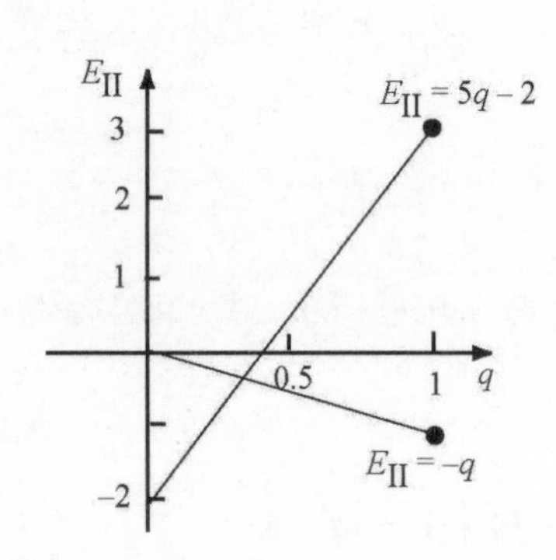

If player I chooses row 1, player II expects to win

$$E_{II} = 3q - 2(1-q) = 5q - 2 \quad (1)$$

If player I chooses row 2, player II expects to win

$$E_{II} = -1q + 0(1-q) = -q \quad (2)$$

These 2 lines are graphed on the left. The intersection of the lines represents player II's maximum payoff and optimal strategy. We solve (1) and (2) simultaneously.

$$E_{II} = E_{II}$$
$$5q - 2 = -q$$
$$6q = 2$$
$$q = \frac{1}{3} \quad \text{and} \quad 1 - q = \frac{2}{3}$$

Player II's optimal strategy is $Q = \begin{bmatrix} \frac{1}{3} \\ \frac{2}{3} \end{bmatrix}$

Using equations (6) to check the results we get $Q = \begin{bmatrix} q_1 \\ q_2 \end{bmatrix}$

$$q_1 = \frac{0-2}{-3+0-2-1} = \frac{-2}{-6} = \frac{1}{3} \qquad q_2 = \frac{-3-1}{-3+0-2-1} = \frac{-4}{-6} = \frac{2}{3}$$

The expected payoff of the game is $E = PAQ = \begin{bmatrix} \frac{1}{6} & \frac{5}{6} \end{bmatrix} \begin{bmatrix} -3 & 2 \\ 1 & 0 \end{bmatrix} \begin{bmatrix} \frac{1}{3} \\ \frac{2}{3} \end{bmatrix} = \frac{1}{3}$

7. $A = \begin{bmatrix} 2 & -1 \\ -1 & 4 \end{bmatrix}$

Player I's strategy is $[p \quad 1-p]$.

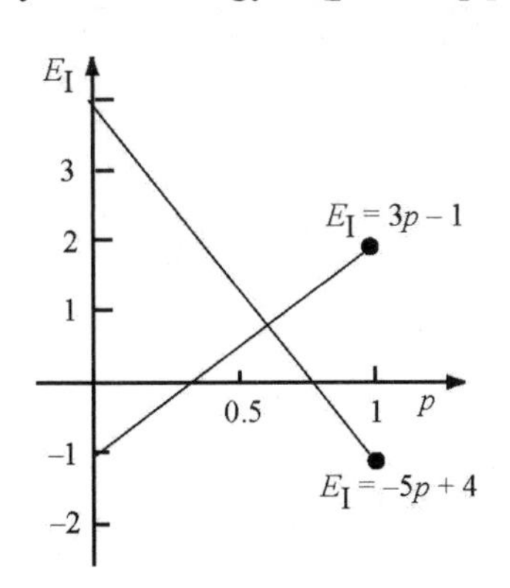

If player II chooses column 1, player I expects to win

$$E_I = 2p - 1(1-p) = 3p - 1 \quad (1)$$

If player II chooses column 2, player I expects to win

$$E_I = -p + 4(1-p) = -5p + 4 \quad (2)$$

These 2 lines are graphed on the left. The intersection of the lines represents player I's maximum payoff and optimal strategy. We solve (1) and (2) simultaneously.

$$E_I = E_I$$
$$3p - 1 = -5p + 4$$
$$8p = 5$$
$$p = \frac{5}{8} \quad \text{and} \quad 1-p = \frac{3}{8}$$

Player I's optimal strategy is $P = \begin{bmatrix} \frac{5}{8} & \frac{3}{8} \end{bmatrix}$.

Using equation (5) to check the results, we get $P = [p_1 \quad p_2]$

$$p_1 = \frac{4+1}{2+4+1+1} = \frac{5}{8} \qquad p_2 = \frac{2+1}{2+4+1+1} = \frac{3}{8}$$

Player II's strategy is $\begin{bmatrix} q \\ 1-q \end{bmatrix}$.

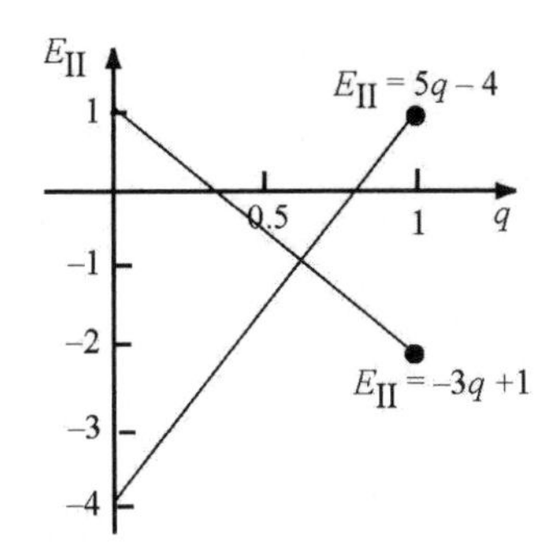

If player I chooses row 1, player II expects to win

$$E_{II} = -2q + 1(1-q) = -3q + 1 \quad (1)$$

If player I chooses row 2, player II expects to win

$$E_{II} = 1q - 4(1-q) = 5q - 4 \quad (2)$$

These 2 lines are graphed on the left. The intersection of the lines represents player II's maximum payoff and optimal strategy. We solve (1) and (2) simultaneously.

$$E_{II} = E_{II}$$
$$-3q + 1 = 5q - 4$$
$$-8q = -5$$
$$q = \frac{5}{8} \quad \text{and} \quad 1-q = \frac{3}{8}$$

Player II's optimal strategy is $Q = \begin{bmatrix} \frac{5}{8} \\ \frac{3}{8} \end{bmatrix}$

Using equations (6) to check the results we get $Q = \begin{bmatrix} q_1 \\ q_2 \end{bmatrix}$

$$q_1 = \frac{4+1}{2+4+1+1} = \frac{5}{8} \qquad q_2 = \frac{2+1}{2+4+1+1} = \frac{3}{8}$$

The expected payoff of the game is $E = PAQ = \begin{bmatrix} \frac{5}{8} & \frac{3}{8} \end{bmatrix} \begin{bmatrix} 2 & -1 \\ -1 & 4 \end{bmatrix} \begin{bmatrix} \frac{5}{8} \\ \frac{3}{8} \end{bmatrix} = \frac{7}{8}$

9. (a) $A = \begin{bmatrix} 4 & -1 \\ 0 & 3 \end{bmatrix}$ We will use equations (5) and (6) to determine the Democrat's (who plays rows) and the Republican's (who plays columns) best strategies.

$$p_1 = \frac{3-0}{4+3+1-0} = \frac{3}{8} \qquad p_2 = 1 - p_1 = \frac{5}{8}$$

The Democrat's best strategy is to spend $\frac{3}{8}$ = 37.5 of the time on domestic issues and $\frac{5}{8}$ = 62.5% of the time on foreign issues.

$$q_1 = \frac{3+1}{4+3+1-0} = \frac{4}{8} = \frac{1}{2} \qquad q_2 = 1 - q_1 = \frac{1}{2}$$

The Republican's best strategy is to spend half the time on domestic issues and half on foreign.

(b) The expected payoff of the game is $PAQ = \begin{bmatrix} \frac{3}{8} & \frac{5}{8} \end{bmatrix} \begin{bmatrix} 4 & -1 \\ 0 & 3 \end{bmatrix} \begin{bmatrix} \frac{1}{2} \\ \frac{1}{2} \end{bmatrix} = \frac{3}{2}$

The game favors the Democrat who expects to gain 1.5 units when optimal strategies are used.

11. First we need to construct the game matrix. We will let the spy play rows.

$$\begin{array}{ll} & \text{Opponent} \\ & \begin{array}{cc}\text{Busy} & \text{Deserted}\end{array} \\ \text{Spy : } \begin{array}{l}\text{Busy} \\ \text{Deserted}\end{array} & \begin{bmatrix} -2 & 10 \\ 30 & -100 \end{bmatrix} \end{array}$$

We will use equations (5) and (6) to find the optimal strategies of the spy (P) and the opponent (Q).

$$P = [p_1 \ \ p_2] : \ p_1 = \frac{-100-30}{-2-100-10-30} = \frac{-130}{-142} = 0.915 \quad p_2 = 1 - p_1 = 0.085$$

The spy should select the busy exit 91.5% of the time and the deserted exit 8.5% of the time.

$$Q = \begin{bmatrix} q_1 \\ q_2 \end{bmatrix} : \quad q_1 = \frac{-100-10}{-2-100-10-30} = \frac{-110}{-142} = 0.775 \qquad q_2 = 1 - q_1 = 0.225$$

The opponent should select the busy exit 77.5% of the time and the deserted exit 22.5% of the time.

The expected payoff of the game is given by $PAQ = \dfrac{(-2)(-100)-(10)(30)}{-2-100-10-30}$

$$= \frac{-100}{-142} = 0.704$$

The game favors the spy whose optimal strategy is [0.915 0.085].

13. We define a game matrix for which Player I plays rows

$$\text{Player I}\ \overset{\text{Player II}}{\begin{bmatrix} a_{11} & a_{12} \\ a_{21} & a_{22} \end{bmatrix}}$$

First we prove equations (5). If Player II plays column 1 then Player I expects to win

$$E_{\text{I}} = a_{11}p + a_{21}(1-p) = a_{11}p + a_{21} - a_{21}p \quad (1)$$

If Player II plays column 2 then Player I expects to win

$$E_{\text{I}} = a_{12}p + a_{22}(1-p) = a_{12}p + a_{22} - a_{22}p \quad (2)$$

Solving the system of equations for p we get

$$E_{\text{I}} = E_{\text{I}}$$
$$a_{11}p + a_{21} - a_{21}p = a_{12}p + a_{22} - a_{22}p$$
$$a_{11}p - a_{21}p - a_{12}p + a_{22}p = a_{22} - a_{21}$$
$$(a_{11} - a_{21} - a_{12} + a_{22})p = a_{22} - a_{21}$$
$$p = \frac{a_{22} - a_{21}}{a_{11} + a_{22} - a_{12} - a_{21}} = p_1$$
$$p_2 = 1 - p_1$$
$$= 1 - \frac{a_{22} - a_{21}}{a_{11} + a_{22} - a_{12} - a_{21}}$$
$$= \frac{a_{11} + a_{22} - a_{12} - a_{21}}{a_{11} + a_{22} - a_{12} - a_{21}} - \frac{a_{22} - a_{21}}{a_{11} + a_{22} - a_{12} - a_{21}}$$
$$p_2 = \frac{a_{11} - a_{12}}{a_{11} + a_{22} - a_{12} - a_{21}}$$

Next we prove equations (6). If Player 1 plays row 1, then Player II expects to win

$$E_{\text{II}} = -a_{11}q - a_{12}(1-q) = -a_{11}q - a_{12} + a_{12}q \quad (1)$$

If Player I plays row 2 then Player I expects to win

$$E_{\text{II}} = -a_{21}q - a_{22}(1-q) = -a_{21}q - a_{22} + a_{22}q \quad (2)$$

Solving the system of equations for p we get

$$E_{\text{II}} = E_{\text{II}}$$
$$-a_{11}q - a_{12} + a_{12}q = -a_{21}q - a_{22} + a_{22}q$$
$$-a_{11}q - a_{22}q + a_{12}q + a_{21}q = a_{12} - a_{22}$$
$$-(a_{11} + a_{22} - a_{12} - a_{21})q = a_{12} - a_{22}$$
$$(a_{11} + a_{22} - a_{12} - a_{21})q = a_{22} - a_{12}$$
$$q = \frac{a_{22} - a_{12}}{a_{11} + a_{22} - a_{12} - a_{21}} = q_1$$
$$q_2 = 1 - q_1$$
$$= 1 - \frac{a_{22} - a_{12}}{a_{11} + a_{22} - a_{12} - a_{21}}$$
$$= \frac{a_{11} + a_{22} - a_{12} - a_{21}}{a_{11} + a_{22} - a_{12} - a_{21}} - \frac{a_{22} - a_{12}}{a_{11} + a_{22} - a_{12} - a_{21}}$$
$$q_2 = \frac{a_{11} - a_{21}}{a_{11} + a_{22} - a_{12} - a_{21}}$$

Chapter 10 Review Exercises

1. P is not regular; the product of row 1 and column 2 is always 0. So $p_{12} = 0$ for every power of P.

3. P is not regular; the product row 2 and column 1 is always 0. So $p_{21} = 0$ for every power of P.

5. P is regular; $P^3 = \begin{bmatrix} \frac{5}{12} & \frac{77}{288} & \frac{91}{288} \\ \frac{1}{3} & \frac{5}{12} & \frac{1}{4} \\ \frac{15}{64} & \frac{455}{1536} & \frac{721}{1536} \end{bmatrix}$ has all positive entries.

7. To find the fixed probability vector $\mathbf{t} = [t_1 \quad t_2]$ we use $t_1 + t_2 = 1$ and $\mathbf{t}P = \mathbf{t}$.

$$[t_1 \quad t_2]\begin{bmatrix} \frac{1}{2} & \frac{1}{2} \\ \frac{2}{3} & \frac{1}{3} \end{bmatrix} = [t_1 \quad t_2] \rightarrow \begin{cases} \frac{1}{2}t_1 + \frac{2}{3}t_2 = t_1 \\ \frac{1}{2}t_1 + \frac{1}{3}t_2 = t_2 \end{cases} \text{ or } \begin{cases} -\frac{1}{2}t_1 + \frac{2}{3}t_2 = 0 \\ \frac{1}{2}t_1 - \frac{2}{3}t_2 = 0 \end{cases}$$

These 2 equations are equivalent, so we have a system of two equations.

$$\begin{cases} -\frac{1}{2}t_1 + \frac{2}{3}t_2 = 0 & (1) \\ t_1 + \quad t_2 = 1 & (2) \end{cases} \quad \text{multiply (1) by 2:} \quad \begin{array}{rl} -t_1 + \frac{4}{3}t_2 = 0 & (1) \\ t_1 + \quad t_2 = 1 & (2) \\ \hline \end{array}$$

$$\text{add:} \quad \frac{7}{3}t_2 = 1 \quad \text{or} \quad t_2 = \frac{3}{7}$$

Substituting t_2 into equation (1) we get $t_1 = \frac{4}{7}$. So $\mathbf{t} = \begin{bmatrix} \frac{4}{7} & \frac{3}{7} \end{bmatrix}$.

9. To find the fixed probability vector $\mathbf{t} = [t_1 \quad t_2 \quad t_3]$ we use $t_1 + t_2 + t_3 = 1$and $\mathbf{t}P = \mathbf{t}$.

$$[t_1 \quad t_2 \quad t_3]\begin{bmatrix} \frac{1}{2} & 0 & \frac{1}{2} \\ \frac{1}{4} & \frac{1}{2} & \frac{1}{4} \\ \frac{1}{3} & \frac{1}{3} & \frac{1}{3} \end{bmatrix} = [t_1 \quad t_2 \quad t_3] \rightarrow \begin{cases} \frac{1}{2}t_1 + \frac{1}{4}t_2 + \frac{1}{3}t_3 = t_1 \\ \frac{1}{2}t_2 + \frac{1}{3}t_3 = t_2 \\ \frac{1}{2}t_1 + \frac{1}{4}t_2 + \frac{1}{3}t_3 = t_3 \end{cases} \text{ or } \begin{cases} -\frac{1}{2}t_1 + \frac{1}{4}t_2 + \frac{1}{3}t_3 = 0 \\ -\frac{1}{2}t_2 + \frac{1}{3}t_3 = 0 \\ \frac{1}{2}t_1 + \frac{1}{4}t_2 - \frac{2}{3}t_3 = 0 \end{cases}$$

The 1st equation is equivalent to the sum of the 2nd and 3rd equations so we have a system of 3 equations to be solved.

$$\begin{cases} t_1 + \quad t_2 + \quad t_3 = 1 & (1) \\ -\frac{1}{2}t_2 + \frac{1}{3}t_3 = 0 & (2) \\ \frac{1}{2}t_1 + \frac{1}{4}t_2 - \frac{2}{3}t_3 = 0 & (3) \end{cases} \quad \text{multiply (3) by } -2: \quad \begin{array}{rl} t_1 + \quad t_2 + \quad t_3 = 1 & (1) \\ -t_1 - \frac{1}{2}t_2 + \frac{4}{3}t_3 = 0 & (3) \\ \hline \end{array}$$

$$\text{Add:} \quad \frac{1}{2}t_2 + \frac{7}{3}t_3 = 1 \quad (3)$$

Add equations (2) and (3) to get

$$
\begin{array}{rl}
-\frac{1}{2}t_2 + \frac{1}{3}t_3 = 0 & (2) \\
\frac{1}{2}t_2 + \frac{7}{3}t_3 = 1 & (3) \\
\hline
\frac{8}{3}t_3 = 1 \quad \text{or} \quad t_3 = \frac{3}{8} &
\end{array}
$$

Back-substituting into equation (2) gives $-\frac{1}{2}t_2 + \frac{1}{3}\left(\frac{3}{8}\right) = 0$ or $t_2 = \frac{1}{4}$.

Back-substituting both t_2 and t_3 into (1) gives $t_1 = 1 - \frac{1}{4} - \frac{3}{8} = \frac{3}{8}$.

So the fixed probability vector $\mathbf{t} = \begin{bmatrix} \frac{3}{8} & \frac{1}{4} & \frac{3}{8} \end{bmatrix}$

11. To find the fixed probability vector $\mathbf{t} = [t_1 \quad t_2 \quad t_3]$ we use $t_1 + t_2 + t_3 = 1$ and $\mathbf{t}P = \mathbf{t}$.

$$
[t_1 \quad t_2 \quad t_3] \begin{bmatrix} 0.7 & 0.1 & 0.2 \\ 0.6 & 0.1 & 0.3 \\ 0.4 & 0.2 & 0.4 \end{bmatrix} = [t_1 \quad t_2 \quad t_3]
$$

$$
\rightarrow \begin{cases} 0.7t_1 + 0.6t_2 + 0.4t_3 = t_1 \\ 0.1t_1 + 0.1t_2 + 0.2t_3 = t_2 \\ 0.2t_1 + 0.3t_2 + 0.4t_3 = t_3 \end{cases} \text{ or } \begin{cases} -0.3t_1 + 0.6t_2 + 0.4t_3 = 0 \\ 0.1t_1 - 0.9t_2 + 0.2t_3 = 0 \\ 0.2t_1 + 0.3t_2 - 0.6t_3 = 0 \end{cases}
$$

The 1st equation is equivalent to the sum of the 2nd and 3rd equations so we have a system of 3 equations to be solved. We will solve the system by elimination, but first we will multiply equations (2) and (3) by 10 to remove the decimals.

$$
\begin{cases} t_1 + t_2 + t_3 = 1 & (1) \\ 0.1t_1 - 0.9t_2 + 0.2t_3 = 0 & (2) \\ 0.2t_1 + 0.3t_2 - 0.6t_3 = 0 & (3) \end{cases} = \begin{cases} t_1 + t_2 + t_3 = 1 & (1) \\ t_1 - 9t_2 + 2t_3 = 0 & (2) \\ 2t_1 + 3t_2 - 6t_3 = 0 & (3) \end{cases}
$$

$$
\begin{array}{lrl}
 & t_1 + t_2 + t_3 = 1 & (1) \\
\text{multiply (2) by } -1: & -t_1 + 9t_2 - 2t_3 = 0 & (2) \\
\hline
\text{add:} & 10t_2 - t_3 = 1 &
\end{array}
\qquad
\begin{array}{lrl}
\text{multiply (1) by } -2: & -2t_1 - 2t_2 - 2t_3 = -2 & (1) \\
 & 2t_1 + 3t_2 - 6t_3 = 0 & (3) \\
\hline
\text{add:} & t_2 - 8t_3 = -2 &
\end{array}
$$

These two equations can be solved by substitution. Substituting $t_2 = 8t_3 - 2$ in the first equation we get $10(8t_3 - 2) - t_3 = 1$ or $80t_3 - 20 - t_3 = 1$ or $79t_3 = 21$ or $t_3 = \frac{21}{79}$.

$t_3 = \frac{21}{79}$ allows us to solve for $t_2 = 8\left(\frac{21}{79}\right) - 2 = \frac{10}{79}$. Finally back-substituting both t_2 and t_3 in equation (1) give us $t_1 = 1 - \frac{10}{79} - \frac{21}{79} = \frac{48}{79}$.

The fixed probability matrix $\mathbf{t} = \begin{bmatrix} \frac{48}{79} & \frac{10}{79} & \frac{21}{79} \end{bmatrix} \approx [\,0.608 \quad 0.127 \quad 0.266]$

13. (a) This situation forms a Markov chain because the shifts in market shares can be thought of as a sequence of experiments each of which results in one of a finite number of states, and the movement to the next state depends only on the current state.

The transition matrix P is given by

$$P = \begin{array}{c} \\ A \\ B \\ C \end{array}\begin{array}{c} \begin{array}{ccc} A & B & C \end{array} \\ \begin{bmatrix} 0.50 & 0.20 & 0.30 \\ 0.40 & 0.40 & 0.20 \\ 0.50 & 0.25 & 0.25 \end{bmatrix} \end{array}$$

(b) We are told that currently each has $\frac{1}{3}$ of the market, so $v^{(0)} = \begin{bmatrix} \frac{1}{3} & \frac{1}{3} & \frac{1}{3} \end{bmatrix}$.

We are looking for $v^{(1)}$, the probability distribution after one year.

$$v^{(1)} = v^{(0)}P$$

$$v^{(1)} = \begin{bmatrix} \frac{1}{3} & \frac{1}{3} & \frac{1}{3} \end{bmatrix}\begin{bmatrix} 0.50 & 0.20 & 0.30 \\ 0.40 & 0.40 & 0.20 \\ 0.50 & 0.25 & 0.25 \end{bmatrix} = \begin{bmatrix} \frac{7}{15} & \frac{17}{60} & \frac{1}{4} \end{bmatrix} \approx [0.4667 \quad 0.2833 \quad 0.25]$$

If the trend continues, after one year distributor A has 46.7% of the market, distributor B has 28.3% of the market, and distributor C has 25% of the market.

(c) The distribution after 2 years is given by $v^{(2)} = v^{(1)}P$.

$$v^{(2)} = \begin{bmatrix} \frac{7}{15} & \frac{17}{60} & \frac{1}{4} \end{bmatrix}\begin{bmatrix} 0.50 & 0.20 & 0.30 \\ 0.40 & 0.40 & 0.20 \\ 0.50 & 0.25 & 0.25 \end{bmatrix} = [0.47167 \quad 0.26917 \quad 0.25917]$$

After 2 years distributor A will have 47.2% of the market, distributor B will have 26.9% of the market and distributor C will have 25.9% of the market.

(d) The long run distribution of the market share is given by the fixed probability vector **t.** We use $t_1 + t_2 + t_3 = 1$, and $\mathbf{t}P = \mathbf{t}$.

$$[t_1 \quad t_2 \quad t_3]\begin{bmatrix} 0.50 & 0.20 & 0.30 \\ 0.40 & 0.40 & 0.20 \\ 0.50 & 0.25 & 0.25 \end{bmatrix} = [t_1 \quad t_2 \quad t_3]$$

$$\rightarrow \begin{cases} 0.5t_1 + 0.4t_2 + \ \ 0.5t_3 = t_1 \\ 0.2t_1 + 0.4t_2 + 0.25t_3 = t_2 \\ 0.3t_1 + 0.2t_2 + 0.25t_3 = t_3 \end{cases} = \begin{cases} -0.5t_1 + 0.4t_2 + \ \ 0.5t_3 = 0 \\ 0.2t_1 - 0.6t_2 + 0.25t_3 = 0 \\ 0.3t_1 + 0.2t_2 - 0.75t_3 = 0 \end{cases}$$

The 1st equation is equivalent to the sum of the 2nd and 3rd equations so we have a system of 3 equations to be solved.

$$\begin{cases} t_1 + \quad t_2 + \qquad t_3 = 1 & (1) \\ 0.2t_1 - 0.6t_2 + 0.25t_3 = 0 & (2) \\ 0.3t_1 + 0.2t_2 - 0.75t_3 = 0 & (3) \end{cases}$$

The solutions are $t_1 = 0.4734$, $t_2 = 0.2663$, and $t_3 = 0.2604$, so the fixed probability vector is $\mathbf{t} = [0.4734 \quad 0.2663 \quad 0.2604]$.

In the long run, distributor A will have 47.3% of the market, distributor B will have 26.6% of the market and distributor C will have 26.0% of the market.

15. (a) This situation forms a Markov chain because her service calls can be thought of as a sequence of experiments each of which results in one of a finite number of states, and the movement to the next state depends only on the current state. The transition matrix P is given by

$$P = \begin{array}{c} \\ U_1 \\ U_2 \\ U_3 \end{array} \begin{array}{c} \begin{array}{ccc} U_1 & U_2 & U_3 \end{array} \\ \begin{bmatrix} 0 & 1 & 0 \\ 0.75 & 0 & 0.25 \\ 0.75 & 0.25 & 0 \end{bmatrix} \end{array}$$

(b) The next month's probability vector is $v^{(1)} = v^{(0)}P$.

$$v^{(1)} = \begin{bmatrix} \frac{1}{3} & \frac{1}{3} & \frac{1}{3} \end{bmatrix} \begin{bmatrix} 0 & 1 & 0 \\ 0.75 & 0 & 0.25 \\ 0.75 & 0.25 & 0 \end{bmatrix} = [0.5 \quad 0.4167 \quad 0.0833] = \begin{bmatrix} \frac{1}{2} & \frac{5}{12} & \frac{1}{12} \end{bmatrix}$$

(c) The fixed probability vector **t** gives the long run probability distribution. It satisfies the equation $\mathbf{t}P = \mathbf{t}$, and can be found from P^n for a large value of n. P^{40} is large enough to find **t**. $\mathbf{t} = [0.4286 \quad 0.4571 \quad 0.1143]$.

In the long run she sells at university 1, 42.9% of the time, at university 2, 45.7% of the time, and at university 3, 11.4% of the time.

17. (a) The Markov chain is not absorbing.

19. (a) The Markov chain is absorbing.
(b) State 2 is the absorbing state.
(c)

$$\begin{array}{c} \\ 2 \\ 1 \\ 3 \end{array} \begin{array}{c} \begin{array}{ccc} 2 & 1 & 3 \end{array} \\ \left[\begin{array}{c|cc} 1 & 0 & 0 \\ \hline \frac{1}{3} & 0 & \frac{2}{3} \\ \frac{1}{4} & \frac{3}{8} & \frac{3}{8} \end{array}\right] \end{array}$$

21. (a) The Markov chain is not absorbing.

23. (a) The Markov chain is absorbing.
(b) States 1 and 4 are absorbing states.
(c)

$$\begin{array}{c} \\ 1 \\ 4 \\ 2 \\ 3 \end{array} \begin{array}{c} \begin{array}{cccc} 1 & 4 & 2 & 3 \end{array} \\ \left[\begin{array}{cc|cc} 1 & 0 & 0 & 0 \\ 0 & 1 & 0 & 0 \\ \hline 0.3 & 0.1 & 0.5 & 0.1 \\ 0.25 & 0.2 & 0.35 & 0.2 \end{array}\right] \end{array}$$

25. (a)

$$P = \begin{array}{c} \\ 0 \\ 1 \\ 2 \\ 3 \\ 4 \\ 5 \end{array} \begin{array}{c} \begin{array}{cccccc} 0 & 1 & 2 & 3 & 4 & 5 \end{array} \\ \begin{bmatrix} 1 & 0 & 0 & 0 & 0 & 0 \\ 0.55 & 0 & 0.45 & 0 & 0 & 0 \\ 0 & 0.55 & 0 & 0.45 & 0 & 0 \\ 0 & 0 & 0.55 & 0 & 0.45 & 0 \\ 0 & 0 & 0 & 0.55 & 0 & 0.45 \\ 0 & 0 & 0 & 0 & 0 & 1 \end{bmatrix} \end{array} \to \begin{array}{c} \\ 0 \\ 5 \\ 1 \\ 2 \\ 3 \\ 4 \end{array} \begin{array}{c} \begin{array}{cccccc} 0 & 5 & 1 & 2 & 3 & 4 \end{array} \\ \left[\begin{array}{cc|cccc} 1 & 0 & 0 & 0 & 0 & 0 \\ 0 & 1 & 0 & 0 & 0 & 0 \\ \hline 0.55 & 0 & 0 & 0.45 & 0 & 0 \\ 0 & 0 & 0.55 & 0 & 0.45 & 0 \\ 0 & 0 & 0 & 0.55 & 0 & 0.45 \\ 0 & 0.45 & 0 & 0 & 0.55 & 0 \end{array}\right] \end{array}$$

(b) Since the gambler began with \$2, we use the second row of the fundamental matrix T which was given in the text. He expects to have \$1 on the average of 1.298 times, \$2 on the average of 2.361 times, \$3 on the average of 1.412 times and \$4 on the average of 0.635 time.

(c) Since he started with \$2 the expected length of the game is the sum of the entries in the 2nd row.

$$E = 1.298406 + 2.360738 + 1.411737 + 0.635282 = 5.706163$$

The gambler should expect to play 5.71 games before absorption.

(d) To find the probability the gambler loses all his money or wins \$5, we look at the product of matrices T and S.

$$TS = \begin{bmatrix} 1.584 & 1.062 & 0.635 & 0.286 \\ 1.298 & 2.361 & 1.412 & 0.635 \\ 0.949 & 1.725 & 2.361 & 1.062 \\ 0.522 & 0.949 & 1.298 & 1.584 \end{bmatrix} \begin{bmatrix} 0.55 & 0 \\ 0 & 0 \\ 0 & 0 \\ 0 & 0.45 \end{bmatrix} = \begin{bmatrix} 0.871 & 0.129 \\ 0.714 & 0.286 \\ 0.522 & 0.478 \\ 0.287 & 0.713 \end{bmatrix}$$

The probability the gambler loses all his money is 0.714. This is the entry in row 2 (he started out with \$2), column 1 (he ended up with \$0) of the product TS. The probability the gambler wins \$5 is 0.286. We can get this either from finding the difference of 1 and 0.714 or from row 2, column 2 of the product TS.

27. A game is strictly determined when there is an entry that is the smallest element in its row and the largest element in its column. That entry, if it exists, is called the value of the game.
The game is not strictly determined; no entry satisfies the criteria.

29. A game is strictly determined when there is an entry that is the smallest element in its row and the largest element in its column. That entry, if it exists, is called the value of the game.
The game is strictly determined; 9 is smallest element in row 2 and largest element in column 1. The value of the game is 9.

31. A game is strictly determined when there is an entry that is the smallest element in its row and the largest element in its column. That entry, if it exists, is called the value of the game.
The game is strictly determined; 12 is the smallest element in row 3 and the largest element in column 3. The value of the game is 12.

33. The expected payoff of player 1 is given by PAQ where P is player I's strategy and Q is player II's strategy.

$$E = PAQ = \begin{bmatrix} \frac{1}{3} & \frac{2}{3} \end{bmatrix} \begin{bmatrix} -1 & 1 \\ 1 & -1 \end{bmatrix} \begin{bmatrix} 1 \\ 0 \end{bmatrix} = \begin{bmatrix} \frac{1}{3} & -\frac{1}{3} \end{bmatrix} \begin{bmatrix} 1 \\ 0 \end{bmatrix} = \frac{1}{3}$$

Player I should expect to win $\frac{1}{3}$.

35. The expected payoff of player 1 is given by PAQ where P is player I's strategy and Q is player II's strategy.

$$E = PAQ = [0.5 \quad 0.5] \begin{bmatrix} 4 & 2 \\ -2 & 3 \end{bmatrix} \begin{bmatrix} 0.5 \\ 0.5 \end{bmatrix} = [1 \quad 2.5] \begin{bmatrix} 0.5 \\ 0.5 \end{bmatrix} = 1.75$$

Player I should expect to win 1.75.

37. The expected payoff of player 1 is given by PAQ where P is player I's strategy and Q is player II's strategy.

$$E = PAQ = \begin{bmatrix} \frac{1}{3} & \frac{1}{3} & \frac{1}{3} \end{bmatrix} \begin{bmatrix} 1 & 4 & -3 \\ 2 & -5 & -1 \\ 1 & -2 & 3 \end{bmatrix} \begin{bmatrix} 1 \\ 0 \\ 0 \end{bmatrix} = \begin{bmatrix} \frac{4}{3} & -1 & -\frac{1}{3} \end{bmatrix} \begin{bmatrix} 1 \\ 0 \\ 0 \end{bmatrix} = \frac{4}{3}$$

Player I should expect to win $\frac{4}{3}$.

39. We will use equations (5) and (6) from section 10.6 to find the optimal strategy. The optimal strategy for player I is $P = [p_1 \quad p_2]$ where

$$p_1 = \frac{a_{22} - a_{21}}{a_{11} + a_{22} - a_{12} - a_{21}} = \frac{-1-1}{-1-1-1-1} = \frac{-2}{-4} = \frac{1}{2} \quad \text{and} \quad p_2 = \frac{a_{11} - a_{12}}{a_{11} + a_{22} - a_{12} - a_{21}} = \frac{-1-1}{-1-1-1-1} = \frac{-2}{-4} = \frac{1}{2}$$

So player I's optimal strategy is $P = \begin{bmatrix} \frac{1}{2} & \frac{1}{2} \end{bmatrix}$.

The optimal strategy for player II is $Q = \begin{bmatrix} q_1 \\ q_2 \end{bmatrix}$ where

$$q_1 = \frac{a_{22} - a_{12}}{a_{11} + a_{22} - a_{12} - a_{21}} = \frac{-1-1}{-1-1-1-1} = \frac{-2}{-4} = \frac{1}{2} \quad \text{and} \quad q_2 = \frac{a_{11} - a_{21}}{a_{11} + a_{22} - a_{12} - a_{21}} = \frac{-1-1}{-1-1-1-1} = \frac{-2}{-4} = \frac{1}{2}$$

So player II's optimal strategy is $Q = \begin{bmatrix} \frac{1}{2} \\ \frac{1}{2} \end{bmatrix}$.

41. (a) We will assume the economy is an active opponent working against the investor. We will use equation (5) to get the investor's optimal strategy $P = [p_1 \quad p_2]$.

$$p_1 = \frac{a_{22} - a_{21}}{a_{11} + a_{22} - a_{12} - a_{21}} = \frac{0-18}{-5+0-20-18} = \frac{-18}{-43} = \frac{18}{43} \quad \text{and} \quad p_2 = \frac{a_{11} - a_{12}}{a_{11} + a_{22} - a_{12} - a_{21}} = \frac{-5-20}{-5+0-20-18} = \frac{-25}{-43} = \frac{25}{43}$$

Player I's optimal strategy is $P = \begin{bmatrix} \frac{18}{43} & \frac{25}{43} \end{bmatrix} \approx [0.4186 \quad 0.5814]$. The investor should invest in A 41.9% of the time and in investment B 58.1% of the time.

(b) The expected payoff corresponding to the optimal strategy is given by the product PAQ. However, since we did not need to find Q in this problem, we will use formula (7) to calculate the expected payoff.

$$E = PAQ = \frac{a_{11}a_{22} - a_{12}a_{21}}{a_{11} + a_{22} - a_{12} - a_{21}} = \frac{(-5)(0)-(20)(18)}{-5+0-20-18} = \frac{-360}{-43} = 8.372$$

The investor should expect a gain 8.37% on the investment.

43. Assuming the county is an active opponent working against the investor, we will use equation (5) to get the investor's optimal strategy $P = [p_1 \quad p_2]$.

$$p_1 = \frac{a_{22} - a_{21}}{a_{11} + a_{22} - a_{12} - a_{21}} = \frac{0.80 - 0.20}{0.70 + 0.80 - 0.30 - 0.20} = \frac{0.60}{1} = 0.60 \qquad \text{and} \qquad p_2 = \frac{a_{11} - a_{12}}{a_{11} + a_{22} - a_{12} - a_{21}} = \frac{0.70 - 0.3}{0.70 + 0.80 - 0.30 - 0.30} = \frac{0.40}{1.0} = 0.40$$

The real estate developer's optimal strategy is $P = \begin{bmatrix} \frac{2}{3} & \frac{1}{3} \end{bmatrix}$. The developer should use 60% of the land for a shopping center and 40% of the land for houses.

Chapter 10 Project

1. If we denote each state (location on the map) by the number associated with it, the transition matrix is

$$P = \begin{array}{c} \\ 1 \\ 2 \\ 3 \\ 4 \\ 5 \\ 6 \\ 7 \\ 8 \\ 9 \\ 10 \end{array} \begin{array}{c} \begin{array}{cccccccccc} 1 & 2 & 3 & 4 & 5 & 6 & 7 & 8 & 9 & 10 \end{array} \\ \begin{bmatrix} 0 & \frac{1}{2} & 0 & \frac{1}{2} & 0 & 0 & 0 & 0 & 0 & 0 \\ \frac{1}{2} & 0 & \frac{1}{2} & 0 & 0 & 0 & 0 & 0 & 0 & 0 \\ 0 & \frac{1}{2} & 0 & 0 & 0 & \frac{1}{2} & 0 & 0 & 0 & 0 \\ \frac{1}{3} & 0 & 0 & 0 & \frac{1}{3} & 0 & 0 & \frac{1}{3} & 0 & 0 \\ 0 & 0 & 0 & \frac{1}{2} & 0 & \frac{1}{2} & 0 & 0 & 0 & 0 \\ 0 & 0 & \frac{1}{4} & 0 & \frac{1}{4} & 0 & \frac{1}{4} & 0 & \frac{1}{4} & 0 \\ 0 & 0 & 0 & 0 & 0 & \frac{1}{2} & 0 & 0 & \frac{1}{2} & 0 \\ 0 & 0 & 0 & \frac{1}{3} & 0 & 0 & 0 & 0 & \frac{1}{3} & \frac{1}{3} \\ 0 & 0 & 0 & 0 & 0 & \frac{1}{4} & \frac{1}{4} & \frac{1}{4} & 0 & \frac{1}{4} \\ 0 & 0 & 0 & 0 & 0 & 0 & 0 & \frac{1}{2} & \frac{1}{2} & 0 \end{bmatrix} \end{array}$$

3. To decide if this is a regular Markov chain, we will look at powers of P. If all the entries are positive for some power of P, we will say the chain is regular.

This is a regular Markov chain. P^{10} has all positive entries.

To find the fixed probability vector, we either solve the system of equations

$$\begin{cases} \mathbf{t}P = \mathbf{t} \\ t_1 + t_2 + t_3 + t_4 + t_5 + t_6 + t_7 + t_8 + t_9 + t_{10} = 1 \end{cases}$$

or raise P to a sufficiently large power so that each row of the resulting transition matrix is the same. We will raise P to powers. P^{80} has 10 identical rows. The fixed probability matrix is

$$\mathbf{t} = [0.077 \quad 0.077 \quad 0.077 \quad 0.115 \quad 0.077 \quad 0.154 \quad 0.077 \quad 0.115 \quad 0.154 \quad 0.077]$$

5. Since we are interested in knowing whether we reach the White House or the Capitol first and nothing else, we make them absorbing states. The subdivided transition matrix is

$$P = \begin{array}{c} \\ 7 \\ 10 \\ 1 \\ 2 \\ 3 \\ 4 \\ 5 \\ 6 \\ 8 \\ 9 \end{array} \begin{array}{c} \begin{array}{cccccccccc} 7 & 10 & 1 & 2 & 3 & 4 & 5 & 6 & 8 & 9 \end{array} \\ \left[\begin{array}{cc|cccccccc} 1 & 0 & 0 & 0 & 0 & 0 & 0 & 0 & 0 & 0 \\ 0 & 1 & 0 & 0 & 0 & 0 & 0 & 0 & 0 & 0 \\ \hline 0 & 0 & 0 & \frac{1}{2} & 0 & \frac{1}{2} & 0 & 0 & 0 & 0 \\ 0 & 0 & \frac{1}{2} & 0 & \frac{1}{2} & 0 & 0 & 0 & 0 & 0 \\ 0 & 0 & 0 & \frac{1}{2} & 0 & 0 & 0 & \frac{1}{2} & 0 & 0 \\ 0 & 0 & \frac{1}{3} & 0 & 0 & 0 & \frac{1}{3} & 0 & \frac{1}{3} & 0 \\ 0 & 0 & 0 & 0 & 0 & \frac{1}{2} & 0 & \frac{1}{2} & 0 & 0 \\ \frac{1}{4} & 0 & 0 & 0 & \frac{1}{4} & 0 & \frac{1}{4} & 0 & 0 & \frac{1}{4} \\ 0 & \frac{1}{3} & 0 & 0 & 0 & \frac{1}{3} & 0 & 0 & 0 & \frac{1}{3} \\ \frac{1}{4} & \frac{1}{4} & 0 & 0 & 0 & 0 & 0 & \frac{1}{4} & \frac{1}{4} & 0 \end{array}\right] \end{array}$$

7. This question, "How many moves will it take you on average before you reach either the White House or the Capitol?" is really asking how many moves do you expect before absorption. The answer is the sum of the entries in row 1 (since we started at #1, West Potomac Park) of the Fundamental matrix T.

$$2.817 + 2.117 + 1.417 + 2.275 + 1.117 + 1.433 + 0.925 + 0.667 = 12.768$$

We expect to make about 12.768 moves before reaching either the White House or the Capitol.

Chapter 11

Logic and Logic Circuits

11.1 Propositions

1. Proposition

3. False. The inclusive disjunction of two statements p and q is true provided at least one of the propositions p, q is true.

5. A proposition is a declarative sentence that can be meaningfully classified as either true or false.
"The cost of shell egg futures was up on June 18, 1990." is a declarative sentence. It is either true or false. The statement is a proposition.

7. A proposition is a declarative sentence that can be meaningfully classified as either true or false.
"What a portfolio!" is not a declarative sentence. It is not a proposition.

9. A proposition is a declarative sentence that can be meaningfully classified as either true or false.
"The earnings of XYZ Company doubled last year." is a declarative sentence. It is either true or false. The statement is a proposition.

11. A proposition is a declarative sentence that can be meaningfully classified as either true or false.
"Jones is guilty of murder in the first degree." is a declarative sentence. It is either true or false. The statement is a proposition.

13. The negation of "A fox is an animal." is "A fox is not an animal."

15. The negation of, "I am buying stocks," is "I am not buying stocks."

17. The negation of, "No one wants to buy my house," is "Someone wants to buy my house."

19. The negation of "Some people have no car," is "Everybody has a car."

21. John is an economics major, or John is a sociology minor (or both).

23. John is an economics major and a sociology minor.

25. John is not an economics major, or John is not a sociology minor (or both).

27. John is not an economics major or John is a sociology minor.

11.2 Truth Tables

3. True

5. $p \wedge p \equiv p$ and $q \vee q \equiv q$

7.

p	q	$\sim q$	$p \vee \sim q$
T	T	F	T
T	F	T	T
F	T	F	F
F	F	T	T

9.

p	q	$\sim p$	$\sim q$	$\sim p \wedge \sim q$
T	T	F	F	F
T	F	F	T	F
F	T	T	F	F
F	F	T	T	T

11.

p	q	$\sim p$	$\sim p \wedge q$	$\sim(\sim p \wedge q)$
T	T	F	F	T
T	F	F	F	T
F	T	T	T	F
F	F	T	F	T

13.

p	q	$\sim p$	$\sim q$	$\sim p \vee \sim q$	$\sim(\sim p \vee \sim q)$
T	T	F	F	F	T
T	F	F	T	T	F
F	T	T	F	T	F
F	F	T	T	T	F

15.

p	q	$\sim q$	$p \vee \sim q$	$(p \vee \sim q) \wedge p$
T	T	F	T	T
T	F	T	T	T
F	T	F	F	F
F	F	T	T	F

17.

p	q	$\sim q$	$p \veebar q$	$p \wedge \sim q$	$(p \veebar q) \wedge (p \wedge \sim q)$
T	T	F	F	F	F
T	F	T	T	T	T
F	T	F	T	F	F
F	F	T	F	F	F

19.

p	q	$\sim p$	$\sim q$	$p \wedge q$	$\sim p \wedge \sim q$	$(p \wedge q) \vee (\sim p \wedge \sim q)$
T	T	F	F	T	F	T
T	F	F	T	F	F	F
F	T	T	F	F	F	F
F	F	T	T	F	T	T

21.

p	q	r	$\sim q$	$p \wedge \sim q$	$(p \wedge \sim q) \veebar r$
T	T	T	F	F	T
T	T	F	F	F	F
T	F	T	T	T	F
T	F	F	T	T	T
F	T	T	F	F	T
F	T	F	F	F	F
F	F	T	T	F	T
F	F	F	T	F	F

23.

p	q	$\sim p$	$q \wedge \sim p$	$p \wedge (q \wedge \sim p)$
T	T	F	F	F
T	F	F	F	F
F	T	T	T	F
F	F	T	F	F

25.

p	q	$\sim p$	$\sim q$	$p \wedge q$	$\sim p \wedge \sim q$	$(p \wedge q) \vee (\sim p \wedge \sim q)$	$[(p \wedge q) \vee (\sim p \wedge \sim q)] \wedge p$
T	T	F	F	T	F	T	T
T	F	F	T	F	F	F	F
F	T	T	F	F	F	F	F
F	F	T	T	F	T	T	F

27. The Idempotent Properties: $p \wedge p \equiv p$ and $p \vee p \equiv p$

p	$p \wedge p$	$p \vee p$
T	T	T
F	F	F

Since the first two columns are the same $p \wedge p \equiv p$.

Since the first and third columns are the same $p \vee p \equiv p$.

29. The Associative Properties: $(p \wedge q) \wedge r \equiv p \wedge (q \wedge r)$ and $(p \vee q) \vee r \equiv p \vee (q \vee r)$

p	q	r	$p \wedge q$	$q \wedge r$	$(p \wedge q) \wedge r$	$p \wedge (q \wedge r)$	$p \vee q$	$q \vee r$	$(p \vee q) \vee r$	$p \vee (q \vee r)$
T	T	T	T	T	T	T	T	T	T	T
T	T	F	T	F	F	F	T	T	T	T
T	F	T	F	F	F	F	T	T	T	T
T	F	F	F	F	F	F	T	F	T	T
F	T	T	F	T	F	F	T	T	T	T
F	T	F	F	F	F	F	T	T	T	T
F	F	T	F	F	F	F	F	T	T	T
F	F	F	F	F	F	F	F	F	F	F

Since the columns $(p \wedge q) \wedge r$ and $p \wedge (q \wedge r)$ are the same, $(p \wedge q) \wedge r \equiv p \wedge (q \wedge r)$.

Since the columns $(p \vee q) \vee r$ and $p \vee (q \vee r)$ are the same, $(p \vee q) \vee r \equiv p \vee (q \vee r)$.

31. The Absorption Properties: $p \vee (p \wedge q) \equiv p$ and $p \wedge (p \vee q) \equiv p$

p	q	$p \vee q$	$p \wedge q$	$p \wedge (p \vee q)$	$p \vee (p \wedge q)$
T	T	T	T	T	T
T	F	T	F	T	T
F	T	T	F	F	F
F	F	F	F	F	F

Since columns p and $p \vee (p \wedge q)$ are the same, $p \vee (p \wedge q) \equiv p$.
Since columns p and $p \wedge (p \vee q)$ are the same, $p \wedge (p \vee q) \equiv p$.

33.

p	q	$\sim q$	$\sim q \vee q$	$p \wedge (\sim q \vee q)$
T	T	F	T	T
T	F	T	T	T
F	T	F	T	F
F	F	T	T	F

Since the columns p and $p \wedge (\sim q \vee q)$ are the same, $p \equiv p \wedge (\sim q \vee q)$.

35.

p	$\sim p$	$\sim(\sim p)$
T	F	T
F	T	F

Since the columns p and $\sim(\sim p)$ are the same $\sim(\sim p) \equiv p$.

37. The compound proposition, "Smith is an ex-convict and Smith is rehabilitated," is equivalent to the compound proposition, "Smith is rehabilitated and Smith is an ex-convict."

The compound proposition, "Smith is an ex-convict or Smith is rehabilitated," is equivalent to the compound proposition, "Smith is rehabilitated or Smith is an ex-convict."

39. Show: $(p \vee q) \wedge r \equiv (p \wedge r) \vee (q \wedge r)$

$$(p \vee q) \wedge r \equiv r \wedge (p \vee q) \quad \text{(commutative property)}$$
$$\equiv (r \wedge p) \vee (r \wedge q) \quad \text{(distributive property)}$$
$$\equiv (p \wedge r) \vee (q \wedge r) \quad \text{(commutative property)}$$

41. Mike cannot hit the ball well, or he cannot pitch strikes.

43. The baby is not crying, and the baby is not talking all the time.

11.3 Implications; The Biconditional Connective; Tautologies

1. Hypothesis; conclusion

3. False. The converse of an implication is not logically equivalent to the implication.

5. $\sim p \Rightarrow q$
Converse: $q \Rightarrow \sim p$ Contrapositive: $\sim q \Rightarrow p$ Inverse: $p \Rightarrow \sim q$

7. $\sim q \Rightarrow \sim p$
Converse: $\sim p \Rightarrow \sim q$ Contrapositive: $p \Rightarrow q$ Inverse: $q \Rightarrow p$

9. If it is raining, the grass is wet.
Converse: If the grass is wet, then it is raining.
Contrapositive: If the grass is not wet, then it is not raining.
Inverse: If it is not raining, then the grass is not wet.

11. If it is not raining, it is not cloudy.
Converse: If it is not cloudy, then it is not raining.
Contrapositive: If it is cloudy, then it is raining.
Inverse: If it is raining, then it is cloudy.

13. Rain is sufficient for it to be cloudy. (Same as, "If it is raining, then it is cloudy.")
Converse: If it is cloudy, then it is raining.
Contrapositive: If it is not cloudy, then it is not raining.
Inverse: If it is not raining, then it is not cloudy.

15.

p	q	$\sim p$	$p \wedge q$	$\sim p \vee (p \wedge q)$
T	T	F	T	T
T	F	F	F	F
F	T	T	F	T
F	F	T	F	T

17.

p	q	$\sim p$	$\sim p \wedge q$	$p \vee (\sim p \wedge q)$
T	T	F	F	T
T	F	F	F	T
F	T	T	T	T
F	F	T	F	F

19.

p	q	$\sim p$	$\sim p \Rightarrow q$
T	T	F	T
T	F	F	T
F	T	T	T
F	F	T	F

21.

p	$\sim p$	$\sim p \vee p$
T	F	T
F	T	T

23.

p	q	$p \Rightarrow q$	$p \wedge (p \Rightarrow q)$
T	T	T	T
T	F	F	F
F	T	T	F
F	F	T	F

25.

p	q	r	$q \wedge r$	$p \wedge (q \wedge r)$	$p \wedge q$	$(p \wedge q) \wedge r$	$p \wedge (q \wedge r) \Leftrightarrow (p \wedge q) \wedge r$
T	T	T	T	T	T	T	T
T	T	F	F	F	T	F	T
T	F	T	F	F	F	F	T
T	F	F	F	F	F	F	T
F	T	T	T	F	F	F	T
F	T	F	F	F	F	F	T
F	F	T	F	F	F	F	T
F	F	F	F	F	F	F	T

Since the last column has only true statements, $p \wedge (q \wedge r) \Leftrightarrow (p \wedge q) \wedge r$ is a tautology.

27.

p	q	$p \vee q$	$p \wedge (p \vee q)$	$p \wedge (p \vee q) \Leftrightarrow p$
T	T	T	T	T
T	F	T	T	T
F	T	T	F	T
F	F	F	F	T

Since the last column contains only true statements, $p \wedge (p \vee q) \Leftrightarrow p$ is a tautology.

29. $p \Rightarrow q$

31. $\sim q \Leftrightarrow \sim p$

33. $q \Rightarrow p$

35. Show $p \Rightarrow q \equiv \sim q \Rightarrow \sim p$

$p \Rightarrow q \equiv \sim p \vee q$ $\quad (p \Rightarrow q \equiv \sim p \vee q)$

$\equiv q \vee \sim p$ $\quad$ (commutative property)

$\equiv \sim q \Rightarrow \sim p$ $\quad (p \Rightarrow q \equiv \sim p \vee q)$

37. Show $(p \wedge q) \Rightarrow r \equiv (p \wedge \sim r) \Rightarrow \sim q$

(a)

p	q	r	$\sim q$	$\sim r$	$p \wedge q$	$(p \wedge q) \Rightarrow r$	$p \wedge \sim r$	$(p \wedge \sim r) \Rightarrow \sim q$	$[(p \wedge q) \Rightarrow r] \Leftrightarrow [(p \wedge \sim r) \Rightarrow \sim q]$
T	T	T	F	F	T	T	F	T	T
T	T	F	F	T	T	F	T	F	T
T	F	T	T	F	F	T	F	T	T
T	F	F	T	T	F	T	T	T	T
F	T	T	F	F	F	T	F	T	T
F	T	F	F	T	F	T	F	T	T
F	F	T	T	F	F	T	F	T	T
F	F	F	T	T	F	T	F	T	T

Since $(p \wedge q) \Rightarrow r \Leftrightarrow (p \wedge \sim r) \Rightarrow \sim q$ is a tautology, $(p \wedge q) \Rightarrow r \equiv (p \wedge \sim r) \Rightarrow \sim q$.

(b) $(p \wedge q) \Rightarrow r \equiv \sim (p \wedge q) \vee r$ $\quad (p \Rightarrow q \equiv \sim p \vee q)$

$\equiv (\sim p \vee \sim q) \vee r$ $\quad$ (De Morgan's property)

$\equiv \sim p \vee (\sim q \vee r)$ $\quad$ (associative property)

$\equiv \sim p \vee (r \vee \sim q)$ $\quad$ (commutative property)

$\equiv (\sim p \vee r) \vee \sim q$ $\quad$ (associative property)

$\equiv \sim (p \wedge \sim r) \vee \sim q$ $\quad$ (De Morgan's property)

$\equiv (p \wedge \sim r) \Rightarrow \sim q$ $\quad (p \Rightarrow q \equiv \sim p \vee q)$

11.4 Arguments

1. Direct proof; Indirect or Proof by Contradiction

3. Let p and q be the statements,

p: It rains. q: John goes to school.

We are assuming q and $p \Rightarrow \sim q$ and q are true statements, and we want to prove $\sim p$ is a true statement.

Direct Proof: $q \Rightarrow \sim p$ is true by the Law of the Contrapositive. With both $q \Rightarrow \sim p$ and q true, then by the Law of Detachment, $\sim p$ must also be true.

We have proved, "It does not rain."

Indirect Proof: We make the additional assumption that $\sim p$ is false.
Then by the Law of Contradiction, p is true. Then since both p and $p \Rightarrow \sim q$ are true, by the Law of Detachment $\sim q$ is true, but this is a contradiction. q is true, meaning $\sim q$ is false. So our additional assumption that $\sim p$ is false is incorrect, and $\sim p$ is true.

We have proved, "It does not rain."

5. Let p, q, and r be the statements, p: Smith is elected president,
q: Kuntz will be elected secretary, and
r: Brown is elected treasurer.

We assume the premises $p \Rightarrow q$, $q \Rightarrow \sim r$, and p are true propositions, and we want to prove $\sim r$ is a true proposition.

Direct Proof: Using the premises $p \Rightarrow q$ and $q \Rightarrow \sim r$, by the Law of Syllogism we get $p \Rightarrow \sim r$ is true. Since $p \Rightarrow \sim r$ and p are true, by the Law of Detachment $\sim r$ is true.

We have shown, "Brown is not elected treasurer."

Indirect Proof: We make the additional assumption that $\sim r$ is false.
Then by the Law of contradiction r is true. $q \Rightarrow \sim r$ is true so its contrapositive $r \Rightarrow \sim q$ is true. Similarly, since $p \Rightarrow q$ is true $\sim q \Rightarrow \sim p$ is true. The Law of Detachment states that since r and $r \Rightarrow \sim q$ are true, then $\sim q$ must be true. Using the Law of Detachment again with $\sim q$ and $\sim q \Rightarrow \sim p$, we conclude that $\sim p$ is true. But this is a contradiction, p is true.
So our additional assumption that $\sim r$ is false is wrong; $\sim r$ is true.

We have shown, "Brown is not elected treasurer."

7. An argument is valid if the conclusion follows logically from the premises.

Let p and q be the propositions, p: Students study,
q: They receive good grades.

We assume the premises $p \Rightarrow q$ and $\sim p$ are true.
Does $\sim q$ follow?

This is an invalid argument. Since $\sim p$ is true, then p is false. In an implication, if p is false, then q can be either true or false. So the conclusion $\sim q$ can be either true or false.

9. An argument is valid if the conclusion follows logically from the premises.

Let p, q, and r be the propositions, p: Tami studies,
q: Tami fails the course,
r: Tami plays with her dolls too often.

We assume the premises $p \Rightarrow \sim q$, $\sim r \Rightarrow p$, and q are true.
Does r follow?

This is a valid argument. Since $p \Rightarrow \sim q$, and $\sim r \Rightarrow p$ are true, then their contrapositives $q \Rightarrow \sim p$ and $\sim p \Rightarrow r$ also true. Using the Law of the Contrapositives and the Law of Syllogism, we have $q \Rightarrow r$ is true. We assumed the premise q is true, so by the Law of Detachment, we conclude r is true.

11.5 Logic Circuits

1. The output of the circuit pictured in Problem 1 is $pq \oplus [\sim p(\sim q \oplus r)]$. We make a truth table of the output.

p	q	r	$\sim p$	$\sim q$	pq	$\sim q \oplus r$	$\sim p(\sim q \oplus r)$	$pq \oplus [\sim p(\sim q \oplus r)]$
1	1	1	0	0	1	1	0	1
1	1	0	0	0	1	0	0	1
1	0	1	0	1	0	1	0	0
1	0	0	0	1	0	1	0	0
0	1	1	1	0	0	1	1	1
0	1	0	1	0	0	0	0	0
0	0	1	1	1	0	1	1	1
0	0	0	1	1	0	1	1	1

We find that the output is 1 when $(p, q, r) = (1,1,1), (1,1,0), (0,1,1), (0,0,1)$ or $(0,0,0)$.

3. The output of the circuit pictured in Problem 3 is $\{[p(\sim p \oplus q)] \oplus \sim q\}q$. We make a truth table of the output.

p	q	$\sim p$	$\sim q$	$\sim p \oplus q$	$p(\sim p \oplus q)$	$\sim q \oplus [p(\sim p \oplus q)]$	$q\{\sim q \oplus [p(\sim p \oplus q)]\}$
1	1	0	0	1	1	1	1
1	0	0	1	0	0	1	0
0	1	1	0	1	0	0	0
0	0	1	1	1	0	1	0

We find that the output is 1 when $p = 1$ and $q = 1$.

5.

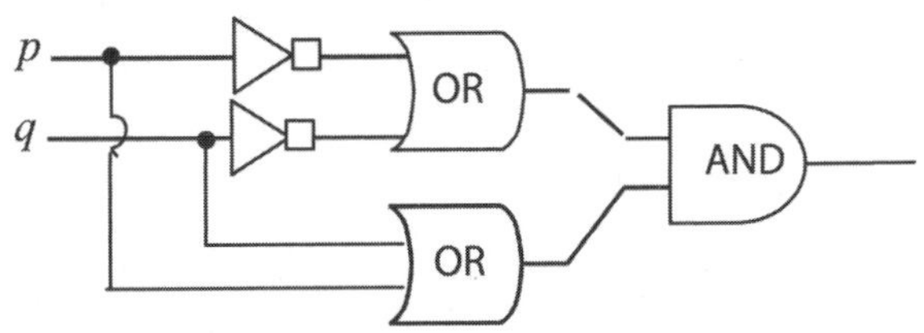

7.

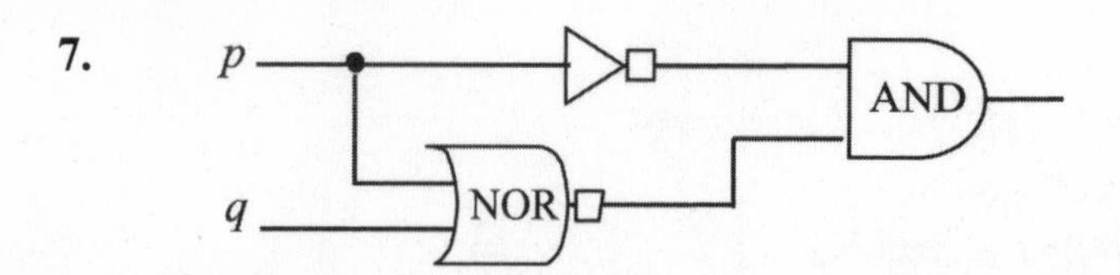

9. To simplify the circuits, we refer back to the truth table corresponding to each circuit.

For problem 1:

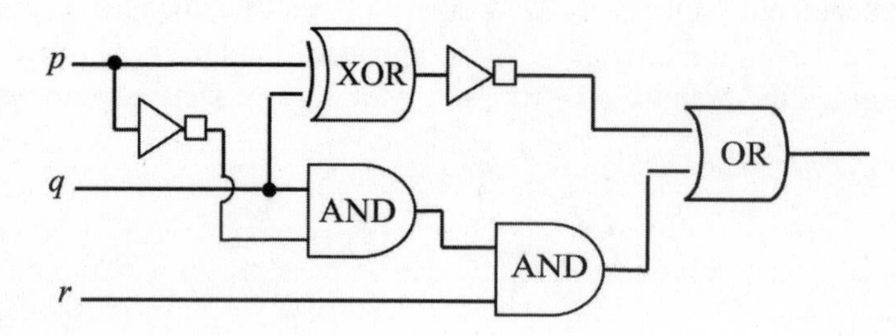

For problem 3: $\{[p(\sim p \oplus q)] \oplus \sim q\}q \equiv pq$

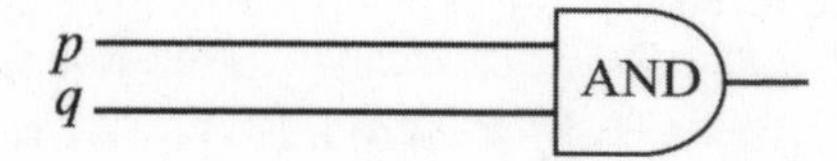

For problem 5: $(\sim p \oplus \sim q)(p \oplus q) = p \veebar q$

For problem 7: $[(\sim p)(\sim q)](\sim p) \equiv (\sim p)(\sim q)$

11. The truth tables and two possible designs for the circuit described are:

p	q	$\sim(p \veebar q)$
1	1	1
1	0	0
0	1	0
0	0	1

p	q	$p \veebar q$
1	1	0
1	0	1
0	1	1
0	0	0

13. $p(\sim p) \equiv 0$

15. Show $pq \oplus pr \oplus q(\sim r) \equiv pr \oplus q(\sim r)$

$$\begin{aligned} pq \oplus pr \oplus q(\sim r) &\equiv pq(r \oplus \sim r) \oplus pr \oplus q(\sim r) \\ &\equiv pqr \oplus pq(\sim r) \oplus pr \oplus q(\sim r) \\ &\equiv pqr \oplus pr \oplus pq(\sim r) \oplus q(\sim r) \\ &\equiv pr(q \oplus 1) \oplus q(\sim r)(p \oplus 1) \\ &\equiv pr(1) \oplus q(\sim r)(1) \\ &\equiv pr \oplus q(\sim r) \end{aligned}$$

Chapter 11 Review

1. A proposition is a declarative sentence that can be meaningfully classified as either true or false.

"A cow is a mammal." is a declarative sentence. It is either true or false. The statement is a proposition.

3. A proposition is a declarative sentence that can be meaningfully classified as either true or false. "Did you see the game?" is not a declarative sentence. It is not a proposition.

5. A proposition is a declarative sentence that can be meaningfully classified as either true or false.

"Fifteen people were injured in the fire." is a declarative sentence. It is either true or false. The statement is a proposition.

7. A proposition is a declarative sentence that can be meaningfully classified as either true or false.

"Get the car." is not a declarative sentence. It is not a proposition.

9. $p \vee q$ I go to the math learning center, or I complete my math homework.

11. $p \Rightarrow q$ If I go to the math learning center, then I complete my math homework.

13. $\sim p \Rightarrow \sim q$ If I do not go to the math learning center, then I do not complete my math homework.

15. $\sim p \wedge \sim q$ I do not go to the math learning center, and I do not complete my math homework.

17. No people are rich.

19. Either Danny is tall, or Mary is not short.

21. (c) Since not rich and poor are not equivalent, (c) is the only answer

23. Answer: (a). We prepare a truth table and compare column labeled "Example" to columns labeled "Choice (a), Choice (b), Choice (c), and Choice (d)." We find only columns labeled "Choice (a)" and "Example" match exactly and so are logically equivalent.

					Example		Choice (a)		Choice (b)
p	q	r	$\sim p$	$\sim p \vee q$	$(\sim p \vee q) \wedge r$	$p \Rightarrow q$	$(p \Rightarrow q) \wedge r$	$q \wedge r$	$\sim p \vee (q \wedge r)$
T	T	T	F	T	T	T	T	T	T
T	T	F	F	T	F	T	F	F	F
T	F	T	F	F	F	F	F	F	F
T	F	F	F	F	F	F	F	F	F
F	T	T	T	T	T	T	T	T	T
F	T	F	T	T	F	T	F	F	T
F	T	T	T	T	T	T	T	F	T
F	T	F	T	T	F	T	F	F	T

	Choice (c)		Choice (b)
$\sim p \Rightarrow q$	$(\sim p \Rightarrow q) \wedge r$	$\sim p \vee r$	$(\sim p \vee r) \wedge (q \wedge r)$
T	T	T	T
T	F	F	F
T	T	T	F
T	F	F	F
T	T	T	T
T	F	T	F
F	F	T	F
F	F	T	F

25.

p	q	$\sim p$	$p \wedge q$	$(p \wedge q) \vee \sim p$
T	T	F	T	T
T	F	F	F	F
F	T	T	F	T
F	F	T	F	T

27.

p	q	$p \vee q$	$(p \vee q) \wedge p$
T	T	T	T
T	F	T	T
F	T	T	F
F	F	F	F

29. p: I will pass the course
q: I will do homework regularly.

I will pass the course if I do homework regularly.

$$q \Rightarrow p$$

31. p: I will pass the course
q: I will do homework regularly.

I will pass this course if and only if I do homework regularly.

$$p \Leftrightarrow q$$

33. Define the statements p: The temperature outside is below 30°, q: I wear gloves.

(a) $p \Rightarrow q$

(b) $q \Rightarrow p$; If I wear gloves, then the temperature outside is below 30°.

(c) $\sim q \Rightarrow \sim p$; If I do not wear gloves, then the temperature outside is not below 30°.

(d) $\sim p \Rightarrow \sim q$; If the temperature is not below 30°, then I do not wear gloves.

35. Define the statements p: Stu works on the project, q: Julie helps.

(a) $q \Rightarrow p$

(b) $p \Rightarrow q$; If Stu works on the project, then Julie helps.

(c) $\sim p \Rightarrow \sim q$; If Stu does not work on the project, then Julie does not help.

(d) $\sim q \Rightarrow \sim p$; If Julie does not help, then Stu does not work on the project.

37. Define the statements p: Kurt will go to the club, q: Jessica comes to town.

(a) $q \Rightarrow p$

(b) $p \Rightarrow q$; If Kurt goes to the club, then Jessica comes to town.

(c) $\sim p \Rightarrow \sim q$; If Kurt does not go to the club, then Jessica does not come to town.

(d) $\sim q \Rightarrow \sim p$; If Jessica does not come to town, then Kurt will not go to the club.

39. Define the statements p: Brian comes to the gym, q: Mike works out.

(a) $q \Rightarrow p$

(b) $p \Rightarrow q$; If Brian comes to the gym, then Mike works out.

(c) $\sim p \Rightarrow \sim q$; If Brian does not come to the gym, then Mike does not work out.

(d) $\sim q \Rightarrow \sim p$; If Mike does not work out, then Brian does not come to the gym.

41. Define the statements p: Patrick goes to practice. q: Patrick starts the game.

Show $p \Rightarrow q \equiv \sim p \vee q$

p	q	$\sim p$	$p \Rightarrow q$	$\sim p \vee q$	$(p \Rightarrow q) \Leftrightarrow (\sim p \vee q)$
T	T	F	T	T	T
T	F	F	F	F	T
F	T	T	T	T	T
F	F	T	T	T	T

Since the last two columns of the truth table are the same, $p \Rightarrow q \equiv \sim p \vee q$

43.

p	q	$\sim p$	$p \Rightarrow q$	$\sim p \vee q$	$\sim p \vee q \Leftrightarrow p \Rightarrow q$
T	T	F	T	T	T
T	F	F	F	F	T
F	T	T	T	T	T
F	F	T	T	T	T

45. Define the propositions p: I paint the house, q: I go bowling.
We are assuming $\sim p \Rightarrow q$ and $\sim q$ are true, and we want to show p is a true proposition.

Direct Proof: Since $\sim p \Rightarrow q$ is true, its contrapositive $\sim q \Rightarrow p$ is true. So $\sim q \Rightarrow p$ and $\sim q$ are true and by the Law of Detachment, p is true, I paint the house.

Indirect Proof: We make the additional assumption that p is false.
Then by the Law of Contradiction $\sim p$ is true. Since $\sim p \Rightarrow q$ and $\sim p$ are true, by the Law of Detachment, q is true, but this is a contradiction, q is false. So our additional assumption that p is false is incorrect. p is true; I paint the house.

47. Define the statements p: John is in town, q: Mark gets tickets, and r: We go to the game.
Assume the premises $p \Rightarrow q$, $\sim r \Rightarrow \sim q$, and p are true. We want to show r.

Direct Proof: Since $\sim r \Rightarrow \sim q$ is true, its contrapositive, $q \Rightarrow r$ is true. Since $p \Rightarrow q$ and $q \Rightarrow r$ are true, $p \Rightarrow r$ is true by Law of Syllogism. Finally, since $p \Rightarrow r$ and p are true, by the Law of Detachment, r is true. We went to the game.

Indirect Proof: We make the additional assumption that. r is false.
Since $\sim r \Rightarrow \sim q$ its contra positive $q \Rightarrow r$ is also true. When $p \Rightarrow q$ and $q \Rightarrow r$ are true, then so is $p \Rightarrow r$ by the Law of Syllogism. Finally, the Law of Detachment states that if p and $p \Rightarrow r$ are true, so is r true. But this is a contradiction. Therefore, our additional assumption stated that r is false, is incorrect. We go to the game.

49. Define the propositions p: I pay a finance charge, q: My payment is late, r: Colleen sends the mail.
Assume the premises $q \Rightarrow p$, $r \vee q$, and $\sim r$ are true. Prove p is true.

Direct Proof: By the Law of Contradiction r is false. The disjunction $r \vee q$ is true provided at least one of its components is true, so q is true. By the Law of Detachment, since $q \Rightarrow p$ and q are true, p is true. I pay a late charge.

Indirect Proof: We make the additional assumption that p is false.
By the Law of Contradiction $\sim p$ is true and r is false. The disjunction $r \vee q$ is true provided at least one of its components is true, so q is true. By the Law of Detachment, since $q \Rightarrow p$ and q are true, p is true. Therefore, the additional assumption stated that p is false, is incorrect; I pay a late charge.

51. Define the statements p: Rob is a bad boy, q: Danny is crying, r: Laura is a good girl.

Assume the premises $p \vee q$, $r \Rightarrow \sim p$, and $\sim q$ are true. We want to prove r (or $\sim r$).

$p \vee q$ is true whenever at least one of its components is true. $\sim q$ is true, and so by the Law of Contradiction q is false. Therefore, p is true. Since $r \Rightarrow \sim p$ is true, its contrapositive, $p \Rightarrow \sim r$, is true, and by the Law of Detachment $\sim r$ is true.

We have shown that $\sim r$ is true. So we conclude: Laura is not a good girl.

53. $\sim(p \oplus q) \equiv (\sim p)(\sim q)$

55.

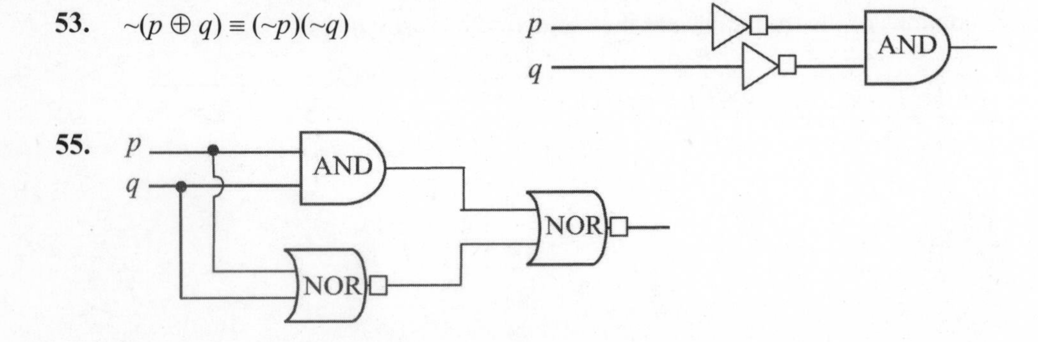

57. $(p \oplus q)[\sim(pq)] \equiv (p \oplus q)(\sim p \oplus \sim q)$

$$\equiv p(\sim p) \oplus p(\sim q) \oplus q(\sim p) \oplus q(\sim q)$$
$$\equiv 0 \oplus p(\sim q) \oplus q(\sim p) \oplus 0$$
$$\equiv p(\sim q) \oplus q(\sim p)$$

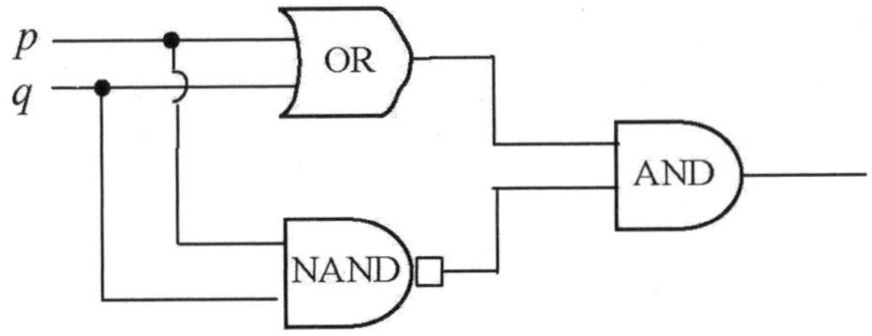

Chapter 11 Project

1. The race:

Person	Place in Race	Color Worn	Number Worn
Alan	1	red	2
Key	2	yellow	3
Steve	3	green	1
John	4	blue	4

3. To do this problem make a 6 by 6 matrix, enter the elements in the first row and the first column as shown. Then make five lists: one of the ages, one of the weights, one of the heights, one of the blood types, and one of the positions. As you read through the statements, mark the lists with the names and fill in the appropriate entries in the matrix. When only one blank entry is remaining in a column of the matrix fill it in using the appropriate list and the process of elimination.

Name	Age	Weight	Height	Blood Group	Position
Jason	9 yrs.	96 lbs.	40 in.	AO	center
Alan	30 yrs.	75 lbs.	74 in.	O	2nd
Adam	5 yrs.	165 lbs.	65 in.	A	1st
Kevin	60 yrs.	125 lbs.	48 in.	B	4th
John	46 yrs.	40 lbs.	60 in.	AB	right

Appendix A

Review

A.1 Real Numbers

1. Rational numbers

3. Distributive

5. True

7. False. The least common multiple of 12 and 18 is 36.

9. (a) Natural Numbers: 2, 5

(b) Integers: −6, 2, 5

(c) Rational numbers: $-6, \frac{1}{2}, -1.333\ldots, 2, 5$

(d) Irrational numbers: π

(e) Real numbers: $-6, \frac{1}{2}, -1.333\ldots, \pi, 2, 5$

11. (a) Natural Numbers: 1

(b) Integers: 0, 1

(c) Rational numbers: $0, 1, \frac{1}{2}, \frac{1}{3}, \frac{1}{4}$

(d) Irrational numbers: none

(e) Real numbers: $0, 1, \frac{1}{2}, \frac{1}{3}, \frac{1}{4}$

13. (a) Natural Numbers: none
(b) Integers: none
(c) Rational numbers: none
(d) Irrational numbers: $\sqrt{2}$, π, $\sqrt{2}+1$, $\pi+\frac{1}{2}$
(e) Real numbers: $\sqrt{2}$, π, $\sqrt{2}+1$, $\pi+\frac{1}{2}$

15. Number: 18.9526
Rounded: 18.953
Truncated: 18.952

17. Number: 28.65319
Rounded: 28.653
Truncated: 28.653

19. Number: 0.06291
Rounded: 0.063
Truncated: 0.062

21. Number: 9.9985
Rounded: 9.999
Truncated: 9.998

23. Number: $\frac{3}{7} = 0.428571\ldots$
Rounded: 0.429
Truncated: 0.428

25. Number: $\frac{521}{15} = 34.73333\ldots$
Rounded: 34.733
Truncated: 34.733

27. $3+2=5$

29. $x+2=3\cdot 4$

31. $3y=1+2$

33. $x-2=6$

35. $\frac{x}{2}=6$

37. $9-4+2=5+2=7$

39. $-6+4\cdot 3=-6+12=6$

41. $4+5-8=9-8=1$

43. $4+\frac{1}{3}=\frac{12}{3}+\frac{1}{3}=\frac{13}{3}$

45.
$$\begin{aligned} 6-[3\cdot 5+2\cdot(3-2)] &= 6-[15+2\cdot 1] \\ &= 6-[15+2] \\ &= 6-17 \\ &= -11 \end{aligned}$$

47.
$$\begin{aligned} 2\cdot(3-5)+8\cdot 2-1 &= 2\cdot(-2)+16-1 \\ &= -4+16-1 \\ &= 12-1=11 \end{aligned}$$

49.
$$\begin{aligned} 10-[6-2\cdot 2+(8-3)]\cdot 2 &= 10-[6-4+5]\cdot 2 \\ &= 10-[7]\cdot 2 \\ &= 10-14 \\ &= -4 \end{aligned}$$

51. $(5-3)\,\frac{1}{2}=(2)\frac{1}{2}=1$

53. $\frac{4+8}{5-3}=\frac{12}{2}=6$

55. $\frac{3}{5}\cdot\frac{10}{21}=\frac{30}{105}=\frac{2}{7}$

57. $\frac{6}{25}\cdot\frac{10}{27}=\frac{2\cdot 3\cdot 2\cdot 5}{5\cdot 5\cdot 3\cdot 9}=\frac{4}{45}$

59. $\frac{3}{4}+\frac{2}{5}=\frac{3\cdot 5}{4\cdot 5}+\frac{2\cdot 4}{4\cdot 5}=\frac{15+8}{20}=\frac{23}{20}$

61. $\frac{5}{6}+\frac{9}{5}=\frac{5\cdot 5}{6\cdot 5}+\frac{9\cdot 6}{5\cdot 6}=\frac{25+54}{30}=\frac{79}{30}$

63. $\frac{5}{18}+\frac{1}{12}=\frac{5\cdot 2}{18\cdot 2}+\frac{1\cdot 3}{12\cdot 3}=\frac{10+3}{36}=\frac{13}{36}$

65. $\frac{1}{30}-\frac{7}{18}=\frac{1\cdot 3}{30\cdot 3}-\frac{7\cdot 5}{18\cdot 5}$
$=\frac{3-35}{90}=\frac{-32}{90}=-\frac{16}{45}$

67. $\frac{3}{20}-\frac{2}{15}=\frac{3\cdot 3}{20\cdot 3}-\frac{2\cdot 4}{15\cdot 4}=\frac{9-8}{60}=\frac{1}{60}$

69. $\frac{\frac{5}{18}}{\frac{11}{27}}=\frac{5}{18}\cdot\frac{27}{11}=\frac{5\cdot 9\cdot 3}{9\cdot 2\cdot 11}=\frac{15}{22}$

71. $6(x+4)=6x+24$

73. $x(x-4)=x^2-4x$

75. $(x+2)(x+4)=(x+2)x+(x+2)4$
$=x^2+2x+4x+8$
$=x^2+6x+8$

77. $(x-2)(x+1)=(x-2)x+(x-2)1$
$=x^2-2x+x-2$
$=x^2-x-2$

79. $(x-8)(x-2)=(x-8)x+(x-8)(-2)$
$=x^2-8x-2x+16$
$=x^2-10x+16$

81. $(x+2)(x-2)=(x+2)x+(x+2)(-2)$
$=x^2+2x-2x-4$
$=x^2-4$

83. Answers will vary.

85. Answers will vary.

87. Subtraction is not commutative.
Examples will vary.

89. Division is not commutative.
Examples will vary.

91. True from the symmetry property.

93. There are no real numbers that are both rational and irrational.
There are no real numbers that are neither rational nor irrational.

95. $0.9999\ldots=1$
To show that $0.9999\ldots=1$, we let $n=0.9999\ldots$, then $10n=9.9999\ldots$

$$\begin{array}{rl} 10n=9.9999\ldots & (1) \\ n=0.9999\ldots & (2) \\ \hline 9n=9.0000\ldots & (\textit{subtract } (2) \textit{ from } (1)) \\ n=1 & (\textit{divide both sides by } 9) \end{array}$$

A.2 Algebra Essentials

1. Variable

3. Strict

5. True

7. False. The absolute value of zero is zero.

9. $\{4\}$

11. -2.5 -1 0 $\frac{3}{4}$ $\frac{3}{2}$ 0.25

13. $\frac{1}{2} > 0$

15. $-1 > -2$

17. $\pi > 3.14$

19. $\frac{1}{2} = 0.5$

21. $\frac{2}{3} < 0.67$

23. $x > 0$

25. $x < 2$

27. $x \le 1$

29. $x \ge -2$

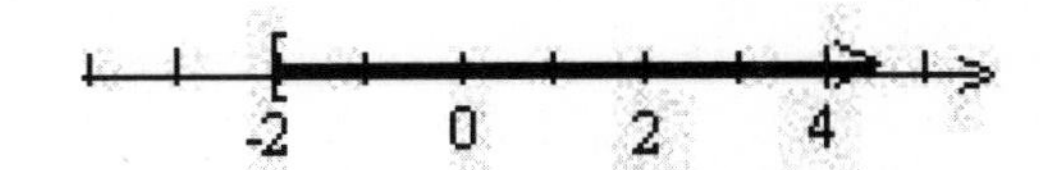

31. $x > -1$

-2 -1 0 2 4

33. $d(C, D) = |C - D| = |0 - 1| = |-1| = 1$

35. $d(D, E) = |D - E| = |1 - 3| = |-2| = 2$

37. $d(A, E) = |A - E| = |-3 - 3| = |-6| = 6$

39. If $x = -2$ and $y = 3$,
then $x + 2y = (-2) + 2(3) = -2 + 6 = 4$

41. If $x = -2$ and $y = 3$,
then $5xy + 2 = 5(-2)(3) + 2 = -30 + 2 = -28$

43. If $x = -2$ and $y = 3$, then $\frac{2x}{x-y} = \frac{2(-2)}{(-2)-3} = \frac{-4}{-5} = \frac{4}{5}$

45. If $x = -2$ and $y = 3$, then $\frac{3x+2y}{2+y} = \frac{3(-2)+2(3)}{2+3} = \frac{-6+6}{5} = 0$

47. If $x = 3$ and $y = -2$, then $|x + y| = |3 + (-2)| = |1| = 1$

49. If $x = 3$ and $y = -2$, then $|x| + |y| = |3| + |-2| = 3 + 2 = 5$

51. If $x = 3$ and $y = -2$, then $\frac{|x|}{x} = \frac{|3|}{3} = \frac{3}{3} = 1$

53. If $x = 3$ and $y = -2$, then $|4x - 5y| = |4(3) - 5(-2)| = |12 - (-10)| = |22| = 22$

55. If $x = 3$ and $y = -2$, then $\left||4x| - |5y|\right| = \left||4(3)| - |5(-2)|\right| = \left||12| - |-10|\right| = |12 - 10| = |2| = 2$

57. We must exclude values of x that would cause the denominator to equal zero.
$x \neq 0$ **(c)**

59. We must exclude values of x that would cause the denominator to equal zero.
$x^2 - 9 \neq 0$
$(x - 3)(x + 3) \neq 0$
$x \neq 3;\ x \neq -3$ **(a)**

61. We must exclude values of x that would cause the denominator to equal zero, but $x^2 + 1$ can never equal zero, so no values are excluded.

63. We must exclude values of x that would cause the denominator to equal zero.
$x^3 - x \neq 0$
$x(x^2 - 1) \neq 0$
$x(x - 1)(x + 1) \neq 0$
$x \neq 0;\ x \neq 1;\ x \neq -1$ **(b), (c),** and **(d)**

65. The domain of the variable x is $\{x \mid x \neq 5\}$.

67. The domain of the variable x is $\{x \mid x \neq -4\}$.

69. $C = \frac{5}{9}(F - 32)$ If $F = 32°$, then $C = \frac{5}{9}(32 - 32) = 0°$.

71. $C = \frac{5}{9}(F - 32)$ If $F = 77°$, then $C = \frac{5}{9}(77 - 32) = \frac{5}{9}(45) = 25°$.

73. $(-4)^2 = -4 \cdot -4 = 16$

75. $4^{-2} = \frac{1}{4^2} = \frac{1}{4 \cdot 4} = \frac{1}{16}$

77. $3^{-6} \cdot 3^4 = 3^{-6+4} = 3^{-2} = \frac{1}{3^2} = \frac{1}{9}$

79. $(3^{-2})^{-1} = 3^{(-2)\cdot(-1)} = 3^2 = 9$

81. $\sqrt{25} = 5$

83. $\sqrt{(-4)^2} = 4$

85. $(8x^3)^2 = 8^2 \cdot (x^3)^2 = 64x^{3\cdot 2} = 64x^6$

87. $(x^2y^{-1})^2 = \left(\frac{x^2}{y}\right)^2 = \frac{x^{2\cdot 2}}{y^2} = \frac{x^4}{y^2}$

89. $\dfrac{x^2y^3}{xy^4}=\dfrac{x^{2-1}}{y^{4-3}}=\dfrac{x}{y}$

91. $2\cdot 2\cdot(-1)^{-1}=\dfrac{4}{-1}=-4$

93. $2^2+(-1)^2=4+1=5$

95. $[2\cdot(-1)]^2=(-2)^2=4$

97. $\sqrt{2^2}=2$

99. $\sqrt{2^2+(-1)^2}=\sqrt{4+1}=\sqrt{5}$

101. $2^{-1}=\dfrac{1}{2}$

103. $\dfrac{(666)^4}{(222)^4}=\left(\dfrac{\cancel{666}^{\,3}}{\cancel{222}_{\,1}}\right)^4=3^4=81$

105. $(8.2)^6=304{,}006.671$

107. $(6.1)^{-3}=0.004$

109. $(-2.8)^6=481.890$

111. $(-3.11)^{-4}=0.011$

113. $x+5=7$
$x=7-5$
$x=2$

115. $6-x=0$
$-x=0-6$
$x=6$

117. $3(2-x)=9$
$2-x=3$ *(divide both sides by 3)*
$-x=3-2$ *(subtract 2 from both sides)*
$x=-1$ *(multiply by −1)*

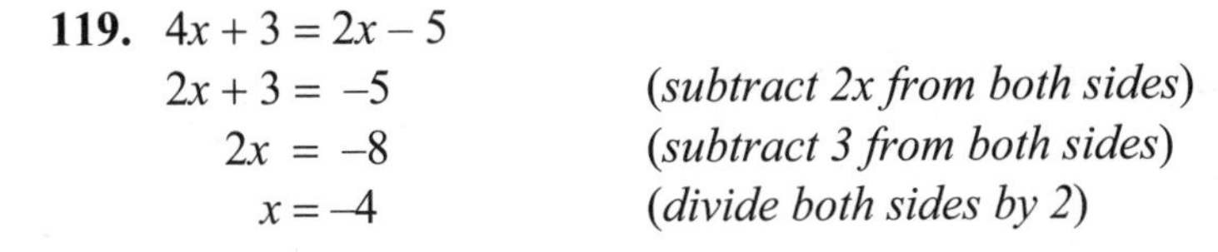

119. $4x+3=2x-5$
$2x+3=-5$ *(subtract 2x from both sides)*
$2x=-8$ *(subtract 3 from both sides)*
$x=-4$ *(divide both sides by 2)*

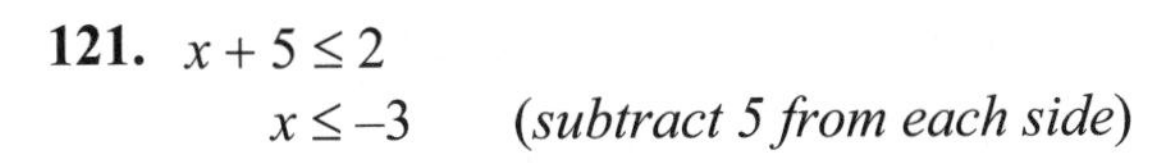

121. $x+5\le 2$
$x\le -3$ *(subtract 5 from each side)*

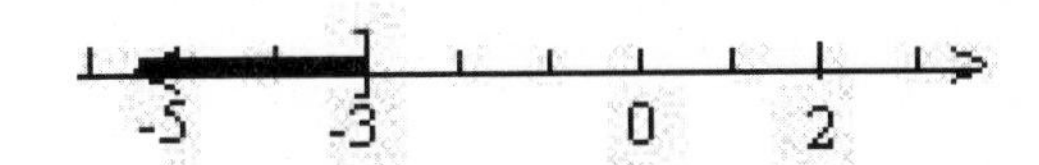

123. $3x+5\ge 2$
$3x\ge -3$ *(subtract 5 from both sides)*
$x\ge -1$ *(divide both sides by 3)*

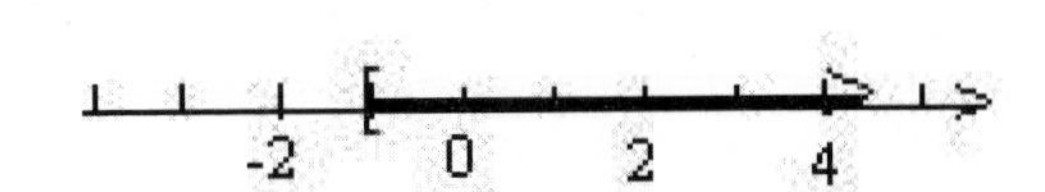

125. $-3x+5\le 2$
$-3x\le -3$ *(subtract 5 from both sides)*
$x\ge 1$ *(divide both sides by −3; change the inequality sign)*

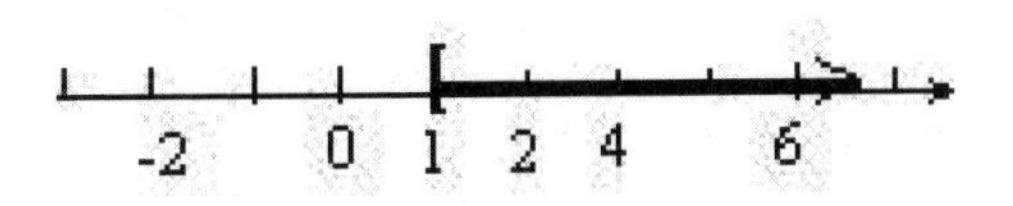

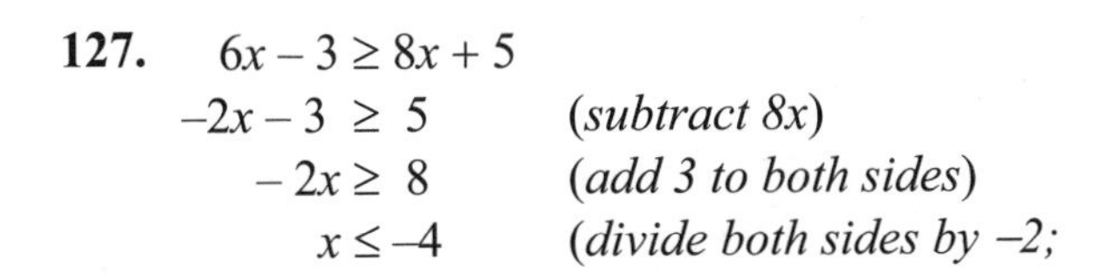

127. $6x-3\ge 8x+5$
$-2x-3\ge 5$ *(subtract 8x)*
$-2x\ge 8$ *(add 3 to both sides)*
$x\le -4$ *(divide both sides by −2;*

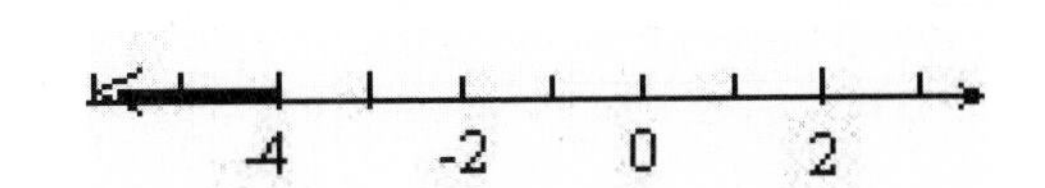

change inequality sign)

129. $A = lw$; domain: $A > 0, l > 0, w > 0$

131. $C = \pi d$; domain: $C > 0, d > 0$

133. $A = \frac{\sqrt{3}}{4} \cdot x^2$; domain: $A > 0, x > 0$

135. $V = \frac{4}{3}\pi r^3$; domain: $V > 0, r > 0$

137. $V = x^3$; domain: $V > 0, x > 0$

139. $C = 4000 + 2x$

(a) If $x = 1000$ watches are produced, it will cost
$C = 4000 + 2(1000) = \$6000$

(b) If $x = 2000$ watches are produced, it will cost
$C = 4000 + 2(2000) = \$8000$

141. (a) If actual voltage is $x = 113$ then
$|113 - 115| = |-2| = 2$
Since $2 < 5$, an actual voltage of 113 is acceptable.

(b) If actual voltage is $x = 109$ then
$|109 - 115| = |-6| = 6$
Since $6 > 5$, an actual voltage of 109 is not acceptable.

143. (a) If the radius is $x = 2.999$, then
$|x - 3| = |2.999 - 3| = |-0.001| = 0.001$
Since $0.001 < 0.010$ the ball bearing is acceptable.

(b) If the radius is $x = 2.89$, then
$|x - 3| = |2.89 - 3| = |-0.11| = 0.11$
Since $0.11 > 0.01$, the ball bearing is not acceptable.

145. $\frac{1}{3} \neq 0.333$; $\frac{1}{3} > 0.333$

$\frac{1}{3} = 0.333\ldots$; $0.333\ldots - 0.333 = 0.000333\ldots$

147. The answer is no. Student answers should justify and explain why not.

A.3 Exponents and Logarithms

3. 3^3

5. $\log_{1.8} x = 4$

7. $\sqrt{16} = 4$

9. $\sqrt[3]{27} = 3$

11. $\sqrt[4]{16} = 2$

13. $8^{2/3} = (8^{1/3})^2 = (2)^2 = 4$

15. $16^{-3/2} = \dfrac{1}{16^{3/2}} = \dfrac{1}{(16^{1/2})^3} = \dfrac{1}{4^3} = \dfrac{1}{64}$

17. $(-8)^{-2/3} = \dfrac{1}{(-8)^{2/3}} = \dfrac{1}{\left[(-8)^{1/3}\right]^2} = \dfrac{1}{(-2)^2} = \dfrac{1}{4}$

19. (a) $3^{2.2} = 11.2116$ (b) $3^{2.23} = 11.5873$ (c) $3^{2.236} = 11.6639$ (d) $3^{\sqrt{5}} = 11.6648$

21. $(1.08)^5 = 1.4693$

23. $\left(1 + \dfrac{0.06}{4}\right)^{12} = (1.015)^{12} = 1.1956$

25. $(1 + 0.11)^{-4} = 1.11^{-4} = 0.6587$

27. $\left(1 + \dfrac{0.04}{12}\right)^{-20} = 0.9356$

29. If $9 = 3^2$, then $\log_3 9 = 2$.

31. If $a^2 = 1.6$, then $\log_a 1.6 = 2$.

33. If $2^x = 7.2$, then $\log_2 7.2 = x$.

35. If $e^x = 8$, then $\ln 8 = x$.

37. If $\log_2 8 = 3$, then $2^3 = 8$.

39. If $\log_a 3 = 6$, then $a^6 = 3$.

41. If $\log_3 2 = x$, then $3^x = 2$.

43. If $\log 4 = x$, then $10^x = 4$.

45. Let $x = \log_2 1$, then $2^x = 1$ or $x = 0$.

47. Let $x = \log_5 25$, then $5^x = 25$ or $x = 2$.

49. Let $x = \log_{1/2} 16$, then

$$\left(\frac{1}{2}\right)^x = 16 = 2^4 = \frac{1}{2^{-4}} = \left(\frac{1}{2}\right)^{-4} \text{ or } x = -4.$$

51. Let $x = \log_{10} \sqrt{10}$, then $10^x = \sqrt{10} = 10^{1/2}$. So $x = \dfrac{1}{2}$.

53. Let $x = \log_{\sqrt{2}} 4$, then $(\sqrt{2})^x = 4$, or $2^{(1/2)x} = 2^2$ or $\dfrac{1}{2}x = 2$ or $x = 4$.

55. Let $x = \log_e \sqrt{e} = \log_e e^{1/2}$, then $e^x = e^{1/2}$ or $x = \dfrac{1}{2}$.

57. $\log \dfrac{5}{3} = 0.222$

59. $\dfrac{\log \frac{10}{3}}{0.04} = 13.072$

61. $\dfrac{\log 5}{\log 1.02} = 81.274$

63. $\dfrac{\log 4}{\log 1.06} = 23.791$

65. $\log_{1.1} 200 = \dfrac{\log 200}{\log 1.1} = 55.590$

67. $\log_{1.005} 1000 = \dfrac{\log 1000}{\log 1.005} = 1385.002$

69. $\log_{1.002} 20 = \dfrac{\log 20}{\log 1.002} = 1499.364$

71. $\log_{1.0005} 500 = \dfrac{\log 500}{\log 1.0005} = 12{,}432.323$

73. $\log_{1.003} 500 = \dfrac{\log 500}{\log 1.003} = 2074.642$

A.4 Recursively Defined Sequences; Geometric Sequences

1. 3; 15

3. $\{n\} = \{1, 2, 3, 4, 5\}$

5. $\left\{\dfrac{n}{n+2}\right\}$: $\dfrac{1}{1+2} = \dfrac{1}{3}$

$\dfrac{2}{2+2} = \dfrac{2}{4} = \dfrac{1}{2}$

$\dfrac{3}{3+2} = \dfrac{3}{5}$

$\dfrac{4}{4+2} = \dfrac{4}{6} = \dfrac{2}{3}$

$\dfrac{5}{5+2} = \dfrac{5}{7}$

$\left\{\dfrac{1}{3}, \dfrac{1}{2}, \dfrac{3}{5}, \dfrac{2}{3}, \dfrac{5}{7}\right\}$

7. $\{(-1)^{n+1} n^2\}$: $(-1)^{(1+1)} \cdot 1^2 = 1$

$(-1)^{(2+1)} \cdot 2^2 = -4$

$(-1)^{(3+1)} \cdot 3^2 = 9$

$(-1)^{(4+1)} \cdot 4^2 = -16$

$(-1)^{(5+1)} \cdot 5^2 = 25$

$\{1, -4, 9, -16, 25\}$

9. $\left\{\dfrac{2^n}{3^n+1}\right\}$: $\dfrac{2^1}{3^1+1}=\dfrac{2}{4}=\dfrac{1}{2}$

$\dfrac{2^2}{3^2+1}=\dfrac{4}{10}=\dfrac{2}{5}$

$\dfrac{2^3}{3^3+1}=\dfrac{8}{28}=\dfrac{2}{7}$

$\dfrac{2^4}{3^4+1}=\dfrac{16}{82}=\dfrac{8}{41}$

$\dfrac{2^5}{3^5+1}=\dfrac{32}{244}=\dfrac{8}{61}$

$\left\{\dfrac{1}{2}, \dfrac{2}{5}, \dfrac{2}{7}, \dfrac{8}{41}, \dfrac{8}{61}\right\}$

11. $\left\{\dfrac{(-1)^n}{(n+1)(n+2)}\right\}$: $\dfrac{(-1)^1}{(1+1)(1+2)}=-\dfrac{1}{6}$

$\dfrac{(-1)^2}{(2+1)(2+2)}=\dfrac{1}{12}$

$\dfrac{(-1)^3}{(3+1)(3+2)}=-\dfrac{1}{20}$

$\dfrac{(-1)^4}{(4+1)(4+2)}=\dfrac{1}{30}$

$\dfrac{(-1)^5}{(5+1)(5+2)}=-\dfrac{1}{42}$

$\left\{-\dfrac{1}{6}, \dfrac{1}{12}, -\dfrac{1}{20}, \dfrac{1}{30}, -\dfrac{1}{42}\right\}$

13. $\left\{\dfrac{n}{e^n}\right\}$: $\left\{\dfrac{1}{e^1}, \dfrac{2}{e^2}, \dfrac{3}{e^3}, \dfrac{4}{e^4}, \dfrac{5}{e^5}\right\}$

15. $\dfrac{1}{2}, \dfrac{2}{3}, \dfrac{3}{4}, \dfrac{4}{5}, \ldots$ Since both the numerator and the denominator increase by one with each term, and the first numerator is one, $a_n=\dfrac{n}{n+1}$

17. $1, \dfrac{1}{2}, \dfrac{1}{4}, \dfrac{1}{8}$ Think of this sequence as $\dfrac{1}{2^0}, \dfrac{1}{2^1}, \dfrac{1}{2^2}, \dfrac{1}{2^3}, \ldots$, then we get $a_n=\dfrac{1}{2^{n-1}}$.

19. $1, -1, 1, -1, 1, -1, \ldots$ Think of this sequence as $(-1)^0, (-1)^1, (-1)^2, (-1)^3, (-1)^4, (-1)^5, \ldots$, then $a_n=(-1)^{n+1}$.

21. $1, -2, 3, -4, 5, -6, \ldots$ This sequence is the natural numbers with alternating signs. Take care of the alternating signs with $(-1)^n$. The general term of this sequence is $a_n=(-1)^{n+1}\cdot n$.

23. $a_1 = 1$
$a_2 = 2 + a_1 = 2 + 1 = 3$
$a_3 = 2 + a_2 = 2 + 3 = 5$
$a_4 = 2 + a_3 = 2 + 5 = 7$
$a_5 = 2 + a_4 = 2 + 7 = 9$

25. $a_1 = -2$
$a_2 = 1 + a_1 = 1 + (-2) = -1$
$a_3 = 2 + a_2 = 2 + (-1) = 1$
$a_4 = 3 + a_3 = 3 + (1) = 4$
$a_5 = 4 + a_4 = 4 + (4) = 8$

27. $a_1 = 5$
$a_2 = 2a_1 = 2 \cdot 5 = 10$
$a_3 = 2a_2 = 2 \cdot 10 = 20$
$a_4 = 2a_3 = 2 \cdot 20 = 40$
$a_5 = 2a_4 = 2 \cdot 40 = 80$

29. $a_1 = 3$
$$a_2 = \frac{a_1}{1} = 3$$
$$a_3 = \frac{a_2}{2} = \frac{3}{2}$$
$$a_4 = \frac{a_3}{3} = \frac{\frac{3}{2}}{3} = \frac{1}{2}$$
$$a_5 = \frac{a_4}{4} = \frac{\frac{1}{2}}{4} = \frac{1}{8}$$

31. $a_1 = 1$
$a_2 = 2$
$a_3 = a_1 \cdot a_2 = 1 \cdot 2 = 2$
$a_4 = a_2 \cdot a_3 = 2 \cdot 2 = 4$
$a_5 = a_3 \cdot a_4 = 2 \cdot 4 = 8$

33. $a_1 = A$
$a_2 = a_1 + d = A + d$
$a_3 = a_2 + d = (A + d) + d = A + 2d$
$a_4 = a_3 + d = (A + 2d) + d = A + 3d$
$a_5 = a_4 + d = (A + 3d) + d = A + 4d$

35. $a_1 = \sqrt{2}$
$$a_2 = \sqrt{2 + a_1} = \sqrt{2 + \sqrt{2}}$$
$$a_3 = \sqrt{2 + a_2} = \sqrt{2 + \sqrt{2 + \sqrt{2}}}$$
$$a_4 = \sqrt{2 + a_3} = \sqrt{2 + \sqrt{2 + \sqrt{2 + \sqrt{2}}}}$$
$$a_5 = \sqrt{2 + a_4} = \sqrt{2 + \sqrt{2 + \sqrt{2 + \sqrt{2 + \sqrt{2}}}}}$$

37. $\{2^n\}$
(a) First term: $a = 2$; common ratio: $r = 2$
(b) $2,\ 2^2,\ 2^3,\ 2^4$ or $2,\ 4,\ 8,\ 16$
(c) $S_n = 2\left(\frac{1 - 2^n}{1 - 2}\right) = -2\left(1 - 2^n\right) = 2^{n+1} - 2$

39. $\left\{-3\left(\frac{1}{2}\right)^n\right\}$

(a) First term: $a = -3\left(\frac{1}{2}\right) = -\frac{3}{2}$; common ratio: $r = \frac{1}{2}$

(b) $-\frac{3}{2},\ -\frac{3}{4},\ -\frac{3}{8},\ -\frac{3}{16}$

(c) $S_n = \left(-\frac{3}{2}\right)\left[\frac{1-\left(\frac{1}{2}\right)^n}{1-\frac{1}{2}}\right] = -3\left[1-\left(\frac{1}{2}\right)^n\right] = -3 + \frac{3}{2^n}$

41. $\left\{\frac{2^{n-1}}{4}\right\}$

(a) First term: $a = \frac{2^0}{4} = \frac{1}{4}$; common ratio: $r = 2$

(b) $\frac{1}{4},\ \frac{2}{4} = \frac{1}{2},\ \frac{2^2}{4} = \frac{4}{4} = 1,\ \frac{2^3}{4} = \frac{8}{4} = 2$

(c) $S = \frac{1}{4}\left[\frac{1-2^n}{1-2}\right] = -\frac{1}{4}\left(1-2^n\right) = 2^{n-2} - \frac{1}{4}$

43. $\left\{2^{\frac{n}{3}}\right\}$

(a) First term: $a = 2^{\frac{1}{3}}$; common ratio: $r = 2^{\frac{1}{3}}$

(b) $2^{\frac{1}{3}},\ 2^{\frac{2}{3}},\ 2^{\frac{3}{3}} = 2,\ 2^{\frac{4}{3}}$

(c) $S = 2^{\frac{1}{3}}\left(\frac{1-2^{\frac{n}{3}}}{1-2^{\frac{1}{3}}}\right)$

45. $\left\{\frac{3^{n-1}}{2^n}\right\}$

(a) First term: $a = \frac{3^0}{2^1} = \frac{1}{2}$; common ratio: $r = \frac{3}{2}$

(b) $\frac{1}{2},\ \frac{3^{2-1}}{2^2} = \frac{3}{4},\ \frac{3^{3-1}}{2^3} = \frac{9}{8},\ \frac{3^{4-1}}{2^4} = \frac{27}{16}$

(c) $S = \frac{1}{2}\left[\frac{1-\left(\frac{3}{2}\right)^n}{1-\frac{3}{2}}\right] = -1\left[1-\left(\frac{3}{2}\right)^n\right]$

47. If $B_0 = \$3000$ and $B_n = 1.01B_{n-1} - 100$, then John's balance after 1 payment is

$$B_1 = 1.01(\$3000) - \$100 = \$2930$$

49. If $B_0 = \$18{,}500$ and $B_n = 1.005B_{n-1} - 534.47$, then Phil's balance after 1 payment is

$$B_1 = 1.005(\$18{,}500) - \$534.47 = \$18{,}058.00$$

51. Since we are only interested in the pairs of mature rabbits, the offspring are only important because they mature.
Month 0: 1 pair mature rabbits – we began with these
Month 1: 1 pair mature rabbits – the same pair from before, but now they have 2 offspring.
Month 2: 2 pairs mature rabbits – the original pair (and their second set of offspring) and the pair that grew up.
Month 3: 3 pairs mature rabbits – The 2 pairs from month 2 (they each have a pair of offspring) and the pair born in month 2.
Month 4: 5 pairs of mature rabbits – The 3 pairs from month 3 (these each have a pair of offspring) and the two pairs born in month 3 that grew up.
Month 5: 8 pairs of mature rabbits – the 5 pairs from month 4 (each have a pair of offspring) and the 3 pairs that grew up.
Month 6: 13 pairs of mature rabbits – the 8 pairs from month 5 (each have a pair of offspring) and the 5 pairs that were born last month and have grown up.
Month 7: 21 pairs of mature rabbits – the 13 pairs from month 6 and the 8 pairs of children that have matured.

53. Diagonal 1: 1
Diagonal 2: 1
Diagonal 3: $1 + 1 = 2$
Diagonal 4: $1 + 2 = 3$
Diagonal 5: $1 + 3 + 1 = 5$
Diagonal 6: $1 + 4 + 3 = 8$
Diagonal 7: $1 + 5 + 6 + 1 = 13$
The sums of the diagonals are 1, 1, 2, 3, 5, 8, 13, … The sums form a Fibonacci Sequence.

Appendix B

Using LINDO to Solve Linear Programming Problems

1. Enter on the LINDO blank window.

max $3x1 + 2x2 + x3$
subject to
$3x1 + x2 + x3 < 30$
$5x1 + 2x2 + x3 < 24$
$x1 + x2 + 4x3 < 20$
end

Click solve;
click no, when asked for reports;
click close.
Move aside the window on which you entered the problem to see the solution on the right.

LP OPTIMUM FOUND AT STEP 1

OBJECTIVE FUNCTION VALUE

1) 24.00000

VARIABLE	VALUE	REDUCED COST
X1	0.000000	2.000000
X2	12.000000	0.000000
X3	0.000000	0.000000

ROW	SLACK OR SURPLUS	DUAL PRICES
2)	18.000000	0.000000
3)	0.000000	1.000000
4)	8.000000	0.000000

NO. ITERATIONS = 1

The maximum value of $P = 24$, and it is attained when $x_1 = 0$, $x_2 = 12$, and $x_3 = 0$.

3. Enter on the LINDO blank window:

max 3x1 + x2 + x3
subject to
x1 + x2 + x3 < 6
2x1 + 3x2 + 4x3 < 10
end

Click solve;
click no, when asked for reports;
click close.
Move aside the window on which you entered the problem to see the solution on the right.

LP OPTIMUM FOUND AT STEP 1

OBJECTIVE FUNCTION VALUE

1) 15.00000

VARIABLE	VALUE	REDUCED COST
X1	5.000000	0.000000
X2	0.000000	3.500000
X3	0.000000	5.000000

ROW	SLACK OR SURPLUS	DUAL PRICES
2)	1.000000	0.000000
3)	0.000000	1.500000

NO. ITERATIONS = 1

The maximum value of $P = 15$, and it is attained when $x_1 = 5$, $x_2 = 0$, and $x_3 = 0$.

5. Enter on the LINDO blank window.

max 2x1 + x2 + 3x3
subject to
x1 + x2 − x3 < 10
x2 + x3 < 4
end

Click solve;
click no, when asked for reports;
click close.
Move aside the window on which you entered the problem to see the solution on the right.

LP OPTIMUM FOUND AT STEP 1

OBJECTIVE FUNCTION VALUE

1) 40.00000

VARIABLE	VALUE	REDUCED COST
X1	14.000000	0.000000
X2	0.000000	6.000000
X3	4.000000	0.000000

ROW	SLACK OR SURPLUS	DUAL PRICES
2)	0.000000	2.000000
3)	0.000000	5.000000

NO. ITERATIONS = 1

The maximum value of $P = 40$, and it is attained when $x_1 = 14$, $x_2 = 0$, and $x_3 = 4$.

7. Enter on the LINDO blank window,

max $2x1 + x2 + x3$
subject to
$x1 + x2 + x3 < 6$
$4x1 + x2 > 12$
end

Click solve;
click no, when asked for reports;
click close.
Move aside the window on which you entered the problem to see the solution on the right.

LP OPTIMUM FOUND AT STEP 1

OBJECTIVE FUNCTION VALUE

1) 12.000000

VARIABLE	VALUE	REDUCED COST
X1	6.000000	0.000000
X2	0.000000	1.000000
X3	0.000000	1.000000

ROW	SLACK OR SURPLUS	DUAL PRICES
2)	0.000000	2.000000
3)	12.000000	0.000000

NO. ITERATIONS = 1

The maximum value of $P = 6$ and it is attained when $x_1 = 6$, $x_2 = 0$, and $x_3 = 0$.

9. Enter on the LINDO blank window

max $2x1 + x2 + 3x3$
subject to
$5x1 + 2x2 + x3 < 20$
$6x1 + x2 + 4x3 < 24$
$x1 + x2 + 4x3 < 16$
end

Click solve;
click no, when asked for reports;
click close.
Move aside the window on which you entered the problem to see the solution on the right.

LP OPTIMUM FOUND AT STEP 3

OBJECTIVE FUNCTION VALUE

1) 15.20000

VARIABLE	VALUE	REDUCED COST
X1	1.600000	0.000000
X2	4.800000	0.000000
X3	2.400000	0.000000

ROW	SLACK OR SURPLUS	DUAL PRICES
2)	0.000000	0.142857
3)	0.000000	0.114286
4)	0.000000	0.600000

NO. ITERATIONS = 3

The maximum value of $P = 15.2$, and it is attained when $x_1 = 1.6$, $x_2 = 4.8$, and $x_3 = 2.4$.

11. Enter on the LINDO blank window.

max 2x1 + 3x2 + x3
subject to
x1 + x2 + x3 < 50
3x1 + 2x2 + x3 < 10
end

Click solve;
click no, when asked for reports;
click close.
Move aside the window on which you entered the problem to see the solution on the right.

LP OPTIMUM FOUND AT STEP 1

OBJECTIVE FUNCTION VALUE

1) 15.00000

VARIABLE	VALUE	REDUCED COST
X1	0.000000	2.500000
X2	5.000000	0.000000
X3	0.000000	0.500000

ROW	SLACK OR SURPLUS	DUAL PRICES
2)	45.000000	0.000000
3)	0.000000	1.500000

NO. ITERATIONS = 1

The maximum value of $P = 15$, and it is attained when $x_1 = 0$, $x_2 = 5$, and $x_3 = 0$.

13. Enter on the LINDO blank window.

max 2x1+ x2 + x3
subject to
−2x1 + x2 − 2x3 < 4
x1 − 2x2 + x3 < 2
end

Click solve;
An error message appears:
Unbounded solution at STEP 1.

Close the boxes to look at the report on the right

UNBOUNDED VARIABLES ARE:
X3
SLK 3
X2

OBJECTIVE FUNCTION VALUE

1) 0.9999990E+08

VARIABLE	VALUE	REDUCED COST
X1	0.000000	10.000000
X2	4.000000	0.000000
X3	99999904.000000	7.000000

ROW	SLACK OR SURPLUS	DUAL PRICES
2)	0.000000	−3.000000
3)	10.000000	−2.000000

NO. ITERATIONS = 1

This problem is unbounded and has no maximum solution.

15. Enter on the LINDO blank window,

```
max 2x1 + x2 + 3x3
subject to
x1 + 2x2 + x3 < 25
3x1 + 2x2 + 3x3 < 30
end
```

Click solve;
click no, when asked for reports;
click close.
Move aside the window on which you entered the problem to see the solution on the right.

```
LP OPTIMUM FOUND AT STEP    1

        OBJECTIVE FUNCTION VALUE

        1)    30.00000

 VARIABLE        VALUE          REDUCED COST
       X1        0.000000          1.000000
       X2        0.000000          1.000000
       X3       10.000000          0.000000

      ROW   SLACK OR SURPLUS     DUAL PRICES
       2)        4.000000          0.000000
       3)        0.000000          1.000000

NO. ITERATIONS =  1
```

The maximum value of $P = 30$, and it is attained when $x_1 = 0$, $x_2 = 0$, and $x_3 = 10$.

17. Enter on the LINDO blank window.

```
max 2x1 + 4x2 + x3 + x4
subject to
2x1 + x2 + 2x3 + 3x4 < 12
2x2 + x3 + 2x4 < 20
2x1 + x2 + 4x3 < 16
end
```

Click solve;
click no, when asked for reports;
click close.
Move aside the window on which you entered the problem to see the solution on the right.

```
LP OPTIMUM FOUND AT STEP    2

        OBJECTIVE FUNCTION VALUE

        1)    42.00000

 VARIABLE        VALUE          REDUCED COST
       X1        1.000000          0.000000
       X2       10.000000          0.000000
       X3        0.000000          2.500000
       X4        0.000000          5.000000

      ROW   SLACK OR SURPLUS     DUAL PRICES
       2)        0.000000          1.000000
       3)        0.000000          1.500000
       4)        4.000000          0.000000

NO. ITERATIONS =  2
```

The maximum value of $P = 42$ and it is attained when $x_1 = 1$, $x_2 = 10$, $x_3 = 0$, and $x_4 = 0$.

19. Enter on the LINDO blank window.

max $2x1 + x2 + x3$
subject to
$x1 + 2x2 + 4x3 < 20$
$2x1 + 4x2 + 4x3 < 60$
$3x1 + 4x2 + x3 < 90$
end

Click solve;
click no, when asked for reports;
click close.
Move aside the window on which you entered the problem to see the solution on the right.

LP OPTIMUM FOUND AT STEP 1

OBJECTIVE FUNCTION VALUE

1) 40.00000

VARIABLE	VALUE	REDUCED COST
X1	20.000000	0.000000
X2	0.000000	3.000000
X3	0.000000	7.000000

ROW	SLACK OR SURPLUS	DUAL PRICES
2)	0.000000	2.000000
3)	20.000000	0.000000
4)	30.000000	0.000000

NO. ITERATIONS = 1

The maximum value of $P = 40$ and it is attained when $x_1 = 20$, $x_2 = 0$, and $x_3 = 0$.

21. Enter on the LINDO blank window.

max $x1 + 2x2 + 4x3 - x4$
subject to
$5x1 + 4x3 + 6x4 < 20$
$4x1 + 2x2 + 2x3 + 8x4 < 40$
end

Click solve;
click no, when asked for reports;
click close.
Move aside the window on which you entered the problem to see the solution on the right.

LP OPTIMUM FOUND AT STEP 2

OBJECTIVE FUNCTION VALUE

1) 50.00000

VARIABLE	VALUE	REDUCED COST
X1	0.000000	5.500000
X2	15.000000	0.000000
X3	5.000000	0.000000
X4	0.000000	6.000000

ROW	SLACK OR SURPLUS	DUAL PRICES
2)	0.000000	0.500000
3)	0.000000	1.000000

NO. ITERATIONS = 2

The maximum value of $P = 50$ and it is attained when $x_1 = 0$, $x_2 = 15$, $x_3 = 5$, and $x_4 = 0$.

23. Enter on the LINDO blank window.

min $x1 + x2 + x3 + x4 + x5 + x6 + x7$
subject to
$4x1 + 2x2 + x3 + 2x5 + x6 > 75$
$x2 + 2x3 + 3x4 + x6 > 110$
$x5 + x6 + 2x7 > 50$
end

Click solve; click no, when asked for reports; click close.
Move aside the window on which you entered the problem to see the solution below.

LP OPTIMUM FOUND AT STEP 3

OBJECTIVE FUNCTION VALUE

1) 76.25000

VARIABLE	VALUE	REDUCED COST
X1	6.250000	0.000000
X2	0.000000	0.166667
X3	0.000000	0.083333
X4	20.000000	0.000000
X5	0.000000	0.083333
X6	50.000000	0.000000
X7	0.000000	0.166667

ROW	SLACK OR SURPLUS	DUAL PRICES
2)	0.000000	–0.250000
3)	0.000000	–0.333333
4)	0.000000	–0.416667

NO. ITERATIONS = 3

The minimum value of $P = 76.25$ and it is attained when $x_1 = 6.25$, $x_4 = 20$, $x_6 = 50$, and $x_2 = x_3 = x_5 = x_7 = 0$.

Appendix C

Graphing Utilities

C.1 The Viewing Rectangle

1. Using the window shown, X-scale = 1, and Y-scale = 2, we get the point (–1, 4) in Quadrant II.

3. Using the window shown, X-scale = 1, and Y-scale = 1, we get the point (3, 1) in Quadrant I.

5. $X\text{min} = -6 \qquad Y\text{min} = -4$

$X\text{max} = 6 \qquad Y\text{max} = 4$

$X\text{scl} = 2 \qquad Y\text{scl} = 2$

7. $X\text{min} = -6 \qquad Y\text{min} = -1$

$X\text{max} = 6 \qquad Y\text{max} = 3$

$X\text{scl} = 2 \qquad Y\text{scl} = 1$

9. $X\text{min} = 3 \qquad Y\text{min} = 2$

$X\text{max} = 9 \qquad Y\text{max} = 10$

$X\text{scl} = 1 \qquad Y\text{scl} = 2$

11. Answers will vary, but an appropriate viewing window would be

$X\text{min} = -12 \qquad Y\text{min} = -4$

$X\text{max} = 6 \qquad Y\text{max} = 8$

$X\text{scl} = 1 \qquad Y\text{scl} = 1$

13. Answers will vary, but an appropriate viewing window would be

$X\min = -30 \qquad Y\min = 0$
$X\max = 50 \qquad Y\max = 100$
$X\text{scl} = 10 \qquad Y\text{scl} = 10$

15. Answers will vary, but an appropriate viewing window would be

$X\min = -10 \qquad Y\min = -20$
$X\max = 110 \qquad Y\max = 180$
$X\text{scl} = 10 \qquad Y\text{scl} = 20$

C.2 Using a Graphing Utility to Graph Equations

1. (a)

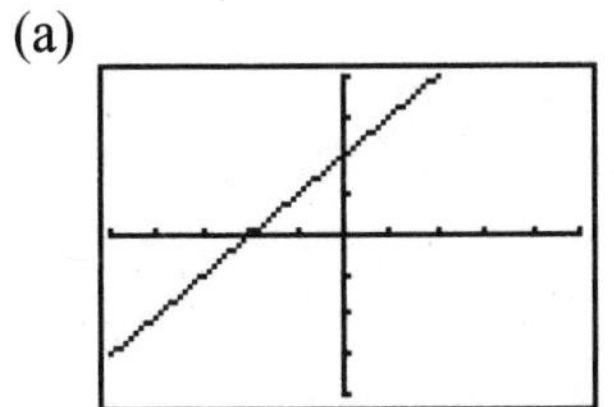

(b)

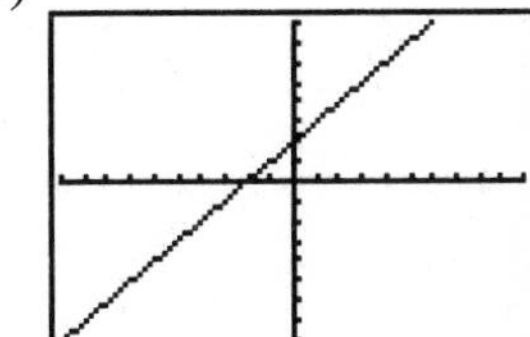

(c)

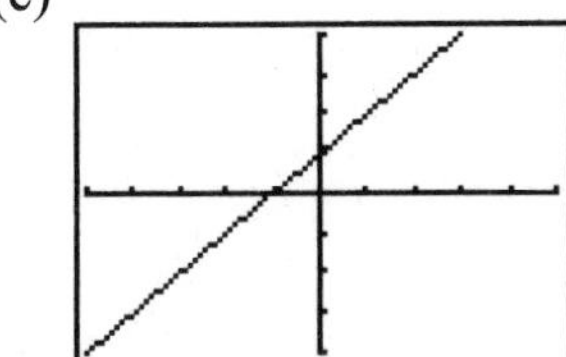

(d)

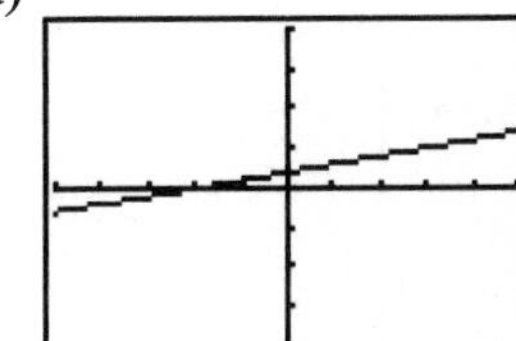

3. (a)

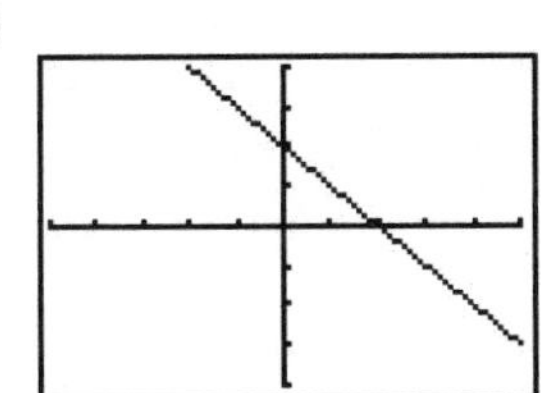

(b)

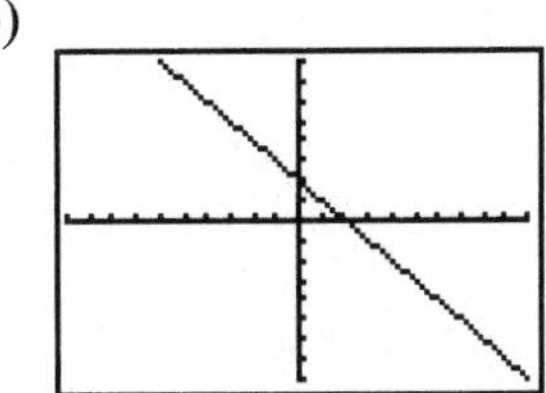

(c)

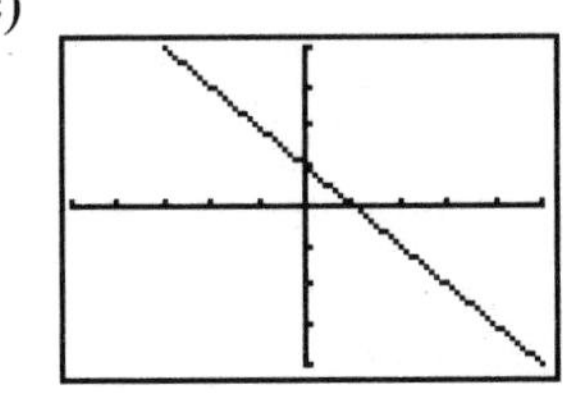

(d)

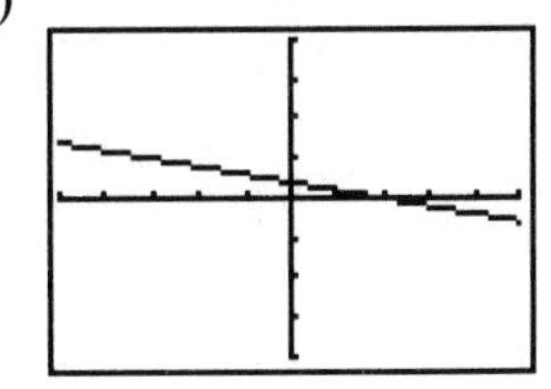

5. (a)

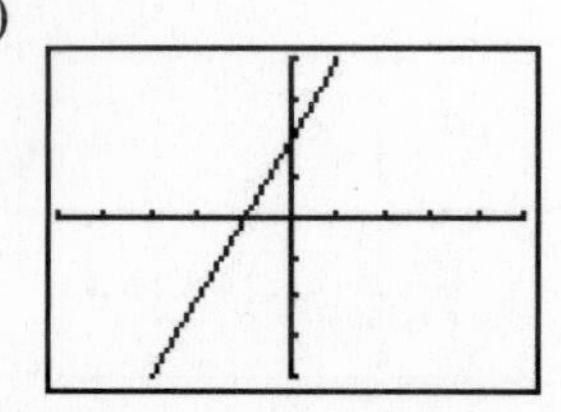

(b)

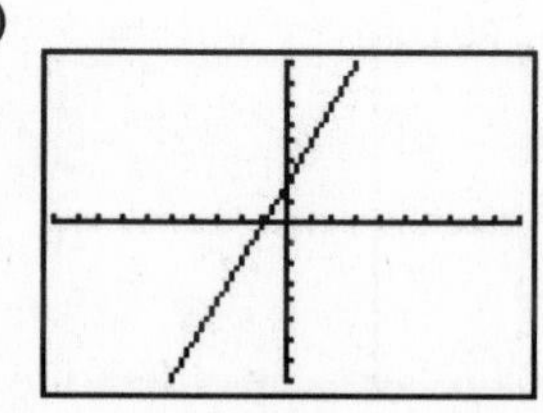

(c)

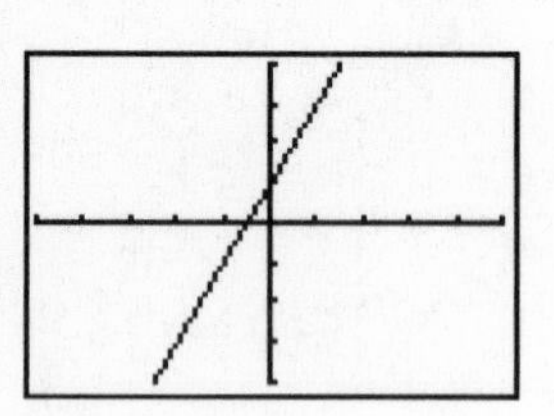

(d)

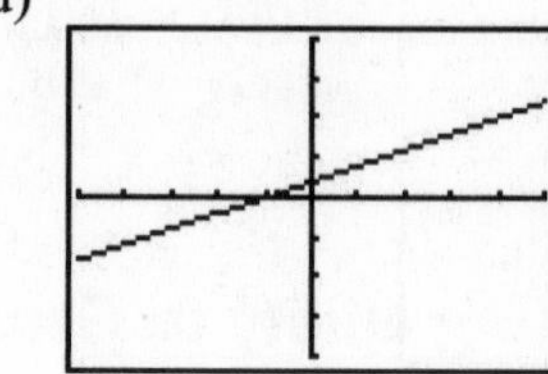

7. (a)

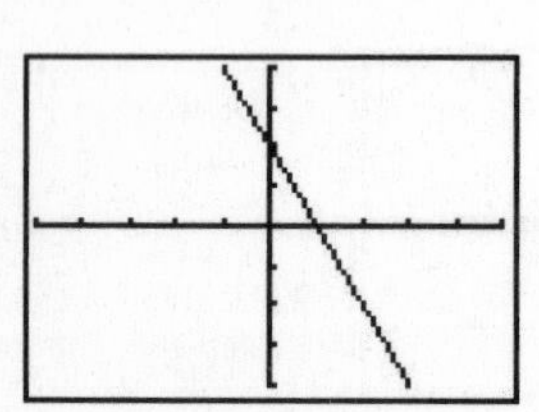

(b)

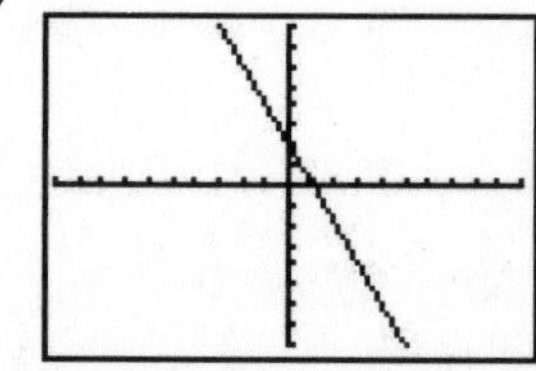

(c)

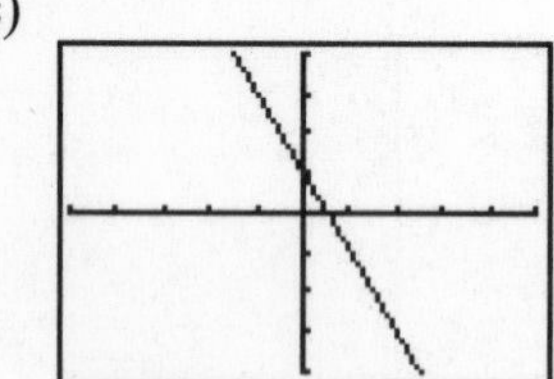

(d)

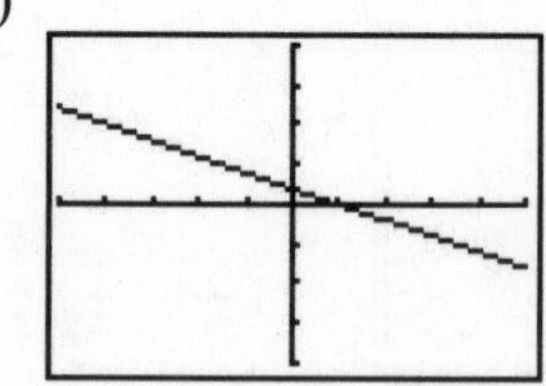

9. (a)

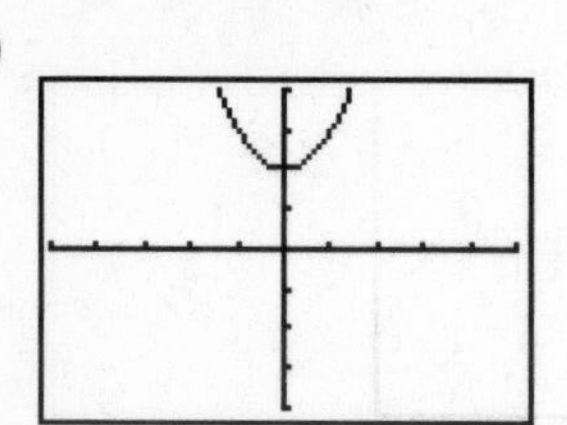

(b)

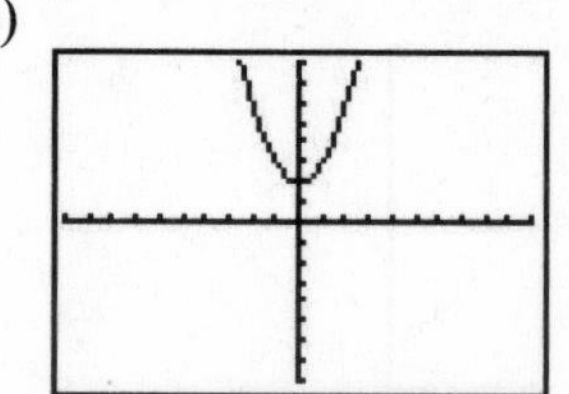

(c)

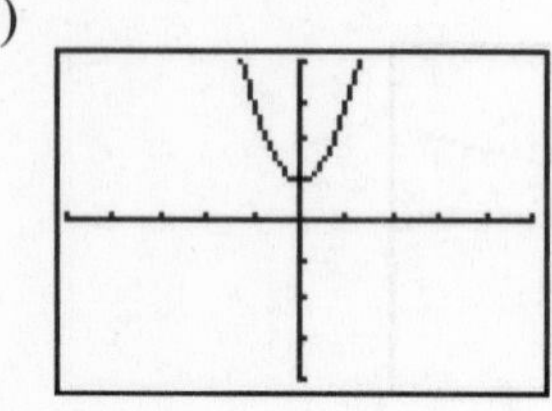

(d)

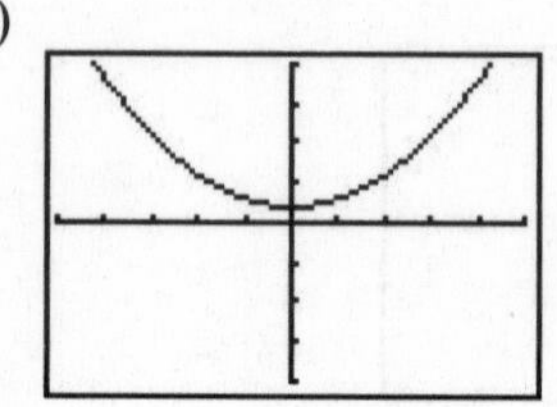

11. (a)

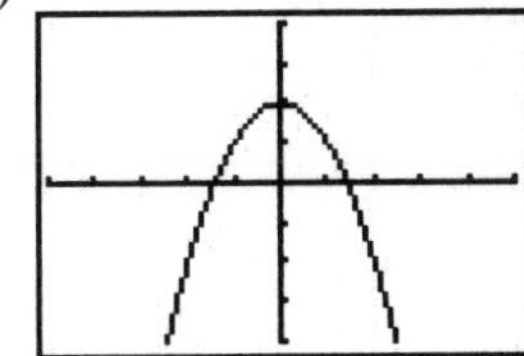

(b)

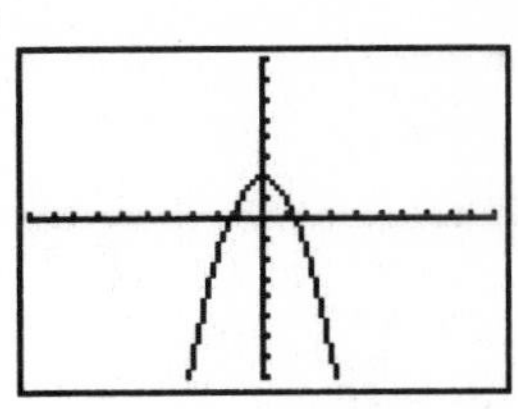

(c)

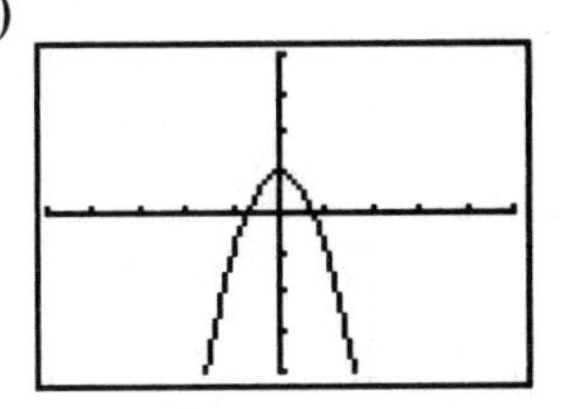

(d)

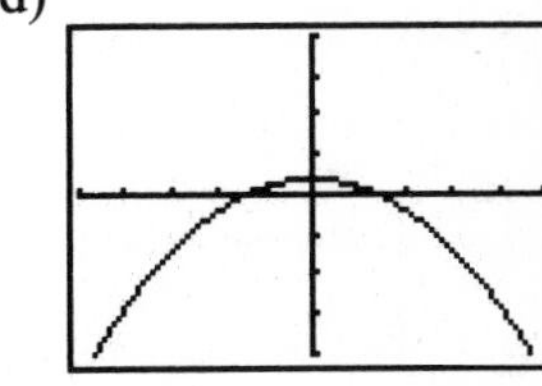

13. (a)

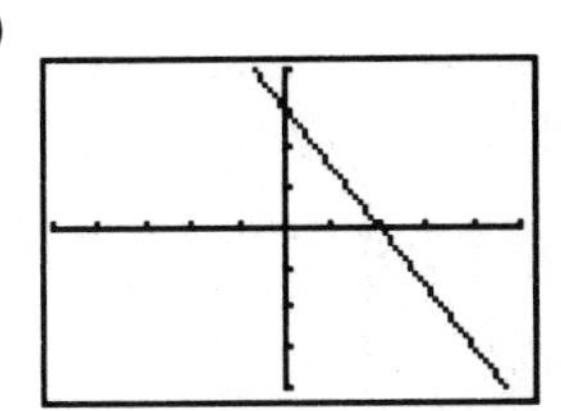

(b)

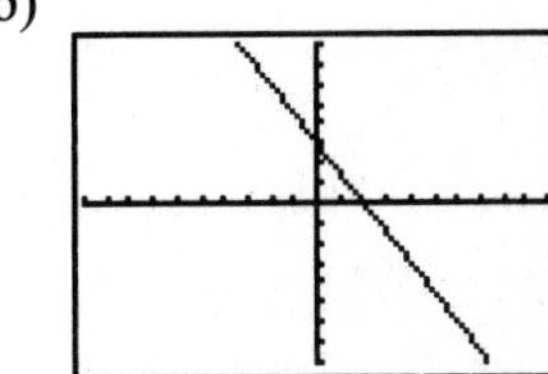

(c)

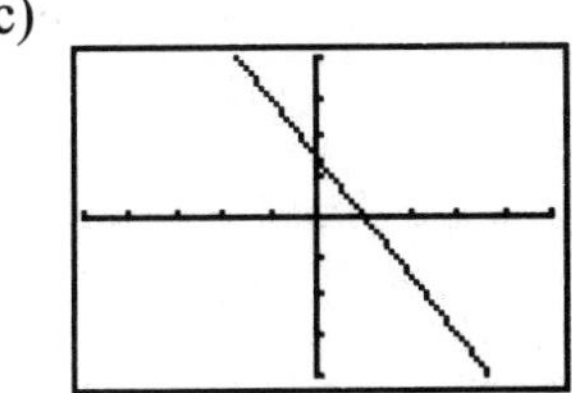

(d)

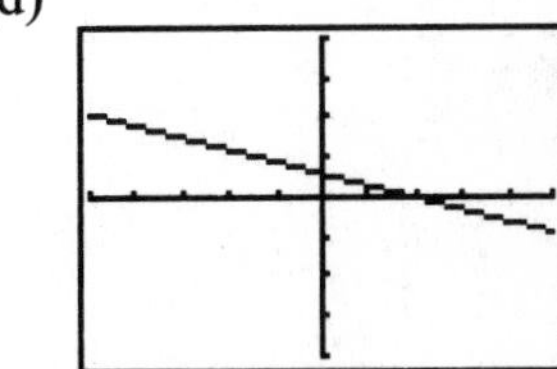

15. (a)

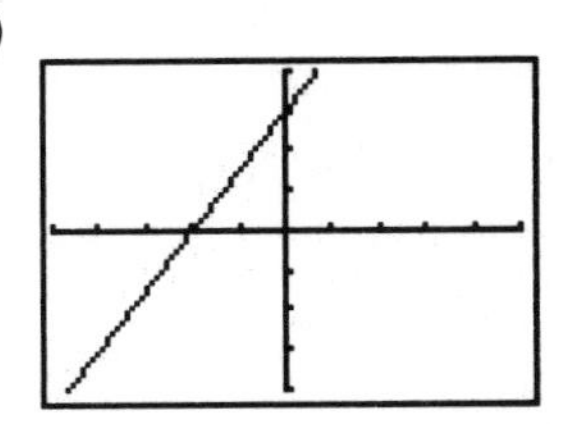

(b)

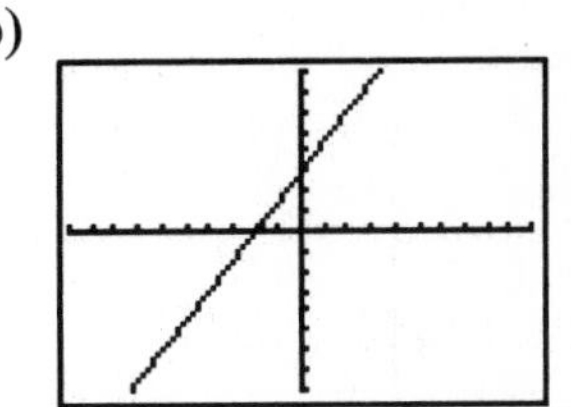

(c)

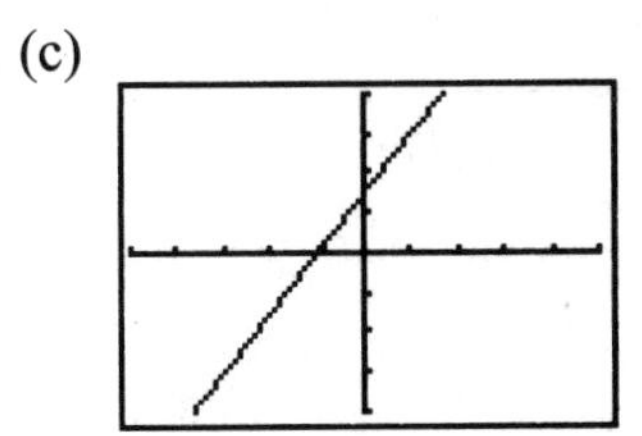

(d)

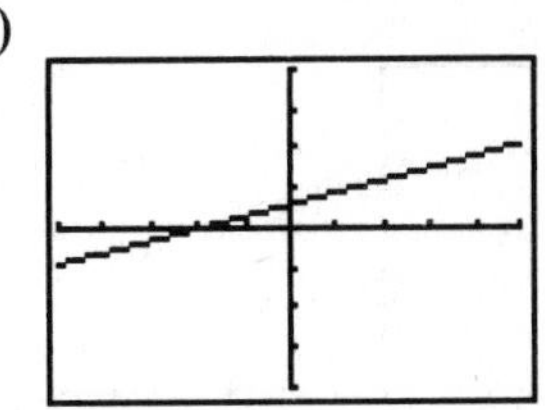

17. $y = x + 2$

X	Y1	
-3	-1	
-2	0	
-1	1	
0	2	
1	3	
2	4	
3	5	

X= -3

19. $y = -x + 2$

X	Y3	
-3	5	
-2	4	
-1	3	
0	2	
1	1	
2	0	
3	-1	

X= -3

21. $y = 2x + 2$

X	Y_5	
-3	-4	
-2	-2	
-1	0	
0	2	
1	4	
2	6	
3	8	

X=-3

23. $y = -2x + 2$

X	Y_7	
-3	8	
-2	6	
-1	4	
0	2	
1	0	
2	-2	
3	-4	

X=-3

25. $y = x^2 + 2$

X	Y_9	
-3	11	
-2	6	
-1	3	
0	2	
1	3	
2	6	
3	11	

X=-3

27. $y = -x^2 + 2$

X	Y_1	
-3	-7	
-2	-2	
-1	1	
0	2	
1	1	
2	-2	
3	-7	

X=-3

29. $3x + 2y = 6$

X	Y_3	
-3	7.5	
-2	6	
-1	4.5	
0	3	
1	1.5	
2	0	
3	-1.5	

X=-3

31. $-3x + 2y = 6$

X	Y_5	
-3	-1.5	
-2	0	
-1	1.5	
0	3	
1	4.5	
2	6	
3	7.5	

X=-3

C.3 Square Screens

1. A square screen results if 2(*X*max – *X*min) = 3(*Y*max – *Y*min)
In this window we test

$$\begin{aligned} 2(3-(-3)) &\ ?\ 3(2-(-2)) \\ 2\cdot 6 &\quad 3\cdot 4 \\ 12 &= 12 \end{aligned}$$

The window is square.

3. A square screen results if 2(*X*max – *X*min) = 3(*Y*max – *Y*min)
In this window we test

$$\begin{aligned} 2(9-0) &\ ?\ 3(4-(-2)) \\ 2\cdot 9 &\quad 3\cdot 6 \\ 18 &= 18 \end{aligned}$$

The window is square.

5. A square screen results if 2(*X*max – *X*min) = 3(*Y*max – *Y*min)
In this window we test

$$\begin{aligned} 2(6-(-6)) &\ ?\ 3(2-(-2)) \\ 2\cdot 12 &\quad 3\cdot 4 \\ 24 &\neq 12 \end{aligned}$$

The window is not square.

7. A square screen results if $2(X\text{max} - X\text{min}) = 3(Y\text{max} - Y\text{min})$
In this window we test

$$\begin{array}{ccc} 2(9-(0)) & ? & 3(4-(-2)) \\ 2 \cdot 9 & & 3 \cdot 6 \\ 18 & = & 18 \end{array}$$

The window is square.

9. Answers may vary.
If $X\text{min} = -4$ and $X\text{max} = 8$, then $2(X\text{max} - X\text{min}) = 2(8 - (-4)) = 2 \cdot 12 = 24$.
To make the screen square, the difference between *Y*min and *Y*max must be $24 \div 3 = 8$.
To include the point (4, 8), the *y*-axis must go at least as high as 8. A possible choice is $Y\text{min} = 1$ and $Y\text{max} = 9$.